GRAPH THEORY WITH APPLICATIONS TO ALGORITHMS AND COMPUTER SCIENCE

GRAPH THEORY WITH APPLICATIONS TO ALGORITHMS AND COMPUTER SCIENCE

Edited by

Y. ALAVI
G. CHARTRAND
L. LESNIAK
D. R. LICK
C. E. WALL

A Wiley-Interscience Publication
JOHN WILEY & SONS
New York • Chichester • Brisbane • Toronto • Singapore

The Theory and Applications of Graphs
with Special Emphasis on Algorithms
and Computer Science Applications

Fifth International Conference
June 4–8, 1984
Western Michigan University
Kalamazoo, Michigan

Library of Congress Cataloging in Publication Data

Main entry under title:

Graph theory with applications to algorithms
and computer science.

"Proceedings of the Fifth Quadrennial International
Conference on the Theory and Applications of Graphs
with special emphasis on Algorithms and Computer
Science Applications, held at Western Michigan
University in Kalamazoo, June 4–8, 1984"—Pref.
"A Wiley-Interscience publication."

1. Graph theory—Congresses. 2. Algorithms—
Congresses. 3. Graph theory—Data processing—
Congresses. I. Alavi, Y. II. International
Conference on the Theory and Applications of
Graphs (5th : 1984 : Western Michigan University)

QA166.G733 1985 511′.5′024 85-9565
ISBN 0-471-81635-3

Printed in the United States of America

10 9 8 7 6 5 4 3 2 1

The Fifth Conference and these Proceedings are dedicated to

Paul Erdös

and

Ronald L. Graham

and the editors herewith recognize and laud their outstanding contributions to mathematics and their promotion of Graph Theory around the world.

PREFACE

This Volume constitutes the Proceedings of the Fifth Quadrennial International Conference on the Theory and Applications of Graphs with special emphasis on Algorithms and Computer Science Applications, held at Western Michigan University in Kalamazoo, Michigan, June 4-8, 1984. Conference participants included research mathematicians and computer scientists from colleges, universities, and the industry, as well as graduate and undergraduate students. Altogether 29 states and 15 countries were represented. The contributions to this Volume include many topics in current research in both the theory and applications in the areas of graph theory and computer science.

ACKNOWLEDGMENTS

The Editors take special pleasure in thanking the many people who contributed to the success of the Fifth Conference as well as the preparation of these Proceedings. In addition to the Conference speakers and participants, and the contributors and the referees for this Volume, we gratefully acknowledge

The outstanding support of Western Michigan University, Dr. John T. Bernhard, President and Dr. Philip S. Denenfeld, Vice President and Dr. L. Michael Moskovis, Associate Vice President, Academic Affairs

The excellent and overall support of the College of Arts and Sciences, Dr. A. Bruce Clarke, Dean and Dr. Clare Goldfarb, Associate Dean

The generous support of the Graduate College, Dr. Laurel Grotzinger, Dean and Chief Research Officer

The enthusiastic and overall support of the Department of Mathematics, Dr. James H. Powell, Chairman and Dr. Joseph T. Buckley, Associate Chairman

The continuing encouragement and fine assistance of our esteemed Graph Theory colleagues Professors Donald L. Goldsmith, S.F. Kapoor, and Arthur T. White

The extraordinary efforts with these Proceedings of many of our graph theory colleagues from around the world, particularly F.R.K. Chung, R.J. Gould, S.T. Hedetniemi, S.F. Kapoor, A.J.Schwenk and A.T. White.

The special administrative and overall assistance of Darlene Lard and general secretarial assistance of Margo Johnson

The outstanding work of the latter two and Karen Schaaf and Myrl Helwig for their skillful typing and assistance with the manuscript

The extensive work of Vicky Koski, senior Mathematics and Computer Science student, as Conference Secretary covering a span of two years

The dedicated and superb work of our Conference Assistants: Afsaneh Behdad, Virginia Mangoyan, Nacer Hedroug, Susan Hollar, Ortrud Oellermann, Nancy Otten, Joan Rahn, Reza Rashidi, Farrokh Saba, Farhad Shahrokhi, Siu-Lung Tang, Zahra Tavakoli, Barbara Treadwell, Mary Wovcha, Hung-Bin Zou, Ek Leng Chua, Andrea Kempher, and Kim Higgins.

Finally, we wish to thank Ken MacLeod, editor, and in particular Rose Ann Campise, Senior Production Supervisor, John Wiley & Sons, for their interest and outstanding assistance with these Proceedings.

The Editors apologize for any oversights in the acknowledgments or any errors in the manuscript,

Y.A.
G.C.
L.L.
D.L.
C.W.

CONTENTS

J. Akiyama*, M. Kano, and M.-J. Ruiz
Tiling Finite Figures Consisting of Regular Polygons 1

N. Alon
Eigenvalues, Geometric Expanders and Sorting in Rounds 15

T. Asano* and S. Sato
Long Path Enumeration Algorithms for Timing Verification on Large Digital Systems 25

K.S. Bagga
On Upsets in Bipartite Tournaments 37

K.S. Bagga and L.W. Beineke*
Some Results on Binary Matrices Obtained via Bipartite Tournaments 47

E.R. Barnes
Partitioning the Nodes of a Graph 57

J.-C. Bermond* and G. Memmi
A Graph Theoretical Characterization of Minimal Deadlocks in Petri Nets 73

G.S. Bloom* and D.F. Hsu
On Graceful Directed Graphs that are Computational Models of Some Algebraic Systems 89

F.T. Boesch
The Cut Frequency Vector 103

J. Bond and C. Peyrat*
Diameter Vulnerability in Networks 123

I. Broere* and C.M. Mynhardt
Generalized Colorings of Outerplanar and Planar Graphs 151

---

*Indicates the speaker at the Conference.

S.A. Burr, P. Erdős, R.J. Faudree, C.C. Rousseau and R.H. Schelp*
The Ramsey Number for the Pair Complete Bipartite Graph-Graph of Limited Degree . . . 163

F.R.K. Chung*, F.T. Leighton, and A.L. Rosenberg
Embedding Graphs in Books: A Layout Problem with Applications to VLSI Design. 175

I.J. Dejter
Hamilton Cycles and Quotients of Bipartite Graphs 189

P. Erdős
Problems and Results on Chromatic Numbers in Finite and Infinite Graphs 201

V. Faber* and A.B. White, Jr.
Supraconvergence and Functions that Sum to Zero on Cycles. 215

R.J. Faudree*, C.C. Rousseau and R.H. Schelp
Edge-Disjoint Hamiltonian Cycles. 231

R.J. Faudree, C.C. Rousseau* and R.H. Schelp
Studies Related to the Ramsey Number $r(K_5 - e)$ 251

N.E. Fenton
The Structral Complexity of Flowgraphs. 273

J.F. Fink* and M.S. Jacobson
n-Domination in Graphs. 283

J.F. Fink and M.S. Jacobson*
On n-Domination, n-Dependence and Forbidden Subgraphs 301

M.A. Fiol*, I. Alegre, J.L.A. Yebra and J. Fábrega
Digraphs with Walks of Equal Length between Vertices. 313

S. Ganesan* and M.O. Ahmad
Application of Numbered Graphs in the Design of Multi-stage Telecommand Codes. 323

*Indicates the speaker at the Conference.

J. Gimbel
The Cochromatic Number of Graphs in a Switching Sequence. 343

R.J. Gould* and R.L. Roth
A Recursive Algorithm for Hamiltonian Cycles in the (1,j,n)-Cayley Graph of the Alternating Group 351

S.L. Hakimi
Further Results on a Generalization of Edge-coloring 371

Y.O. Hamidoune and M. Las Vergnas*
The Directed Shannon Switching Game and the One-way Game. 391

F. Harary
Graph Theoretic Approaches to Finite Mathematical Structures 401

H. Harborth
Drawings of Graphs and Multiple Crossings . . . 413

S. Hedetniemi, S. Hedetniemi* and R. Laskar
Domination in Trees: Models and Algorithms. . . 423

J.P. Hutchinson* and L.B. Krompart
Connected Planar Graphs with Three or More Orbits 443

M. Johnson
Relating Metrics, Lines and Variables Defined on Graphs to Problems in Medicinal Chemistry 457

M. Kano
[a,b]-Factorizations of Nearly Bipartite Graphs. 471

M.M. Klawe
The Complexity of Pebbling for Two Classes of Graphs 475

---

*Indicates the speaker at the Conference.

C.P. Kruskal and D.B. West*
Compatible Matchings in Bipartite Graphs. . . . 489

E.L. Lawler* and P.J. Slater
A Linear Time Algorithm for Finding an Optimal Dominating Subforest of a Tree. . . . 501

M.J. Lipman* and R.E. Pippert
Toward a Measure of Vulnerability II. The Ratio of Disruption 507

F. Loupekine and J.J. Watkins*
Cubic Graphs and the Four-Color Theorem 519

V.V. Malyshko
An Effective Approach to Some Practical Capacitated Tree Problems 531

D.W. Matula
Concurrent Flow and Concurrent Connectivity in Graphs 543

Z. Miller
A Linear Algorithm for Topological Bandwidth in Degree Three Trees 561

C.M. Mynhardt* and I. Broere
Generalized Colorings of Graphs 583

T. Nishizeki* and K. Kashiwagi
An Upper Bound on the Chromatic Index of Multigraphs 595

A.M. Odlyzko and H.S. Wilf*
Bandwidths and Profiles of Trees. 605

J. Pach* and L. Surányi
2-Super-universal Graphs. 623

B. Peruničić* and Z. Durić
An Efficient Algorithm for Embedding Graphs in the Projective Plane 637

---

*Indicates the speaker at the Conference.

R.E. Pippert* and M.J. Lipman
Toward a Measure of Vulnerability I. The Edge-connectivity Vector. 651

F.S. Roberts* and Z. Rosenbaum
Some Results on Automorphisms of Ordered Relational Systems and the Theory of Scale Type in Measurement 659

R.W. Robinson
Counting Strongly Connected Finite Automata . . 671

E. Schmeichel* and D. Hayes
Some Extensions of Ore's Theorem. 687

S. Schuster
Packing a Tree of Order p with a (p,p) Graph. . 697

A.J. Schwenk
How Many Rinds can a Finite Sequence of Pairs Have? 713

D.R. Shier
Iterative Algorithms for Calculating Network Reliability 741

R.E. Tarjan
Shortest Path Algorithms. 753

J. Wang*, S. Tian and J. Liu
The Binding Number of Lexicographic Products of Graphs 761

R.W. Whitty
Spanning Trees in Program Flowgraphs. 777

R.J. Wilson
Analysis Situs. 789

P.M. Winkler
On Graphs which are Metric Spaces of Negative Type 801

---

*Indicates the speaker at the Conference.

GRAPH THEORY WITH APPLICATIONS TO ALGORITHMS AND COMPUTER SCIENCE

TILING FINITE FIGURES CONSISTING OF REGULAR POLYGONS

Jin AKIYAMA
Tokai University

Mikio KANO
Akashi Technological College

Mari-Jo RUIZ
Ateneo de Manila University

ABSTRACT

We give some sufficient conditions for a finite plane figure consisting of squares or hexagons to be covered with tiles of specific shapes.

1. Introduction

Tiling problems examine the possibility of covering plane figures with tiles of specific shapes, where covering a figure with tiles means to lay tiles over the figure so that it is completely covered, but such that no tiles are stacked on each other and no tiles exceed the edges of the figure. For example, the problem of deciding whether the defective chessboard of Figure 1.1 (a) can be covered with dominoes (Figure 1.1 (b)) is a tiling problem. In this paper, we shall present some results on tiling plane figures consisting of a finite number of squares or regular hexagons.

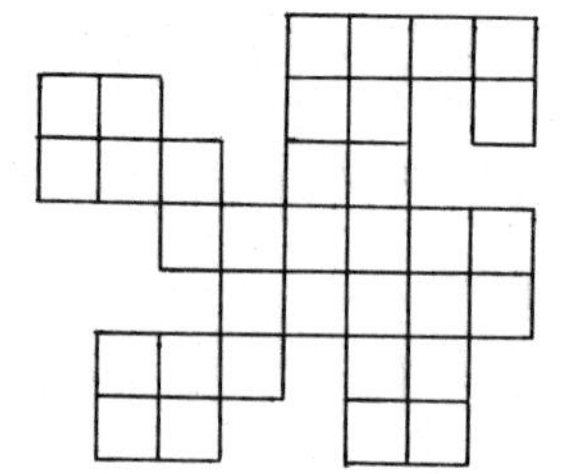

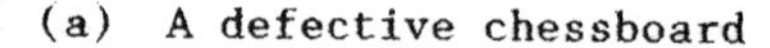

(a) A defective chessboard (b) A domino

Figure 1.1

Consider a plane figure consisting of some polygons. Two polygons in the figure are said to be adjacent if they have a common edge. For any figure, we can associate a graph as follows: Represent each polygon in the figure by a vertex and join two vertices by an edge if the polygons represented by these vertices are adjacent. There is a close relationship between the solution of a tiling problem and the existence of a certain component factor (i.e. a spanning subgraph with a given component) in the graph associated with the figure. For example, the graph associated with the defective chessboard of Figure 1.1 (a) is shown in Figure 1.2 (a). Moreover, it is clear that the existence of tiling a defective chessboard with dominoes is equivalent to the existence of a 1-factor in the graph associated with the defective chessboard. Note that the graph given in Figure 1.2 (b) cannot be the graph associated with any defective chessboard since two vertices x and y are not joined by an edge.

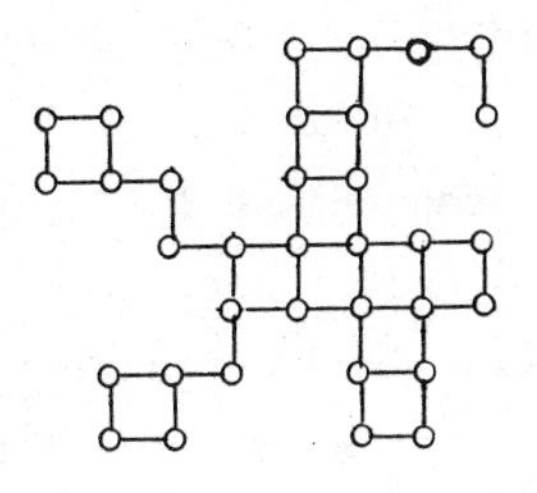

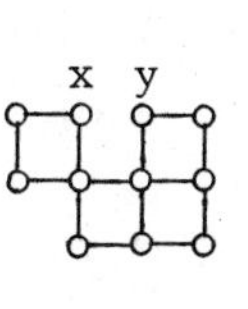

(a) (b)

Figure 1.2

We use the following notation. By P_n, C_n and $K_{1,n-1}$, we denote the path, the cycle and the star with n vertices, respectively. For a graph G and a set $\{A,B,\cdots,K\}$ of graphs, a spanning subgraph F of G is called an $\{A,B,\cdots.K\}$-factor of G if each component of F is isomorphic to one of $\{A,B,\cdots,K\}$. In particular, each component of an H-factor (i.e. an $\{H\}$-factor) is H.

We conclude this section by mentioning that some papers on tiling problem are concerned with the infinite plane, the infinite half plane and others, rather than finite planes. Golomb [7] studied the problem of tiling the infinite plane, the half plane, a quadrant of the plane, infinite strips, and other infinite figures with polyominoes of one shape. He also gave a classification of the capability of each polyomino, consisting of up to six unit squares, tiling each of the figures mentioned above. The papers [5], [6], [8] discuss various ways of tiling the plane.

2. Tiling defective chessboards

A figure obtained from an $m \times n$ chessboard (m rows and n columns) by removing a certain number of unit squares is called a defective chessboard, where m and n are arbitrary integers. The order of a defective chessboard is the number of unit squares in it. A defective chessboard B is said to be tough if every pair of adjacent dominoes in B is contained in a 2×2 square of B. Observe that the defective chessboard in Figure 2.1 (a) is tough, while the one in Figure 2.1 (b) is not. The graph associated with a defective chessboard is called a square graph, and a square graph is said to be tough if its associated defective chessboard is tough, that is, a square graph is tough if every edge is contained in some C_4.

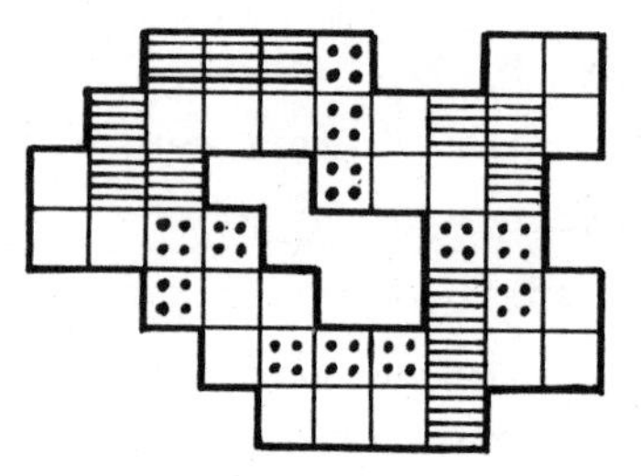

(a) A tough defective chessboard that covered with triminoes.

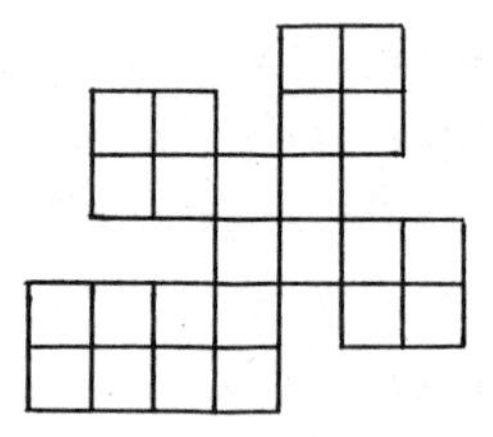

(b) A defective chessboard is not tough and cannot be covered with triominoes.

Figure 2.1

We say that a defective chessboard is connected if the corresponding square graph is connected. Generally, a plane figure is said to be connected if the corresponding graph is connected.

Tutte's 1-factor theorem [11] provides a solution to tiling a defective chessboard with dominoes. Moreover, Edmonds [4] gives a polyominal time algorithm for determining the existence of tiling a defective chessboard with dominoes. We now turn our attention from dominoes to other tiles. There are two kinds of triominoes, namely L-shaped and I-shaped triominoes (Figure 2.2 (a) and (b)).

Theorem 2.1 [2] Every connected tough defective chessboard of order 3p can be covered with triomiones (Figure 2.1 (a)).

We omit the proof of this theorem, which is similar to that of Theorem 2.2, presented next. Note that both I-shaped and L-shaped triomiones are represented by P_3 in the corresponding graph, and so the above theorem also gives a local sufficient condition for a square graph to have a P_3-factor.

There exists a connected tough square graph of order 4p which has neither P_2-factors nor $\{P_4, K_{1,3}\}$-factors for any $p \geq 7$ (Figure 2.3). However, we have the next theorem, which says that every connected tough chessboard of even order can be tiled with dominoes and tetrominoes given in Figure 2.2 (c).

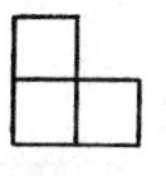

(a) The L-shaped triomino

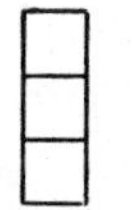

(b) The I-shaped triomino

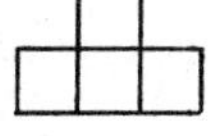

(c) The tetriomino corresponding to $K_{1,3}$

Figure 2.2

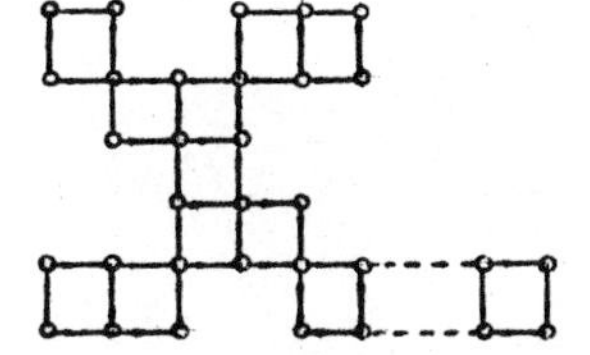

A tough square graph having neither P_2-factors nor $\{P_4, K_{1,3}\}$-factors

Figure 2.3

Theorem 2.2 Every connected tough square graph of even order has a $\{P_2, K_{1,3}\}$-factor (Figure 2.4).

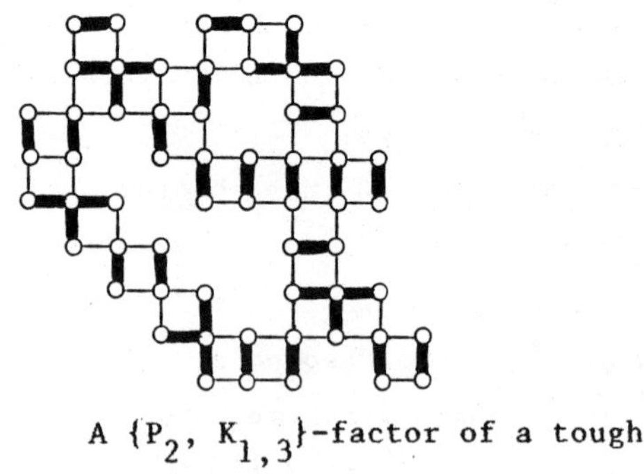

A $\{P_2, K_{1,3}\}$-factor of a tough square graph

<u>Figure 2.4</u>

Proof We prove by induction, the following statement (*) which includes the theorem.

(*) Let G be a connected tough square graph. It follows that if $|V(G)| \equiv 0 \pmod 2$, then G has a $\{P_2, K_{1,3}\}$-factor, and that if $|V(G)| \equiv 1 \pmod 2$, then for any vertex v of degree 2, $G - v$ has a $\{P_2, K_{1,3}\}$-factor.

Let G be a connected tough square graph. We briefly say that G has a <u>factor</u> when G has a $\{P_1, K_{1,3}\}$-factor, and that G has a (v)-semifactor when $G - v$ has a $\{P_2, K_{1,3}\}$-factor, where v is a vertex of G. We denote by capital letters A, B, $\cdots$ cycles of order 4 in G, and by small letters a,b,$\cdots$ vertices of G. We may assume $|V(G)| \geq 10$ since if $|V(G)| < 10$, then (*) holds. We first assume $|V(G)| \equiv 0 \pmod 2$.

(1) If G contains the part in Figure 2.5 (a), then G has a factor.

Proof We use Figure 2.5 (b). If $G \not\supset B$, then $G \supset C$ and so $G - \{a,x,b\}$ has a (c)-semifactor by the induction hypothesis. Hence G has a factor. Therefore we may assume $G \supset B, C$. We consider two cases.

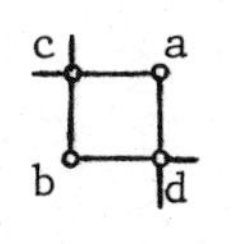

(a)

$\deg_G(a)=\deg_G(b)=2$, $\deg_G(c) \geqq 2$ and $\deg_G(d) \geqq 2$.

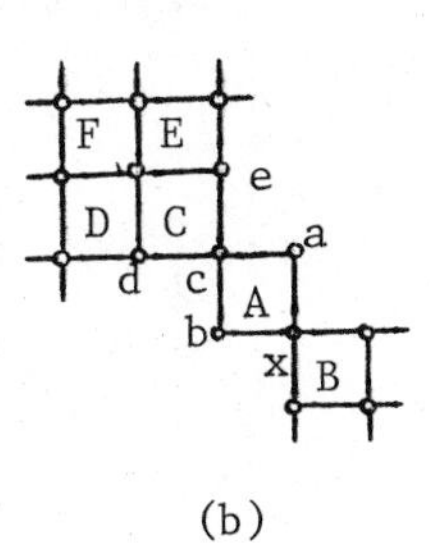

(b)

Figure 2.5

Case 1.1 x is not a cut vertex of G.

If $G \supset D, E$, then $G-\{a,c,b\}$ has an (x)-semifactor by the induction hypothesis, and so G has a factor. If $G \supset D, \not\supset E$, then $G - \{e,c,a,b\}$ has a factor and so G has a factor. If $G \not\supset D, E$, then $G \supset F$ and $G - \{a,b,c,d,e\}$ has an (x)-semifactor. Hence G has a factor.

Case 1.2 x is a cut vertex of G.

Put $G = \{a,b\} = X \cup Y$ such that $X \supset C$ and $Y \supset B$, where X and Y are connected subgraphs of G. If $|V(X)| \equiv 0 \pmod 2$, then X has a factor and $Y + \{a,b,c\}$, which is the subgraph of G induced by $V(Y) \cup \{a,b,c\}$, has a (c)-semifactor. Hence G has a factor. If $|V(X)| \equiv 1 \pmod 2$, then X has a (c)-semifactor and Y has an (x)-semifactor. Thus G has a factor.

(2) If G contains the part in Figure 2.6 (a), then G has a factor.

We use the Figure 2.6 (b) and consider two cases.

Case 2.1 v is a cut vertex of G.

Put $G - v = X \cup Y$ such that $e,f \in X$ and $a,b \in Y$. If $|V(X)| \equiv 0 \pmod 2$, then $X + v$ has a (v)-semifactor and $Y + v$ has a factor. Hence G has a factor. If $|V(X)| \equiv 1 \pmod 2$, then $X + v$ has a factor and $Y + v$ has a (v)-semifactor, and thus G has a factor.

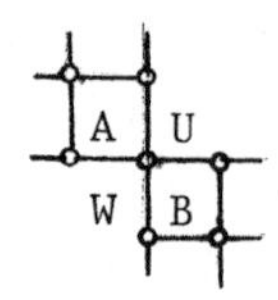

$G \supset A, B$ and $G \not\supset U, W$

(a)

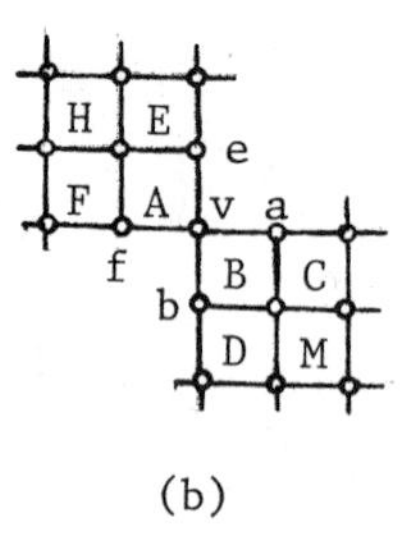

(b)

Figure 2.6

Case 2.2 v is not a cut vertex of G.
If $G \not\supset D, C$, then $G \supset M$ and so G has a factor by (1). Thus we may assume $G \supset C$. Suppose $G \not\supset D$. If $G \supset E, F$, then $G - \{v,b\}$ has a factor and so G has a factor. If $G \supset E$, $\not\supset F$, then $G - \{f,v,b\}$ has an (a)-semifactor and thus G has a factor. If $G \not\supset$, E, F, then $G \supset H$ and it follows from (1) that G has a factor. Consequently we may assume $G \supset D$. If $G \supset E, F$, then $G - v$ has an (a)-semifactor and so G has a factor. If $G \supset E$, $\not\supset F$, then $G-\{v,f\}$ has a factor and so G has a factor. If $G \not\supset E, F$, then $G \supset H$ and thus G has a factor by (1).
(3) If G contains neither the part of Figure 2.5 (a) nor the part of Figure 2.6 (a), then G has a factor. We can take a vertex v of degree 2 as in Figure 2.7. Let A be the C_4 containing v. If $G \not\supset B$, then $G \supset I$ by the assumption of (3). Hence $G-\{v,a\}$ has a factor, and so G has a factor. Therefore we may assume $G \supset B$. If $G \not\supset I$, then $G \not\supset H$ and so $G-\{v,h\}$ has a factor. Thus G has a factor. Therefore we may assume $G \supset I$, and we have that $G - v$ has an (a)-semifactor. Hence G has a factor.

Consequently, the proof of the case that $|V(G)| \equiv 0 \pmod 2$ is complete.

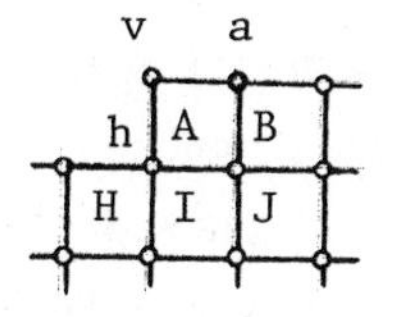

Figure 2.7

We next assume $|V(G)| \equiv 1 \pmod 2$. Let v be a vertex of degree 2, and let $A, B, \cdots$ denote cycles of order 4, if any, as in Figure 2.8. If $G \supset B, C$, then $G - v$ has a factor. Hence we may assume $G \not\supset C$. In each of the following seven cases it can be shown that G has a (v)-semifactor.

Case 1. $G \not\supset B, D$.

Case 2. $G \supset B, \not\supset D$.

Case 3. $G \not\supset B, \supset D$.

Case 4. $G \supset B, D$ and $G \supset I, J$.

Case 5. $G \supset B, D$ and $G \supset I, \not\supset J$.

Case 6. $G \supset B, D$ and $G \not\supset I, \supset J$.

Case 7. $G \supset B, D$ and $G \not\supset I, J$.

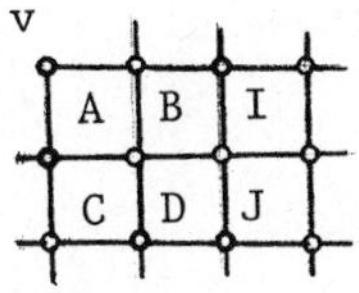

Figure 2.8

We conclude with a conjecture. A defective chessboard is said to be strongly tough if it is tough and every I-shaped triomino in it is contained in a 2×3 rectangle of it.

Conjecture Every connected strongly tough defective chessboard of order np can be covered with all n-omiones (i.e. all connected figures consisting of n unit squares), where $n \geq 3$.

This conjecture is true if $n = 3$ (Theorem 2.1) while the conjecture does not hold if $n = 2$. Furthermore, as we pointed out before, there exists a connected tough defective chessboard of order $4p$ which cannot be covered with triominoes (Figure 2.3).

3. Tiling honeycomb patterns

A honeycomb pattern is a plane figure consisting of unit hexagons (Figure 3.1). Figure 3.2 shows a dihexon and a trihexon. A honeycomb pattern H is said to be tough if every dihexon in H is included in a trihexon of H. For example, the honeycomb pattern in Figure 3.1 is not tough, while the one in Figure 3.3 is. The graph associated with a honeycomb pattern is called a triangle graph (Figure 3.4), and a triangle graph is said to be tough if the corresponding honeycomb pattern is tough. The next theorem implies that a connected tough triangle graph of even order has a 1-factor, and is proved by induction without using Tutte's 1-factor theorem.

Theorem 3.1 [1] Every connected tough honeycomb pattern of even order can be covered with dihexons.

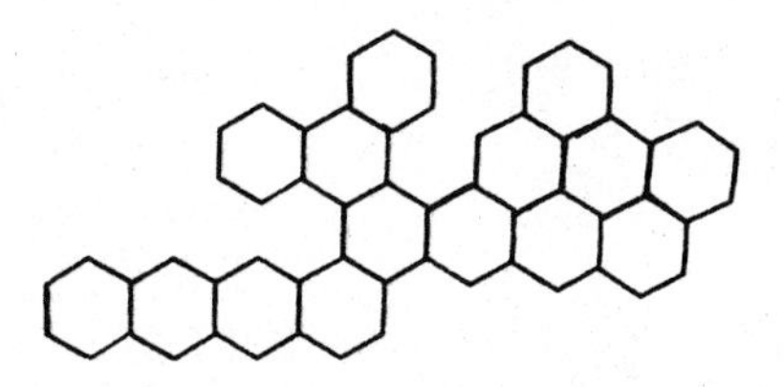

A honeycomb pattern

Figure 3.1

(a) A dihexon (b) A trihexon

Figure 3.2

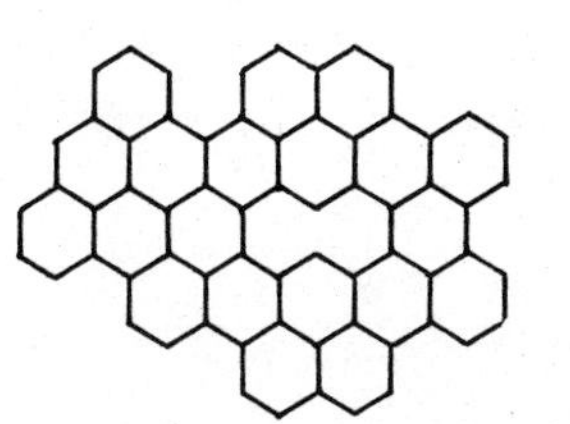

A Tough Honeycomb Pattern

Figure 3.3

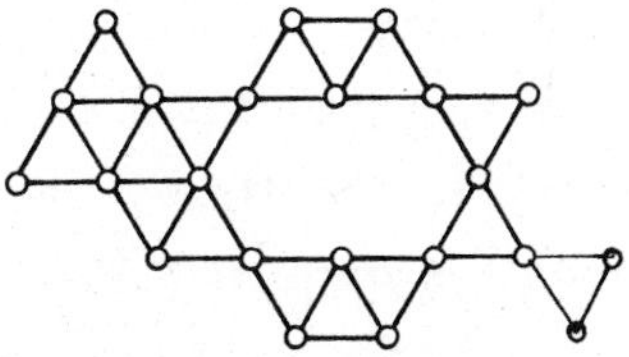

The triangle graph associated with Figure 3.3

Figure 3.4

It seems to be difficult to find a good sufficient condition for a honeycomb pattern to be coverable with trihexons. We next give a necessary condition.

Theorem 3.2 If a honeycomb pattern H of order 3p (p > 1) is coverable with trihexons, then every trihexon in H is contained in one of the following configurations in H:

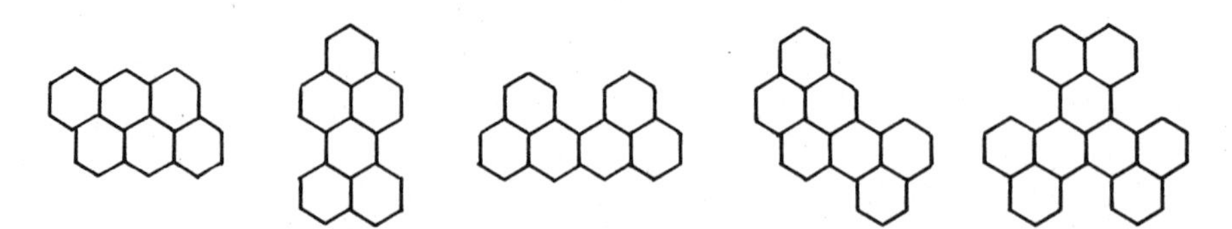

Figure 3.5

We conclude this section with a theorem on graph factors, which yields some results on tiling problems. Let a and b be integers such that $0 \le a \le b$. Then a graph G is called an [a,b]-graph if the degree $\deg_G(x)$ of each vertex x of G satisfies $a \le \deg_G(x) \le b$.

Theorem 3.3 [3], [10] Let k and n be positive integers. Then every [k,nk]-graph has a $\{K_{1,1}, K_{1,2}, \cdots, K_{1,n}\}$-factor.

For example, we obtain the following result by this theorem: A connected tough defective chessboard can be covered with dominoes and triominoes since the graph associated with a tough defective chessboard is a [2,4]-graph, and has a $\{K_{1,1}, K_{1,2}\}$-factor.

REFERENCES

1. J. Akiyama and M. Kano, 1-factors of a triangle graph, to appear.
2. J. Akiyama and M. Kano, Path factors of a graph, *Graphs and its Applications*, Wiley and Sons, New York, 1984.

3. A. Amahashi and M. Kano, On factors with given component, *Discrete Math.* 42 (1982), 1-6.

4. J. Edmonds, Paths, trees and flowers, *Can. J. Math.*, 17 (1965), 449-467.

5. A. Fontaine and G. Martin, Polymorphic prototiles, *J. Combinatorial Theory (A)*, 34 (1983), 119-121.

6. A. Fontaine and G. Martin, Tetramorphic and pentamorphic prototiles, *J. Combinatorial Theory (A)*, 34 (1983), 115-188.

7. S. W. Golomb, Tiling with polyominoes, *J. Combinatorial Theory*, 1 (1966), 280-296.

8. B. Grunbaum and G. C. Shephard, *The Mathematical Gardner* (D. A. Klarner, Ed.), Wadsworth, Belmont, California, (1981).

9. M. Las Vergnas, A note on matchings in graphs, *Cahiers Centre Etudes Recherche Oper.*, 17 (1975), 257-260.

10. C. Payan, Sur quelques problemes de couverture st de couplages en Combinatoire, Thesis, Grenoble, 1977.

11. W. T. Tutte, The factorization of linear graphs, *J. London Math. Soc.*, 22 (1947), 107-111.

EIGENVALUES, GEOMETRIC EXPANDERS AND SORTING IN ROUNDS

Noga Alon

Mathematics Research Center
AT&T Bell Laboratories
Murray Hill, NJ 07974

ABSTRACT

Expanding graphs are relevant to theoretical computer science in several ways. Here we use finite geometries to construct explicitly highly expanding graphs with essentially the smallest possible number of edges. The expansion properties of the graphs are proved using the eigenvalues of their adjacency matrices.

Our graphs enable us to improve significantly previous results on a parallel sorting problem that arises in structural modeling, by describing an explicit algorithm to sort n elements in k time units using $O(n^{\alpha_k})$ parallel processors, where, e.g., $\alpha_2 = 7/4$, $\alpha_4 = 26/17$ and $\alpha_5 = 22/15$.

1. INTRODUCTION

A graph G is called (n, α, β)-expanding, where $0 < \alpha \leqslant \beta \leqslant n$, if it is a bipartite graph on the sets of vertices I (inputs) and O (outputs), where $|I| = |O| = n$, and every set of at least α inputs is joined by edges to at least β different outputs.

Expanding graphs with a small number of edges, which are the subject of an extensive literature, are relevant to theoretical computer science in several ways. Here we merely point out two examples. A family of linear expanders of density k and expansion d is a set $\{G_n\}_{n=1}^{\infty}$ of graphs, where G_n has $\leqslant (k + o(1))n$ edges and is $(n, \alpha, \alpha(1 + d(1 - \alpha/n)))$-expanding for all $\alpha \leqslant n/2$, where $d > 0$ and k are fixed. Such a family is the basic building block used in the constructions of graphs with special connectivity properties and small number of edges (see, e.g., Chung [9]). An example of a graph of this type is an n-superconcentrator, which is a directed acyclic graph with n inputs and n outputs such that for every $1 \leqslant r \leqslant n$ and every two sets A of r inputs and B of r outputs there are r vertex disjoint paths from the vertices of A to the vertices of B. Superconcentrators have been used in the construction of graphs that are hard to pebble (see Lengauer and Tarjan [18], Pippenger [22] and Paul, Tarjan and Celoni [24]), in the study of lower bounds (see Valiant [29]), and in the establishment of time space tradeoffs for computing various functions (Abelson [1], Ja'Ja' [16] and Tompa [27]).

A family of linear expanders is also essential in the recent parallel sorting network of Ajtai, Komlós and Szemerédi [2].

It is not too difficult to prove the existence of a family of linear expanders using probabilistic arguments (see, e.g., Chung [9], Pinkser [20] and Pippenger [21]). However, for applications an explicit construction is desirable. Such a construction is far more difficult and was first given in Margulis [19] and modified in Gaber and Galil [10]. (See also Alon and Milman [4] for a similar but more general construction.)

The expanding graphs used in [10] to construct superconcentrators and those used in the sorting network of [2] are (n, α, β)-expanding for some fixed (independent of n) ratio of β/α, i.e., they are rather weakly expanding. For some applications, however, a higher amount of expansion is necessary and $(n, \alpha(n), \beta(n))$-expanding graphs are needed, where $\beta(n)/\alpha(n) \rightarrow \infty$ as $n \rightarrow \infty$. A possible (and essentially the only known) method to obtain (explicitly) highly expanding graphs with a small number of edges is an "iteration" of the known expander of [10] (see Pippenger [23]). Unfortunately, this method is a poor substitute for the probabilistic construction since it supplies graphs with too many edges. This makes some of the applications impossible.

Here we use finite geometries to explicitly construct highly expanding graphs with essentially the smallest possible number of edges. Specifically, we show using the correspondence (proved in Tanner [26] and in [4]) between the eigenvalues of the adjacency matrix of a graph and its expansion properties, that the points versus hyperplanes incidence graph of a finite geometry of dimension d is an $(n, x, n-n^{1+1/d}/x)$-expanding graph, for all $0 < x < n$. As pointed out by Pippenger [23], the results of Guy and Znam [11] imply that any such graph must have at least $\Omega(n^{2-1/d})$ edges. Our graphs have $(1 + o(1))n^{2-1/d}$ edges; only a constant times the theoretical lower bound. The previous methods were not sufficient to construct graphs with this amount of expansion having $o(n^2)$ edges.

By a theorem of Singer [12, p 128] the edges of our graphs can be defined by a set of $\simeq n^{1-1/d}$ translations modulo n, in contrast to the result of Klawe [17] that asserts that no family of linear expanders can have this form. This reveals a difference between weakly expanding and highly expanding graphs.

Our new expanding graphs enable us to obtain an explicit algorithm for sorting n elements in two time units using $O(n^{7/4})$ parallel processors. This improves results of Bollobás and Rosenfeld [7], Häggvist and Hell [14] and Pippenger [23] who gave explicit algorithms to this problem using $2/5\ n^2 + O(n^{3/2})$, $13/30(n^2 - n)$ and $O(n^{1.943...}(\log n)^{0.943...})$ processors, respectively. It also enables us to improve the best known algorithms for sorting n elements in k time units, for all (fixed) $k \geqslant 4$.

Our paper is organized as follows: In Section 2 we show how the eigenvalues of the adjacency matrix of a graph are related to its expansion properties, and use this to construct our geometric expanders. In Section 3 we describe how our expanders can be applied to the problem of sorting in rounds.

In a subsequent paper [3] we obtain further results using our eigenvalues method. In particular we construct several graphs relevant to Ramsey Theory and obtain a strengthened version of the well known de Bruijn-Erdös Theorem [6].

Due to space limitations we present here only the results, without proofs. The detailed proofs will appear in [3].

2. THE EIGENVALUES METHOD AND THE GEOMETRIC EXPANDERS

Relations between the expansion properties of a graph and the eigenvalues of certain matrices associated with it are proved in Tanner [26] and in [4]. For our purposes here we need a generalization of a result of [26].

Let G be a bipartite graph with classes of vertices I and O, where $|I| = n$, $|O| = m$. Suppose the degree of each $i \in I$ is k and the degree of each $o \in O$ is s. (Thus $k \cdot n = s \cdot m$). Let $A = (a_{io})_{i \in I, o \in O}$ be the nxm adjacency matrix of G defined by

$$a_{io} = \begin{cases} 1 \text{ if } i \text{ and } o \text{ are adjacent} \\ 0 \text{ otherwise} . \end{cases}$$

AA^T is a real symmetric positive semidefinite matrix and thus has real non-negative eigenvalues with orthogonal eigenvectors. Let $\lambda_1 \geqslant \lambda_2 \geqslant \cdots \geqslant \lambda_n$ be these eigenvalues.

For $X \subseteq I$, let $N(X)$ denote the set of all neighbours of X in G. If $X = \{x\}$ we write $N(x)$ instead of $N(\{x\})$.

Theorem 2.1

Suppose $Z \subseteq 0$, $|Z| = z$ and assume $b \leqslant (kz)/(2m)$. Put $X = \{i \in I : |N(i) \cap Z| \leqslant b\}$. Then

$$|X| \leqslant \frac{\lambda_2 n \cdot (m - z)}{ksz - 2kbn + \lambda_2 (m - z) + b^2 n} \left(\leqslant \frac{\lambda_2 n \cdot m}{ksz - 2kbn} \right) .$$

□

Remark 2.2.

For $X \subseteq I$, one can apply Theorem 2.1 with $Z = 0 - N(x)$ and $b = 0$ to obtain that

$$|N(X)| \geqslant \frac{k^2 |X|}{(k \cdot s - \lambda_2)|X|/n + \lambda_2} .$$

This is the main result of [26].

We can now construct our geometric expanding graphs. Let $d, q \geqslant 2$ be integers. Let I and O be, respectively the sets of points and hyperplanes of a finite geometry of dimension d and order q. (As is well known, such as a geometry always exists if q is a prime power, and has an easy explicit description — see [12, p. 128].) Let $G = G(q, d)$ denote the bipartite graph with classes of vertices I and O in which $p \in I$ is joined to $h \in O$ iff p is incident with h. The next theorem shows that $G(q, d)$ is a highly expanding graph.

Theorem 2.3.

Put $n = (q^{d+1} - 1)/(q - 1)$, $k = (q^d - 1)/(q - 1)$.

1. $G = G(q, d)$ is k-regular and $|I| = |O| = n$; thus G has $(1 + o(1))n^{2-1/d}$ edges. (As $q \to \infty$ for fixed d.)

2. If $X \subseteq I$, $|X| = x$ then

$$|N(X)| \geqslant n - \frac{(n - x)(n(q - 1) + 1)}{n(q - 1) + 1 + (n - q - 1)x} \geqslant n - \frac{n^{1+1/d}}{x}.$$

Thus G is $(n,x,n - n^{1+1/d}/x)$-expanding for all $0 < x < n$.

□

Theorem 2.3 follows from Theorem 2.1 by computing the corresponding eigenvalues. We omit the details

Remarks

1. The known results about the distribution of primes (see, e.g., [5,*p. xx*]) clearly imply that for every fixed $d \geqslant 2$ and every integer n there exists a prime p such that $n \leqslant (p^{d+1} - 1)/(p - 1) \leqslant n + O(n^{1-1/(3d)})$. Any induced subgraph of $G(p,d)$ with n inputs and n outputs has $(1 + o(1)) \dfrac{n^{1+1/d}}{x}$ neighbours. Thus we have for every $d \geqslant 2$, an explicit construction of a family of graphs $\{H(n,d)\}_{n=1}^{\infty}$ where $H(n,d)$ has $(1 + o(1))n^{2-1/d}$ edges and is $(n,x,n-(1 + o(1)) \dfrac{n^{1+1/d}}{x})$-expanding for all $0 < x < n$.

2. Theorem 2.3 implies that if $P \subseteq I$, $|P| = q \leqslant n^{1/d}$ then $|N(P)| \geqslant n - \frac{1}{2}(n-q) \geqslant n/2$. Thus $G(q,d)$ is $(n, n^{1/d}, n/2)$-expanding. As noted

by Pippenger [23], the well-known results about the problem of Zarenkiewicz (see, e.g., [11]) supply lower bounds on the number of edges of expanding graphs. Using the results of [11] one can easily show that the number of edges of an $(n, n^{1/d}, n/2)$ expanding graph is at least $(1 + o(1)) \cdot \ell n 2 \cdot n^{2-1/d}$. Note that the number of edges of $G(q, d)$ (or of $H(n,d)$) is $(1 + o(1)) \cdot n^{2-1/d}$ and thus these graphs have (up to a constant of $1/\ell n 2$) the smallest possible number of edges.

3. Let $PG(d,q)$ be the finite geometry of dimension d over the field $GF(q)$ and let $G(q,d)$ be the corresponding expander. Let n,k be as in Theorem 2.3. By Singer's Theorem ([12, p. 128]) there exist $0 \leqslant a_1 < a_2 < \cdots < a_k < n$ such that $G(q,d)$ is isomorphic to the bipartite graph with classes of vertices $A = B = \{0,1,2,...,n-1\}$ in which $a \in A$ is joined to $b \in B$ iff $b = (a + a_i) \ (mod\ n)$ for some $1 \leqslant i \leqslant k$. This contrasts with the result of [17] that implies that no family of linear expanders can have this form and thus shows a difference between highly expanding and weakly expanding graphs.

3. SORTING IN ROUNDS

Suppose we are given n elements with a linear order unknown to us. In the first round we ask m_1 simultaneous questions, each a binary comparison. Having the answers we deduce all implications and ask, in the next round, another m_2 questions, deduce their implications, and so on. A choice of our questions that guarantees that after r rounds we will know the complete order of the elements is an algorithm for sorting in r rounds. The need for such algorithms with fixed r arises in structural modeling (see Häggvist and Hell [15]). Since all comparisons within a round are evaluated simultaneously, such algorithms have obvious connection to parallel sorting, as defined by Valiant [28], and seem to be practical in situations like testing consumer preferences (see Scheele [25]), where the communication between our sorting computer and the consumers is being performed by correspondence. Let $f_r(n)$ denote the minimum possible number of comparisons sufficient to sort n elements in r rounds. Clearly $f_1(n) = \binom{n}{2}$. In Häggvist and Hell [13, 14] and

Bollobás and Thomason [8], probabilistic arguments are used to obtain estimates of $f_r(n)$ for $r \geqslant 2$. In particular it is known that $f_2(n) = O(n^{3/2}\log n)$ and $f_2(n) = \Omega(n^{3/2})$, (see [8]). For practical applications, however, a probabilistic argument is not enough and an explicit sorting algorithm is desirable. Häggvist and Hell observed this fact and in [15] they gave explicit algorithms for sorting in k rounds with $O(n^{s_k})$ comparisons, where $s_k \to 1$ as $k \to \infty$ and, e.g., $s_3 = 8/5$, $s_4 = 20/13$ and $s_5 = 28/19$. It seems more difficult to find an efficient explicit sorting algorithm in two rounds. In [14] such an algorithm with $13/30\ (n^2 - n)$ comparisons is given. A somewhat better algorithm is given in Bollobás and Rosenfeld [7] — with $2/5\ n^2 + O(n^{3/2})$ comparisons. The only construction with $o(n^2)$ comparisons is due to Pippenger [23] — $O(n^{1.943...}(\log n)^{0.943...})$.

Here we use our geometric expanders corresponding to finite geometries of dimension 4 to prove:

Theorem 3.1

By an explicit construction

$$f_2(n) = O(n^{7/4}) \ .$$

□

The first round of our algorithm uses $O(n^{7/4})$ comparisons corresponding to the expanders of finite geometries of dimension 4. We then apply Theorems 2.1 and 2.3 to show that even by deducing only direct implications (i.e., if we find in the first round that $x < y$, $y < z$ and $z < t$ we conclude that $x < z$ and $y < t$ but not necessarily that $x < t$), we will have to compare in the second round only $O(n^{7/4})$ pairs. It is worth noting that in [8] a lower bound of $\Omega(n^{5/3})$ is proved for such an algorithm. Thus our construction is not that far from being best possible.

Theorem 3.1 together with the results of Häggvist and Hell [15, Theorem 3], supply an explicit sorting algorithm in k rounds with $O(n^{\alpha_k})$ comparisons, where $\alpha_1 = 2$, $\alpha_2 = 7/4$ and $\alpha_k = \min(2(2^j - 1)\ \alpha_{k-j} - 2^j)/((2^j - 1)\ \alpha_{k-j} - 1)$, with the minimum taken over all j,

$0 < j < k$, for which $\alpha_{k-j} \geqslant 2^j/(2^j - 1)$. This improves the results of [15] for all $k \geqslant 4$. In particular one can easily check that $\alpha_4 = 26/17$ and $\alpha_5 = 22/15$, slightly better than the corresponding bounds $s_4 = 20/13$ and $s_5 = 28/19$ given in [15].

REFERENCES

[1] H. Abelson, A note on time space tradeoffs for computing continuous functions, Infor. Proc. Letters 8 (1979), 215-217.

[2] M. Ajtai, J. Komlós and E. Szemerédi, Sorting in c log n parallel steps, Combinatorica 3 (1983), 1-19.

[3] N. Alon, Eigenvalues, geometric expanders, sorting in rounds and Ramsey theory, preprint.

[4] N. Alon, and V. D. Milman, λ_1, isoperimetric inequalities for graphs and superconcentrators, preprint.

[5] B. Bollobás, "Extremal Graph Theory," Academic Press, London and New York (1978).

[6] N. G. deBruijn and P. Erdös, On a combinatorial problem, Indagationes Math. 20 (1948), 421-423.

[7] B. Bollobás and M. Rosenfeld, Sorting in one round, Israel J. Math 38 (1981), 154-160.

[8] B. Bollobás and A. Thomason, Parallel sorting, Discrete Appl. Math. 6 (1983), 1-11.

[9] F. R. K. Chung, On concentrators, superconcentrators, generalizers, and nonblocking networks, Bell Sys. Tech. J. 58 (1978), 1765-1777.

[10] O. Gabber and Z. Galil, Explicit construction of linear sized superconcentrators, J. Comp. and Sys. Sci. 22 (1981), 407-420.

[11] R. K. Guy and S. Znam, A problem of Zarenkiewicz, in "Recent Progress in Combinatorics" (W. T. Tutte, ed.) Academic Press, 1969, pp. 237-243.

[12] M. Hall, Jr., "Combinatorial Theory," Wiley and Sons, New York and London, 1967.

[13] R. Häggvist and P. Hell, Graphs and parallel comparison algorithms, Congr. Num. 29 (1980), 497-509.

[14] R. Häggvist and P. Hell, Parallel sorting with constant time for comparisons, SIAM J. Comp. 10, (1981), 465-472.

[15] R. Häggvist and P. Hell, Sorting and merging in rounds, SIAM J. Alg. and Disc. Meth. 3 (1982), 465-473.

[16] J. Ja'Ja', Time space tradeoffs for some algebraic problems, Proc. 12th Ann. ACM Symp. on Theory of Computing, 1980, 339-350.

[17] M. Klawe, Non-existence of one-dimensional expanding graphs, Proc. 22nd Ann. Symp. Found. Comp. Sci. Nashville (1981), 109-113.

[18] T. Lengauer and R. E. Tarjan, Asymptotically tight bounds on time space trade-offs in a pebble game, J. ACM 29 (1982), 1087-1130.

[19] G. A. Margulis, Explicit constructions of concentrators, Prob. Per. Infor. 9 (4), (1973), 71-80 (English translation in problems of Infor. Trans. (1975), 325-332).

[20] M. Pinsker, On the complexity of a concentrator, 7th International Teletraffic Conference, Stockholm, June 1973, 318/1-318/4.

[21] N. Pippenger, Superconcentrators, SIAM J. Computing 6 (1977), 298-304.

[22] N. Pippenger, Advances in pebbling, Internat. Colloq. on Autom. Lang. and Prog. 9 (1982), 407-417.

[23] N. Pippenger, Explicit construction of highly expanding graphs, preprint.

[24] W. J. Paul, R. E. Tarjan and J. R. Celoni, Space bounds for a game on graphs, Math. Sys. Theory 20 (1977), 239-251.

[25] S. Scheele, Final report to office of environmental education, Dept. of Health, Education and Welfare, Social Engineering Technology, Los Angeles, CA 1977.

[26] R. M. Tanner, Explicit construction of concentrators from generalized N-gons, SIAM J. Alg. Discr. Meth., to appear.

[27] M. Tompa, Time space tradeoffs for computing functions, using connectivity properties of their circuits, J. Comp. and Sys. Sci. 20 (1980), 118-132.

[28] L. G. Valiant, Parallelism in comparison networks. SIAM J. Comp. 4 (1975), 348-355.

[29] L. G. Valiant, Graph theoretic properties in computational complexity, J. Comp. and Sys. Sci. 13 (1976), 278-285.

Long Path Enumeration Algorithms for Timing Verification on Large Digital Systems

Tetsuo Asano

Osaka Electro-Communication University, Japan

Shinichi Sato

NEC Corporation, Japan

ABSTRACT

To insure that a digital system will operate at a desired speed, the designer must verify that every path delay is not too long or too short. Thus, an efficient path enumeration technique is needed. In this paper we present two algorithms for enumerating all possible paths in a directed acyclic graph that exceed given delay time. They run in $O(m \log m + N(K))$ and $O(Km)$ time, respectively, and require $O(m + n)$ space, where K, m, and n are the number of such long paths, arcs, and nodes of a graph, and N(K) is the total number of nodes on K such long paths.

1. Introduction

To insure that a digital system will operate correctly and at a desired speed, the designer must verify that the propagation delay along every path is within a specified limit. Previously, circuit simulators and delay simulators (simulators providing timing analysis capability) have been used as timing verification tools. Circuit simulators can produce detailed circuit behavior and are suitable for simulating transistors or small circuits. However, since they require too much computer time, they cannot be applied to circuits containing a large number of gates. Delay simulators also have the limitation that they cannot evaluate all the logic paths, they can only evaluate certain paths which are activated by the simulation input patterns defined by the designers. All other paths will not be checked. As is stated above, simulators are not suitable for timing verification. Our approach is to enumerate all possible logic paths and to examine their validity, that is, to verify that the path delay is not too long or too short for every logic path that may be activated. Thus, an algorithm is needed for enumerating all paths longer (or shorter) than a given limit.

In the field of graph theory, the problem of finding the shortest (or longest) path between two specified nodes has been intensively studied [1-3]. Among several generalizations of the problem, the problem of finding K shortest (or longest) paths between two specified nodes in a graph is important for the above-mentioned purpose. A number of algorithms have been reported for the problem[4-6]. The fastest one [6] proposed by Katoh, Ibaraki, and Mine requires O(Kc(n, m)) time and O(Kn + m) space where c(n, m) is the time to compute shortest paths from one node to the others in an undirected graph with n nodes and m edges, and $c(n, m) < \min[O(n^2), O(m \log n)]$ is known. Note that a graph corresponding to a logic diagram has no cycle. Thus, an efficient algorithm may exist for such restricted graphs. In fact, we can easily construct an O(m) time algorithm for finding a shortest (or longest) path, in other words, c(n, m) = O(m) holds in our case. Thus, we have an O(Km) time and O(Kn + m) space algorithm for enumerating K shortest (or longest) paths. Based on this algorithm, the following approach may be considered: Enumerate paths in the increasing (or decreasing) order of their lengths until the length becomes longer (or shorter) than a given limit. However, how can we estimate the value of K, the number of such paths to be enumerated? Recall that the purpose here is to enumerate all paths shorter (or longer) than a given limit, in other words, we do not have to enumerate them in the increasing or decreasing order of their lengths. We propose two efficient algorithms for enumerating all the paths exceeding a given limit. They run in O(m log m + N(K')) and O(K'm) time, respectively, and require O(m + n) space, where K', m, and n are the number of such long paths, arcs, and nodes, respectively, and N(K') is the total number of nodes on K' such long paths.

2. Graphical Representation of Logic Diagram

Delay analysis is performed by tracing logic paths in combinatorial circuits. Fig. 1 shows the four kinds of logic paths. An example of a logic diagram is shown in Fig. 2 where each box indicates a gate and a figure in it represents its delay time. We can represent such a logic diagram by a directed acyclic graph as shown in Fig. 3. Nodes of the graph may be decomposed into three kinds: starting nodes, ending nodes, and intermediate nodes. Each directed arc is assigned a value corresponding to the delay time. The problem is to check whether each logic path (a directed path from a starting node to an ending node) which may be activated satisfies the predefined design specification. Thus, we need an algorithm for enumerating all possible paths exceeding a given limit in a directed acyclic graph.

3. Path Enumeration Algorithm

Given a directed acyclic graph G corresponding to a logic diagram, we construct another directed acyclic graph G* by adding two nodes s and t, called "source" and "terminal", respectively, and a directed arc (s, v) for every starting node v and (v', t) for every ending node v'. The length of every added arc is defined to be zero. Thus, if P = $(v_1, v_2, \ldots, v_m)$ is a directed path from a stating node v_1 to an ending node v_m, then P* = $(s, v_1, v_2, \ldots, v_m, t)$ is a directed path in G* between s and t, and it has the same length as P, where length of a path P, denoted by length(P), is the sum of the lengths of the arcs on P. Similarly, the length of an arc (v, v') is denoted by length(v, v'). Fig. 4 shows an example of a directed acyclic graph G and its extended graph G*. Throughout this paper for each node v the successors of v are ordered so that the first successor of v is denoted by fs(v) and the next successor of v following from v' is denoted by next(v; v'). If v' is the last successor of v then next(v, v') is defined to be NIL.

We first present a simple exhaustive algorithm which enumerates all possible paths in a graph and tests whether the length of each path exceeds a given limit. In the following algorithm, by v(i) we denote the *i*th node on the current path.

[Algorithm-A]

(Step 1) Specify a limit L_θ on lengths of paths.

(Step 2) Find a path P = (s=v(0), v(1), ... , v(q)=t) between s and t by tracing a pointer fs() to the first successor at each node, starting at the source node s. Computationally, perform the operation v(i) = fs(v(i-1)) for i = 1, 2, ... until v(i) = t.

(Step 3 - Comparison) If length(P) $\geq L_\theta$ then output the path P.

(Step 4 - Backtrack) Find the largest i such that next(v(i), v(i+1)) is not NIL. If there exists no such i then all the possible paths have been enumerated, and thus the algorithm terminates.

(Step 5 - Next path) Set v(i+1) = next(v(i), v(i+1)), and then trace the pointer fs() from v(i+1) to the terminal t to find a path P = (s=v(0), v(1), ... , v(q')=t). Go to (Step 3).

It is easily verify that the above algorithm enumerates all possible paths but the time complexity is $O(2^m)$ in the worst case where m is the number of arcs in a graph. Note that it may run in $O(2^m)$ time even if a given graph contains a small number of paths which exceed a given limit. We can dramatically reduce the time complexity. The basic idea is to test whether the path being formed has any chance to be longer than the limit, and to ignore those which have such a chance as early as possible. We need some preprocessing as follows: First, we calculate at each node v the value dist(v), which is the length of the longest path from v to the terminal node t. Then, at each node v we sort its successors v' in the decreasing order of length(v, v') + dist(v'), which indicates the length of the longest path from v via the arc (v, v'), in other words, v" = next(v, v') means that length(v, v") + dist(v") $\leq$ length(v, v') + dist(v'). It is easily shown that the length of the longest path with an initial subpath P = (s=v(0), v(1), ... , v(i)) is given by

$$\text{dist}(v(i)) + \Sigma_{j=1}^{i} \text{length}(v(j\text{-}1), v(j))$$

or equivalently

$$\text{dist}(v(i))+ \text{length}(v(i\text{-}1), v(i))+ \Sigma_{j}^{i-1}\text{length}(v(j\text{-}1),v(j)).$$

Therefore, given an initial subpath P from s to a node v, if length(P) + dist(v) < L_θ then we can ignore all of those paths with the initial subpath P.

[Algorithm - B]

(Step 0 - Preprocess) (i) For each node v, calculate dist(v), the length of the longest path from v to t. (ii) For each node v, sort its successor v' in the decreasing order of length(v, v') + dist(v'), and define pointers fs(v) and next(v, v'), that is, if

$$\text{length}(v, v^1) + \text{dist}(v^1) \geq \text{length}(v, v^2) + \text{dist}(v^2) \geq \ldots$$

then fs(v) = v^1 and next(v, v^j) = v^{j+1}, j = 1, 2, ...

(Step 1) Specify a limit L_θ on lengths of paths.

(Step 2) Find a path P = (s=v(0), v(1), ... , v(q)=t) between s and t by tracing a pointer fs() at each node. This is the longest path in the graph. Thus, if the length of the path does not exceed the limit L_θ then the search is completed with the answer that there exists no such long path. Otherwise, set r(s) = 0 and r(v(i)) = r(v(i-1)) +

length(v(i-1), v(i)) for each i, $1 \leq i \leq q$, where r(v) is the length of the current initial subpath from s to v.

(Step 3 - Backtrack) Find the largest i such that next(v(i), v(i+ 1)) is not NIL and r(v(i)) + dist(v) + length(v(i), v) $\geq L_\theta$, where v is next(v(i), v(i+ 1)). If there exists no such i then all the possible paths have been enumerated, and thus the algorithm terminates.

(Step 4 - Next path) Set v(i+ 1) = next(v(i), v(i+ 1)), and then trace the pointer fs() from v(i+ 1) to the terminal t to find a path P = (s=v(0), v(1), ... , v(q')=t). For j = i+ 1, i+ 2, ..., q'-1, recompute the value r(v(j)) by r(v(j)) = r(v(j-1)) + length(v(j-1), v(j)). Output the path P since length(P) $\geq L_\theta$ holds, and then go to (Step 3).

Consider the graph shown in Fig. 4(b). The first step of the above algorithm is to compute the value dist(v) at each node v and then sort its successors by the value length(v, v') + dist(v'), the length of the longest path from v via the arc (v, v') to the terminal t. Table 1 summerizes the results.

Suppose the limit is 18. At the step 2 we find a longest path P = (s, A, D, G, I, L, t) by tracing a pointer fs() at each node. Since the length of the path exceeds the limit, we calculate the value r(v), the length of the current initial subpath from s to v. The results are as follows:

$$r(s)=0,\ r(A)=0,\ r(D)=3,\ r(G)=8,\ r(I)=14,\ r(L)=20,\ r(t)=20.$$

The next step is to find the largest i such that next(v(i), v(i+ 1)) is not NIL and r(v(i)) + dist(v) + length(v(i), v) $\geq L_\theta$, where v is next(v(i), v(i+ 1)). So the nodes L, I, G, D, A, and s are examined in this order. At the node L and I, we have next(L, t) =NIL and next(I, L) = NIL. At the node G, next(G, I) is J and the path composed by the initial subpath of P from s to G and the longest path from G via the arc (G, J) to the terminal node t exceeds the limit 18 because the length is calculated as follows:

$$r(G) + \text{length}(G, J) + \text{dist}(J) = 8 + 5 + 5 = 18.$$

Thus, we have found the second path P' = (s, A, D, G, J, M, t). For the node J, M, and t, the values of the r()'s are recomputed:

$$r(J) = r(G) + \text{length}(G, J) = 13,$$
$$r(M) = r(J) + \text{length}(J, M) = 18,$$
$$r(t) = r(M) + \text{length}(M, t) = 18.$$

Once again, we try to find the next long path by backtracking. Thus, we have found the third long path P" = (s, A, E, G, I, L, t) by tracing the pointer fs() from the node E to t. The algorithm produces no other paths.

The space complexity of the algorithm is O(m + n) since we do not have to store long paths enumerated. The time complexity is O(m log m + N(K)) where m is the number of arcs and N(K) is the total number of nodes on K paths which exceed a given limit. The algorithm checks the inequality

$$r(v(i)) + \text{length}(v(i), v) + \text{dist}(v) \geq L_\theta,$$
$$\text{where } v = \text{next}(v(i), v(i+1)),$$

for each node v(i), i = q-1, q-2, ... , 0. Thus it takes at most O(q) time. If the inequality holds for any i, the algorithm finds the next long path P' = (s=v(0), v(1), ... , v(i), v'(i+1), ... , v'(q')=t) by tracing the pointer fs(), which takes at most O(q') time. Recomputation of r() is also performed in O(q') time. Thus, if we denote by q_k the number of nodes on the *k*th long path found by the algorithm then the time consumed is given by

$$2q_1 + (q_1 + 2q_2) + \dots + (q_{k-1} + 2q_k) + q_k = 3N(K),$$

where $N(K) = \Sigma\, q_i$, i.e., the total number of nodes on K paths longer than the given limit. Therefore we can conclude that the time complexity of the algorithm is O(m log m + N(K)).

The above algorithm may be inefficient for some graphs. An extreme case is that the longest path in a graph is shorter than a given limit. In this case the preprocessing operation is meaningless. Generally speaking, if a graph contains a small number of long paths then the O(m log m) time needed by the preprocessing operation becomes a dominating factor in the time complexity of the algorithm. In the following we present another algorithm which does not require much preprocessing. For this purpose, at each node v of a graph we divide the set of successors into two subsets S(v) and U(v), where v' ∈ S(v) (v" ∈ U(v), resp.) means that the successor v' (v", resp.) of v belongs to a set of sorted successors (unsorted successors, resp.). Initially, at each node v we find a successor v' to maximize the value length(v, v') + dist(v') among successors of v and then set S(v) = {v'} and U(v) = {all other successors of v}. Also initially we set next(v, v') = NIL for every edge (v, v') of a graph. The preprocessing operation is just the above stated, and thus takes only O(m + n) time. The algorithm proceeds in the same way as Algorithm B. The only exception is when next(v, v') finds NIL. We must check whether next(v, v') is really NIL or not. The check is easy. We first examine the set U(v). If it is empty the next(v, v') is really NIL. Otherwise, next(v, v') should be such a node v" that it maximize the value of length(v, v") + dist(v") among the nodes in U(v). Thus, we set next(v, v') = v" and remove v" from the set U(v).

[Algorithm - C]

(Step 0 - Preprocessing) (i) For each node v, calculate dist(v). (ii) For each arc (v, v'), set next(v, v') = NIL. (iii) For each node v, find a successor v' to maximize the value of length(v, v') + dist(v'), and set fs(v) = v' and U(v) = {all other successors of v}.

(Step 1) Specify a limit L_θ.

(Step 2) Find a path P = (s=v(0), v(1), ... v(q)=t) by tracing a pointer fs() at each node. If the length of the path is shorter than the given limit then return the answer that there exists no such long path. Otherwise, set r(s) = 0 and r(v(i)) = r(v(i-1)) + length(v(i-1), v(i)) for i = 1, 2, ... , q.

(Step 3 - Backtrack) For each i, i = q-1, q-2, ... , 1, perform the following operation:

(a) If next(v(i), v(i+ 1)) is NIL and U(v(i)) is not empty then find a successor v of v(i) to maximize the value length(v(i), v) + dist(v) among the set U(v(i)) and then set next(v(i), v(i+ 1)) = v and remove v from the set U(v(i)).

(b) Check the current value of next(v(i), v(i+ 1)). If it is not NIL, say v, and the inequality r(v(i)) + length(v(i), v) + dist(v) $\geq L_\theta$ holds, then go to (Step 4).

If the condition in (b) does not hold for any i, then return the answer that all the possible paths have been enumerated.

(Step 4 - Next path) Set v(i+ 1) = next(v(i), v(i+ 1)), and then trace the pointer fs() from v(i+ 1) to the terminal node t to find a path P = (s=v(0), v(1), ... , v(q')=t). For j = i+ 1, i+ 2, ... , q'-1, recompute the value r(v(j)) by r(v(j)) = r(v(j-1)) + length(v(j-1), v(j)). Output the path P and then go to (Step 3).

The time complexity of the above algorithm is O(Km) since each iteration takes at most O(m) time.

4. Conclusions

In this paper we have given two algorithms for enumerating those paths in a directed acyclic graph which exceed a given limit: One runs in O(m log m + N(K)) time and the other in O(Km) time. The space required is O(m + n). N(K) is the total number of nodes on such long paths. Since O(N(K)) time is an obvious lower bound on the time complexity, the first algorithm seems to be near optimal. The second algorithm is efficient for small K, i.e., for those graphs containing only a small number of long paths.

These algorithms will be useful for the timing verification of large digital systems. However, the problem of enumerating only those paths that may be activated by some input patterns remains to be solved.

References

[1] Dijkstra, E.W.: "*A note on two problems in connexion with graphs*", Numerishe Mathematik, 1, p.269 (1959).

[2] Dreyfus, S.E.: "*An appraisal of some shortest path algorithms*", Oper. Res., 17, p.395 (1969).

[3] Johnson, D.B.: "*Efficient algorithms for shortest paths in sparse networks*", J. Assoc. Comput. Mach., 24, 1, p.1 (1977).

[4] Yen, J.Y.: "*Finding the K shortest loopless paths in a network*", Manag. Sci., 17, 11, p.712 (1971).

[5] Fox, B.L.: "*Calculating Kth shortest paths*", INFOR, 11, 1, p.66 (1973).

[6] Katoh, N., T. Ibaraki, and H. Mine: "*An Efficient algorithm for K shortest simple paths*", Networks, 12, p.411 (1982).

Table 1. Results of preprocessing operation.

node	dist()	fs()	ordering of successors
s	20	A	A, C, B
A	20	D	D, E
B	16	F	F, I
C	17	E	E, G, F
D	17	G	G, I
E	14	G	G, H, J
F	10	H	H, K
G	12	I	I, J, L
H	8	J	J, M
I	6	L	L
J	5	M	M, L
K	4	L	L, M
L	0	t	t
M	0	t	t

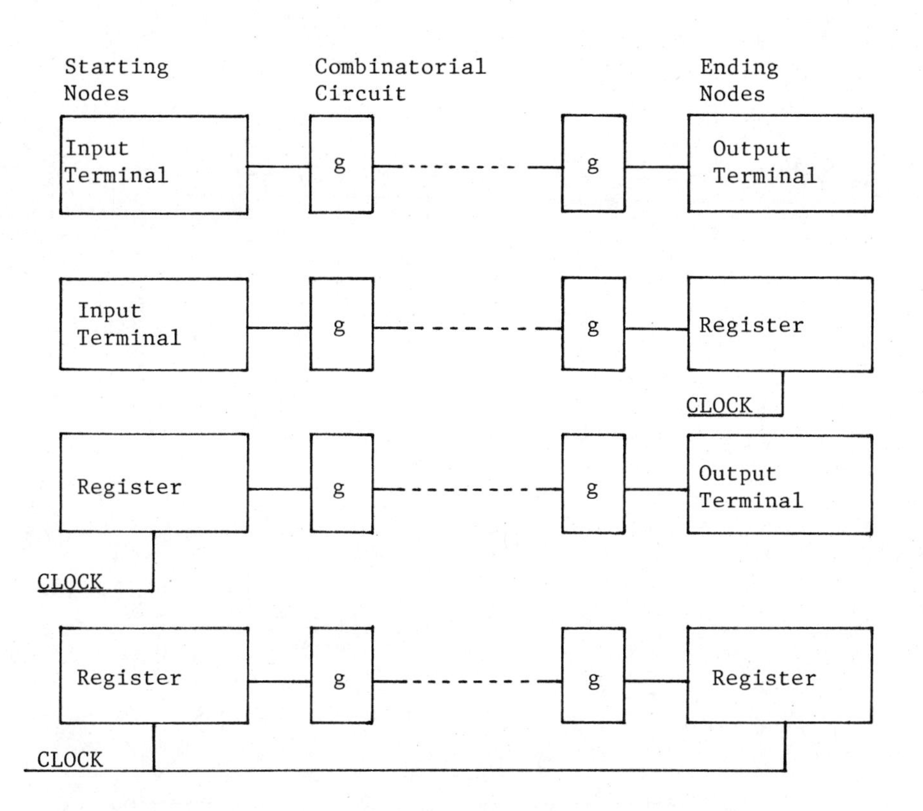

Fig. 1 Logical path.

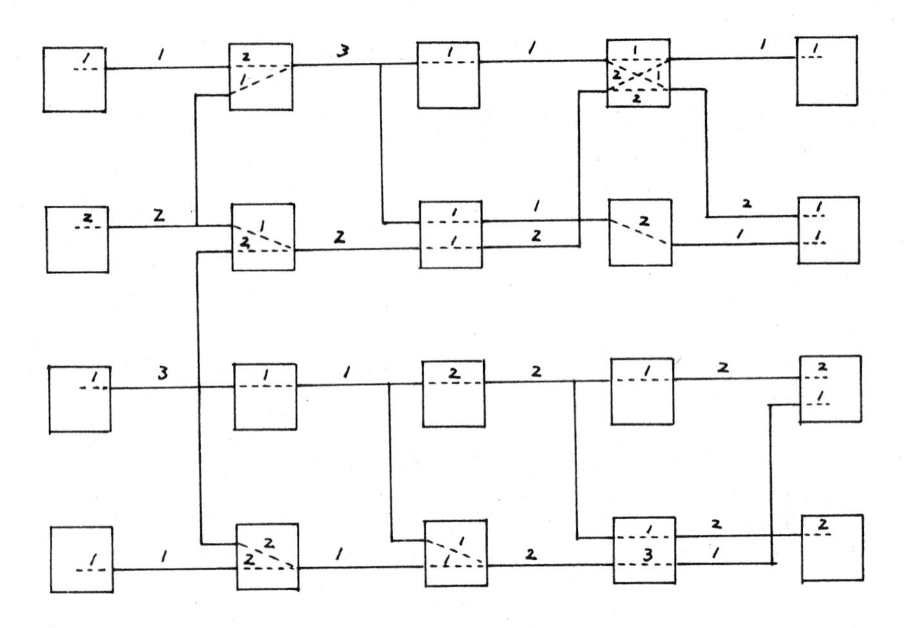

Fig. 2 A logic diagram.

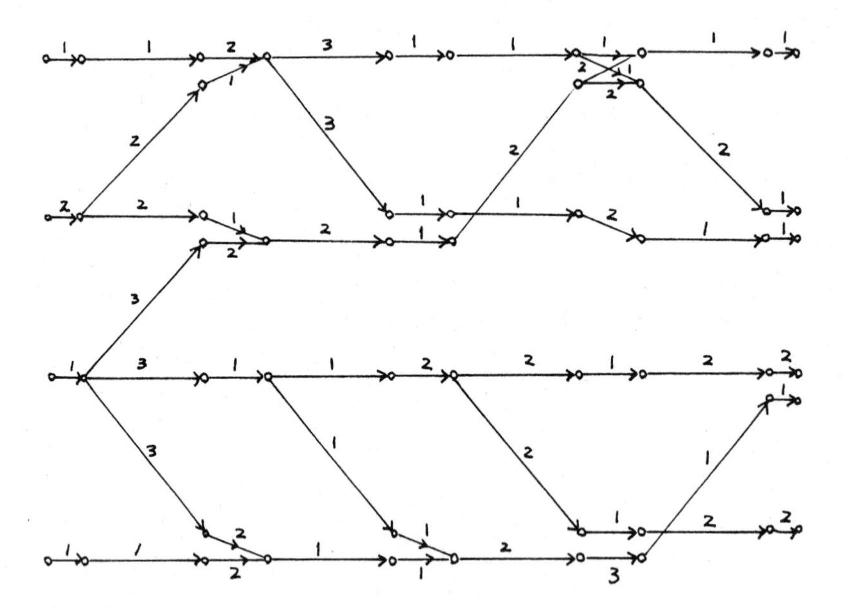

Fig. 3 A directed acyclic graph corresponding to a logic diagram.

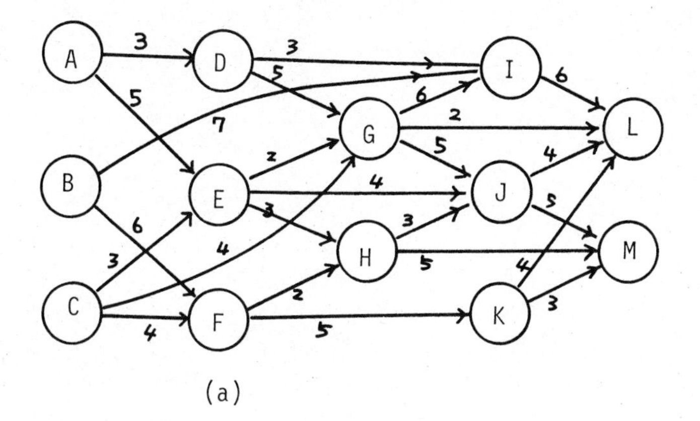

(a)

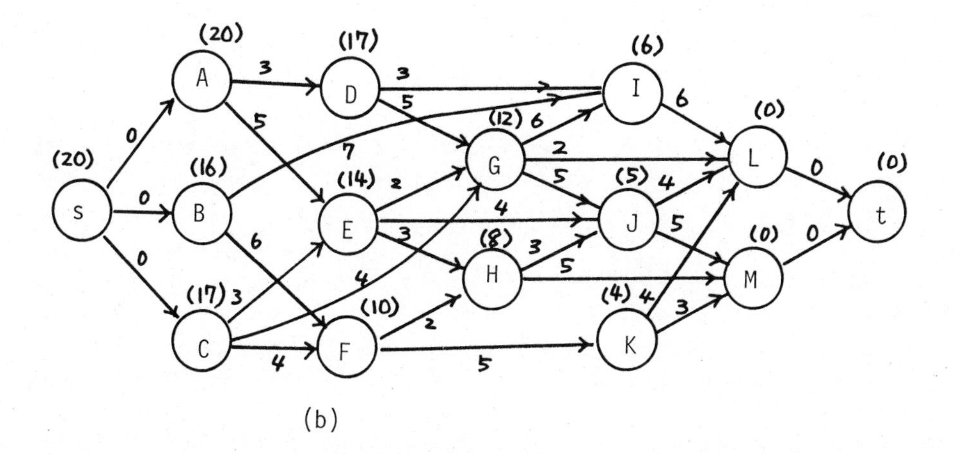

(b)

Fig. 4. Example of a directed acyclic graph.
(a) A directed acyclic graph.
(b) Its extended graph.

ON UPSETS IN BIPARTITE TOURNAMENTS

Kunwarjit S. Bagga
Indiana University-Purdue University at Fort Wayne

ABSTRACT

Given two teams X and Y of m and n players respectively, a bipartite tournament can be thought of as a result of the competition in which each member of X plays against every member of Y and there are no ties. Suppose that the players in each team are ranked according to their scores. We then investigate the occurrence and bounds on the number of upsets in the teams.

Introduction

An _ordinary tournament_ (or _n-tournament_) is an orientation of K_n, the complete graph on n vertices. An _m × n bipartite tournament_ is an orientation of $K_{m,n}$, the complete bipartite graph. As is well-known, ordinary tournaments can be used to represent round robin competitions. Analogously, an m × n bipartite tournament T results when each player of a team $X = \{x_1, x_2, \ldots, x_m\}$ competes against every player of another team $Y = \{y_1, y_2, \ldots, y_n\}$

and there are no ties. We write $x_i \longrightarrow y_j$ when the player x_i beats (or dominates) the player y_j. The <u>score</u> a_i of x_i is the number of players in Y dominated by x_i and b_j = score y_j is similarly defined. The lists $A = [a_1, a_2, \ldots, a_m]$ and $B = [b_1, b_2, \ldots, b_n]$ are the <u>score lists</u> of T. The <u>dominance matrix</u> of T is the $m \times n$ matrix $D = (d_{ij})$ in which $d_{ij} = 1$ if $x_i \longrightarrow y_j$ and $d_{ij} = 0$ if $y_j \longrightarrow x_i$.

The first major study of bipartite tournaments was done by Moon [8]. For terminology and a variety of results on bipartite tournaments and their score lists, we refer the reader to Beineke and Moon [3]. A comparative study of ordinary and bipartite tournaments can be found in Beineke [2].

<u>Upsets</u>

We consider a bipartite tournament T on teams $X = \{x_1, x_2, \ldots, x_m\}$ and $Y = \{y_1, y_2, \ldots, y_n\}$ with score lists $A = [a_1, a_2, \ldots, a_m]$ and $B = [b_1, b_2, \ldots, b_n]$, respectively. We rank players in each team according to their scores. Thus, relabelling if necessary, we may assume $a_1 \geq a_2 \geq \cdots \geq a_m$ and $b_1 \geq b_2 \geq \cdots \geq b_n$ so that in the team X, x_1 is the superiormost player while x_m is the inferiormost player. The players in Y are similarly ranked. We will see below that the order in which players of equal scores are ranked is immaterial for our discussion.

An <u>upset</u> is said to have occurred in the team X if there exist i, j and k with $1 \leq i < j \leq m$ and $1 \leq k \leq n$ such that

$$x_j \longrightarrow y_k \longrightarrow x_i .$$

Thus, an upset results when a player "beats" a superior player via a player of the other team. For $1 \leq i < j \leq m$, let $u_{ij} = |\{y_k | x_j \longrightarrow y_k \longrightarrow x_i\}|$. Then $U_X = \sum_{1 \leq i < j \leq m} u_{ij}$ is the total number of upsets in the team X. We similarly define

$U_y = \sum_{1 \leq k < \ell \leq n} v_{k\ell}$ and let $U = U_x + U_y$ denote the total number of upsets in T.

Using D, the dominance matrix of T, the upsets can be easily displayed and counted as follows: For $1 \leq i < j \leq m$, u_{ij} = number of r such that there is a zero and a one in the rth coordinates of the rows R_i and R_j of D respectively. Similarly $v_{k\ell}$ = number of s such that there is a one and a zero in the sth coordinates of the columns C_k and C_ℓ respectively ($1 \leq k < \ell \leq n$).

From this it can be seen that the number of upsets in each of the teams is independent of the order in which the players of equal scores are ranked.

Now consider two players x_i, x_j with $1 \leq i < j \leq m$. If there exists a y_k with $x_j \longrightarrow y_k \longrightarrow x_i$, then $a_i \geq a_j$ implies that there is a y_ℓ with $x_i \longrightarrow y_\ell \longrightarrow x_j$. In other words, we have a (directed) 4-cycle containing x_i and x_j. Conversely, a 4-cycle containing x_i and x_j gives an upset. However, this correspondence is not in general one-to-one. We thus have the following result.

Proposition. The following statements are equivalent for a bipartite tournament T:

(i) T has no upsets.

(ii) T has no 4-cycles.

(iii) T has no cycles.

It is a well-known fact that for every $n \geq 1$, there is a unique acyclic n-tournament, the transitive tournament on n vertices. This is not true of bipartite tournaments, however. The following result holds; its proof may be found in [4].

Theorem 1. A bipartite $m \times n$ tournament with score lists $A = [a_1, a_2, \ldots, a_m]$ and $B = [b_1, b_2, \ldots, b_n]$ is acyclic iff $[a_1, a_2, \ldots, a_m]$ and $[m - b_n, m - b_{n-1}, \ldots, m - b_1]$ are conjugate partitions.

Now let $u = \min_{1 \le i < j \le m} u_{ij}$. The probability of u being 0 is seen to be at most $\binom{m}{2}(3/4)^n$ and this $\longrightarrow 0$ when $m, n \longrightarrow \infty$ while being not too disparate. Equivalently, the probability that a given pair of vertices in the same team lie on a 4-cycle $\longrightarrow 1$. This can also be proved to hold for any two vertices in different teams.

We now discuss some properties of irreducible tournaments which will be used in obtaining bounds on the number of upsets. Recall that a tournament (ordinary or bipartite) is <u>irreducible</u> (or <u>strong</u>) if given any two vertices, there is a directed path from each to the other. A well-known result of Camion states that every irreducible n-tournament has a cycle of length n. This was generalized by Harary and Moser [7] who showed that there are cycles of all lengths between 3 and n. From this we get the immediate corollary: For any irreducible n-tournament T, there is a vertex v such that T - v is also irreducible. The above results do not hold for bipartite tournaments, however. For example, the bipartite tournament T with dominance matrix

$$D = \begin{bmatrix} 1 & 0 & 0 \\ 0 & 1 & 1 \\ 0 & 1 & 1 \end{bmatrix}$$

is irreducible but non-hamiltonian. Also, for $n \ge 2$, the $n \times n$ bipartite tournament T shown in Figure 1 (where only $X \longrightarrow Y$ arcs are indicated) is irreducible and has the property that T - v is reducible for every vertex v.

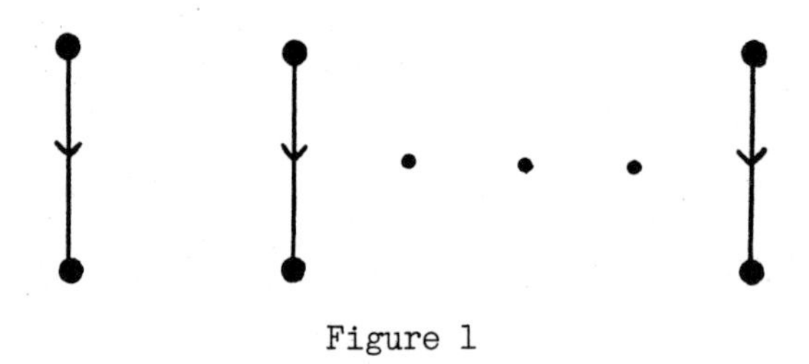

Figure 1

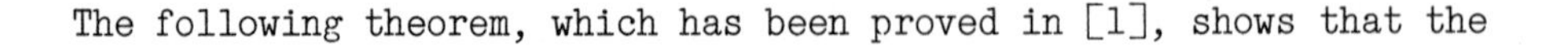

The following theorem, which has been proved in [1], shows that the above example is special.

Theorem 2. If T is an irreducible $m \times n$ bipartite tournament, then there exists a vertex v such that T - v is also irreducible unless $m = n$ and T is the tournament of Figure 1.

In considering a lower bound on the number of upsets, we observe that upsets occur within irreducible components. This follows from our earlier remarks about the correspondence between upsets and 4-cycles. Hence, the following theorem is stated for irreducible bipartite tournaments.

Theorem 3. The minimum number of upsets in the class of all irreducible $m \times n$ bipartite tournaments is $m + n - 2$.

Proof: We use induction on $m + n$, the total number of vertices, to show that the number of upsets is bounded below by $m + n - 2$. The only irreducible bipartite tournament with $m + n = 4$ is the 4-cycle; it has two upsets, one in each team. For $m + n = 5$, there is, up to isomorphism, only one 3×2 (or 2×3) irreducible bipartite tournament with dominance matrix $\begin{bmatrix} 1 & 0 \\ 0 & 1 \\ 0 & 1 \end{bmatrix}$ and it has 3 upsets. The inductive step follows immediately from Theorem 2 and the observation that any vertex in an irreducible bipartite tournament lies on a 4-cycle. For the exceptional $n \times n$ bipartite tournament of Figure 1, the number of upsets is seen to be $n(n - 1)$, which is at least $2n - 2$ for $n \geq 2$. This completes the induction.

Now consider the $m \times n$ bipartite tournament T with score lists $[n - 1, 1, \ldots, 1]$ and $[m - 1, m - 1, \ldots, m - 1, 1]$ and dominance matrix

$$D = \begin{bmatrix} 0 & 1 & \dots & 1 & 1 \\ 0 & 0 & \dots & 0 & 1 \\ \cdot & & & & \\ \cdot & & & & \\ \cdot & & & & \\ 0 & 0 & \dots & 0 & 1 \\ 1 & 0 & \dots & 0 & 0 \end{bmatrix}$$

T is easily seen to be irreducible, and it has $m + n - 2$ upsets. This proves the theorem.

We observe that the minimum numbers of upsets in the two teams are $m - 1$ and $n - 1$, respectively.

In the following theorem, we consider an upper bound.

Theorem 4. The number of upsets in an $m \times n$ bipartite tournament is bounded above by

$$\frac{n}{2}\left[\frac{m^2}{4}\right] + \frac{m}{2}\left[\frac{n^2}{4}\right]$$

Proof: As usual, let T be an $m \times n$ bipartite tournament on teams $X = \{x_1, x_2, \dots, x_m\}$ and $Y = \{y_1, y_2, \dots, y_n\}$ with corresponding score lists $[a_1, a_2, \dots, a_m]$ and $[b_1, b_2, \dots, b_n]$. Consider 2-paths of the type $x_i \longrightarrow y_k \longrightarrow x_j$. The number of such paths is $\sum_{k=1}^{n} b_k(m - b_k)$. Also, for every such path with $a_i \leq a_j$, there exists at least one 2-path $x_j \longrightarrow y_\ell \longrightarrow x_i$. Hence, $U_x \leq \frac{1}{2} \sum_{k=1}^{n} b_k(m - b_k)$. The sum on the right side of this inequality attains its maximum value when b_k and $m - b_k$ are as nearly equal as possible for each k. The theorem follows.

We note that the upper bound is sharp when both m and n are even. An extremal example is the $m \times n$ bipartite tournament with the matrix $D = \begin{bmatrix} J & 0 \\ 0 & J \end{bmatrix}$, where J is the $\frac{m}{2} \times \frac{n}{2}$ matrix of all ones.

The total number of 2 paths of T is

$P = \sum_{i=1}^{m} a_i(n - a_i) + \sum_{k=1}^{n} b_k(m - b_k)$. In the proof of Theorem 4, we found that $U \leq \frac{1}{2}P$. The case when equality holds is treated in the next theorem.

Theorem 5. If $A = [a_1, a_2, \ldots, a_m]$ and $B = [b_1, b_2, \ldots, b_n]$ are the score lists of T, then $U = \frac{1}{2}P$ iff both A and B are constant.

Proof: Given vertices x_i and x_j with scores $a_i \geq a_j$ respectively, the vertices of Y can be partitioned into four classes:

$$\begin{aligned} C_1 &: x_i \longrightarrow y_k \longrightarrow x_j \\ C_2 &: x_i \longleftarrow y_k \longleftarrow x_j \\ C_3 &: x_i \longrightarrow y_k \longleftarrow x_j \\ C_4 &: x_i \longleftarrow y_k \longrightarrow x_j \;. \end{aligned}$$

Then $a_i = |C_1| + |C_3| \geq |C_2| + |C_3| = a_j$ so that $|C_1| \geq |C_2|$ and the equality holds iff $a_i = a_j$. The theorem follows.

It is known that there exist non-isomorphic bipartite tournaments with the same score lists. For example, given the score lists $A = [2, 2, 2, 2] = B$, we have the non-isomorphic bipartite tournaments T_1 and T_2 with their respective matrices

$$D_1 = \begin{bmatrix} 1 & 1 & 0 & 0 \\ 1 & 1 & 0 & 0 \\ 0 & 0 & 1 & 1 \\ 0 & 0 & 1 & 1 \end{bmatrix} \quad \text{and} \quad D_2 = \begin{bmatrix} 1 & 1 & 0 & 0 \\ 0 & 1 & 1 & 0 \\ 0 & 0 & 1 & 1 \\ 1 & 0 & 0 & 1 \end{bmatrix}$$

From the above theorem, it follows that when A and B are constant, the number of upsets depends only on A and B and is independent of their tournament realization. In fact, this is true in general, as we now show. We will need the following result, whose proof may be found in [3].

Theorem 6. If two bipartite tournaments have the same score lists, then each can be transformed into the other by successively reversing the arcs of 4-cycles.

We observe that in the matrix form, 4-cycle arc reversals correspond to replacing one of the following two 2×2 submatrices of D by the other:

$$\begin{bmatrix} 1 & 0 \\ 0 & 1 \end{bmatrix} \qquad \begin{bmatrix} 0 & 1 \\ 1 & 0 \end{bmatrix}.$$

As an immediate corollary of the above theorem we get our next result.

Theorem 7. If two bipartite tournaments have the same score lists, then they have the same number of upsets. In fact, each team has the same number of upsets in the two tournaments.

The next objective, then, is to find a formula which expresses the number of upsets in terms of the scores. This we can obtain by following the reasoning in the proof of Theorem 5.

Theorem 8. For the bipartite tournament T on teams X and Y with the score lists $[a_1 \geq a_2 \geq \cdots \geq a_m]$ and $[b_1 \geq b_2 \geq \cdots \geq b_n]$ respectively, the number of upsets in X is given by

$$U_X = \frac{1}{2}\left[\sum_{k=1}^{n} b_k(m - b_k) - \sum_{1 \leq i < j \leq n} (a_i - a_j)\right].$$

It was remarked earlier that the upper bound of Theorem 4 is sharp when both m and n are even. The formula of Theorem 8 can be used to show that the upper bound is also sharp when $m = n$, and that it is sharp in no other case.

We observe that due to our method of ranking players in each team according to their scores, the upsets also include those between players of equal scores. This also happens for upsets in ordinary tournaments, which have been extensively studied in [5], [6]. That the number of upsets in bipartite tournaments between players of unequal scores is not independent of tournament realizations may be seen from the following example of bipartite tournaments T_1 and T_2 with score lists A = [3, 2, 2, 1] and B = [3, 3, 1, 1] and their corresponding dominance matrices

$$D_1 = \begin{bmatrix} 1 & 1 & 1 & 0 \\ 0 & 0 & 1 & 1 \\ 0 & 0 & 1 & 1 \\ 0 & 0 & 0 & 1 \end{bmatrix} \quad \text{and} \quad D_2 = \begin{bmatrix} 1 & 0 & 1 & 1 \\ 0 & 1 & 1 & 0 \\ 0 & 0 & 1 & 1 \\ 0 & 0 & 0 & 1 \end{bmatrix}$$

The first has five upsets between players of unequal scores and one upset between players of equal scores, while the corresponding numbers for the second are three and three.

ACKNOWLEDGEMENT: The author is indebted to Professor Lowell Beineke for his constant encouragement and advice.

REFERENCES

[1] Bagga, Kunwarjit S., Some structural properties of bipartite tournaments, Ph.D. Thesis, Purdue University, 1984, to appear.

[2] Beineke, L. W., A tour through tournaments or Bipartite and ordinary tournaments: A comparative survey, in Combinatorics, London Mathematical Society Lecture Note Series 52, Cambridge University Press (1981), pp 41-45.

[3] Beineke, L. W. and Moon, J. W., On bipartite tournaments and scores, in The Theory and Applications of Graphs, (ed. G. Chartrand et al), Wiley (1981), pp 55-71.

[4] Bollobás, B., Frank, O. and Karoński, M., On 4-cycles in random bipartite tournaments. Journal of Graph Theory. 7 (1983), 183-194.

[5] Brualdi, R. A. and Li, Qiao, Upsets in round robin tournaments. J. Combin. Theory, Ser. B 35(1983), 62-77.

[6] Fulkerson, D. R., Upsets in round robin tournaments, Canad. J. Math 17(1965), 957-969.

[7] Harary, F. and Moser, L., The theory of round robin tournaments. Amer. Math. Monthly 73(1966), 231-245.

[8] Moon, J. W., On some combinatorial and probabilistic aspects of bipartite graphs, Ph.D. Thesis, University of Alberta, Edmonton, 1962.

SOME RESULTS ON BINARY MATRICES OBTAINED VIA BIPARTITE TOURNAMENTS

Kunwarjit S. Bagga
Indiana University-Purdue University at Fort Wayne

Lowell W. Beineke
Indiana University-Purdue University at Fort Wayne

ABSTRACT

Criteria for determining which pairs of lists of nonnegative integers are realizable as the row and column sums of a matrix of zeros and ones were found by H. J. Ryser. We use bipartite tournaments to establish which pairs of lists are uniquely realizable (a result first found by M. Koren) and which pairs belong to self-complementary matrices.

1. Introduction.

Ryser [7] determined which pairs of lists of nonnegative integers are realizable as the row and column sums of a matrix of zeros and ones. Two related questions which we will answer here are (1) which pairs are uniquely realizable, and (2) which pairs are realizable by "self-complementary" matrices.

Gale [4] (the same year as Ryser) and Moon [6] provided other solutions to Ryser's problem. Gale's solution was constructive in nature and was given in the context of network flows, while Moon's was stated for bipartite tournaments. It is also in the context of bipartite tournaments that we do much of our work here. We note that our work on the first question was done independently of earlier work of Koren, who established the same result in the context of undirected graphs, and we also note that an incomplete form was stated in [2]. Because of the length of the proofs, we shall not attempt to provide them here.

2. Terminology and Notation.

An $m \times n$ _bipartite tournament_ is the result of orienting the edges of the complete bipartite graph $K_{m,n}$. (It is thus a bipartite analogue of an ordinary tournament, which is the result of orienting a complete graph.) We say that vertex v _dominates_ vertex w if there is an arc from v to w, and we write $v \to w$. The _score_ of a vertex is the number of vertices which it dominates.

The two partite sets of vertices in a bipartite tournament T will be denoted by $X = \{x_1, \ldots, x_m\}$ and $Y = \{y_1, \ldots, y_n\}$. If the score of x_i is a_i and that of y_j is b_j, then the lists (collections in which order is immaterial and repetitions are permitted) $A = [a_1, \ldots, a_m]$ and $B = [b_1, \ldots, b_n]$ are the _score lists_ of T.

In general, given a pair of lists $A = [a_1, \ldots, a_m]$ and $B = [b_1, \ldots, b_n]$, we speak of the pair of lists A and B as being _realizable_ if there is a bipartite tournament with A and B as its score lists.

The _dominance matrix_ of an $m \times n$ bipartite tournament T is the $m \times n$ matrix $D = (d_{ij})$ in which

$$d_{ij} = \begin{cases} 1 & \text{if } x_i \to y_j \\ 0 & \text{if } y_j \to x_i \end{cases}$$

Thus, D is a matrix of 0's and 1's in which the score of vertex x_i is the number of 1's in the i'th row, while the score of y_j is the number of 0's in the j'th column. That is, the row sums form the score list A, but the column sums the dual list $\overline{B}$.

It might seem from this that the use of bipartite graphs rather than tournaments would be more reasonable, since then both row and column sums would correspond to valencies. However, there is a good reason for using directed graphs, that being that the concept of reducibility is much more natural there than for undirected graphs or even for matrices.

In the three cases, reducibility takes these forms, all equivalent to each other:

(a) A bipartite tournament (or any directed graph) is <u>reducible</u> if there is a nonempty proper subset of the vertex set which has no incoming arcs (from outside the subset).

(b) A bipartite graph is <u>reducible</u> if there are partitions of the partite sets $X = X_1 \cup X_2$ and $Y = Y \cup Y_2$ (not both degenerate) so that all pairs of vertices from X_1 and Y_2 are adjacent and no pairs from X_2 and Y_1 are.

(c) A matrix D of zeros and ones is <u>reducible</u> if after suitable permutations of rows and columns it can be partitioned into the block form

$$D = \begin{pmatrix} D_{11} & J \\ \underline{0} & D_{22} \end{pmatrix},$$

where J and $\underline{0}$ are matrices of all 1's and all 0's respectively.

Any of these structures is called <u>irreducible</u> if it is not reducible. (For digraphs, irreducibility is also known as strong connectedness.) A maximal irreducible subtournament is called an irreducible component.

3. Ryser's Theorem and Some Consequences.

We state Ryser's theorem in its bipartite tournament form, extended to include irreducibility, because that is the form in which we use it.

Theorem 1. (Ryser) Two lists of nonnegative integers $A = [a_1 \geq \ldots \geq a_m]$ and $B = [b_1 \geq \ldots \geq b_n]$ are realizable by a bipartite tournament if and only if, for $k = 1, \ldots, m$,

$$(R_k) \qquad \sum_{i=1}^{k} a_i \leq \sum_{j=1}^{n} \min \{k_1, m - b_j\},$$

with equality when $k = m$.

Furthermore, a realization is reducible if and only if $b_i = m$, $b_n = 0$, or for some $k < m$, equality holds in (R_k).

For completeness, we also include versions of Gale's and Moon's theorems.

Theorem 2. (Gale) Let A and B be a pair of lists of nonnegative integers, and let A' and B' be obtained by deleting one entry a_i from A and reducing $n - a_i$ highest entries of B by 1 each. Then A and B are realizable if and only if A' and B' are.

Theorem 3. (Moon) A pair of lists $A = [a_1 \leq \ldots \leq a_m]$ and $B = [b_1 \leq \ldots \leq b_n]$ of nonnegative integers are realizable if and only if for $k = 1, \ldots, m$ and $\ell = 1, \ldots, n$,

$$(M_{k,\ell}) \qquad \sum_{i=1}^{k} a_i + \sum_{j=1}^{\ell} b_j \geq k\ell,$$

with equality when $k = m$ and $\ell = n$.

Furthermore, a realization is reducible if and only if $a_1 = 0$, $b_1 = 0$, or equality holds in $(M_{k,\ell})$ for another pair k,ℓ.

Returning to Ryser's theorem, we note that if some scores in a list are equal, then one need not consider all of the inequalities (R_k), but only those for which $a_k > a_{k+1}$ (and $k = m$). A similar statement can also be made for Moon's result.

If a pair of lists is realizable, then the dominance of vertices is determined up to dominance within irreducible components. That is, the number of, the orders of, and the dominance between irreducible components are completely determined by the lists. Therefore, the only difference between two realizations of a given pair of lists can be within irreducible components.

In obtaining our results on uniqueness, we derive some other results on realizability. For example, we show that, given a pair of lists with an irreducible realization, if one entry is increased by 1, then the resulting lists are also realizable (but not necessarily irreducible).

4. Unique Realizability.

A pair of realizable score lists is called <u>uniquely realizable</u> if all realizations are isomorphic. The matrix equivalent of uniqueness means unique up to permutations of rows and columns.

It is easily seen that if one of the lists in a realizable pair is all 1's, then all realizations must be isomorphic. This is indicated in Figure 1, where we assume that A is an all-1 list and only the X-to-Y arcs are shown. Obviously, the same conclusion holds if the dual of either list is all 1's.

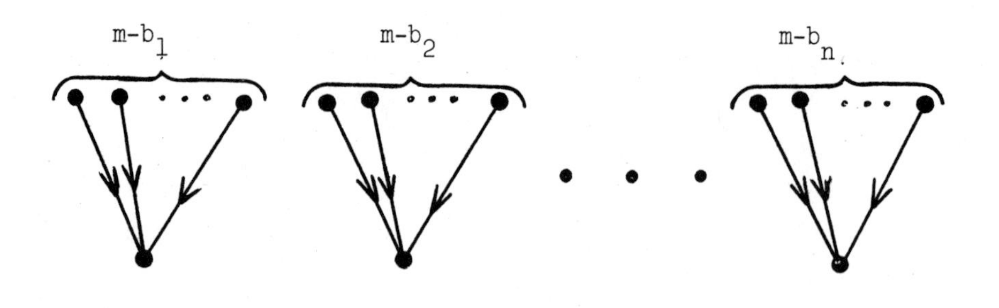

Figure 1

So, we assume that A and B are a uniquely realizable pair of lists belonging to an irreducible bipartite tournament, and that neither is an all-1 list nor the dual of such a list. We are then able to deduce these facts:

(a) One of the lists is constant and the other is not.

(b) The nonconstant list has precisely two values, one of which occurs exactly once.

These facts provide the basis for our proof of the necessity of the conditions for unique realizability.

Theorem 4. An irreducible pair of score lists is uniquely realizable if and only if either they or their duals are in one of the following three classes:

(I) one list is all 1's (the other arbitrary);

(II) one list is [d, 1, ..., 1] and the other constant;

(III) one list is [d, ..., d, 1] and the other all 2's.

We note that the corresponding result for ordinary tournaments was provided by Avery [1].

Theorem 5. The score list of a nontrival irreducible ordinary tournament is uniquely realizable if and only if it is one of the lists [1, 1, 1], [1, 1, 2, 2] or [2, 2, 2, 2, 2].

5. Self-complementarity.

The complement $\overline{P}$ of an $m \times n$ matrix P of zeros and ones has as its i, j entry $1 - p_{m+1-1,n+1-j}$. Informally, we note that $\overline{P}$ can be obtained by rotating the matrix J - P about its "center." If $\overline{P} = P$, we say that P is self-complementary. An example of such a matrix is

$$P = \begin{pmatrix} 1 & 1 & 0 & 1 \\ 1 & 0 & 1 & 0 \\ 0 & 1 & 0 & 0 \end{pmatrix}.$$

Our interest in determining which row and column sums belong to self-complementary matrices was piqued by J. W. Moon, who first raised the corresponding question for bipartite tournaments (the ordinary tournament problem was solved by Eplett [3]). We will show later that in fact the bipartite tournament and matrix questions are not exactly the same in this instance.

Let P be an $m \times n$ self-complementary matrix. Clearly, P must have as many 0's as 1's so that not both m and n can be odd. Furthermore, the lists A and B of row and column sums must be self-dual, in that, for $1 \leq i \leq m$ and $1 \leq j \leq n$, $a_i + a_{m+1-i} = n$ and $b_j + b_{n+1-j} = m$. However, as Moon observed, these conditions are not sufficient. For example, the lists [3, 3, 3, 1, 1, 1] and [4, 4, 2, 2] are self-dual and realizable, but they belong to no self-complementary matrix.

To see this, we introduce some matrix notation. If P is a matrix, we let $P(i_1, \ldots, i_r | j_1, \ldots, j_s)$ denote the $r \times s$ submatrix of P obtained by deleting those rows i_k and columns j_ℓ not listed. Now assume that P realizes the lists [3, 3, 3, 1, 1, 1] and [4, 4, 2, 2], and let x be the number of 1's in $P(1, 2, 3|1, 2)$, y the number in $P(1, 2, 3|3, 4)$ and z the number in $P(4, 5, 6|1, 2)$. Then clearly $x + y = 9$ and $x + z = 8$, so that y and z have different parity. But if P were self-complementary then $y + z = 2 \cdot 3$, which is impossible.

As it happens, this line of reasoning provides us with the condition needed to solve our problem (the proof of sufficiency is quite lengthy). Let P be an $m \times n$ matrix with A and B its lists of row and column sums. Assume that $m = 2r$ and $n = 2s$, and let

$$A_r = \sum_{i=1}^{r} a_i \text{ and } B_s = \sum_{j=1}^{s} b_j.$$

Also, let the numbers of 1's in the $r \times s$ submatrices $P(1, \ldots, r|1, \ldots, s)$, $P(1, \ldots, r|s + 1, \ldots, n)$, and $P(r + 1, \ldots, m|1, \ldots, s)$ be x, y, and z respectively. Then $x + y = A_r$ and $x + z = B_s$. Furthermore, if P is self-complementary, then $y + z = rs$, so that the number

$$\sigma(A, B) := A_r + B_s - rs$$

must be even. This turns out to be the additional condition we require.

<u>Theorem 6</u>. Let $A = [a_1 \geq \ldots \geq a_m]$ and $B = [b_1 \geq \ldots \geq b_n]$ be realizable as the row and column sums of a matrix. Then they are realized by a self-complementary matrix if and only if

(1) both lists are self-dual; and either

(2a) $m + n$ is odd, or

(2b) m, n, and $\sigma(A, B)$ are all even.

We turn now to the consideration of tournaments. The <u>converse</u> T' of a bipartite tournament T is obtained by reversing each arc in T, and T is called <u>self-converse</u> if T and T' are isomorphic (with partite sets fixed). Clearly, if a bipartite tournament T can be labeled so that its dominance matrix is self-complementary, then T must be self-converse. However, the converse statement is not true -- there exist self-converse bipartite tournaments for which no labeling of the vertices yields a self-complementary dominance matrix. An example is given in Figure 2 (where again only the X-to-Y arcs are shown). This tournament is self-converse and has self-dual score lists A = [3, 3, 2, 2, 1, 1] and B = [5, 4, 2, 1]. Since $\sigma(A, B)$ is odd, they cannot be realized by a self-complementary matrix. (We note that T and T' are isomorphic under the mapping which interchanges x_1 and x_6, x_2 and x_5, y_1 and y_4, and y_2 and y_3 but leaves x_3 and x_4 fixed.)

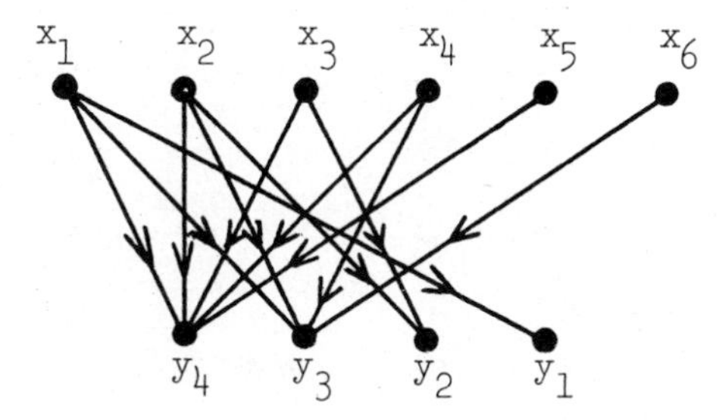

Figure 2

The following result is an extension of Theorem 6 to self-converse bipartite tournaments. In essence, it removes the restriction of evenness on $\sigma(A, B)$ in certain cases.

<u>Theorem 7</u>. A realizable pair of score lists $A = [a_1 \geq \dots \geq a_m]$ and $B = [b_1 \geq \dots \geq b_n]$ are realized by a self-converse bipartite tournament if and only if

(1) both lists are self-dual;

(2) m or n is even; and

(3) whenever m = 2r and n = 2s are both even and $a_r > a_{r+1}$ and

$b_s > b_{s+1}$, then $\sigma(A, B)$ is also even.

In conclusion, we note that again the corresponding result [3] for ordinary tournaments has a much simpler statement -- only "self-duality" is required.

Theorem 8. (Eplett) A tournament score list $S = [s_1 \geq \cdots \geq s_n]$ belongs to a self-converse ordinary tournament if and only if, for $i = 1, \ldots, n$, $s_i + s_{n+1-i} = n - 1$.

REFERENCES

[1] P. Avery, Condition for a tournament score sequence to be simple, J. Graph Theory, 4(1980), 157-164.

[2] L. W. Beineke and J. W. Moon, On bipartite tournaments and scores, Chapter in The Theory and Applications of Graphs (ed. G. Chartrand), Wiley (1981), pp. 55-71.

[3] W. J. Eplett, Self-converse tournaments, Canad. Math. Bull. 22(1979), 23-27.

[4] D. Gale, A theorem on flows in networks, Pacific J. Math. 7(1957), 1073-1082.

[5] M. Koren, Pairs of sequences with a unique realization by bipartite graphs, J. Comb. Theory 21(1976), 224-234.

[6] J. W. Moon, On some combinatorial and probabilistic aspects of bipartite graphs, Ph.D. thesis, University of Alberta, Edmonton, 1962.

[7] H. J. Ryser, Combinatorial properties of matrices of zeros and ones, Canad. J. Math. 9(1957), 371-377.

PARTITIONING THE NODES OF A GRAPH

Earl R. Barnes
IBM Thomas J. Watson Research Center
Yorktown Heights, New York 10598

ABSTRACT:

We prove a separating hyperplane theorem which can be used, in conjunction with a procedure described in [6], to obtain tight lower bounds on the number of edges that must be cut when the nodes of a graph are partitioned into a specified number of groups of specified sizes.

1. **Introduction**. Let G be an undirected graph having nodes $N = \{ 1,...,n \}$ and edge set E. It is often of interest to partition the nodes of G into a given number, say k, of disjoint subsets $S_1,...,S_k$, of specified sizes $|S_1| = m_1 \geq ... \geq |S_k| = m_k$, in such a way as to minimize the number of edges joining nodes in distinct subsets of the partition. For example, such problems arise in laying out computer circuits on chips. See, for example, [1] and [2].

Several heuristic algorithms have been proposed for solving graph partitioning problems. We mention [3], [4],[5] to name a few. In [6] we described a procedure for determining when a partition obtained by a heuristic procedure is optimal or nearly optimal. We gave a procedure for computing a lower bound on the number of edges that must be cut by a partition of the type we have described. When a heuristic partition cuts a number of edges close to this lower bound we know it is nearly optimal. In this paper we show how to compute optimal values for some of the parameters that occured in [6]. This leads to a sharpening of the bounds obtained there.

2. A Procedure for Computing Lower Bounds. Let $N = S_1 \cup \ldots \cup S_k$ be a partitioning of the nodes of G. Let

$$x_j = (x_{1j}, \ldots, x_{nj})^T$$

be an indicator vector for S_j, $j = 1, \ldots, k$. Thus

$$x_{ij} = \begin{cases} 1 \text{ if } i \in S_j \\ 0 \text{ if } i \notin S_j, \quad i = 1, \ldots, n, \; j = 1, \ldots, k. \end{cases}$$

Clearly

$$\sum_{i=1}^{n} x_{ij} = |S_j| = m_j, \; j = 1, \ldots, k,$$

and

$$\sum_{j=1}^{k} x_{ij} = 1, \; i = 1, \ldots, n.$$

Let a_{ij} denote the number of edges connecting nodes i and j and let $A = (a_{ij})$ denote the adjacency matrix for G. The number of edges with both endpoints in S_j is given by

$$\frac{1}{2} \sum_{r \in S_j} \sum_{s \in S_j} a_{rs} = \frac{1}{2} \sum_{r=1}^{n} \sum_{s=1}^{n} a_{rs} \, x_{rj} \, x_{sj} = \frac{1}{2} x_j^T A x_j.$$

Thus the number of edges not cut by our partition is given by

$$E_{nc} = \frac{1}{2} \sum_{j=1}^{k} x_j^T A x_j.$$

Let E_c denote the number of edges cut. Since $E_c + E_{nc} = |E|$, minimizing E_c is equivalent to maximizing E_{nc}. Thus our graph partitioning problem reduces to the quadratic assignment problem

$$
\begin{aligned}
\text{maximize} \quad & \sum_{j=1}^{k} x_j^T A x_j \\
\text{subject to} \quad & \sum_{i=1}^{n} x_{ij} = m_j,\ j = 1,\ldots,k, \\
& \sum_{j=1}^{k} x_{ij} = 1,\ i = 1,\ldots,n, \\
& x_{ij} = 0 \text{ or } 1 \text{ for all } i \text{ and } j.
\end{aligned}
\qquad (2.1)
$$

Consider the problem of finding an upper bound on the value of this maximum. To this end, define $v_j = \frac{1}{\sqrt{m_j}} x_j$, $j = 1,\ldots,k$. The v_j's form an orthonormal set of vectors. Let $\lambda_1 \geq \ldots \geq \lambda_n$ denote the eigenvalues of A and let $u_1,\ldots,u_n$ denote a corresponding orthonormal set of eigenvectors. Then

$$A = \sum_{i=1}^{n} \lambda_i u_i u_i^T.$$

It follows that

$$
\begin{aligned}
\sum_{j=1}^{k} x_j^T A x_j = \sum_{j=1}^{k} m_j v_j^T A v_j \quad & = \sum_{i=1}^{n} \sum_{j=1}^{k} \lambda_i m_j (u_i^T v_j)^2 \\
& = \sum_{i=1}^{n} \sum_{j=1}^{k} \lambda_i m_j s_{ij}
\end{aligned}
$$

where $s_{ij} = (u_i^T v_j)^2 \geq 0$. Note that

$$\sum_{i=1}^{n} s_{ij} = \| v_j \|^2 = 1,\ j = 1,\ldots,k,$$

and

$$\sum_{j=1}^{k} s_{ij} \leq \| u_i \|^2 = 1,\ i = 1,\ldots,n.$$

We can also find numbers γ_{ij} such that

$$\sum_{r=1}^{i} \sum_{t=1}^{j} s_{rt} \leq \gamma_{ij}, \; i = 1,\ldots,n, \; j = 1,\ldots,k. \tag{2.2}$$

Thus we can find an upper bound on the value of the maximum in (2.1) by solving the linear programming problem

$$\begin{aligned}
\text{maximize} \quad & \sum_{i=1}^{n} \sum_{j=1}^{k} \lambda_i \, m_j \, s_{ij} \\
\text{subject to} \quad & \sum_{i=1}^{n} s_{ij} = 1, \; j = 1,\ldots,k \\
& \sum_{j=1}^{k} s_{ij} \leq 1, \; i = 1,\ldots,n, \\
& \sum_{r=1}^{i} \sum_{t=1}^{j} s_{rt} \leq \gamma_{ij}, \; i = 1,\ldots,n, \; j = 1,\ldots,k, \\
& s_{ij} \geq \text{ for all i and j.}
\end{aligned} \tag{2.3}$$

It turns out that the solution of this problem depends on only a few of the eigenvalues (approximately k) of A. Moreover, a solution can be obtained by a greedy algorithm. Thus, the amount of work required to obtain upper bounds by the procedure we are developing is quite modest. It involves computing a few eigenvalues of A and values for the parameters γ_{ij}.

With the introduction of slack variables problem (2.3) becomes a special case of the following linear programming problem studied in [7]. Let (c_{ij}) be an $n \times k$ matrix satisfying

$$c_{ij} + c_{i+1,j+1} \geq c_{i,j+1} + c_{i+1,j} \tag{2.4}$$

for $i = 1,\ldots,n-1$ and $j = 1,\ldots,k-1$. Consider the transportation problem

$$\begin{aligned}
\text{maximize} \quad & \sum_{i=1}^{n} \sum_{j=1}^{k} c_{ij}\, s_{ij} \\
\text{subject to} \quad & \sum_{j=1}^{k} s_{ij} = a_i,\ i = 1,\dots,n, \\
& \sum_{i=1}^{n} s_{ij} = b_j,\ j = 1,\dots,k, \\
& \sum_{r=1}^{i} \sum_{t=1}^{j} s_{rt} \le \gamma_{ij} \quad i = 1,\dots,n,\ j = 1,\dots,k \\
& s_{ij} \ge 0 \qquad i = 1,\dots,n,\ j = 1,\dots,k
\end{aligned} \tag{2.5}$$

where $\sum_{i=1}^{n} a_i = \sum_{j=1}^{k} b_j$.

Condition (2.4) for problem (2.3) requires that

$$\lambda_i\, m_j + \lambda_{i+1}\, m_{j+1} \ge \lambda_i\, m_{j+1} + \lambda_{i+1}\, m_j$$

which holds since

$$(\lambda_i - \lambda_{i+1})(m_j - m_{j+1}) \ge 0.$$

Consider the transportation problem (2.5) and the following greedy algorithm from [7].

Let

$$s_{11} = \min \{a_1, b_1, \gamma_{11}\}. \tag{2.6a}$$

If s_{rt} has been defined for $r \le i$, $t \le j$, $(r,t) \ne (i,j)$, define

$$s_{ij} = \min \left\{ a_i - \sum_{t=1}^{j-1} s_{it},\ b_j - \sum_{r=1}^{i-1} s_{rj},\ \gamma_{ij} - \sum_{\substack{r \le i\, t \le j \\ (r,t) \ne (i,j)}} \sum s_{rt} \right\} \tag{2.6b}$$

Theorem 2.1 . If the $n \times k$ matrix (s_{ij}) defined by (2.6) is feasible for problem (2.5), it is also optimal.

Theorem 2.2. A necessary and sufficient condition for problem (2.5) to have a feasible solution is that

$$\gamma_{ij} \geq \max \left\{ 0, \sum_{r=1}^{i} a_r - \sum_{t=j+1}^{k} b_t \right\} \tag{2.7}$$

for i = 1,...,n and j = 1,...,k-1.

Moreover, if (2.7) holds, and if

$$\gamma_{ij} + \gamma_{i+1,j+1} \geq \gamma_{i,j+1} + \gamma_{i+1,j}$$

and

$$\gamma_{i+1,j} \geq \gamma_{ij}, \gamma_{i,j+1} \geq \gamma_{ij},$$

for i = 1,...,n-1 and j=1,...,k-1, (2.6) gives a feasible, hence optimal solution of (2.5).

These theorems are proved in [7].

Example. Consider the problem of partitioning the nodes of the following graph into two sets containing 5 nodes each. The partition $N = \{1,2,3,4,5\} \cup \{6,7,8,9,10\}$ cuts 3 edges and appears to be optimal. In order

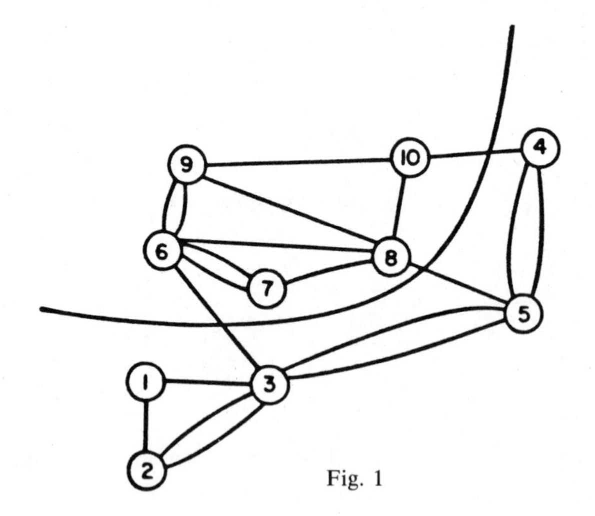

Fig. 1

to prove that it is optimal we must show that for any partition $E_c \geq 3$. Since $|E| = 20$ this amounts to showing that $E_{nc} \leq 17$. We can assign values to the γ_{ij}'s and solve (2.3) in order to obtain an upper bound on E_{nc}. Since

$$\sum_{t=1}^{k} \sum_{r=1}^{n} s_{rt} \leq k$$

we can certainly take each $\gamma_{ij} = k$ in (2.3). If we do this the algorithm (2.6) gives $\sum_{j=1}^{k} \lambda_j m_j$ as the value of the maximum in (2.3). It follows that

$$E_{nc} \leq \frac{1}{2} \sum_{j=1}^{k} \lambda_j m_j. \tag{2.8}$$

This bound was derived by a different approach in [8].

For later use we have computed the first four eigenvalues and eigenvectors of the adjacency matrix for the graph of Fig. 1. They are given by

$$\lambda_1 = 4.418 \quad \lambda_2 = 3.076 \quad \lambda_3 = 2.024 \quad \lambda_4 = .666$$

and

$$u_1 = \begin{matrix} .1362 \\ .2087 \\ .393 \\ .2103 \\ .359 \\ .4646 \\ .2962 \\ .3792 \\ .344 \\ .2113 \end{matrix} \quad u_2 = \begin{matrix} .2829 \\ .3986 \\ .4716 \\ .1878 \\ .3513 \\ -.3319 \\ -.2933 \\ -.2382 \\ -.3339 \\ -.125 \end{matrix} \quad u_3 = \begin{matrix} -.289 \\ -.3626 \\ -.2224 \\ .5737 \\ .4126 \\ -.2611 \\ -.1926 \\ -.1324 \\ -.0266 \\ .3357 \end{matrix} \quad u_4 = \begin{matrix} .2796 \\ .2446 \\ -.0583 \\ -.0709 \\ -.3143 \\ -.1791 \\ -.4639 \\ .0492 \\ .4088 \\ .5813 \end{matrix} \tag{29}$$

Substituting into (2.8) gives

$$E_{nc} \leq \frac{5}{2}(\lambda_1 + \lambda_2) = 18.735.$$

But E_{nc} is an integer. So we must have $E_{nc} \leq 18$. From this we can only conclude that $E_c \geq 2$. In order to show that $E_c \geq 3$ we must begin again with a better choice of the γ_{ij}'s. In Section 4 we will show how to determine optimal values for a few of the γ_{ij}'s and return to prove that $E_c \geq 3$ Section 5.

3. A Separating Hyperplane Theorem. Recall that

$$\sum_{r=1}^{i} \sum_{t=1}^{j} s_{rt} = \sum_{r=1}^{i} \sum_{t=1}^{j} (u_r^T v_t)^2 = \sum_{t=1}^{j} \frac{1}{m_t} x_t^T \Big(\sum_{r=1}^{i} u_r u_r^T \Big) x_t$$
$$= \sum_{t=1}^{j} \frac{1}{m_t} x_t^T Q Q^T x_t$$

where Q is the $n \times i$ matrix whose r-th column is u_r. We define

$$\gamma_{ij} = \max \sum_{t=1}^{j} \frac{1}{m_t} x_t^T QQ^T x_t \qquad (3.1)$$

where the maximum is taken over all vectors $x_1,\ldots,x_j$ satisfying the constraints

$$\sum_{r=1}^{n} x_{rt} = m_t, \; t = 1,\ldots,j,$$
$$\sum_{t=1}^{j} x_{rt} \leq 1, \; r = 1,\ldots,n,$$
$$x_{rt} = 0 \text{ or } 1 \text{ for all } r \text{ and } t.$$

At first glance it may appear that computing the γ_{ij}'s is as hard as our original problem (2.1). But this is not the case. The factored form of the matrix in (3.1) makes the computation of γ_{ij} fairly simple for certain values of i and j. To see this we need the following theorem.

Theorem 3.1. Let q_r denote the r-th row of Q and let $x_1,\ldots,x_j$ be a solution of (3.1). We think of q_r as a point in the i dimensional Euclidean space E_i. Define the sets

$$\hat{S}_t = \{q_r \mid x_{rt} = 1\}, \; t = 1,...,j. \tag{3.2}$$

If j=1 there is a hyperplane of dimension i-1 separating the points in $\hat{S}_1$ from the remainder of the points $q_1,...,q_n$. If $j > 1$, any two of the sets $\hat{S}_1,...,\hat{S}_j$ can be separated by a hyperplane of dimension i-1.

<u>Proof</u>. If all the points $q_1,...,q_n$ are identical the theorem holds trivially. So we assume that some two of these points are distinct.

Let

$$e_t = \frac{1}{m_t} \sum_{r=1}^{n} q_r x_{rt}, \; t = 1,...,j$$

denote the means of the sets $\hat{S}_1,...,\hat{S}_k$, respectively. Let $y_1,...y_j$ be any vectors satisfying the constraints on the vectors in (3.1). We will first prove that

$$\sum_{t=1}^{j} e_t(\sum_{r=1}^{n} q_r x_{rt})^T \geq \sum_{t=1}^{j} e_t(\sum_{r=1}^{n} q_r y_{rt})^T. \tag{3.3}$$

If (3.3) is false for some $y_1,...,y_j$, then

$$\begin{aligned}
\sum_{t=1}^{j} \|\sqrt{m_t} e_t\|^2 &= \sum_{t=1}^{j} \sqrt{m_t}\, e_t(\frac{1}{\sqrt{m_t}} \sum_{r=1}^{n} q_r x_{rt})^T \\
&< \sum_{t=1}^{j} \sqrt{m_t}\, e_t(\frac{1}{\sqrt{m_t}} \sum_{r=1}^{n} q_r y_{rt})^T \\
&\leq \sum_{t=1}^{j} \frac{1}{2}\{\|\sqrt{m_t}\, e_t\|^2 + \|\frac{1}{\sqrt{m_t}} \sum_{r=1}^{n} q_r y_{rt}\|^2\}
\end{aligned}$$

by Minkowski's inequality. We can rewrite this as

$$\sum_{t=1}^{j} \left\| \frac{1}{\sqrt{m_t}} \sum_{r=1}^{n} q_r x_{rt} \right\|^2 = \sum_{t=1}^{j} \frac{1}{m_t} \| Q^T x_t \|^2$$

$$= \sum_{t=1}^{j} \frac{1}{m_t} x_t^T Q Q^T x_t \quad < \sum_{t=1}^{j} \left\| \frac{1}{\sqrt{m_t}} \sum_{r=1}^{n} q_r y_{rt} \right\|^2$$

$$= \sum_{t=1}^{j} \frac{1}{m_t} y_t^T Q Q^T y_t.$$

This contradicts the fact that $x_1,\dots,x_j$ is a solution of (3.1). It follows that (3.3) holds.

Consider the case j=1. (3.3) implies that

$$e_1\Big(\sum_{r=1}^{n} q_r x_{r1}\Big)^T \geq e_1\Big(\sum_{r=1}^{n} q_r y_{r1}\Big)^T \tag{3.4}$$

for any 0-1 vector y_1 satisfying $\sum_{r=1}^{n} y_{r1} = m_1$. Let y be a vector obtained from x_1 by interchanging a 0 and a 1. (3.4) then implies that for some $q_{r_1} \in \hat{S}_1$, and some $q_{r_2} \notin \hat{S}_1$,

$$e_1\, q_{r_1}^T \geq e_1\, q_{r_2}^T. \tag{3.5}$$

Moreover, if we consider all the possible y_1's that can be derived from x_1 in this way, we deduce (3.5) for any $q_{r_1} \in \hat{S}_1$ and any $q_{r_2} \notin \hat{S}_1$. To complete the proof of the theorem for j=1 we need only show that $e_1 \neq 0$.

We have

$$\| \sqrt{m_1}\, e_1 \|^2 = \frac{1}{m_1} \left\| \sum_{r=1}^{n} q_r x_{r1} \right\|^2 = \frac{1}{m_1} x_1^T Q Q^T x_1.$$

Thus, if $e_1 = 0$, the maximum in (3.1) is zero and this implies that

$$\sum_{r=1}^{n} q_r y_{r1} = 0$$

for every vector y, satisfying constraints on x_1 in (3.1). In particular we have

$$
\begin{array}{llllll}
q_1 & +\dots+q_{m_1-1} & +q_{m_1} & & & =0 \\
q_1 & +\dots+q_{m_1-1} & & +q_{m_1+1} & & =0
\end{array}
$$

$$
\begin{array}{lllll}
q_1 & +\dots+q_{m_1-1} & & & +q_n=0 \\
q_1 & & & +q_{m_1+1}+\dots & +q_n=0 \\
& q_2 & & +q_{m_1+1}+\dots & q_n=0
\end{array} \tag{3.6}
$$

$$
q_{m_1} \quad +q_{m_1+1}+\dots \quad +q_n=0.
$$

These equations imply that $q_1 = \dots = q_n = 0$ which contradicts the fact that some two of these vectors are distinct. It follows that $e_1 \neq 0$ and the theorem holds for $j=1$.

To see that the theorem holds for $j > 1$ first consider the case where $j = k = 2$. In this case we have $x_{r2} = 1 - x_{r1}$ and $y_{r2} = 1 - y_{r1}$. Substituting these values into (3.3) gives

$$
(e_1-e_2)\Big(\sum_{r=1}^{n} q_r x_{r1}\Big)^T \geq (e_1-e_2)\Big(\sum_{r=1}^{n} q_r y_{r1}\Big)^T.
$$

The argument used to deduce (3.5) now shows that

$$
(e_1-e_2)q_{r_1}^T \geq (e_1-e_2)q_{r_2}^T
$$

for any $q_r \in \hat{S}_1$ and $q_{r_2} \in \hat{S}_2$. To complete the proof of the theorem we need only show that $e_1 \neq e_2$.

Observe that

$$\| \frac{1}{n} \sum_{r=1}^{n} q_r \|^2 = \| \frac{m_1}{n} e_1 + \frac{m_2}{n} e_2 \|^2$$
$$\leq \frac{m_1}{n} \| e_1 \|^2 + \frac{m_2}{n} \| e_2 \|^2 \tag{3.7}$$
$$= \frac{1}{n} \{ \frac{1}{m_1} x_1^T Q Q^T x_1 + \frac{1}{m_2} x_2^T Q Q^T x_2 \}$$

with equality throughout if and only if $e_1 = e_2$. If $e_1 = e_2$ the maximum in (3.1) is given by the left hand side of (3.7). On the other hand, if there exists vectors y_1, y_2 satisfying the constraints in (2.1) and

$$\frac{1}{m_1} \sum_{r=1}^{n} q_r \, y_{r1} \neq \frac{1}{m_2} \sum_{r=1}^{n} q_r \, y_{r2} \tag{3.8}$$

then the maximum in (3.1) is greater than $\| \frac{1}{n} \sum_{r=1}^{n} q_r \|^2$. This can be seen by substituting the unequal means (3.8) in (3.7). It follows that $e_1 \neq e_2$ if (3.8) holds. So to complete the proof of the theorem for $j = k = 2$ we need only show that there are vectors y_1, y_2 satisfying (2.1) and (3.8).

Suppose that for all vectors y_1, y_2 satisfying the constraints in (2.1) with $k = 2$, we have

$$\frac{1}{m_1} \sum_{r=1}^{n} q_r \, y_{r1} = \frac{1}{m_2} \sum_{r=1}^{n} q_r \, y_{r2}$$

This implies that

$$(\frac{1}{m_1} + \frac{1}{m_2}) \sum_{r=1}^{n} q_r y_{r1} = \frac{1}{m_2} \sum_{r=1}^{n} q_r. \tag{3.9}$$

Since y_1 is arbitrary, subject to $\sum_{r=1}^{n} y_{r1} = m_1$, we can deduce from (3.9) that the points $q_1, \ldots, q_n$ satisfy a system of equations similar to (3.6) with a common right hand side. This implies that $q_1 = \ldots = q_n$ which contradicts our assumption that at least two of these points are distinct. This contradiction establishes the theorem for $j = k = 2$.

The fact that the theorem holds in the general case is now clear. For if we fix j-2 of the vectors $x_1,...,x_j$ at their optimal values, and optimize with respect to the remaining two, we obtain a problem equivalent to the one we just discussed.

4. **Optimal Values for the** γ_{ij}'s. We turn now to the problem of computing the γ_{ij}'s defined by (3.1). First consider γ_{1k}. In this case Q is just the eigenvector u_1. It follows from Theorem 3.1 that there are k-1 real numbers which separate the components of u_1 into k groups $\hat{S}_1,...,\hat{S}_k$ corresponding to the solution $x_1,...,x_k$ of (3.1). Given $\hat{S}_1,...,\hat{S}_k$ the vectors $x_1,...,x_k$ are given by

$$x_{rt} = \begin{cases} 1 \text{ if } u_{r1} \in \hat{S}_t \\ 0 \text{ if } u_{r1} \notin \hat{S}_t, \ r = 1,...,n, \ t = 1,...,k. \end{cases}$$

Since there are only k! ways of separating the components of u_1 into k groups of given sizes by k-1 constants, only k! sets of vectors $x_1,...,x_k$ need be examined to determine the one corresponding to the maximum in (3.1). In a similar fashion we can find the best possible values for γ_{1j}, j = 1,...,k. In fact we can find tight upper bounds on all the partial row sums in the matrix (s_{ij}) for all rows corresponding to known eigenvectors u_i. These bounds were computed in [6]. However, it does not seem possible to use the technique developed there to compute γ_{ij} for other values of i and j.

Now consider the problem of evaluating γ_{i1} for small values of i. γ_{11} has already been evaluated. Consider γ_{21}. In this case the rows of Q are points in the Euclidean plane E_2. According to Theorem 3.1, if x_1 achieves the maximum in (3.1), the m_1 points q_r corresponding to $x_{r1} = 1$ can be separated by a line in E_2 from the $n - m_1$ points corresponding to $x_{r1} = 0$. Since it is a simple matter to find all possible lines in E_2 which separate the points $q_1,...,q_n$ into two sets of sizes m_1 and $n - m_1$ respectively, it is a simple matter to compute γ_{21}.

Now consider γ_{31} given by (3.1). In this case the points q_r lie in E_3 and we must find all hyperplanes which separate m_1 of these points from the remaining $n - m_1$. Any such hyperplane can be chosen to pass through at least three of the points $q_1,...,q_n$, so at most $\binom{n}{3}$ hyperplanes need be considered. This explains how to compute γ_{31}. In principle, we could continue in this way to compute any γ_{i1} for which the eigenvectors $u_1,\ldots,u_i$ are known. However, this calculation becomes very expensive for large values of n and for values of $i > 3$.

It is now clear that we can compute upper bounds on the partial row and column sums of the matrix (s_{ij}) in (2.3). In particular, whenever u_i is known, we can find $a_i \leq 1$ such that $\sum_{j=1}^{k} s_{ij} \leq a_i$. We always replace the corresponding constraint $\sum_{j=1}^{k} s_{ij} \leq 1$ in (2.3) by this one in applying our method. Constraints on partial row and column sums can also be added to those in (2.3) and the resulting problem solved in greedy fashion. We leave the proof of this to the reader.

The reader can also verify that our separating hyperplane technique can be used to compute γ_{i2} for small values of i in the case $k = 2$ when we are partitioning the nodes into two groups. Some examples of this case are given in the next section.

5. **Examples.** 1. Consider the graph in Fig. 1. The eigenvectors in (2.9) can be used to compute $\gamma_{11} = .725$, $\gamma_{12} = .9784$, $\gamma_{21} = .9244$, $\gamma_{22} = 1.838$, $\gamma_{31} = .9244$, $\gamma_{32} = 1.8568$ and $a_1 = .9784$, $a_2 = .9224$, $a_3 = .7602$.

When we solve (2.3) using these bounds we obtain $E_{nc} \leq 17.7$ which implies $E_{nc} \leq 17$ and $E_c \geq 3$ as we suspected.

2. In our second example we give a proof of the following result due to Fiedler [9].

Let $A = (a_{ij})$ be a doubly stochastic symmetric matrix of order n and define

$$\mu(A) = \min \sum_{\substack{i \in M \\ j \notin M}} a_{ij} \tag{4.1}$$

where the minimization is taken over all nonempty subsets M of $N = \{ 1,\ldots,n \}$. Then

$$\left(\frac{n}{n-1}\right)\mu(A) \geq 1-\lambda_2$$

where λ_2 denotes the second largest eigenvector of A.

To see this let G be the graph obtained from a complete graph on n nodes by adding loops to the nodes and assigning weight a_{ij} to each edge (i,j). Let M be a set on which the minimum in (4.1) is achieved and let $m = |M|$. Without loss of generality we assume $n\text{-}m \geq m$. Consider the problem of partitioning the nodes of G into two subsets of sizes n-m and m so as to minimize the sum of the weights on edges cut. The minimum sum is given by $\mu(A)$. Clearly our method for finding lower bounds on E_c can be used to find a lower bound on $\mu(A)$.

Let x_1 and x_2 denote characteristic vectors for the partition corresponding to (4.1). Then clearly the sum of the weight on edges not cut is given by

$$E_{nc} = \frac{1}{2} \sum_{j=1}^{2} x_j^T A x_j + \frac{1}{2}\mathrm{Tr}(A).$$

The first eigenvalue of A is $\lambda_1 = 1$ and the corresponding eigenvector is $u_1 = \frac{1}{\sqrt{n}}(1,\ldots,1)^T$. Thus $x_1^T u_1 = \frac{n-m}{\sqrt{n}}$, $x_2^T u_1 = \frac{m}{\sqrt{n}}$, and we can take $\gamma_{11} = \frac{n-m}{n}$, $\gamma_{12} = \gamma_{21} = 1, \gamma_{22} = 2$. Also, $a_1 = a_2 = 1$. The greedy algorithm then gives $s_{11} = \frac{n-m}{n}$, $s_{12} = s_{21} = \frac{m}{n}$, $s_{22} = \frac{n-m}{n}$. Thus

$$\begin{aligned} E_{nc} &\leq \frac{1}{2}\{\lambda_1(n-m)s_{11} + \lambda_1 m s_{12} + \lambda_2(n-m)s_{21} + \lambda_2 m s_{22}\} + \frac{1}{2}\mathrm{Tr}(A) \\ &= \frac{1}{2}\{n + 2(\lambda_2 - 1)\frac{(n-m)m}{n}\} + \frac{1}{2}\mathrm{Tr}(A). \end{aligned}$$

The total weight on edges of G is $\frac{1}{2}n + \frac{1}{2}\mathrm{Tr}(A)$.Thus

$$\mu(A) \geq \frac{1}{2}n + \frac{1}{2}\mathrm{Tr}(A) - E_{nc} \geq (1-\lambda_2)\frac{(n-m)m}{n}.$$

This implies that

$$1-\lambda_2 \leq \frac{n}{(n-m)m}\mu(A) \leq \left(\frac{n}{n-1}\right)\mu(A)$$

as we claimed.

References

[1] T.C. Raymond, "LSI/VLSI Design Automation", Computer, Vol. 14, No. 7, July 1981, pp. 89-100.

[2] J.R. Tobias, "LSI/VLSI Building Blocks" Computer, Vol. 14, No. 8, August 1981, pp 83-101.

[3] E.R. Barnes, "An Algorithm for Partitioning the Nodes of a Graph", SIAM J. Algebraic and Discrete Methods, Vol. 3, No. 4, Dec. 1982, pp 541-550.

[4] E.R. Barnes, A. Vannelli, J. Q. Walker, "A New Procedure for Partitioning the Nodes of a Graph" IBM Research Report RC 10561, June 1984.

[5] B.W. Kernighan and S. Lin, "An Efficient Heuristic Procedure for Partitioning Graphs", Bell Systems Technical Journal, Vol. 49, Feb. 1970, pp 291-307.

[6] E.R. Barnes and A.J. Hoffman, "Partitioning, Spectra and Linear Programming", Progress in Combinatorial Optimization, W.R. Pulleyblank (Editor), Academic Press, 1984.

[7] E.R. Barnes and A.J. Hoffman, "On Transportation Problems with Upper Bounds on Leading Rectagles", to appear in SIAM J. Algebraic and Discrete Methods.

[8] W.E. Donath and A.J. Hoffman, Lower Bounds for the Partitioning of Graphs, IBM J. Research and Development, Vol. 17, 1973, pp 420-425.

[9] M. Fiedler, "Bounds for Eigenvalues of Doubly Stochastic Matrices", Linear Algebra and Applications, Vol. 5, 1972, pp 299-310.

A GRAPH THEORETICAL CHARACTERIZATION OF MINIMAL DEADLOCKS IN PETRI NETS

J-C. BERMOND

L.R. I., U. A. 410 C.N.R. S., INFORMATIQUE, bat 490,

Universite Paris-Sud, 91405 ORSAY, FRANCE.

G. MEMMI

E.N. S. T. 46 rue Barrault,

75634 PARIS Cedex 13, FRANCE.

ABSTRACT

Petri Nets are now a well known formal model of parallel computation. They can be modeled by weighted bipartite digraphs, with two set of vertices: places and transitions. Here we study the properties of minimal deadlocks, which are important for liveness properties of the net. A deadlock D is a set of places such that $\Gamma^{-}(D) \subseteq \Gamma(D)$. *We give a characterization of minimal deadlocks in terms of path properties and show their application to special classes of Petri Nets.*

0 INTRODUCTION

Petri Nets were introduced in 1962 to model the dynamic evolution of discrete systems. Since then they have been widely used and they constitute now one of the most complete and advanced formal models of parallel computation. Curiously, although a Petri Net is a graph, very few combinatorial studies have been carried out. In this paper we give an answer to such a graph problem.

We first recall some definitions and properties of Petri Nets useful to understand the motivation of the problem; the interested reader will find more details in the recent book of G.W. Brams [2]. In the second part, we study in details the properties of deadlocks. In the third part, we solve a combinatorial problem associated to them. (This part can be read independently of the others).

1 PETRI NETS

In a Petri Net there are two kinds of elements: *places* and *transitions*. Roughly the set of places is defined in order to represent the set of states of the system, the set of transitions represent the set of rules to go from one state to another. We will associate to each state of the system a marking which consists in putting in each place marks or tokens. (One can think that these tokens correspond to available resources). A transition can be fired (i.e. a rule can be executed) if there are enough resources in the places concerned by this transition and the firing of the transition will result in the transfer of some resources of these places to other ones.

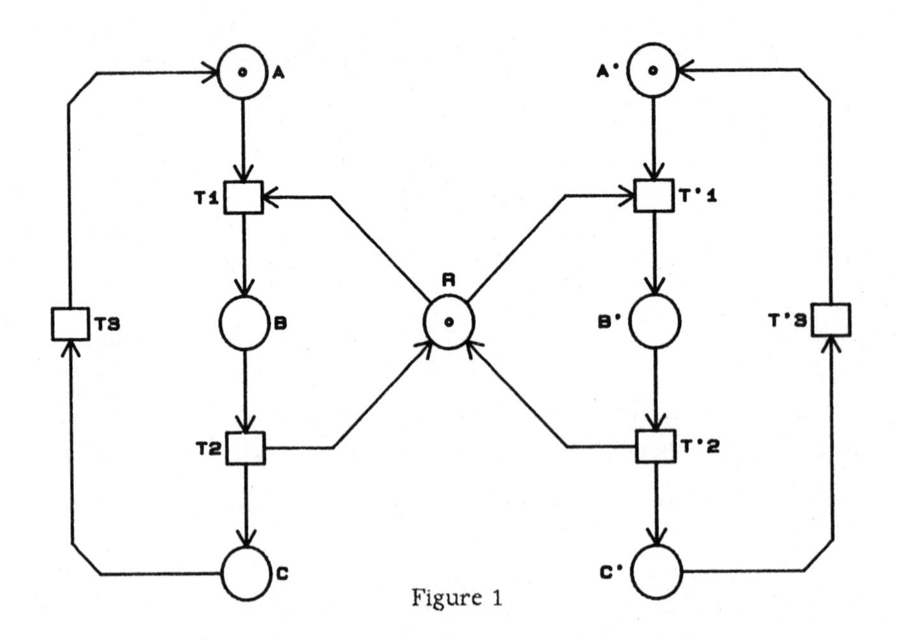

Figure 1

Example 1.1 : The Petri Net of *Figure 1* models the mutual exclusion of two processes sharing a common resource. The place R corresponds to this resource. In place A (resp A') process 1 (resp 2) is waiting for the resource. In place B (resp B') the process 1 (resp 2) uses the resource. In place C (resp C') process 1 (resp 2) is working. Transition T_1 can be fired if there is a token in place A and one in place R, that is if process 1 is waiting and the resource is available. When this transition is fired, we go to place B; we delete the tokens in A and R and add one in B. The functioning of T_1 ' is the same. When T_1 is fired, it is not possible to fire T_1 ' until we have fired T_2, which corresponds to put the first process in state C and to give back the resource in R; again we model this by deleting the token in B and adding one in C and one in R.

Definition 1.2 : A marked Petri Net is a pair $<R,M_0>$ where:
R = < P,T,Γ,V> is a finite weighted bipartite directed graph with vertex set $P \cup T$ where P is the set of places and T the set of transitions and $P \cap T = \phi$;
Γ is the function successor which associates to a vertex of P (resp. T) a subset of T (resp. P); (x,y) is an arc of R if and only if y belongs to Γ(x).
V: $PxT \cup TxP \rightarrow N$,where N is the set of natural numbers is a mapping such that V(x,y) = 0 if and only if $y \notin \Gamma(x)$. V is called the valuation of the Petri Net and consists to associate to each arc of the net a positive integer. (we shall write V(x,y) for the valuation of the arc (x,y));

M_0 is a mapping from P to N which is called ***the initial marking*** and if $M_0(p) = k$, we will say that the place p contains k tokens (or marks).

Remarks: Usually places are pictured with circles and transitions with bars, squares or rectangles.In graph theory, Γ(x) it is also noted $\Gamma^+(x)$,and we denote by $\Gamma^-(x)$ the set $\{y / x \in \Gamma(x)\}$.

Petri Nets can also be defined in a purely algebraic way [8].
The dynamic behavior of a Petri Net is taken into account by considering at any time the marking of the net, where the marking of a place is the number of tokens in this place. The way of passing from one marking to another one is the firing of a transition defined as follows.

Definition 1.3 : A transition t is ***fireable*** at a marking M if and only if:

$$\text{for every p of P, } M(p) \geq V(p,t).$$

If t is fireable at M, we can ***fire*** t and ***reach*** M ' with:

for every p of P, $M'(p) = M(p) - V(p,t) + V(t,p)$.

In the example 1.1. P = {A,B,C,R,A',B',C'} ; T = { $T_1, T_2, T_3, T_1', T_2', T_3'$ } and the valuation of each arc is 1. $M_o(A) = 1$, $M_o(A') = 1$, $M_o(R) = 1$, $M_o(p) = 0$ otherwise. When T_1 is fired, we reach M' with M'(B) = 1, M'(A') = 1, and M'(p) = 0 otherwise.

Example 1.4: In the example of Figure 2, P = {A,B}; T = {t,r}; V(A,r) =2; V(t,A) = 3; and V(x,y) = 1 for the other arcs. $M_o(A) = 2$; $M_o(B) = 0$.

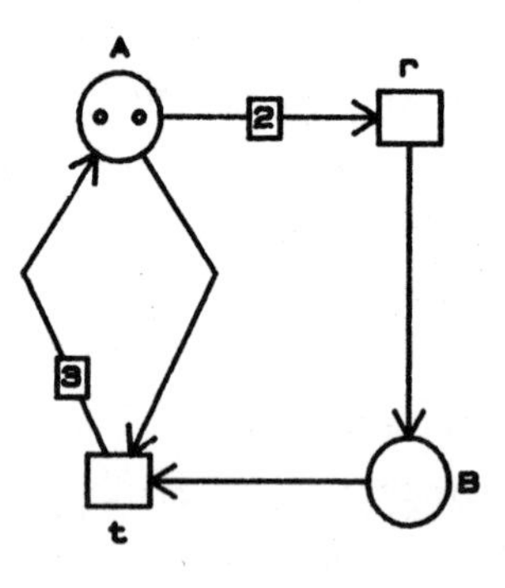

Figure 2

In what follows, we will denote that t is fireable at M by M(t> , and that M' is reached by M(t>M' .We recall the following definitions for a Petri Net <R,M_0>.

Definition 1.5: Let $s = t_1, t_2, \ldots, t_i$ be a sequence of transitions; s is a *firing sequence* fireable at M_o if and only if there exists a sequence of markings $M_1, \ldots, M_i$ such that for every j, $1 \leq j \leq i$, $M_{j-1}(t_j > M_j$ and we will write $M_o(s > M_i$

Notation 1.6: The set of markings reachable from a marking M is denoted by

$$Acc(R, M_o) = \{m \mid \exists s \in T^*, M_o(s > m\}$$

where T^* is the set of all finite sequences on T .

Definition 1.7: A transition t is *live* for M_o if and only if for every M of $Acc(R,M_o)$, there exists s of T^* such that M(st> . $<R;M_o>$ is said to be live if and only if every transition of T is live.

The net of example 1.1 is live for the initial marking M_o. In example 1.4, it can be seen that if $M(A) + 2M(B) = 2k+1$, with k a positive integer, then the net is live; otherwise it is not.

The problem of deciding if a Petri Net is live is exp-space hard [9]. Therefore it is interesting to find classes of Petri Nets for which the problem is easier to solve.

2 DEADLOCKS, TRAPS and COMMONER'S PROPERTY.

We can imagine the firing of transitions as a move of tokens along paths in the Petri Net . A subset of places, such that a token cannot enter in it if it does not contain already a certain amount of tokens, will be called a deadlock and similarly a subset of places such that some tokens can never quit it will be called a trap. These two notions have been introduced in [3] and [6]. Let us give precise definitions.

Definition 2.1 A place p is *deficient in tokens* (shortly deficient) for a marking M if and only if:

$$\Gamma(p) \neq \phi \text{ and } M(p) < \underset{t \in \Gamma(p)}{Min} \{V(p,t)\}.$$

A subset of places Q is deficient for M if and only if each place of Q is deficient for M.

When a subset of places Q is deficient for a marking M, then no transition of $\Gamma(Q)$ is fireable at M.

Definition 2.2 Let D be a subset of places, D is a *deadlock* if and only if $\Gamma^-(D) \subseteq \Gamma(D)$.

Definition 2.3 Let S be a subset of places, S is a *trap* if and only if for every t in $\Gamma(S)$, there exists p in $\Gamma(t) \cap S$ such that either $\Gamma(p) = \phi$, or $V(t,p) \geq \underset{t' \in \Gamma(p)}{Min} (V(p,t'))$.

Remark 2.4: We also have that S is not deficient after the firing of a transition of $\Gamma(S)$. If S is a trap, we have $\Gamma(S) \subseteq \Gamma^{-}(S)$. The union of two deadlocks (or two traps) is a deadlock (resp. a trap). So, we can speak about the maximal deadlock or trap of a given subset. But the intersection of two deadlocks (or two traps) is not necessarily a deadlock (resp. a trap).

Example 2.5 : In the net of Figure 3, {A,B} and {A,C} are both deadlocks and traps. $\{A,B\} \cap \{A,C\} = \{A\}$ is neither a deadlock ($t \in \Gamma^{-}(\{A\})$ and $t \notin \Gamma(\{A\})$) nor a trap ($r \in \Gamma(\{A\})$ and $V(r,A) \leq Min(V(A,r) , V(A,s))$).

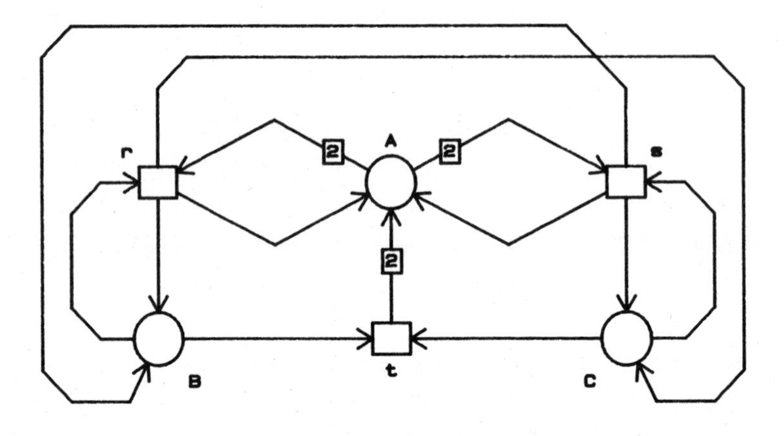

Figure 3

The fundamental property of a deadlock D is the following : if D is deficient for M, it remains deficient for any marking reachable from M.

Property 2.6 : Let D be a subset of places such that there does not exist a place p with $\Gamma(p) = \phi$. Then D is a deadlock if and only if every marking M for which D is deficient is such that D remains deficient for any marking of

Acc(R,M).

Proof : Let D be a deadlock and let M_1 be a marking for which D is deficient. Suppose that there exists a marking M_n of Acc(R,M_1) for which D is not deficient, and let $s = t_2,...,t_n$ be such that $M_1(t_2>M_2(t_3>M_3,...,(t_n>M_n$. Let M_i be the first marking in the sequence for which D is not deficient, necessarily $t_i \in \Gamma^-(D)$; but D is deficient for M_{i-1} then $t_i \notin \Gamma(D)$. This is in contradiction with $\Gamma^-(D) \subseteq \Gamma(D)$.

Conversely, suppose there exists $t \in \Gamma^-(D) - \Gamma(D)$. We can find a marking M such that D is deficient for M (for example by putting no tokens in D; as $\Gamma(p) \neq \phi$ every place is deficient). We can also choose M such that for a given n $M(t^n>$ (where t^n is the sequence where t is fired n times) belongs to Acc(R,M) (it suffices to put enough tokens in the places of $\Gamma^-(t)$). But if n is large enough after n firings of the transition t, a place of D is no more deficient (choose a place p of $\Gamma(t) \cap D$ and $n > V(p,r)$ for one transition r of $\Gamma(p)$). Therefore we have a contradiction and $\Gamma^-(D) \subseteq \Gamma(D)$. □

Property 2.7 : Let A be a subset of places such that there does not exist a place p with $\Gamma(p) = \phi$. Then A is a trap if and only if every marking M for which A is non deficient remains non deficient for any marking of Acc(R,M).

Proof: It is similar to that of property 2.6.

We are now ready to explain the basic idea of Commoner. If a deadlock D is deficient for a marking M, it remains deficient and that means that any transition of $\Gamma(D)$ is not live. On the other hand if a trap is not deficient, it will remain not deficient. Therefore if we want a net to be live, it is necessary that no deadlock becomes deficient. A step to achieve that is that every deadlock contains a non deficient trap. This is the so-called *deadlock-trap property* and it depends of the structure of the Petri Net if this property is a necessary, a sufficient or a necessary and sufficient condition of liveness for a given marked Petri Net (see [2],[7],[10]).

A first restriction on the structure of a Petri Net consists in giving constraints on the valuation V of the net.

Definition 2.8 : A valuation V of a Petri Net is said to be *homogeneous* if and only if for every place p and for every pair of transitions t and t' of $\Gamma(p)$, $V(p,t) = V(p,t')$

Definition 2.9 : A valuation V of a net is said to be *non blocking* if and only if for every transition t, for every $p \in \Gamma(t)$ either $\Gamma(p) = \phi$ or $V(t,p) \geq \underset{t' \in \Gamma(p)}{Min} (V(p,t'))$.

If V is non blocking, a subset of places S is a trap if and only if: $\Gamma(S) \subseteq \Gamma^-(S)$.
We slightly generalize a result of [7] for Petri Net with an homogeneous valuation.

Property 2.10 : Let <R;M_0> be a marked Petri Net with an homogeneous valuation. If every deadlock contains a non deficient trap for M_0, then there does not exist a marking M of $Acc(R,M_0)$ such that no transition is fireable at M.

Proof : Suppose in contrary that such a marking M exists, and consider $D = \{p \in P \mid p$ is deficient for $M\}$. Let t be a transition of $\Gamma^-(D)$, t is not fireable; then there exists p in $\Gamma^-(t)$ such that $M(p) < V(p,t)$. As V is homogeneous, p is deficient for M and $t \in \Gamma(D)$. Then D is a deadlock with no non deficient trap for M. □

Remark 2.11 : This property does not hold in general.

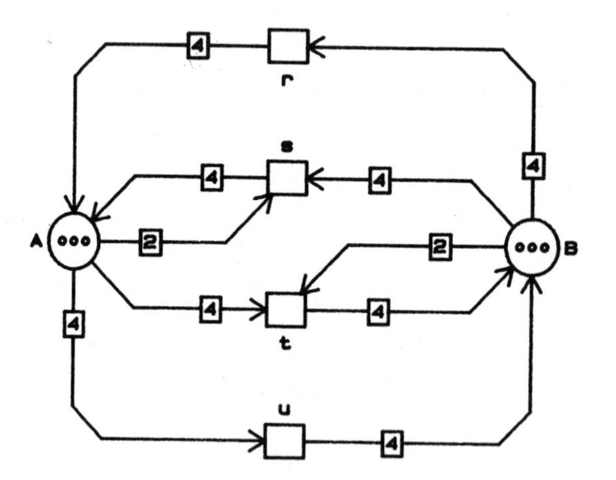

Figure 4

In the example of Figure 4, there is only one deadlock {A,B} which is also a non deficient trap for M_o and there is no transition fireable at M_o.

Now let us consider Petri Nets with non blocking valuations. The following proposition generalizes a result of Commoner in [3] given without proof. Furthermore, that proof gives a straightforward algorithm to get the maximal trap of a given deadlock.

Property 2.12 : Let R =<P,T,Γ,V> be a net such that V is a non blocking valuation. Let D be a deadlock and S_D its maximal trap. If S_D is deficient for a marking M, then there exists a marking M' such that for every p in P, $M'(p) \geq M(p)$ and such that no transition of $\Gamma(D)$ is live in <R;M'>

Proof : First, let us construct a trap of D maximal for the inclusion. Let $D_0 = \phi$ We successively construct :

$$T_i = \Gamma\left(D - \bigcup_{j=0}^{i-1} D_j\right) - \Gamma^-\left(D - \bigcup_{j=0}^{i-1} D_j\right)$$

$D_i = \Gamma^-(T_i) \cap \left(D - \bigcup_{j=0}^{i-1} D_j\right)$ for $i > 0$ until $T_i = \phi$. We claim that when $T_i = \phi$, then $S_D = D - \bigcup_{j=0}^{i-1} D_j$ is a maximal trap of D.

Note that such an i exists because D and T are finite. We have : $\Gamma(S_D) - \Gamma^-(S_D) = \phi$; so $\Gamma(S_D) \subseteq \Gamma^-(S_D)$. As the valuation is non blocking, S_D is a trap. Suppose that S_D is not maximal and let $S \supset S_D$ be a maximal trap. Let k be the first index such that there exists p in $D_k \cap (S-S_D)$. By the construction there exists t in $T_k \cap \Gamma(p)$ (in particular $t \in \Gamma(S)$), such that $\Gamma(t) \cap D \subseteq \bigcup_{j=0}^{k-1} D_j$; therefore $t \notin \Gamma^-(S)$ contradicting the definition of a trap.

Now let M be a marking for which S_D is deficient. Let us construct M', from M by adding tokens in such a manner that starting at M' we can fire a sequence of transitions which uses all the tokens in $D - S_D$ and add no tokens in S_D. To do that consider the places of D_{i-1} and let m be the maximum number of tokens of a place of D_{i-1}; then add tokens to all the places of $\Gamma^-(T_{i-1})$ in order to fire each transition of $\Gamma^-(T_{i-1})$ exactly m times. The marking M_1 obtained is such that $M_1(p) \geq M(p)$ for every place p. If we fire each transition of T_{i-1} n times, then we obtain a marking where the places of D_{i-1} have no more marks. In this construction and firing the places of S_D have not been affected. Repeat the construction for D_{i-2} and so on until D_1. We finally obtain a marking M' and a firing of transition of $\bigcup_{j=1}^{i-1} T_i$ such that for the marking obtained each place of D - S_D has no token and S_D is deficient. Therefore no transition of Γ(D) is live.

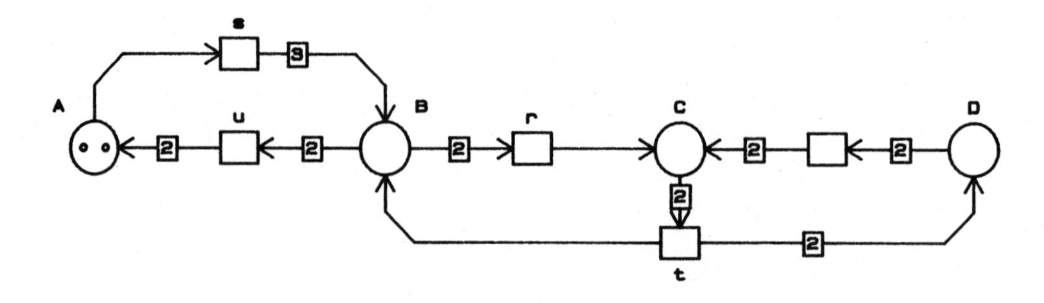

Figure 5

Remark 2.13 : This result does not hold in general. In the example of Figure 5, {C,D} is the maximal trap of {A,B,C,D}, and for each marking $M' \geq M_0$ the net is not bounded.

Problem : It will be interesting to find for what classes of Petri Nets the deadlock-trap property is either sufficient or necessary for the liveness of the Petri Net.

3. A CHARACTERIZATION OF MINIMAL DEADLOCKS

As we have seen in § 2 it is interesting to study deadlocks and traps. Here we restrict our attention to deadlocks; similar results can be obtained in exactly the same manner for traps. In particular, in order to check if every deadlock contains a non deficient trap it suffices to verify it for every minimal deadlock. We present a characterization of minimal deadlocks in terms of path properties. This result was also presented at the conference on Petri Nets held in Toulouse [1].

Recall that a deadlock D is a subset of places such that $\Gamma^{-}(D) \subseteq \Gamma(D)$. D is minimal if and only if it does not contain any deadlock except D and ϕ.

Example 3.1 :

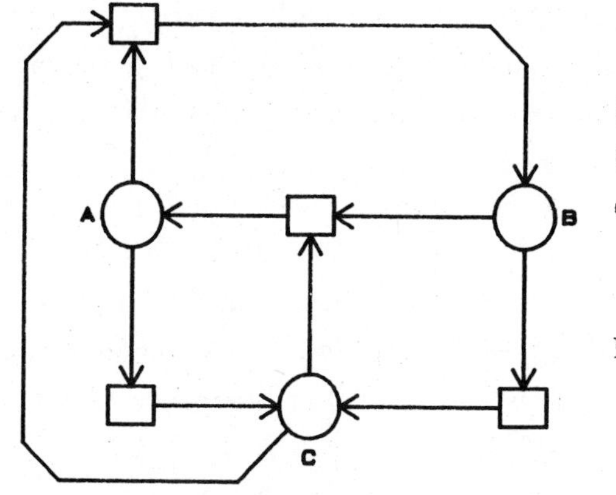

{A,B,C} is a deadlock,
{A,B} and {A,C} are minimal deadlocks

Figure 6

Theorem 3.2 : Let D be a set of places. Then D is a minimal deadlock if and only if:

a) For every set **C** of paths in $D \cup \Gamma^{-}(D)$ beginning by a place, and for every pair (p,q) of places of D, there exists an elementary path C_p from p to q in $D \cup \Gamma^{-}(D)$ such that for each edge (x,t) of C_p such that t belongs to T we have : either (x,t) is an edge of a path of **C**, or t does not appear in any path of **C**.

b) $D \cup \Gamma^{-}(D)$ generates a strongly connected subgraph.

Proof : First, let us consider a minimal deadlock D.

Point a) directly follows from the proposition :

Proposition 3.3 : Let D be a minimal deadlock; let **C** be a set of paths in $D \cup \Gamma^{-}(D)$ beginning by a place of D and let p,q be two places of D. Then there exists an elementary path C_p from p to q such that :

i) $\{p,q\} \subseteq C_p \subseteq D \cup \Gamma^{-}(D)$

ii) No transition t of C_p has an input place in the subpath of C_p from t to q.

iii) For each edge (x,t) of C_p such that t belongs to T we have :

either (x,t) is an edge of a path of **C** or t does not appear in any path of **C**.

Proof : Let R = { x ∈ D such that there exists an elementary path C_x from x to q satisfying i), ii), iii) as in the proposition.} R ≠ ϕ as q ∈ R.

We are going to show that R is a deadlock, as R ⊆ D and D is a minimal deadlock, then R = D and p will belong to R.

Let t ∈Γ⁻ (R), t ∈ Γ⁻(D); therefore there exists u ∈ D ∩ Γ⁻(t). If u ∈ R, then t belongs to Γ(R) and we are done.

Otherwise, let r ∈Γ(t) ∩ R; r ∈ R, thus there exists a path C_r from r to q satisfying i), ii), iii). Let C_u be the path obtained by the concatenation of (u,t,r) and C_r. If C_u is non elementary, then t belongs to C_r and therefore as any place of C_r is in R, t belongs to Γ(R).

C_u satisfies by construction i); as u does not belong to R, it does not satisfy ii) or iii). If C_u does not satisfy ii), Γ⁻(t) ∩ C_r ≠ ϕ and t ∈ Γ(R). If C_u satisfies ii) but not iii), as C_r satisfies itself iii), there exists (π,t) in **C**, with π ≠ u and (u,t) does not appear in **C**; π ∈ **C** ∩ Γ⁻(t), but the path consisting in the concatenation of (π,t,r) and C_r satisfies i), ii), iii) and therefore t belongs to Γ(R) a contradiction.

Point b) is proved by using point a) and the following lemma :

Lemma 3.4: If D is any set of places such that for every pair of vertices x and y, there exists a path from x to y whose places all belong to D, then the subgraph generated by D ∪ (Γ(D) ∩ Γ⁻(D)) is strongly connected

Proof : To prove the lemma, let H be the subgraph generated by D ∪ (Γ(D) ∩ Γ⁻(D)) and let x and y be two vertices of H.

Case 1 : x ∈ D : If y belongs to D, by hypothesis there exists a path from x to y in H. If y belongs to (Γ(D) ∩ Γ⁻(D)), there exists z in Γ⁻(y) ∩ D and by hypothesis a path from x to z in H and therefore one from x to y.

Case 2 : x ∈ Γ(D) ∩ Γ⁻(D) : there exists z in Γ(x) ∩ D ; by case 1 there exists a path from z to y in H and therefore a path from x to y in H.

Conversely let us note that if D ∪ Γ⁻(D) is a strongly connected graph (condition b) then, D is a deadlock. Indeed it is clear that Γ⁻(D) ⊆ Γ(D). D does not need to be minimal. Suppose that D satisfies condition a) and is not minimal. Let D_o be properly included in D and let **C** be a set of paths (p_o,t,q_o)

where $\{p_o, q_o\} \subseteq D_o$, such that every transition of $\Gamma^-(D_o)$ is in at least one path. Let $p \in D - D_o$ and $q \in D_o$. Then the path C_p from p to q uses necessarily a transition t of $\Gamma^-(D_o)$ and an edge (x,t) with $x \notin D_o$; therefore there exists a path (p_o, t, q_o) in **C**, contradicting a). ▫

We think that this last result gives a powerful tool to construct new classes or to better understand the structure of existing ones. Here we apply our theorem for three classes of Petri Nets. These applications are straightforward and we are looking for more complex cases.

Definitions 3.5: A Petri Net $< P,T,\Gamma >$ is said to be *simple* [3] if and only if : $\forall p, q \in \Gamma(p) \cap \Gamma(q) \neq \phi \Rightarrow$ either $\Gamma(p) \subseteq \Gamma(q)$ or $\Gamma(q) \subseteq \Gamma(p)$

A Petri Net $< P,T;\Gamma >$ is said to be *N.S.K. (nicht-selbst-Kontrollierend)* [4] if and only if there does not exist a place p in P such that : $\Gamma(p) = \{t_1, \ldots, t_m\}$ and there exists i of $\{1,\ldots,m\}$ with :

i) there exists an elementary path from p to a transition t_j, containing t_i and $i \neq j$.

ii) there exists a circuit containing p and t_i.

In such classes, we can easily deduce from our theorem the following proposition :

Corollary 3.6 : Let D be a minimal deadlock in a simple Petri Net or a N.S.K. Petri Net, then : $\forall t \in \Gamma(D) : |\Gamma^-(t) \cap D| = 1$.

From this last corollary, we easily derive a theorem on liveness and safeness for F.C. nets in [5], N.S.K. in [4] (see [10]).

Definition 3.7 : G is said to be backward-conflict-free (b.c.f.) if and only if every place has one and only one input transition :

$$\forall p \in P, \sum_{t \in T} V(t,p) = 1$$

Corollary 3.8 : Let G be a backward-conflict-free net, D is a minimal deadlock if and only if $D \cup \Gamma^-(D)$ generates an elementary circuit.

Proof : If $D \cup \Gamma^-(D)$ generates an elementary circuit, then D is clearly a minimal deadlock.

Conversely, let D be a minimal deadlock. Then $D \cup \Gamma^{-}(D)$ generates a strongly connected subgraph. $\Gamma^{-}(D) \neq \phi$ by definition of a b.c.f.; there exists an elementary circuit B in the subgraph generated by $D \cup \Gamma^{-}(D)$. We apply the theorem with **C** consisting of two paths whose union is B. Suppose that $D - B \neq \phi$ and let p be in D -B, q in $D \cap B$: any path from p to q meets the circuit B for the first time necessarily in a transition t from a place x and the edge (x,t) does not satisfy condition a) for **C**. ▫

From this property it is easy to deduce that the deadlock-trap property of Commoner is a necessary and sufficient condition of liveness for the b.c.f. nets (see [10]).

REFERENCES

[1] : J-C Bermond and G. Memmi : " A graph theoretical characterization of minimal deadlocks."
lecture at the 4th Eur. Work. on Applications and Theory of Petri Nets. Toulouse, 1983, (no publication).

[2] : G.W. Brams : "*Reseaux de Petri : Theorie et Pratique (tome 1)*"
Masson - Paris ,1982.

[3] : F. Commoner : "Deadlocks in Petri Nets "
Applied data Research Inc. Wakefield Mass. CA. 7206-2311 ,1972.

[4] : W. Griese : "Lebendikeit in NSK - Petri - Netzen "
Tec. Univ. Munchen - TUM - INFO - 7906 ,1979.

[5] : M. Hack : "Analysis of production schemata by Petri Nets "
Technical Report TR-94 Project MAC M.I.T. ,1972.

[6] : A. W. Holt and F. Commoner : "Events and conditions "
Applied Data Research Inc. New York ,1970.

[7] : M. Jantzen and R. Valk : " Formal properties of place transition nets"
Proc. of Advanced Course on General Net Theory of Processes Systems
- Hamburg (1979), Springer Verlag, L.N.C.S. 84, 1981, 165-212.

[8] : R.M. Karp and R.E. Miller : "Parallel Program Schemata"
J.C.S.S. 3 (1969), 147-195.

[9] : E.W. Mayr : " An Algorithm for the general Petri Net Reachability problem"
SIAM. J. of Computing 13 (1984), 441-460.

[10] : G. Memmi : " Methode d'analyse de reseaux de Petri, reseaux a files, et applications aux systemes temps reels"
These de doctorat d'Etat. Universite Pierre et Marie Curie. Paris, 1983.

ON GRACEFUL DIRECTED GRAPHS THAT ARE COMPUTATIONAL MODELS OF SOME ALGEBRAIC SYSTEMS

G. S. Bloom
The City College of New York

D. F. Hsu
Fordham University

ABSTRACT

A generalization of the process of numbering directed graphs is formulated in which the nodes and arcs of a digraph are labelled with the elements of a finite group Γ. Equivalence is established between the existence of a finite group Γ with a particular property and the existence of a Γ-graceful numbering of its characterizing unidirectional digraph. It is established specifically that each sequenceable group gracefully labels a path; each R-sequenceable group gracefully labels one cycle; each group starter gracefully labels a collection of 2-cycles; and each (K,1) near complete mapping gracefully labels the union of an appropriate collection of disjoint cycles with a single path. It is seen that the existence of certain classes of gracefully numbered graphs depends upon the existence of certain cyclic neofields. Significantly, the existence of some algebraic structures with complicated descriptions can be simply understood in terms of the

Γ-graceful numbering of appropriate digraphs.

1. Introduction

Combinatorial techniques have a history of salutary symbiosis with algebraic applications. It has often been the case, for example, that algebraic structures have been used to construct combinatorial designs and related objects. Perhaps the best known example of this type is Singer's Theorem which uses the structure of finite fields to construct symmetric block designs and finite difference sets (see [9], page 129).

Recently, the authors have used algebraic techniques to explore a certain graph numbering problem [2, 3, 4] and have discovered several startling equivalences between the existence of graceful numberings for specific classes of directed graphs and the existence of classes of corresponding algebraic structures.

At first consideration the discovery of this correspondence was surprising. For example, although the graceful numbering of digraphs is closely related to the graceful numbering of (undirected) graphs as well as to harmonious and elegant numberings of graphs, there seems not to be any similar striking algebraic connnections to those objects (see [1, 5, 8, 11] for example). On the other hand, this work generalized some previous results of Hsu [10] which used concepts of graph numberings in the study of cyclic neofields.

In this paper the authors explain exactly what correspondences have been found. In Section 2 the basic concepts of Γ-graceful numberings are introduced, and in the following sections there are given several theorems that explicitly present the equivalences between the particular algebraic structures and the corresponding graceful digraph numbering problems.

2. Basic Concepts

Let D be a directed graph with n nodes and m arcs, no self-loops and no more than one arc directed from any one node x to any other node y. Let Γ be a finite group of order m+1 with identity element e. A numbering α of D comprises both a mapping from the nodes of D to Γ and the subsequently induced arc numbering (in which arc (x,y) is numbered by group element $(\alpha(x))^{-1} * \alpha(y)$ where $\alpha(x)$ and $\alpha(y)$ are the elements assigned to x and y respectively).

Definition 1. A numbering α of a directed graph D using elements of a finite group Γ is a Γ-graceful numbering of D if the node labels are distinct (hence $n \leq m+1$), and if the arc labels are distinct (so that the mapping from the arcs to Γ is bijective). A digraph which admits a Γ-graceful numbering is called Γ-graceful.

Consistent with previous terminology, if α is a Z_{m+1} graceful numbering of D, then α is simply said to be a graceful numbering of D; and D itself is simply said to be graceful. Graceful digraphs have recently been studied in [2, 3, 4].

It should be emphasized that a digraph D with m arcs may be Γ_1-graceful but not be Γ_2 graceful for groups Γ_1 and Γ_2 that are unequal but are both of order m+1. For example the unidirectional path, $\vec{P}_{21}$ is not graceful, but $(\vec{P}_{21}, \Gamma_{21})$ is graceful where

$$\Gamma_{21} = gp\{a,b:a^3 = b^7 = e, a^{-1}ba = b^2\},$$

a non-abelian group. A graceful numbering of the successive nodes of $\vec{P}_{21}$ follows:

$e\ b\ b^2\ b^3\ b^5\ a\ b^4\ ab^4\ b^6\ ab^3\ a^2b\ a^2b^2\ ab^6\ a^2\ ab^2\ a^2b^4\ ab^5\ a^2b^6\ ab^4$
$a^2b^3\ a^2b^5$

3. Sequencing and R-Sequencing

Definition 2. A sequencing of a finite group Γ of order k is an arrangement of the elements of Γ: $a_0 = e, a_1, a_2, \ldots, a_{k-1}$; such that the partial products,

$$\begin{aligned} b_0 &= a_0, \\ b_1 &= a_0 a_1, \\ b_2 &= a_0 a_1 a_2, \\ &\vdots \\ b_{k-1} &= a_0 a_1 a_2 \ldots a_{k-1} \end{aligned}$$

are all distinct. A finite group which admits a sequencing is called a <u>sequenceable group</u>.

<u>Theorem 1</u>. A finite group Γ of order n is sequenceable if and only if the unidirectional path $\vec{P}_n$ is Γ-graceful.

<u>Proof</u>. If Γ is sequenceable with sequencing $a_0 = e, a_1 \ldots, a_{n-1}$, the partial product, $b_i = \sum_{k=0}^{i} a_k$, can be used as the node label for the i-th node of $\vec{P}_n$. Then the i-th arc label

$$\begin{aligned} b_i^{-1} * b_{i+1} &= (a_0 a_1 \ldots a_{i-1} a_i)^{-1} (a_0 a_1 \ldots a_{i-1} a_i a_{i+1}) \\ &= (a_i^{-1} a_{i-1}^{-1} \ldots a_1^{-1} a_0^{-1})(a_0 a_1 \ldots a_{i-1} a_i a_{i+1}) \\ &= a_{i+1} \end{aligned}$$

is distinct from the other arc labels and is not equal to the identity element. Hence $(\vec{P}_n, \Gamma)$ is graceful.

Conversely, let $g_0, g_1\ g_2, \ldots, g_{n-1}$ be a Γ-graceful numbering of the successive nodes of $\vec{P}_n$. Define $a_0 = e$ and $a_i = g_{i-1}^{-1} g_i$ for $i=1,2,\ldots,n-1$. Since for $i>0$, the set of a_i are the arc numbers of a Γ-graceful digraph, then the numbers of the set $\{a_i: 0 \leq i \leq n-1\}$ are all distinct. This set of arc values of $\vec{P}_n$ defines a sequencing of Γ, since the partial products are readily seen to be distinct:

$$b_0 = a_0 = e,$$

$$b_1 = a_0 a_1 = e g_0^{-1} g_1 = g_0^{-1} g_1,$$

$$b_2 = a_0a_1a_2 = (g_0^{-1}g_1)(g_1^{-1}g_2) = g_0^{-1}g_2,$$
$$\vdots$$
$$b_{n-1} = a_0a_1\ldots a_{n-1} = (g_0^{-1}g_1)(g_1^{-1}g_2)\ldots(g_{n-2}^{-1}g_{n-1}) = g_0^{-1}g_{n-1}. \blacksquare$$

Several examples of the use of Theorem 1 follows:

Example 1. The cyclic group H_8 with generator a can be sequenced by

$$e\ a^7\ a^2\ a^5\ a^4\ a^3\ a^6\ a.$$

The corresponding graceful node numbering of $\vec{P}_8$ is

$$e\ a^7\ a\ a^6\ a^2\ a^5\ a^3\ a^4.$$

The structure of this sequencing can be realized using modular arithmetic in Z_8, the set of exponents of the generator. The corresponding numbering of the nodes of $\vec{P}_8$, 0 7 1 6 2 5 3 4, is graceful in the sense of our earlier work [2, 3, 4].

Example 2. The dihedral group

$$D_6 = gp\{a,b:\ a^6 = b^2 = e;\ ab = b^{-1}a\}$$

may be sequenced in the following way:

$$e\ a\ a^2\ b\ a^3\ ba^2\ ba\ ba^4\ ba^3\ a^4\ a^5\ ba^5.$$

The corresponding graceful node numbering of $\vec{P}_{12}$ is

$$e\ a\ a^3\ ba^3\ b\ a^2\ ba^5\ a^5\ ba^4\ ba^2\ ba\ a^4.$$

Example 3. A sequencing of Γ_{21} was used to gracefully number $\vec{P}_{21}$ as given at the end of Section 2.

Gordon [7] has shown that a finite group is sequenceable if and only if its Sylow 2-subgroup is cyclic; nevertheless, the problem of generally determining sequencings of groups is unsolved (see, for example, [14] Although R-sequenceable groups have been defined by Friedlander, Gordon, and Miller [6] in a manner similar to that of sequenceable groups, it has been shown [12] that in their algebraic properties, they differ considerably.

Definition 3. An R-sequencing of a finite group Γ of order n is an arrangement of the elements of Γ: $a_0 = e, a_1, a_2, \ldots, a_{n-1}$, such that the first n-1 partial products are distinct; i.e. $b_0 = a_0 = e$, $b_1 = a_0a_1 = a_1$, $b_2 = a_0a_1a_2, \ldots, b_{n-2} = a_0a_1a_2\ldots a_{n-2}$; and such that the maximal partial product $b_{n-1} = a_0a_1\ldots a_{n-1} = e$. A finite group that admits an R-sequencing is called an <u>R-sequenceable group</u>.

<u>Theorem 2.</u> A finite group Γ of order n is R-sequenceable if and only if $\vec{C}_{n-1}$, the unidirectional cycle with n-1 nodes, is Γ-graceful.

<u>Proof.</u> Let Γ be an R-sequenceable group with the R-sequencing $a_0 = e, a_1, a_2, \ldots, a_{n-1}$ with the distinct partial products $b_i = \prod_{k=0}^{i} a_k$ for $0 \leq i \leq n-2$; and finally with $b_{n-1} = \prod_{k=0}^{n-1} a_k = e$. Let d be the element of Γ not present in the set of partial products. Since the identity element has not been accounted for, $d^{-1} \neq e$, and $d^{-1}, d^{-1}b_1, d^{-1}b_2, \ldots, d^{-1}b_{n-2}$ give a Γ-graceful numbering for the successive nodes of $\vec{C}_{n-1}$.

Conversely, if $g_1, g_2, \ldots, g_{n-2}$ is a Γ-graceful numbering of $\vec{C}_{n-1}$ using group Γ of order n, then define $a_{i+1} = g_i^{-1}g_{i+1}$ for i = 1,2,...,n-2 and $a_i = g_{n-1}^{-1}g_i$. Then $e, a_1, a_2, \ldots, a_{n-1}$ is an R-sequencing of Γ since

(i) the first n-1 partial products are distinct:

$$b_0 = a_0 = e,$$

$$b_1 = a_0a_1 = ea_1 = g_{n-1}^{-1}g_1,$$

$$b_2 = a_0a_1a_2 = (g_{n-1}^{-1}g_1)(g_1^{-1}g_2)(g_2^{-1}g_3) = g_{n-1}^{-1}g_3,$$

$$\vdots$$

$$b_{n-2} = a_0a_1a_2\ldots a_{n-2} = (g_{n-1}^{-1}g_1)(g_1^{-1}g_2)\ldots(g_{n-3}^{-1}g_{n-2})$$

$$= g_{n-1}^{-1}g_{n-2}; \text{ and}$$

(ii) $b_{n-1} = a_0 a_1 a_2 \ldots a_{n-1} = (g_{n-1}^{-1} g_1) \ldots (g_{n-2}^{-1} g_{n-1}) = e.$ ∎

Example 4. $\vec{C}_6$ is H_7-graceful where $H_7 = gp\{a: a^7 = e\}$. The nodes of $\vec{C}_6$ can be numbered by $a^2\ a\ a^3\ a^6\ a^4\ a^5$. The first non-identity element in the corresponding R-sequencing is $(a^5)^{-1}a^2 = a^4$. The entire corresponding R-sequencing of H_7 is $e\ a^4\ a^6\ a^2\ a^3\ a^5\ a$, the successive arc labels for the H_7-graceful numbering.

Example 5. The dihedral group D_6 in Example 2 is not only sequenceable, but also is R-sequenceable and gives the following corresponding D_6-graceful numbering for $\vec{C}_{11}$: $a^2\ a^4\ a^3\ ba^2\ ba^3\ ba\ ba^4\ a^5\ ba^5\ a\ b$

It was proved in [12] that if a group Γ is both sequenceable and R-sequenceable, then Γ is not abelian. Consequently, if Γ is abelian and of order n, it cannot give a Γ-graceful numbering of both $\vec{P}_n$ and $\vec{P}_{n-1}$. Thus, in considering Z_{12} we note that while $\vec{P}_{12}$ is graceful, $\vec{C}_{11}$ is not.

Not much is yet known about the sequenceability and R-sequenceability of non-abelian groups, but as Examples 2 and 5 clearly illustrate, those properties are not mutually exclusive for that class of groups.

4. Group Starters

Definition 4. A starter of a finite group Γ of order 2s+1 is a partitioning of the non-identity elements of Γ into $\{a_1, b_1\}$, $\{a_2, b_2\}, \ldots, \{a_s, b_s\}$ such that $\bigcup_i \{a_i^{-1} b_i, b_i^{-1} a_i\}$ consists of the non-identity elements of Γ.

In Section 2 we presented finite group properties corresponding to Γ-graceful numberings of single component digraphs. In this section the groups correspond to Γ-gracefully numbering highly disjoint digraphs with isomorphic components. It is easy to show the following:

Theorem 3. A group Γ has a starter if and only if D is Γ-graceful where $D = \bigcup_{i=1}^{s} \vec{C}_2$ and $\vec{C}_2$ is the unidirectional 2-cycle.

Starters have had extensive combinatorial application in the study of Howell designs, Room squares, orthogonal Latin squares, etc. See, for example, [13].

5. Neofields and Generalized Complete Mapings

In this section we establish relationships between the Γ-graceful numberings of unions of disjoint (not necessarily isomorphic), unidirectional cycles and certain generalized complete mappings, as well as between the Γ-graceful numberings of the union of disjoint, unidirectional cycles with a unidirectional path and other generalized complete mappings. Because of their intrinsic algebraic connections, these Γ-graceful numberings are also associated with the existence of appropriate neofields. It will be shown here how the existence of algebraic structures with complicated descriptions can be simply understood in terms of Γ-graceful numberings of appropriate digraphs.

Definition 5. A finite left neofield N_v is a set B of v elements with two binary operations (+) and (*) such that

(i) (B,+) is a loop with 0 as its identity,

(ii) (B-{0},*) is a group with e as its identity,

and

(iii) (*) is distributive over (+) on the left-hand side.

A left neofield which also satisfies the right distributive law is called a neofield.

Definition 6. A neofield (or left neofield) whose multiplicative group is cummutative is called an abelian neofield (or an abelian left neofield). It is a cyclic neofield if its multiplicative group is cyclic.

A Galois field is a finite cyclic neofield which has associative and commutative addition. Neofields were introduced and studied by Paige (see [10]). Cyclic neofields were extensively studied by Hsu [10].

Recently Hsu and Keedwell [12] defined (K, λ) complete mappings and (K, λ) near-complete mappings which include respectively the concepts of sequencing and R-sequencing. The classes also led to the first complete characterization of left neofields.

It is well-known that because of its left distribution law $((a+b) = a(e+a^{-1}b))$, a left neofield N_v is completely described by the second row $e + x$ of its addition table. This row is designated as the presentation function $\Pi(x) = e + x$ of N_v.

Definition 7. A (K, λ) complete mapping of a group Γ, where $K = \{k_1, k_2, \ldots, k_s\}$ and the k_i are integers such that $\sum_{i=1}^{s} k_i = \lambda(|\Gamma|-1)$, is an arrangement of the non-identity elements of Γ (each used λ times) into s cyclic sequences of lengths $k_1, k_2, \ldots, k_s$ say

$$(g_{11}g_{12}\ldots g_{1k_1})(g_{21}g_{22}\ldots g_{2k_2})\ldots(g_{s1}g_{s2}\ldots g_{sk_s}),$$

such that the elements $g_{ij}^{-1}g_{i,j+1}$ (where $i=1,2,\ldots,s$; and the second suffix j is added modulo k_i) comprise the non-identity elements of Γ each counted λ times.

A (K, λ) near complete mapping of a group Γ, where $K = \{h_1, h_2, \ldots, h_r; k_1, k_2, \ldots, k_s\}$ and the h_i and k_j are integers such that $\sum_{i=1}^{r} h_i + \sum_{j=1}^{s} k_j = \lambda|\Gamma|$, is an arrangement of the elements of Γ (each used λ times) into r sequences with lengths $h_1, h_2, \ldots, h_r$ and s cyclic sequences with lengths $k_1, k_2, \ldots, k_s$, say $[g'_{11}g'_{12}\ldots g'_{1h_1}]\ldots[g'_{r1}g'_{r2}\ldots g'_{rh_r}](g_{11}g_{12}\ldots g_{1k_1})\ldots(g_{s1}g_{s2}\ldots g_{sk_s})$ such that the elements $(g'_{ij})^{-1}g'_{i,j+1}$ and $g_{ij}^{-1}g_{i,j+1}$ together with the elements $g_{ik_i}^{-1}g_{i1}$ comprise the non-identity elements of Γ each counted λ times. We have $\Sigma(h_i-1) + \Sigma k_j = \lambda(|\Gamma|-1)$ so it is immediate from the definition itself that $r = \lambda$.

A <u>generalized complete mapping</u> is either a (K, λ) complete mapping or a (K, λ) near-complete mapping.

<u>Theorem 4.</u> (Hsu and Keedwell). (i) There exists a $(K,1)$ complete mapping of the finite group Γ of order $v-1$ if and only if there exists a left neofield $(N_v, +, *)$ whose multiplicative group is Γ and $e + e = 0$.

(ii) There exists a $(K,1)$ near-complete mapping of the finite group Γ of order $v-1$ if and only if there exists a left neofield $(N_v, +, *)$ whose multiplicative group is Γ and $e + e \neq 0$.

The basic relationship between a generalized $\lambda = 1$ complete mapping of a group Γ and the corresponding Γ-graceful numbering of the associated directed graph follows. By $\bigcup_{i=1}^{s} \vec{C}_{k_i}$ we denote the union of s unidirectional cycles whose lengths are respectively $k_1, k_2, \ldots, k_s$.

<u>Theorem 5.</u> (i) There exists a $(K,1)$ complete mapping of the finite group Γ of order n for which $K = \{k_1, k_2, \ldots, k_s\}$ with $\sum_{i=1}^{s} k_i = n-1$ if and only if $\bigcup_{i=1}^{s} \vec{C}_{k_i}$ is Γ-graceful.

(ii) There exists a $(K,1)$ near-completing mapping of the finite group Γ of order n for which $K = \{h; k_1, k_2, \ldots, k_s\}$ with $\sum_{i=1}^{s} k_i + h = n$ if and only if $\vec{P}_h \cup (\bigcup_{i=1}^{s} \vec{C}_{k_i})$ is Γ-graceful.

<u>Proof.</u> (i) Suppose that

$$(g_{11}g_{12}\cdots g_{1k_1})(g_{21}g_{22}\cdots g_{2k_2})\cdots(g_{s1}g_{s2}\cdots g_{sk_s})$$

is a $(K,1)$ complete mapping of Γ. Then $\sum_{i=1}^{s} k_i = n-1$. Moreover, the elements $g_{ij}^{-1} g_{i,j+1}$ (where $i = 1,2,\ldots,$ s and the second index j is added modulo k_i) are all distinct and non-identity. Hence, this $(K,1)$ complete mapping gives a Γ-graceful node numbering for $\bigcup_{i=1}^{s} k_i$ (which has $m = n-1$ arcs). Conversely, the node numbers of the Γ-graceful numbering of $\bigcup_{i=1}^{s} \vec{C}_{k_i}$ gives a $(K,1)$ complete mapping of Γ.

(ii) The proof follows that of (i) except the single (r=1) non-cyclic sequence $[g'_{11}g'_{12}\cdots g'_{1h}]$ is identified with the sequence of node numbers of the single unidirectional path $\vec{P}_h$. Inasmuch as the digraph consists of the disjoint union of s unidirectional cycles and a single path, $\sum_{i=1}^{s} k_i + h = n$ which is also the order of Γ. ∎

As a consequence of Theorems 4 and 5, we are able to relate the existence of neofield, a non-associative algebraic system, with the possibility of producing a generalized graceful numbering of appropriate digraphs.

Corollary 6. (i) $\bigcup_{i=1}^{s} \vec{C}_{k_i}$ with $\sum_{i=1}^{s} k_i = n-1$ is Γ-graceful for a finite group Γ of order n-1 if and only if there exists a left-neofield $(N_n, +, *)$ whose multiplicative group is Γ and for which $e + e = 0$.

(ii) $\vec{P}_h U(\bigcup_{i=1}^{s} \vec{C}_{k_i}$ with $\sum_{i=1}^{s} k_i + h = n-1$ is Γ-graceful for a finite group Γ of order n-1 if and only if there exists a left neofield $(N_n, +, *)$ whose multiplicative group is Γ and for which $e + e \neq 0$.

Example 6. $[e\ a^4](a\ a^2\ a^7)(a^3\ a^6\ a^5)$ is a (K,1) near-complete mapping of the cyclic group $H_8 = \text{gp}\{a:\ a^8 = e\}$. The corresponding left neofield is

x	0	e	a	a^2	a^3	a^4	a^5	a^6	a^7
$\Pi = e + x$	e	a^4	a^2	a^7	a^6	0	a^3	a^5	a

which is equivalent to an H_8-graceful numbering of the directed graph $\vec{P}_i U(\bigcup_{i=1}^{2} \vec{C}_3)$.

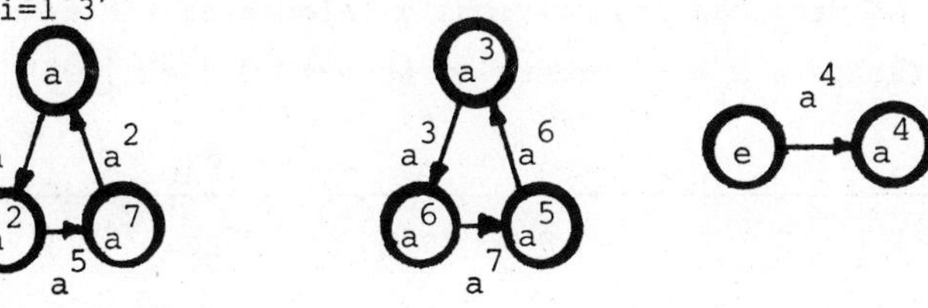

In Section 2 we noted that an abelian group cannot be both sequenceable and R-sequenceable. More generally, the following is true:

Theorem 7. An abelian group Γ of order n cannot Γ-gracefully number $\bigcup_{i=1}^{s} \vec{C}_{k_i}$ (where $\sum_{i=1}^{s} k_i = n-1$) and $\vec{P}_h \cup (\bigcup_{i=1}^{s} \vec{C}_{k_i})$ where $h+\sum_{i=1}^{s} k_i = n$) even if all other constraints on path and cycle sizes are met.

Proof. In [12] it was proved that an abelian group Γ has a (K,1) complete mapping if and only if e is the product of all the elements of Γ. Similarly, Γ has a (K,1) near-complete mapping if and only if it contains a unique element of order 2 which also equals the product of all the elements of Γ. By Theorem 5, the proof is complete. ∎

Example 7. Using Theorem 7 and Example 6 we note that

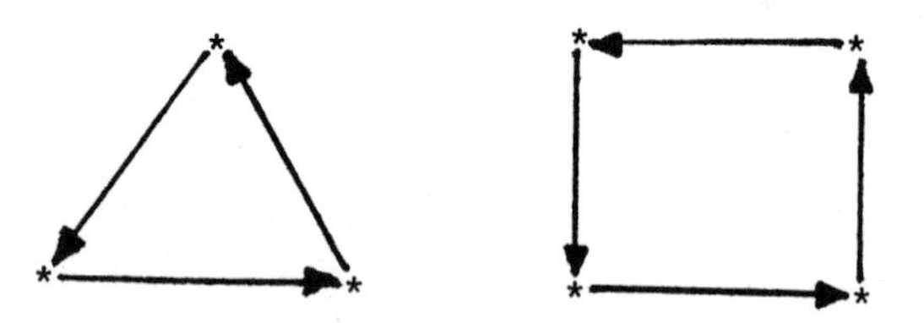

is not H_8-graceful.

6. Remarks

The relations between neofields and Γ-graceful numberings of certain digraphs provide us with a powerful tool to construct the members of specified classes of neofields. A program run at the University of Michigan has previously calculated the number of cyclic neofields $\#(N_v)$ of orders up to v = 12 (See [10]).

v	4	5	6	7	8	9	10	11	12
$\#(N_v)$	1	2	3	8	19	64	225	928	3441

The simplifications described in this paper in representing neofields by graceful digraphs greatly reduce both the computer data structures for representing neofields and the complexity of the algorithms for calculating them. This analysis will appear later.

From among the plethora of questions raised by this work, we cite two here either of whose answers would be significant both combinatorially and computationally.

Question 1. What is the behavior of $\#(N_v)$, where N_v is a cyclic neofield of order v?

Question 2. Is $\bigcup_{i=1}^{18} \vec{C}_3$ H_{55}-graceful, where $H_{55} = gp\{a: a^{55} = e\}$?

Acknowledgements

This research was conducted under the partial support of PSC-CUNY Grant #6-63230 and Fordham University Faculty Research Grant #071225.

References

1. J. C. Bermond, Graceful graphs, radio antennae, and French windmills, in Graph Theory and Combinatorics (ed. R. Wilson). London: Pittman, 1979, 18-37.

2. G. S. Bloom, & D. F. Hsu, On graceful graphs and a problem in network addressing, in Congressus Numerantium, 35. Winnipeg: Utilitas Mathematica, (1982), 91-103.

3. G. S. Bloom, & D. F. Hsu, Graceful directed graphs, Technical Report #CCNY-CS283, The City College of New York, 1983.

4. G. S. Bloom, & D. F. Hsu, On graceful directed graphs, to appear in SIAM J. Algebraic and Discrete Methods, 1985.

5. G. S. Bloom, A chronology of the Ringel-Kotzig conjecture and the continuing quest to call all trees graceful, in Topics in Graph Theory (Annals of the New York Academy of Sciences, 328, ed. F. Harary). New York: NYAS, 1979, 32-51.

6. R. J. Friedlander, B. Gordon and M. D. Miller, On a group sequencing problem of Ringel, Congressus Numerantium XXI, Utilitas Math., (1978), 307-321.

7. B. Gordon, Sequences in groups with distinct partial products, Pacific J. Math., II, (1961), 1309-1313.

8. R. L. Graham, & N. J. A. Sloane, On Additive bases and harmonious graphs, SIAM J. Algebraic and Discrete Methods, 4, (1980), 382-404.

9. M. Hall, Jr., Combinatorial Theory. Waltham, Mass., Blaisdell, (1967).

10. D. F. Hsu, Cyclic Neofields and Combinatorial Designs (Lecture Notes in Mathematics, 824). Berlin: Springer-Verlag, (1980).

11. D. F. Hsu, Harmonious labellings of windmill graphs and related graphs, J. Graph Theory, 6, 1982, 85-87.

12. D. F. Hsu, & A. D. Keedwell, Generalized complete mappings, neofields, sequenceable groups, and block designs, I, Pacific J. Math., III, (1984), 317-332.

13. F. K. Hwang, Strong starters, balanced starters and partitionable starters, Bull of Institute of Mathematics, Academia Sinica, VII (1983), 561-572.

14. A. D. Keedwell, Sequenceable groups: a survey, in Finite Geometries and Designs. Cambridge: Cambridge University Press, 1981, 205-215.

THE CUT FREQUENCY VECTOR

F. T. Boesch*
Stevens Institute of Technology

ABSTRACT

The problem of finding networks with the smallest probability of disconnection is used to motivate a new definition of vectors associated with a graph. They are called the edge-cut and point-cut frequency vectors and have their i-th components equal to the total number of i-th order cuts. In this context a cut is any set of edges (points) whose removal creates a disconnected, trivial, or empty graph. The problem of finding (p,q) graphs that minimize these cut frequency vectors is considered herein. It is shown that this problem includes a variety of connectivity and spanning tree extremal problems as special cases. Several new results and conjectures regarding the solution to these problems are presented.

1.0 Introduction

This work will introduce and study some properties of vectors which are related to the connectivity of a graph. Herein the known results concerning these vectors are summarized, and several new results are presented. Unless otherwise stated the notation and terminology follows Harary's book [16]. All theorems which have

---

* The author's research was supported by NSF Grant ECS-8100652

been published elsewhere are denoted by Theorem n[m]. The other theorems are new results. By way of motivation consider first the following classical model for network reliability. A network is a graph G together with a probability of failure ρ associated with each edge. It is assumed that the edges of G fail independently. A set of edges E is said to be an <u>edge-cut</u> if G-E is disconnected or trivial. If m_i denotes the total number of edge-cuts or order i then it is easy to verify that the probability $P(G,\rho)$ of the network failing can be espressed as

$$P(G,\rho) = \sum_{i=1}^{q} m_i \rho^i (1-\rho)^{q-1}$$

Hence the following definition captures the network reliability properties in terms of properties of the graph G.

<u>Definition 1.</u> For a p-point, q-edge graph G, let $\underset{\sim}{m}(G) = (m_1, m_2, \ldots, m_q)$, where m_i is the total number of edge-cuts of order i, be called the edge-cut frequency vector.

Note that $m_i = 0$ for $i<\lambda$ the edge connectivity.

The definition is illustrated in Figure 1 which shows the m vector for two 5 point, 6 edge graphs. From this example it is clear that the m_i are not monotone in i. Furthermore, in contrast to the degree frequency vector, the sum of the m_i can not be expressed as a function that depends only on p and q. However one elementary property of the $\underset{\sim}{m}$ vector is given by Bauer, Boesh, Suffel, and Van Slyke [4]. Their result is stated here as Theorem 1.

<u>Theorem 1</u> [4] The terms $m_i / \binom{q}{i}$ are monotone non-decreasing in i.

2.0 A connectivity extremal problem

Motivated by the question of trying to find a most reliable (p,q) graph, one might consider the following extremal problem. To this end we say that $\underset{\sim}{m}(G_1) \leq \underset{\sim}{m}(G_2)$ if $m_i(G_1) \leq m_i(G_2)$ for each $i (1 \leq i \leq q)$.

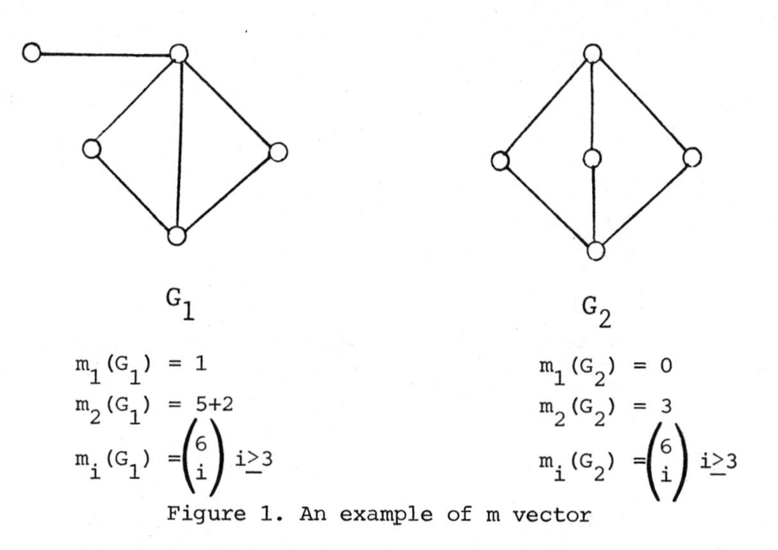

Figure 1. An example of m vector

Definition 2. A p-point, q-edge graph $\hat{G}$ is said to be a minimum edge-cut frequency graph or simply a minimum $\underset{\sim}{m}$ graph if $\underset{\sim}{m}(\hat{G}) \leq \underset{\sim}{m}(G)$ for all p-point, q-edge graphs G.

Of course it might be the case that minimum edge-cut frequency graphs do not exist for all possible values of p and q. However if they do, then they will correspond to networks that have largest reliability for any value of ρ. Note that if $\hat{G}$ is a minimum $\underset{\sim}{m}$ graph then $\underset{\sim}{m}(\hat{G})$ must have the maximum possible number of zeros, i.e., λ must be maximum over the class of (p,q) graphs. For convenience it is assumed that 2q/p is an integer δ. In which case, it is well known that G has maximum λ if and only if G is a regular degree δ graph with $\lambda=\delta$. Many classes of regular graphs have this property. For example it is easy to verify that any edge-disjoint union (on the same point set) of hamiltonian cycles has $\lambda=\delta$. A very general, but little known, result due to Mader [20] is stated here as Theorem 2.

Theorem 2 [20] Every connected point symmetric graph has $\lambda=\delta$.

3.0 Special classes of point-symmetric graphs

When the number of points is prime, Turner [27] showed that point-symmetric graphs belong to a very special class of graphs.

These graphs, known as circulants are defined as follows. To this end we assume the points of a graph are labelled 0,1,...,p-1, and we refer to point i instead of saying the point labelled i. The <u>circulant graph</u> $C_p\langle a_1,a_2,\ldots,a_k\rangle$ or briefly $C_p\langle a_i\rangle$, where $0<a_1<a_2<\ldots<a_k<(p+1)/2$, has $i+a_1, i+a_2,\ldots,i+a_k$ (mod p) adjacent to each point i. The sequence $\langle a_1\rangle$ is called the <u>jump sequence</u> and the a_i are called the <u>jumps.</u> Notice that our definition precludes jumps of size >p/2 as such jumps would produce the same result as a jump of size (p-a) where p-a<p/2. Also note that if $a_k \neq p/2$ then the circulant is always regular of degree 2k. When p is even we have allowed $a_k = p/2$ (called a <u>diagonal</u> jump), and when $a_k = p/2$ the circulant has degree 2k-1.

An example of a circulant is shown in Figure 2. In the event that the jump sequence is $\langle 1,2,\ldots,k\rangle$ the circulant becomes C_p^k;

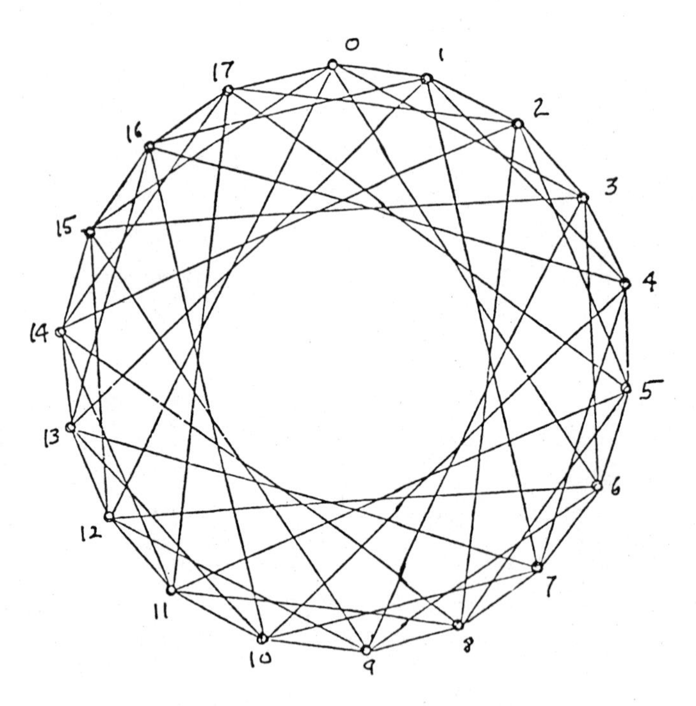

Figure 2. $C_{18}\langle 1,3,6\rangle$

an example is shown in Figure 3. Another special

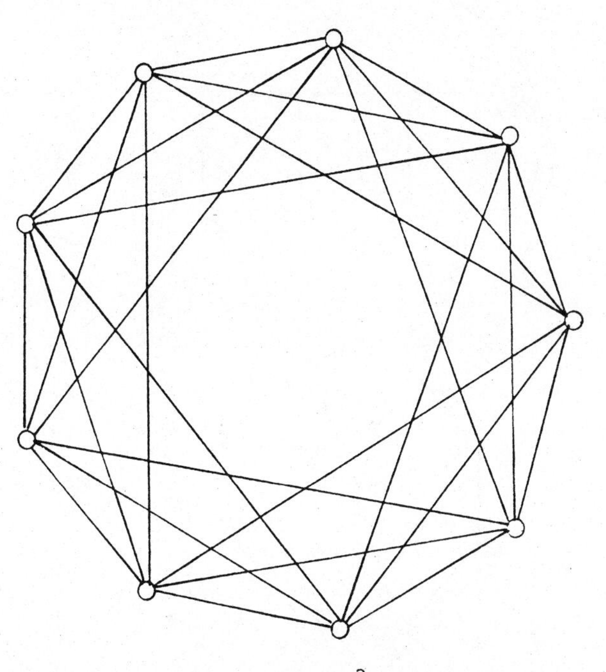

Figure 3. $C_9^{\,3}$

circulant called a Moebius ladder M_n was introduced by Harary [17]. These are the circulants given by $C_{2n}\langle 1,n\rangle$; an example is shown in Figure 4, which also illustrates the labelling that identifies the two representations as being isomorphic. A slightly less obvious circulant is the regular, complete, n-partite graph K(m,m,...m) which can be abbreviated as $K(m^n)$. An example of the complement $\overline{K}(m^n)$ is shown in Figure 5. Since the complement of a circulant is also a circulant, one can express $K(m^n)$ for $n \geq 2$ as follows

$$\overline{K}(m^n) = C_{n.m}\langle n, 2n, 3n, \ldots, \lfloor m/2 \rfloor n\rangle$$

where $\lfloor x \rfloor$ denotes the largest integer that does not exceed x.

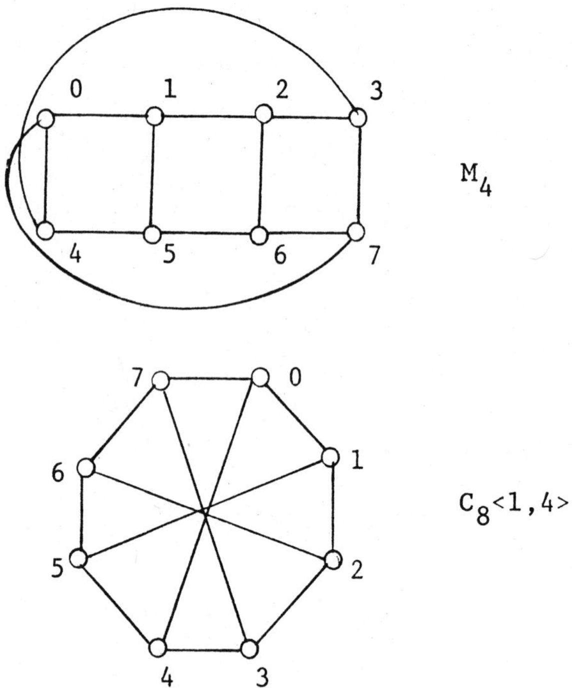

Figure 4. The Moebius Ladder

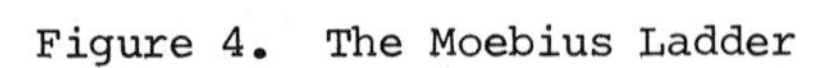

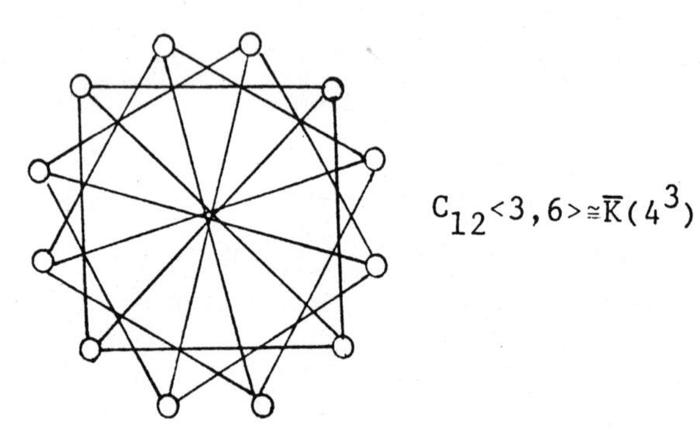

Figure 5. The regular complete n-partite graph

Two other special classes of point-symmetric graphs are K_2xK_n and K_2xC_n (sometimes called the prism); see Figure 6. Notice that Figure 6 also shows K_2xC_5 represented as a circulant. The next theorem specifies when this is possible.

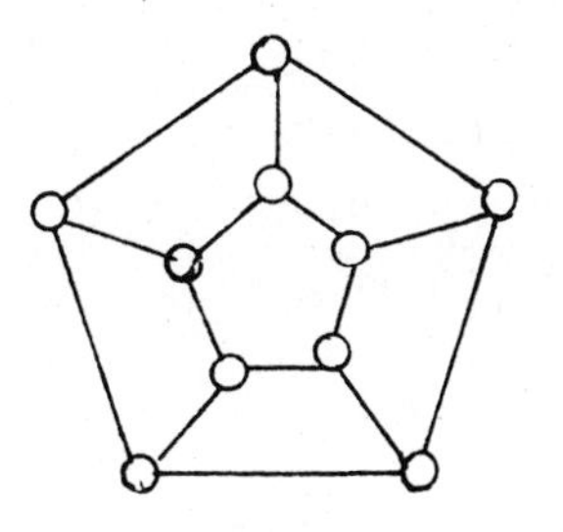

THE
PRISM
K_2xC_5

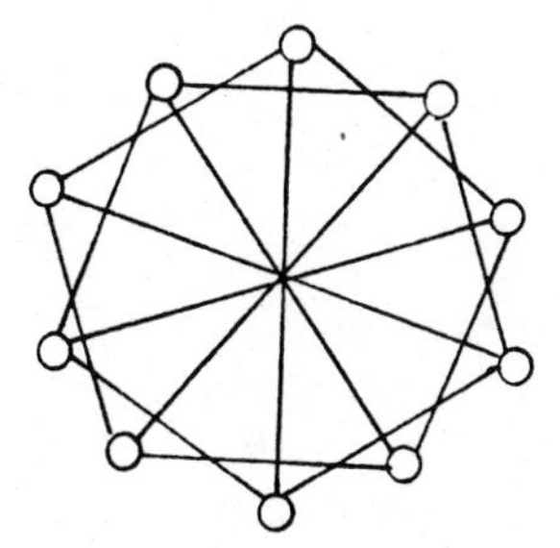

$C_{10}<2,5>\cong K_2xC_5$

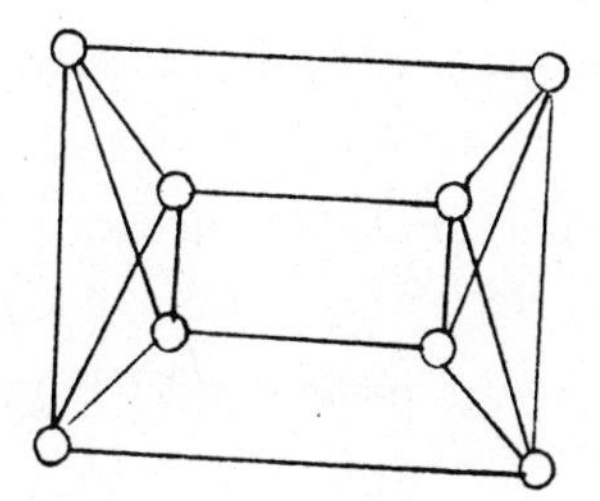

K_2xK_4

Figure 6. K_2xC_n and K_2xK_n

Theorem 3 The prism K_2xC_n is a circulant if and only if n is odd. If n is odd, K_2xC_n is $C_{2n}<2,n>$.

Proof The fact that $C_{2n}<2,n>$ is K_2xC_n when n is odd is easily seen to be valid. Now the only degree 3 circulant is $C_{2n}<\alpha,n>$ for some α. However this is connected if and only if $\gcd(2n,\alpha,n) = 1$. Thus if n is even α must be the odd number $2\beta+1$. Furthermore, it is easily verified that $C_{2n}<1,n>$ is isomorphic to $C_{2n}<1\times(2\beta+1),n\times(2\beta+1) (\mathrm{mod}\ 2n)> = C_{2n}<2\beta+1,n>$. But $C_{2n}<1,n>$ always contains the odd cycle defined by points $0,1,2,\ldots,n,0$ when n is even, and K_2xC_n has no odd cycles as it is easily shown to be bipartite.

As a corollary to this it might be noted that when n is odd, K_2xK_n is $C_{2n}<2,4,\ldots,n-1,n>$ but when n is even, K_2xK_n can never be a circulant as it would contain K_2xC_n. However, it is easily shown that both K_mxC_n and K_mxK_n are point-symmetric.

For further information on circulants see Boesch and Tindell [9,10].

4.0 Minimizing the first non-zero term in $\underset{\sim}{m}$

The first non-zero term of $\underset{\sim}{m}$ is m_λ. In order to minimize m_λ over the class of (p,q) graphs with $\lambda = 2q/p = \delta$, an integer, observe that every point incidence set is a different edge cut of order λ. Thus $m_\lambda \geq p$. The following definition and theorems were given by Bauer, Boesch, Suffel, and Tindell [2].

Definition 3 A regular graph is super-λ if every minimum edge-cut is the incidence set of a point.

Clearly a regular, super-λ graph has $\lambda=\delta$ (the maximum), and $m_\lambda = p$ (the minimum).

Theorem 4 [2] If $\kappa=\lambda$ then every minimum edge-cut is a point incidence set or a matching.

Theorem 5 [2] The circulants $C_p<1,2,\ldots,k>$ and $C_{2n}<1,2,\ldots,k,n>$, dnown as Harary graphs, are super-λ.

Theorem 6 [2] The circulant $C_p<a_1,a_2,\ldots,a_k>$ is super-λ if $a_1=1$.

The last theorem suggests that connected circulants might all be super-λ. It is easily seen, however, that the cycle C_p and $K_2 x K_n$ (for n odd) are not super-λ. Theorem 7, a recent result of Boesch and Wang [11], states that these are the only exceptions.

Theorem 7 [11] The only connected circulants that are not super-λ are $C_p\langle a\rangle$ and $C_{2n}\langle 2,4,\ldots,n-1,n\rangle$ for n odd.

For a complete solution to the problem of minimizing m_λ when 2q/p is not an integer, see Bauer, Boesch, Suffel, and Tindell [3].

5.0 Minimizing m_i for $i>\lambda$

The preceeding sections included results which show that a large class of circulants minimize the first λ terms of the m vector over the class of (p,q) graphs where 2q/p is an integer. Very little is known about the minimum values of m_i for all possible i. One result by Boesch and Wang [11] is stated here as Theorem 8.

Theorem 8 [11] The Harary graph C_p^k, which has $\lambda=\delta=2k$, minimizes m_i for $2k \leq i \leq 4k-3$.

Results are also available regarding the properties of m_i for large i. For example, if i is large enough, every set of i edges will be an edge-cut. Specifically if $q-i<p-1$ then $m_i = \binom{q}{i}$.

Now let μ denote q-p+1 and consider m_μ. Clearly the subgraph which remains after removing μ edges from G is connected if and only if it is a spanning tree of G. Hence $m_\mu = \binom{q}{\mu} - t(G)$ where t(G) is the number of spanning trees of G. Thus if G is to be a minimum $\tilde{m}$ graph over the class of (p,q) graphs having $2q/p = \bar{\delta}$ an integer, then G must be super-λ and have the maximum t(G). Now although one could always calculate t(G) by using Kirchhoff's Matrix Tree Theorem, it would be useful to have explicit formulas giving t(G) for the special class of super-λ graphs.

One can obtain some insight into the question of max t(G) for regular graphs via some special cases. For example, it is easy to show that a connected, cubic, 8 point graph has no cut-points.

Thus one can apply Jackson's Theorem [18] to verify that all connected, cubic, 8 point graphs are hamiltonian. This fact enables one to enumerate all the possible cases. Figure 7 shows all the non-isomorphic 8 point, cubic, connected graphs and gives the number of spanning trees for each such graph. In this case it can be verified that the only circulant is the Moebius ladder M_4.

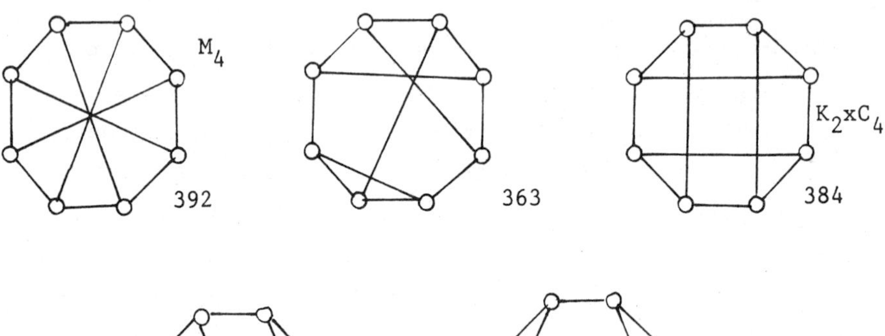

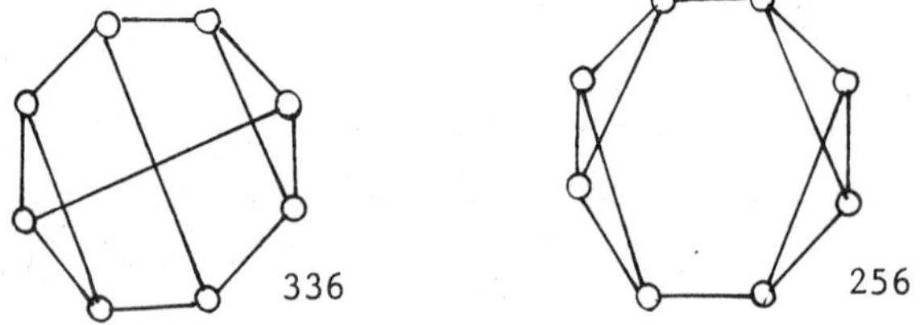

Figure 7. All connected, 8 point, cubic graphs and their number of spanning trees

Figure 8 gives a table listing the known cases where there is an explicit formula for t(G). The cases which use the notation G_1-G_2 always refer to removing the edges of G_2 from the graph G_1 under the assumption that G_1 contains G_2 as a subgraph.
It is natural to conjecture that Harary graphs are minimum m graphs. However, the explicit formulas show that

G	t(G)	Reference
C_p	p	-
K_p	p^{p-2}	Cayley [13]
W_n	$\left(\frac{3+\sqrt{5}}{2}\right)^n + \left(\frac{3-\sqrt{5}}{2}\right)^n -2$	Sedláček [22]
$K(n,m)$	$n^{m-1}\, m^{n-1}$	Berge [5]
$K(m^n)$	$[m(n-1)]^{(m-1)n}\,(mn)^{n-2}$	Schwenk [23]
$K_m x K_n$	$m^{m-2} n^{n-2} (m+n)^{(n-1)(m-1)}$	Schwenk [23]
$C_p^{\ 2}$	$pF_p^{\ 2}$ (F_p is the pth Fibonacci #)	Baron, Boesch Prodinger, Tichy Wang [1]
$C_{2n}\langle 1,n\rangle = M_n$	$\frac{n}{2}[(2+\sqrt{3})^n + (2-\sqrt{3})^n +2]$	Moon [21]
$K_2 x C_n$	$t(M_n) -2n$	Boesch Bogdanowicz [6]
$\overline{C}_{2n}\langle n\rangle = K_{2n}-nK_2$	$(2n)^{n-2}\,(2n-2)^n$	Berge [5]
$K_n-K(1,m)$	$n^{n-m-2}(n-1)^{m-1}(n-m-1)$	Berge [5]
$L_n = K_2 x P_n$	$\frac{1}{2\sqrt{3}} \lvert (2+\sqrt{3})^n - (2-\sqrt{3})^n \rvert$	Moon [21]
$K_p - K_m$	$p^{p-m-1}(p-m)^{m-1}$	Berge [5]
$K_2^{\ n}$	$\frac{1}{2^n} \prod_{i=1}^{n} (2i)^{\binom{n}{i}}$	Schwenk [23]

Figure 8. A Table of Tree Formulas

$t(C_8^{\ 2}) = 3528$, $t(C_8\langle 1,3\rangle) = t(K(4,4)) = 4096$, $t(K_2 x K_4) = 3456$ - Hence the Harary graphs do not always minimize m. The case of degree 4 graphs on 8 points is interesting because $C_8\langle 1,3\rangle = K(4,4)$ and $C_8^{\ 2} = C_8\langle 1,2\rangle$; it is easily verified that all other circulants of degree 4 or 8 points are isomorphic to either of these two. Notice that by the observation following Theorem 3, $K_2 x K_4$ is not a circulant.

There is very little known about extremal spanning tree graphs. The problem of maximizing t(G) over the class of (p,q) graphs is clearly related to the minimum m vector problem. Certainly the case where 2q/p is an integer and G is restricted to be regular is of interest. Furthermore if an explicit formula for the minimum value of t(G) were known it could be used together with Theorem 1 to establish upper bounds of m_i. Specifically let t_0 denote the minimum value of t(G) over the class of (p,q) graphs. Then for any (p,q) graph G,

$$m_i(G) \leq \frac{(q-p+1)!(p-1)!}{i!(q-i)!}\left[\binom{q}{p-1} - t_0\right]$$

$$\text{for } \lambda \leq i \leq q-p+1$$

The known extremal results for t(G) are easily summarized as follows. The maximum and minimum values of t(G) over the class of regular degree 3 and regular degree 4 connected graphs on 8 points are known by exhaustive search. They are shown in Figure 7 for degree 3 and are K(4,4) and $K_2 x K_4$ for degree 4. For the case of regular degree 2n-1 graphs on 2n points there is only one such graph; it is $K_{2n} - nK_2$. Finally we note that M_n does not maximize t(G) over the class of 2n point regular degree 3 graphs since the Peterson graph has t(G) =2000 while $t(M_5)$=1815.

The known results for maximum and minimum t(G) over the class of (p,q) graphs are summarized in the next four theorems. Theorem 9 is due to Cheng [14]. Theorem 10 is by Shier [24] and independently by Kelmans and Chelnokov [19]. Theorem 11 is from [19]. Theorem 12 was conjectured in [4] and proven by Boesch and Bogdanowicz [7].

Theorem 9*[14] $K(m^n)$ is a unique maximum of $t(G)$ over the class of mn point graphs having $m^2n(n-1)/2$ edges.

Theorem 10*[19,24] $K_{2n} - mK_2$ for $m \leq n$ maximizes $t(G)$ over the class of (p,q) graphs having 2n points and $2n(2n-1)/2 - m$ edges.

Theorem 11*[19] $K_{2n} - K(1,m)$ for $m \leq 2n-1$ minimizes $t(G)$ over the class of (p,q) graphs having 2n points and $2n(2n-1)/2 - m$ edges.

Theorem 12*[7] If $q \geq p-1$, then the minimum value of $t(G)$ over the class of (p,q) graphs is achieved by the graph H which is formed by adding a path of p-k-1 edges to the point of degree ℓ in $G=K_{k+1}-K(1,k-\ell)$,
where k is the largest positive integer, such that
$q-p+k+1+1-k(k-1)/2 \geq 1$ and $\ell = q-p+k+1-k(k-1)/2$.

To conclude this section, it should be noted that the solution to the tree extremal problem for all (p,q) graphs may not resolve the question of the existence of a minimum m graph. Namely the values of m_i for $\lambda<i<\mu$ would also have to be minimized. The only possibility for a negative resolution would be a situation where it was shown that for some p and q, no graph which maximizes $t(G)$ is also a minimum m_λ, maximum λ graph. The experience gained from known results regarding the tree extremal problem indicates that such a possibility is highly unlikely. Certainly in the case that 2q/p is an integer, it is reasonable to conjecture that some G which maximizes $t(G)$ will also be a regular super-λ graph. Finally, it is noted that the results of Cheng [14] and Boesch and Felzer [8] indicate that it is reasonable to conjecture that $K(m^n)$ would actually minimize all m_i values.

6.0 The Point Cut Frequency Vector

The network realiability model corresponding to point failure

---

* The extremal values of $t(G)$ for the cases of Theorems 9,10,11 and 12 can be found in Figure 8. In case of Theorem 12, $t(H)=t(G)$

motivates a similar definition of a point-cut or simply a cut frequency vector. In this case one says that a cut is any set of points where revoval makes G disconnected, trivial, or empty. The inclusion of the posibility of being empty must be allowed for, as one must include the case where all points fail in the reliability model.

Definition 4 For a p-point, q-edge graph G let $\underset{\sim}{\ell}(G) = (\ell_1, \ell_2, \ldots, \ell_p)$ where ℓ_i is the total number of cuts of order i, be called the cut frequency vector.

Note that $\ell_i = 0$ for $i<\kappa$ the connectivity. Again one can compare graphs G_1 and G_2 by saying $\underset{\sim}{\ell}(G_1) \leq \underset{\sim}{\ell}(G_2)$ if $\ell_i(G_1) \leq \ell_i(G_2)$ for each i $(1 \leq i \leq p)$.

As before, a minimum $\underset{\sim}{\ell}$ graph must have the maximum possible number of zero elements in $\underset{\sim}{\ell}$. Hence κ must be a maximum. In the case of κ it is not true that all point-symmetric graphs have maximum κ. In fact, even for the case where $2q/p = \delta$ an integer, the maximum possible value of κ is δ, but not all circulants have $\kappa = \delta$. Necessary and sufficient conditions for a circulent to have $\kappa = \delta$ are given in [10].

As in the case of the $\underset{\sim}{m}$ vector, the following material will be restricted to the case where $2q/p$ is an integer.

In order to consider the problem of minimizing ℓ_κ, it is natural to make the following definition.

Definition 6 A regular graph is said to be super-κ if every minimum cut is the adjacency set of a point.

Note that super-κ graphs have $\kappa = \delta$. Furthermore it is easy to show that any maximum κ regular graph has maximum λ. It is natural to ask whether the same is true of super-κ. The next theorem answers this question.

Theorem 13 For regular graphs, super-κ implies super-λ except for C_4, C_5, and $C_6\langle 2,3\rangle$.

Proof If G is super-κ but not super-λ, one can apply Theorem 4 as $\kappa = \delta$ implies $\lambda = \delta$. Let the edge set of the matching be denoted by M and the disjoint point sets of the partition of V created by G - M be V_1 and V_2; see Figure 9. Suppose the endpoints of M that are in V_1 are called $\mu_1,\mu_2,\ldots,\mu_\lambda$ and in V_2 are $v_1,v_2,\ldots,v_\lambda$. Now if $\lambda = \delta \geq 3$, the set of points W, where

$$W = \{\mu_1,\mu_2,v_3,v_4,\ldots,v_\lambda\}$$

will be a minimum disconnecting set. Furthermore W can never be the adjacency set of any point of G if $\lambda \geq 4$, as

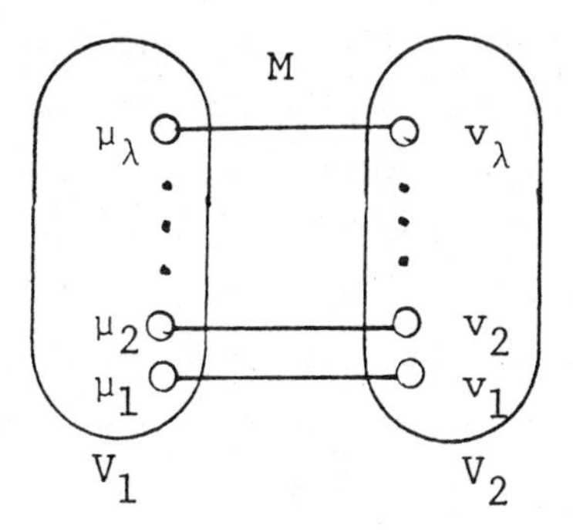

Figure 9. A minimum edge-cut for a super-κ graph

no point adjacency set of G can contain μ_1,μ_2,v_1, and v_2 when M is a matching. The case $\delta = \lambda = 3$ would allow the possibility of $\{\mu_1,v_2,v_3\}$ and $\{v_1,\mu_2,\mu_3\}$ to be adjacency sets. However this implies that G is $K_2 x K_3$ which is C_6<2,3>. The cases of $\delta = 1$ or 2 are obvious.

Unfortunately it is not possible to mimic the techniques used to minimize m_λ when dealing with ℓ_κ. Specifically ℓ_κ is not bounded below by p, as adjacency sets for different points need not be distinct. Thus one might define $\gamma(G)$ to be the total number of adjacency sets of G. Note that $\ell_\kappa(G) \geq \gamma(G)$, where equality is attained by super-κ graphs. Define min ℓ_κ and min γ over the class of regular (p,q) graphs G.

It is easy to see that if G is super-κ and $\gamma(G) = \min \gamma$, then $\ell_\kappa(G) = \min \ell_\kappa$. Thus one is tempted to study the structure of min γ graphs. There are some known results for these cases due to Hakimi and Amin [15], Smith [25], and Boesch and Wang [12]. However it has been shown via a collection of counter examples by

Hakimi and Amin [15] that "min γ graphs do not solve the min ℓ_κ problem". To be precise they verified the following result, which is designated here as a theorem because the generation of the required examples is not trivial.

Theorem 14 [15] Let $S(p,r)$ be the class of all regular degree r graphs on p points. Then:

a) No min γ graph in $S(12,4)$ is a super-κ graph

b) No min γ graph in $S(8,3)$ is a min ℓ_κ graph

c) No min ℓ_κ graph in $S(12,4)$ is duper-κ.

There are some results regarding other classes with the properties indicated in Theorem 14, and some structural properties of such classes were obtained by Tindell [26]. However such results do not aid in the solution of the min ℓ_κ problem.

This work will be concluded with a result that seems to indicate a major distinction between the minimum $\underset{\sim}{m}$ and minimum $\underset{\sim}{\ell}$ graph problems.

Theorem 15 Minimum $\underset{\sim}{\ell}$ graphs can fail to exist even when $2q/p$ is an integer.

Proof Let $p = 8$, $q = 12$, and $\delta = 2q/p = 3$. If $\Delta_3(G)$ denotes the number of 3 point induced subgraphs of G that are connected, then

$$\ell_{p-3} = \binom{p}{p-3} - \Delta_3$$

If G is a regular degree 3, 8 point, triangle free graph, then the connected 3 point subgraphs are always 3 point paths P_3, and

$$\Delta_3 = \binom{3}{2} \times 8 = 24.$$

Furthermore if such a graph has triangles then $\Delta_3 \leq 24$. Now since any maximum κ graph $\hat{G}$ for $p = 8$, $q = 12$ will be regular of degree 3, it follows that

$$\ell_{p-3}(\hat{G}) \geq \binom{8}{3} - 24 = 32.$$

However $K(2,6)$ has 8 points and 12 edges, but it is easy to see that

$$\ell_{p-3}(K(2,6)) = \binom{6}{3} = 20.$$

Because $\kappa(K(2,6)) = 2$, it follows that there is no graph that minimizes all the terms of the $\underset{\sim}{\ell}$ vector for $p = 8$, $q = 12$.

Bauer and Suffel * have generalized the construction given in the above theorem to show that there are classes of (p,q) graphs for which no minimum $\underset{\sim}{\ell}$ vector exists. However, it is not known whether there is any class of regular graphs for which there is no minimum $\underset{\sim}{\ell}$ graph.

We conclude by noting that there are some limited results for the minimum ℓ_κ problem. They can be found in Hakimi and Amin (15), Smith (25), and Boesch and Felzer (8).

REFERENCES

1. G. Baron, F. Boesch, H. Prodinger, R. Tichy, and J. Wang, The number of spanning trees in the square of a cycle. Fibonacci Quart. (to appear).

2. D. Bauer, F. Boesch, C. Suffel, and R. Tindel, Connectivity extremal problems and the design of reliable probabilistic networks. The Theory and Applications of Graphs (G. Chartrand, Y. Alavi, D. Goldsmith, L. Lesniak-Foster, and D. Lick, Eds.) Wiley, New York, (1981, 45-54.

3. D. Bauer, F. Boesch, C. Suffel, and R. Tindell, Combinatorial optimization problems in the analysis and design of probabilistic networks. Networks (to appear).

4. D. Bauer, F. Boesch, C. Suffel, and R. Van Slyke, On the validity of reduction of reliable network design to a graph extremal problem. Operations Research (submitted).

5. C. Berge, Principles of Combinatorics, Academic Press, NY, (1971).

6. F. Boesch and Z. Bogdanowicz, The number of spanning trees in the Prism. Utilitas Mathematica (submitted).

7. F. Boesch and Z. Bogdanowicz, The minimum number of spanning trees in a (p,q) graph. J. Graph Theory (submitted).

* In unpublished work

8. F. T. Boesch and A. P. Felzer, On the invulnerability of the regular complete k-partite graphs. SIAM J. APPL. MATH 20 (1971), 176-182.

9. F. Boesch and R. Tindell, Connectivity and Symmetry in Graphs. Proc. First Colorado Graph Theory and Applications Symposium (to appear).

10. F. Boesch and R. Tindell, Circulants and their connectivities. J. Graph Theory, (to appear).

11. F. Boesch and J. Wang, Super Line-Connectivity Properties of Circulant Graphs. SIAM J. Algebraic and Discrete Methods (submitted).

12. F. Boesch and J. Wang, On a Class of Graphs which Optimize Several Point Reliability Measures. Discrete Appl. Math. (submitted)

13. A. Cayley, A theorem on trees. Quart. J. Math. 23 (1889, 376-378.

14. C. Cheng, Maximizing the total number of spanning trees in a graph: Two related problems in graph theory and optimum design theory. J. Combin. Theory Series B (1981), 240-248.

15. S. L. Hakimi and A. T. Amin, On the design of reliable networks, Networks 3 (1973), 241-260.

16. F. Harary, Graph Theory, Addison-Wesley, Reading (1969).

17. F. Harary, The maximum connectivity of a graph. Proc. Nat. Acad. of Sci. USA 48 (1962), 1142-1146.

18. B. Jackson, Hamilton cycles in regular graphs. J. Graph Theory 2 (1978), 363-365.

19. A. K. Kelmans and V. M. Chelnokov, A Certain Polynomial of a graph and graphs with an extremal number of trees. J. Combinatorial Theory (B) 16, (1974), 197-214.

20. W. Mader, Minimale n-fach kantenzusammenhängende graphen Math. Ann. 191 (1971), 21-28.

21. J. W. Moon, Counting labelled trees. University of Alberta Mathematics Department Report, Edmonton (1969).

22. J. Sedláček, Lucas numbers in graph theory. Mathematics (geometry and Graph Theory) (Czech.), Univ. Karlova, Prague, (1970), 111-115.

23. A. Schwenk, (private communication).

24. D. R. Shier, Maximizing the number of spanning trees in a graph with n nodes and m edges. Journal of Research NBS 78B (1974), 193-196.

25. D. Smith, Graphs with the smallest number of minimum cut-sets. Networks 14 (1984), 47-62.

26. R. Tindell, (private communication).

27. J. Turner, Point-symmetric graphs with a prime number of points. J. Combinatorial Theory 3 (1967), 136-145.

DIAMETER VULNERABILITY IN NETWORKS

J. Bond & C. Peyrat
L. R. I., Universite de Paris Sud
bat 490, F-91405 ORSAY Cedex, France

Abstract :

Interconnection Networks (telecommunication or microprocessors networks) are modeled by graphs. An important property is that of vulnerability (or fault tolerance). One wants that the centers can communicate in case of link or node failures and in particular with a small message delay. Here we consider graphs of diameter D, maximum degree Δ such that after deletion of s vertices, the diameter is at most D'. We give bounds on D' and on the maximum order of such graphs and tables of their greatest known order for $s = 1$ and $D' = D, D+1, D+2$.

0. Introduction

It is well known that telecommunication networks or interconnection networks can be modeled by graphs. Recent advances in technology, especially in the advent of very large scale integrated (VLSI) circuit technology have enabled very complex interconnection networks to be constructed. Thus, it is of great interest to study the topologies of interconnection networks,and, in particular, their associated graphical properties. If there are point-to-point connections, the computer network is modeled by a graph in which the nodes or vertices correspond to the computer centers in the network and the edges correspond to the communication links.

In the design of these networks, several parameter are very important, for example message delay, message traffic density, reliability or fault tolerance, existence of efficient algorithms for routing messages, cost of the networks,... All these properties correspond to properties of the associated graph, for example, message delay corresponds to the diameter. Furthermore, for technical reasons, a node may not be connected to too many others, which means that the maximum degree of the associated graph is bounded.

For a survey on this problem, see [2] and for a more general study of networks architectures see [13] where the importance of graph studies is outlined.

Here, we focus our attention on some vulnerability problem : we want to design networks with a small diameter D, a given maximum degree Δ (this is a well known problem, with a great literature on it, see [2]) and furthermore such that in case of node (or link) failures the message delay remains still small, that is the diameter of the graph resulting after the deletion of vertices (or edges) is small.

Definitions and notation

We recall here some of the most important definitions and notation. Definitions not given here can be found in [1]. Let G be a simple connected graph. The *distance* $d_G(x,y)$ between x and y is the length of a shortest path between x and y. The *diameter* $D(G)$ (or D if there is no ambiguity) is the maximum of $d_G(x,y)$ over all pairs of vertices $x,y \in V(G)$. If $x \in V(G)$, we note $\Gamma(x)$ the set of the vertices adjacent to x. More generally, we note $\Gamma_i(x)$ the set of vertices y, whose distance from x is i. We note $d(x)$ the degree of the vertex x, and by Δ the maximum degree of the graph.

A (Δ,D)*-graph* will be a graph of maximum degree Δ and diameter at most D. We note by $n(\Delta,D)$ the maximum number of vertices a (Δ,D)-

graph can have. We call (Δ, D, D', s)*–graph* a (Δ, D)-graph such that the diameter of the graph is at most D' after deletion of s vertices (note that this implies that G is (s+1)-connected); $n(\Delta, D, D', s)$ will denote the maximum number of vertices of a (Δ, D, D', s)-graph.

F.R.K. Chung and *M.R. Garey* [9], have shown that the deletion of t edges in a (t+1)-edge-connected graph results in a graph of diameter at most $(t+1)D + t$. These results were refined for D=2,3 and extended for digraphs in [11]. *F.R.K. Chung* and *M.R. Garey* have also shown that after the deletion of vertices the diameter is not necessarily bounded in terms of D. In the first part of the article, we show that it is bounded by a function of Δ and D. We give also upper bounds on $n(\Delta, D, D', s)$-graph and in particular we give tables of the largest known order of (Δ, D, D', s)-graphs for $s = 1$, $D' = D, D+1, D+2$.

Upper bounds for D' and $n(\Delta, D, D', s)$

1.1. Upper bounds for D'

Let G be a λ-connected graph of diameter D and order n. In [9], *F. R. K. Chung* and *M.R. Garey* have shown that after deletion of s vertices, (with $s < \lambda$), the diameter of the resulting graph is at most $\left\lfloor \frac{n-s-2}{\lambda-s} \right\rfloor + 1$ and that this bound is as sharp as possible. This bound is very large, but if we add the constraint that G is of maximum degree Δ, then we have the following result :

Proposition : *Let G be a $(\Delta, D, D', 1)$-graph then :*

$$D' \leq \Delta \left(2 \left\lceil \frac{D}{2} \right\rceil - 1 \right) + D$$

if D *is even and* $\Delta \geq 5$ *then*, $D' \leq s\Delta(D-1) - 1$

if D *is odd then*, $D' \leq s\Delta(D-2) + 6$

Proof : Let z be a vertex of G whose deletion give a graph of diameter D'. Let x_0 and $x_{D'}$ be two vertices of $G' = G-z$ such that $d_{G'}(x_0,x_{D'}) = D'$ and let P be a shortest path between x_0 and $x_{D'}$.

$$P = (x_0, x_1, \ldots, x_i, \ldots, x_{D'})$$

Let $A = \left\{ i \mid x_i \in P \text{ and } d(x_i,z) > \left\lceil \frac{D}{2} \right\rceil \right\}$

Let $\Gamma = \Gamma(z)$, and for every y belonging to Γ, let

$$I(y) = \left\{ i \mid x_i \in P \text{ and } d(x_i,y) < d(x_i,z) \right\}$$

$I(y)$ is the set of indices of the vertices of P whose shortest path to z includes y.

Let $a(y) = \inf I(y)$ and $b(y) = \sup I(y)$.

We will first prove $D' \leq \Delta \left(2 \left\lceil \frac{D}{2} \right\rceil - 1 \right) + D$

Let x_i be any vertex of P, then x_i belongs to A or to at least one of the $I(y)$. Therefore :

$$|V(P)| \leq |A| + \sum_{y \in \Gamma} |I(y) - A| \qquad (1)$$

Indeed ,let i and j be two indices of $I(y)-A$, then :

$$d_{G'}(y,x_i) \leq \left\lceil \frac{D}{2} \right\rceil - 1 \text{ and } d_{G'}(y,x_j) \leq \left\lceil \frac{D}{2} \right\rceil - 1$$

therefore :

$$d_{G'}(x_i,x_j) \leq 2 \left\lceil \frac{D}{2} \right\rceil - 2$$

and

$$|I(y) - A| \leq 2 \left\lceil \frac{D}{2} \right\rceil - 1$$

To prove $|A| \leq D+1$, suppose $|A| > D + 1$.

In this case, there exist two indices i and j of A, such that $|j-i| > D$.

In G', we have :

$$|j-i| = d_{G'}(x_i,x_j) > D$$

furthermore $d_G(x_i,z) + d_G(x_j,z) > D$
and therefore $d_G(x_i,x_j) > D$, a contradiction.

Consequently $D' + 1 = |V(P)| \leq D + 1 + \Delta\left(2\left\lfloor\frac{D}{2}\right\rfloor - 1\right)$

We will now prove the second part of the proposition.

* if $A = \phi$

  By the relation (1), we have : $D' \leq \Delta\left(2\left\lfloor\frac{D}{2}\right\rfloor - 1\right) - 1$

* if $A \neq \phi$

  Let $j = \inf A$ and $k = \sup A$. Obviously, $k - j \leq D$.

  Let $S = \left\{y \mid b(y) \leq k-D-1 \text{ or } a(y) \geq j+D+1\right\}$

  and $\overline{S} = \Gamma - S$. Note $\overline{S} \neq \phi$.

  As above, it can be shown that :

i) For all y such that $b(y) < j$ or $a(y) > k$, we have :

$$b(y) - a(y) \leq 2\left\lfloor\frac{D}{2}\right\rfloor - 2$$

In particular, every y belonging to S satisfies the condition, whence, we have :

ii) If $y \in S$, then $b(y)-a(y) \leq 2\left\lfloor\frac{D}{2}\right\rfloor - 2$

Furthermore, if D is even, then we have :

iii) For all y belonging to S and for all D even, $b(y) - a(y) \leq D - 4$

Indeed, it is sufficient to show that, for all y in S and for all i in $I(y)$, we have : $d(x_i,y) \leq \frac{D}{2} - 2$.

Suppose for instance : $a(y) \geq j+D+1$, then $d_{G'}(x_i,x_j) > D$ for all i in $I(y)$.

As $d_G(x_i,x_j) \leq D$, necessarily we have :

$$d_G(x_i,x_j) = d_G(x_i,z) + d_G(x_j,z)$$

As $d_G(x_j,z) > \frac{D}{2}$ (j belongs to A), we have $d_G(x_i,z) \leq \frac{D}{2} - 1$.

From that, it follows $d_G(x_i,y) \leq \frac{D}{2} - 2$ for all i in $I(y)$.

Finally, we have :

iv) for all y in Γ, $b(y) - a(y) \leq D$

Indeed, if $b(y) - a(y) \geq D + 1$, then in G, the shortest path between $x_{a(y)}$ and $x_{b(y)}$ contains z.

Therefore :

$$\begin{aligned} d_G(x_{a(y)},x_{b(y)}) &= d_G(x_{a(y)},z) + d_G(x_{b(y)},z) \\ &= d_G(x_{a(y)},y) + d_G(x_{b(y)},y) + 2 \\ &\leq D \end{aligned}$$

But as we have :

$$d_{G'}(x_{b(y)},y) = d_G(x_{b(y)},y)$$

and

$$d_{G'}(x_{a(y)},y) = d_G(x_{a(y)},y),$$

we obtain :

$$D + 1 \leq d_{G'}(x_{a(y)},y) + d_{G'}(x_{b(y)},y) \leq D - 2$$

a contradiction, and we need have $b(y) - a(y) \leq D$.

To achieve the demonstration, we will distinguish three cases according to the cardinality of $\overline{S}$.

Case a) $|\overline{S}| \geq 4$

Let y in $\overline{S}$, we have : $k - D \leq b(y)$ and $a(y) \leq j + D$

- if $b(y) < j$ then $b(y) - a(y) \leq 2\left\lfloor\frac{D}{2}\right\rfloor - 2$ (from (i)), whence :

$$a(y) \geq b(y) - 2\left\lfloor\frac{D}{2}\right\rfloor + 2 \geq k - D - 2\left\lfloor\frac{D}{2}\right\rfloor + 2.$$

- if $b(y) \geq j$, as $b(y) - a(y) \leq D$, from (iv), we have :
$a(y) \geq b(y) - D \geq j - D$.

Similarly:

- if $a(y) > k$, then $b(y) \leq a(y) + 2\left\lceil \frac{D}{2} \right\rceil - 2 \leq j + D + 2\left\lceil \frac{D}{2} \right\rceil - 2$.
- if $a(y) \leq k$ then, $b(y) \leq a(y) + D$ (from (iv)), and
$b(y) \leq k + D$.

We will now majorize

$$M = \underset{y \in \overline{S}}{Sup}\, b(y) - \underset{y \in \overline{S}}{Inf}\, a(y)$$

α) If $k - j \geq 2\left\lceil \frac{D}{2} \right\rceil - 2$, then $k + D \geq j + D + 2\left\lceil \frac{D}{2} \right\rceil - 2$,
and for all y in $\overline{S}$, we always have :

$$b(y) \leq k + D$$

We also have : $j - D \leq k - D - 2\left\lceil \frac{D}{2} \right\rceil + 2$ and then
$a(y) \geq j - D$.
In this case, we have : $M \leq k + D - j + D \leq 3D$
- if D is even, from (iii), for all y in S we have :
$b(y) - a(y) \leq D - 4$ and $|I(y)| \leq D - 3$
as $|S| \leq \Delta - 4$, we have :

$$D' \leq (\Delta - 4)(D - 3) + M \leq (\Delta - 4)(D - 3) + 3D$$

$$\leq \Delta(D - 1) - 1 \quad \text{as soon as } \Delta \geq 5.$$

- if D is odd, from (ii), we have for all y in S :
$b(y) - a(y) \leq D - 3$ and $|I(y)| \leq D - 2$.
As $|\overline{S}| \leq \Delta - 4$, we finally obtain :

$$D' \leq (\Delta - 4)(D - 2) + 3D$$

$$\leq \Delta(D - 2) + 5 \quad \text{as soon as } D \geq 3.$$

β) If $k - j < 2\left\lfloor \frac{D}{2} \right\rfloor - 2$, then for all y in $\overline{S}$, we have :

$$b(y) \leq j + D + 2\left\lfloor \frac{D}{2} \right\rfloor - 2$$

$$a(y) \geq k - D - 2\left\lfloor \frac{D}{2} \right\rfloor + 2$$

Therefore,we have : $M \leq 2D + 4\left\lfloor \frac{D}{2} \right\rfloor - 4 + j - k$

As $j - k \leq 0$, we obtain : $M \leq 2D + 4\left\lfloor \frac{D}{2} \right\rfloor - 4$

- if D is even, from (iii), for all y in S, we have

$$|I(y)| \leq b(y) - a(y) + 1 \leq D-3$$

and, as $|S| \leq \Delta - 4$, we have :
$D' \leq (\Delta - 4)(D - 3) + M$
Hence the result :

$$D' \leq (\Delta - 4)(D - 3) + 2D + 4\left\lfloor \frac{D}{2} \right\rfloor - 4$$

$$\leq \Delta(D - 3) + 8$$

$$\leq \Delta(D - 1) - 1$$

as soon as $\Delta \geq 5$.
-if D is odd, then (from (ii)), for all y in S, we have :
$|I(y)| \leq b(y) - a(y) + 1 \leq D - 2$.
Consequently :

$$D' \leq (\Delta - 4)(D - 2) + 4D - 6 = \Delta(D - 2) + 2$$

Case b) : $|\overline{S}| = 3$

$\overline{S} = \{y_1, y_2, y_3\}$

We now have :

$|V(P)| = D' + 1 \leq \sum_{y \in S} |I(y)| + |I(y_1) \cup I(y_2) \cup I(y_3)|$

We will prove :

$|I(y_1) \cup I(y_2) \cup I(y_3)| \leq 3D + 1$ which will imply :

- if D is even : as from (iii), $|I(y)| \leq D - 3$ for all y in S, and as $|S| \leq \Delta - 3$, we have

$$
\begin{aligned}
D' &\leq (\Delta - 3)(D - 3) + 3D \\
&\leq \Delta (D - 1) + 9 - 2\Delta \\
&\leq \Delta (D - 1) - 1
\end{aligned}
$$

as soon as $\Delta \geq 5$.

- if D is odd : from (iv), for all y in S, we have :

$|I(y)| \leq D - 2$ and as $|S| \leq \Delta - 3$, we have :

$$D' \leq (\Delta - 3)(D - 2) + 3D = \Delta (D - 2) + 6$$

α) If $I(y_i) \cap [j,k] \neq \phi$ for all i, $1 \leq i \leq 3$, then $b(y_i) \leq k + D$ and $a(y_i) \geq j - D$. It results :

$$
\begin{aligned}
|I(y_1) \cup I(y_2) \cup I(y_3)| &\leq \operatorname{Sup}_{y_i \in S} b(y_i) - \operatorname{Inf}_{y_i \in S} a(y_i) + 1 \\
&\leq k + D - j + D + 1 \\
&\leq 3D + 1
\end{aligned}
$$

β) if there exists i, $1 \leq i \leq 3$, such that $I(y_i) \cap [j,k] = \phi$, then, from (ii), $|I(y_i)| \leq 2\left\lceil \frac{D}{2} \right\rceil - 1$ and therefore, (from (iv)),

$$|I(y_1) \cup I(y_2) \cup I(y_3)| \leq 2(D + 1) + 2\left\lceil \frac{D}{2} \right\rceil - 1 \leq 3D + 1$$

Case c) $|\bar{S}| = 2$ or $|\bar{S}| = 1$,

We still have :

$|V(P)| = D' + 1 \leq \sum_{y \in S} |I(y)| + \sum_{y \in \bar{S}} |I(y)|$.

- if D is even, it follows from (iii) that :

$|I(y)| \leq D - 3$ for all y in S and consequently,

$$
\begin{aligned}
D' &\leq (\Delta - |\bar{S}|)(D - 3) + |\bar{S}|(D + 1) \\
&= \Delta(D - 1) - 2\Delta + 4|\bar{S}| - 1
\end{aligned}
$$

$$\leq \Delta(D-1)-1$$

as soon as $\Delta \geq 5$.

- if D is odd, it results from (ii) :
$|I(y)| \leq D-2$ for all y in S and consequently.

$$D' \leq (\Delta - |\overline{S}|)(D-2) + 3|\overline{S}| - 1$$

$$\leq \Delta(D-2)+5.$$

Remark 1: *there exist 2-connected graphs of diameter D, maximum degree $\Delta \geq 5$ and a vertex z of $V(G)$ such that:*
$D(G-z) = D'$ and $D' = \Delta(D-1)-1$ if D is even, $D' = \Delta(D-2)$ if D is odd.

Here, we will give only an example for $\Delta = 5$ and $D = 4$ In [6] or [12], we give a general construction.

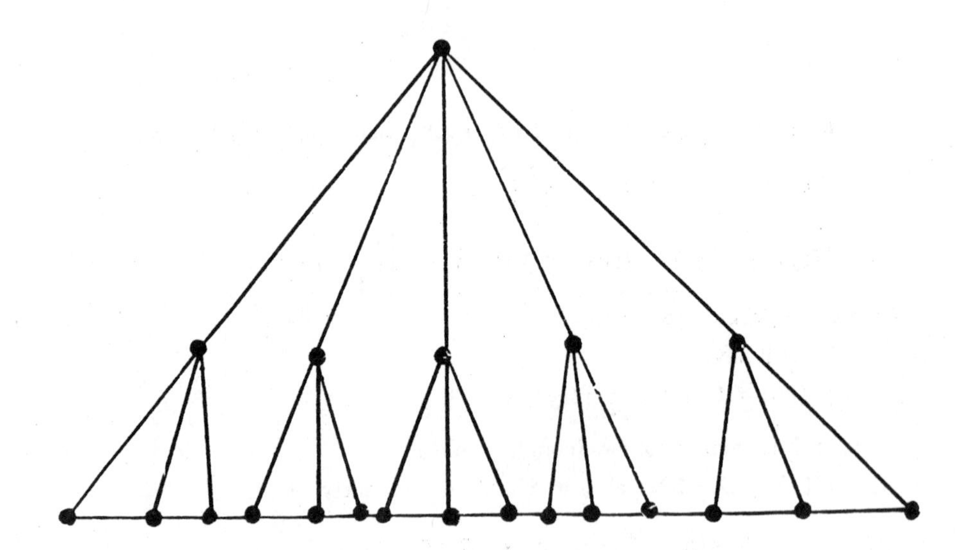

Remark 2: *With a similar proof, it can be shown that if G is a* (Δ, D, D', s)*-graph, then :*

$D' \leq s\,\Delta(D-1) - 1$ if D *is even and* $\Delta > 5$

$D' \leq s\,\Delta(D-2) + 6$ if D *is odd.*

Proposition : *Let G be a* $(\Delta, D, D', 1)$ *graph, such that there exists a vertex z with* $D(G-z) = D'$ *then :*

if $|V(G)| > 1 + \Delta \dfrac{\left[(\Delta-1)^{D-1} - 1\right]}{(\Delta - 2)}$ *then* $D' \leq 2D$.

Proof : Suppose $D' > 2D$

Let x and y be two vertices of $G-z$ such that :

$$d_{G-z}(x,y) = D' \geq 2D + 1.$$

Let v be any vertex of $G-z$, then :

$$d_{G-z}(x,v) \geq D + 1 \text{ or } d_{G-z}(y,v) \geq D+1.$$

Suppose, for instance $d_{G-z}(x,v) \geq D + 1$.

As G has diameter D, this implies that the shortest path between x and v contains z and, therefore :

$$d_G(z,v) < d_G(v,x) \leq D.$$

Hence :

$$\forall v \in V(G-z), \text{ we have : } d_G(z,v) \leq D - 1.$$

Now, if we count the vertices according to their distance from z, we obtain :

$$|V(G)| \leq 1 + \Delta \frac{\left[(\Delta-1)^{D-1} - 1\right]}{\Delta - 2}$$

1.2. Upper bounds for $n(\Delta, D, D', s)$

Proposition : *Let G be a (Δ, D, D, s)-graph ($D \geq 3$) then ,*

$$|V(G)| \leq 1 + \Delta\frac{\left[(\Delta - 1)^{D-1} - 1\right]}{(\Delta - 2)} + \frac{(\Delta - s - 1)\,\Delta\,(\Delta - 1)^{D-2}}{s+1}$$

Proof : Let z be any vertex of G, we have : $|V(G)| = \sum_{i=0}^{D} |\Gamma_i(z)|$

For all i, $i \leq D - 1$, we have : $|\Gamma_i(z)| \leq \Delta(\Delta - 1)^{i-1}$. Let x be any vertex of $\Gamma_D(z)$, x has at least $s + 1$ neighbours in $\Gamma_{D-1}(z)$, otherwise, the deletion of the neighbours of x in $\Gamma_{D-1}(z)$ would yield a graph of diameter greater than D. On the other hand, every vertex x of $\Gamma_{D-1}(z)$, has at least $s + 1$ neighbours in $\Gamma_{D-1}(z) \cup \Gamma_{D-2}(z)$. Suppose x has only λ ($\lambda < s + 1$) neighbours, say $y_1, \ldots, y_\lambda$, in $\Gamma_{D-1}(z) \cup \Gamma_{D-2}(z)$. As $d_{G-\{y_1,\ldots,y_\lambda\}}(x,z) \leq D$, x has necessarily at least one other neighbour in $\Gamma_{D-1}(z) \cup \Gamma_{D-2}(z)$, this prove that x has at least $s + 1$ neighbours in $\Gamma_{D-1}(z) \cup \Gamma_{D-2}(z)$ and therefore at most $\Delta - s - 1$ neighbours in $\Gamma_D(z)$

So, we have :

$$|\Gamma_D(z)| \leq \frac{(\Delta - s - 1)}{s+1}|\Gamma_{D-1}(z)| \leq \frac{(\Delta - s - 1)\Delta(\Delta - 1)^{D-2}}{s+1}$$

and

$$|V(G)| \leq 1 + \Delta\frac{\left[(\Delta - 1)^{D-1} - 1\right]}{(\Delta - 2)} + \frac{(\Delta - s - 1)\,\Delta\,(\Delta - 1)^{D-2}}{s+1}$$

Remark : This bound is asymptotically equal to the Moore bound over s+1. Therefore it is very huge and can not be easily used as soon as D is large, because we do not know enough large graphs. It is quite difficult to obtain better bounds in the general case, but in the particular case where $D = 2$, we have the following result :

Proposition : *Let G be a $(\Delta,2,D',1)$-graph, such that there exist a vertex z of G such that $D(G-z) = D'$. We then have*

(i) *If* $D' = 2$ $|V(G)| \leq \frac{\Delta^2 + \Delta + 2}{2}$

(ii) *If* $D' = 4$ $|V(G)| \leq \frac{\Delta^2}{2} + 1$

(iii) *If* $D' \geq 5$ $|V(G)| \leq \Delta + 1$

Proof : (iii) has already be proved.

(i) the proof is the same as that of the last proposition.

(ii) Let z be a vertex of G such that $D(G-z) = 4$. Let x_0 and x_4 be two vertices of G such that $d_{G-z}(x_0,x_4) = 4$ and let P be a shortest path between x_0 and x_4 in $G-z$

$$P = (x_0,x_1,x_2,x_3,x_4)$$

As G has diameter 2, x_0,x_1,x_3 and x_4 are adjacent with z.
Let x be any vertex of $\Gamma(x_0) - z$. Because G has diameter 2, $d_G(x,x_4) \leq 2$. If $d_G(x,x_4) = 1$, then $d_{G-z}(x_0,x_4) = 2$ a contradiction and necessarily, $d_G(x,x_4) = 2$.
Let (x,y,x_4) be a shortest path between x and x_4.
$y = z$, because else $d_{G-z}(x_0,x_4) = 3$. This proves $x \in \Gamma(z)$
Therefore, we have : $\Gamma(x_0) - z \subseteq \Gamma(z)$ and $\Gamma(x_4) - z \subseteq \Gamma(z)$
As $\Gamma(x_0) \cap \Gamma(x_4) = z$, we obtain

$$|\Gamma(x_0)| + |\Gamma(x_4)| \leq |\Gamma(z)| + 2 \leq \Delta + 2$$

Therefore: $|\Gamma(x_0)| \leq \left\lfloor \frac{\Delta}{2} \right\rfloor + 1$ or $|\Gamma(x_4)| \leq \left\lfloor \frac{\Delta}{2} \right\rfloor + 1$

Suppose for instance $|\Gamma(x_0)| \leq \left\lfloor \frac{\Delta}{2} \right\rfloor + 1$

We will now count the vertices according to their distance from x_0 :

- x_0 has p neighbours (with $p \leq \left\lfloor \frac{\Delta}{2} \right\rfloor + 1$) and one of these neighbours is z.
- z has at most $\Delta - p$ vertices not taken into account.

The other neighbours of x_c have at most $\Delta - 2$ new neighbours.
Hence the result :

$$|V(G)| \leq 1 + p + (\Delta - p) + (p - 1)(\Delta - 2)$$

$$\leq 1 + \Delta + (\Delta - 2)\left\lfloor \frac{\Delta}{2} \right\rfloor$$

$$\leq 1 + \frac{\Delta^2}{2}$$

Remark 1 : The bound of the relation (iii) is as sharp as possible. The graph constructed from a path of length Δ by adding a vertex adjacent to all the other vertices show it.

Remark 2 : The best known $(\Delta,2)$-graphs are $(\Delta,2,3,1)$-graphs. Indeed, they have about Δ^2 vertices.

Construction of $(\Delta,D,D',1,)$-graphs

We will give now constructions of $(\Delta,D,D',1)$-graphs, using known construction of (Δ,D)-graphs.
Here, we will not give proofs of the propositions, these proofs can be found in [6] or [12].

2.1. WBDQ construction

We will denote these graphs by WBDQ, because these constructions were given by *Wegner* in [14] and by *Bermond, Delorme and Quisquater* in [5].

Let G be a $(\Delta - 1, D - 2)$-graph of order n.
We construct G' from G in the following way : We take Δ copies of G, and then add n extra vertices, called central vertices. Every central vertex is adjacent with all the copies of a vertex of G.

Proposition : *G' is a $(\Delta, D, D+2, 1)$-graph.*

Corollary 1 : $n(\Delta, D, D+2, 1) \geq (\Delta+1)\ n(\Delta-1, D-2)$.

Corollary 2 : $n(\Delta, D, D+1, 1) \geq (\Delta-1)\ n(\Delta-1, D-2, D+1, 1)$.

Let G be a $(\Delta-1, D-2, D-2, 1)$-graph of order n. Let $p = \sum_{x \in V(P)} (\Delta-1-d(x))$.
If p is null, the following construction is possible only if n is even.
Else, let $n' = n + p$ or $n + p - 1$ such that n' is even.
We construct G' from G in the following way:

Take $\Delta - 1$ copies of G, then add n' extra vertices, called central vertices, each of these central vertices is adjacent with all the copies of one vertex of G. Then, add also a perfect matching on the central vertices without creating any triangle in G (this can always be done provided $p \ll n$).

Proposition : *If $\Delta \geq 3$, then G' is a $(\Delta, D, D, 1)$-graph.*

2.2. The construction $G_1[G_2]$ where G_2 is a complete graph

The $G_1[G_2]$ construction is defined in the general case by *Bermond, Delorme* and *Quisquater* in [5].

Here, we will consider the particular case where G_1 is a (Δ_1, D_1)-graph of order n_1 and G_2 is a complete graph of order n_2 with :

$$n_2 = n_1 \Delta - \sum_{x \in V(G_1)} (d_1(x) + 1) \text{ with } \Delta > \Delta_1$$

$G = G_1[G_2]$ is a $(\Delta, 2D_1)$-graph of order $n_1 n_2$ constructed from n_2 copies of G_1 in such a way that there is always at least an edge between two copies of G_1.

Proposition : *If G_1 is a $(\Delta_1, D_1, 2D_1 + 2)$-graph, then $G_1[K_{n_2}]$ is a $(\Delta, 2D_1 + 1, 2D_1 + 2, 1)$-graph.*

2.3. Cartesian sum

The cartesian sum $G_1 + G_2$ of a (Δ_1, D_1)-graph $G_1 = (V_1, E_1)$ by a (Δ_2, D_2)-graph $G_2 = (V_2, E_2)$ is defined by example in [1] :

$V(G_1 + G_2) = V_1 \times V_2$

$((x_1, x_2), (y_1, y_2)) \in E(G_1 + G_2)$ if and only if

$x_1 = y_1$ *and* $(x_2, y_2) \in E_2$ or $(x_1, y_1) \in E_1$ *and* $x_2 = y_2$.

Proposition : *If G_1 is a (Δ_1, D_1)-graph and G_2 is a (Δ_2, D_2)-graph, then $G_1 + G_2$ is a $(\Delta_1 + \Delta_2, D_1 + D_2, D_1 + D_2, 1)$-graph as soon as $D_1 \geq 2$ and $D_2 \geq 2$.*

2.4. A construction of $(\Delta, D, D, 1)$-graphs

Let G be a (Δ, D)-graph of order n. G' is constructed from G by duplicating the vertices of G and replacing the edges of G by $K_{2,2}$ (that is the lexicographical product of G by an independent set of two vertices).

Proposition : *G' is a $(2\Delta, D, D, 1)$-graph.*

If we choose as graph G the graphs P_q' (factor graphs of incidence graphs associated to a projective plane of order q), then, we have proved that $n(2\Delta, D, D, 1)$ is of the order of Δ^2.

2.5. The $k*G$ construction

Let G be a $(\Delta - k + 1, D - 1, D', 1)$-graph, and let G' be the graph constructed from G by taking k copies of G and then add all the possible edges between copies of a given vertex of G

Proposition : *G' is a $(\Delta, D, D'', 1)$-graph where $D'' = Sup(D', D)$.*

2.6. A construction of $(\Delta, D, D, 1)$-graph with Δ odd

When Δ is even, we have a family of $(\Delta, D, D, 1)$-graphs, the Kautz graphs (see [7]).

Let us recall that the vertices of these graphs are words of length D on an alphabet of $(\frac{\Delta}{2}) + 1$ letters, two consecutive letters being different. Two vertices are adjacent if and only if the $D - 1$ first letters of one are the same as the $D - 1$ last letters of the other. These graphs have order $(\frac{\Delta}{2})^D + (\frac{\Delta}{2})^{D-1}$. There exists an extension of these graphs (see [6] or [12]) for $\Delta = 2d$ into graphs of the same diameter and degree 2d+1, such that the deletion of one vertex does not increase the diameter.

2.7. Sequence Graphs

This family of graphs has been exhibited in [10] by *M.A. Fiol, J.L.A. Yebra* and *J. Fabrega*. Let G be a connected graph, $S^q(G)$, the sequence graph of order q of G is the graph whose vertices represent the paths of length q-1 (q vertices and q-1 edges) in G. Two vertices of $S^q(G)$ are adjacent if the union of the two paths they represent is a path of length q. The diameter of $S^q(G)$ is at most $D(G) + q - 1$.

In the particular case where G is a bipartite graph and the parity of $D(G)$ is the same as the parity of q, the diameter of $S^q(G)$ is equal to $D(G) + q - 2$.

Propositon : *$S^q(K_{d+1,d+1})$ is a $(\Delta, D, D+2, 1)$-graph.*

TABLES

Using all the constructions given here and some other given in [6] or [12], we will give tables of maximum order of $(\Delta, D, D', 1)$-graphs for $D' = D, D+1, D+2$.

Symbols and notations used in the tables.

$\otimes$: WBDQ construction (See 2.1)
$G_1[K_n]$: See 2.2
$+$: Cartesian sum (See 2.3)
$G(K_{2,2})$: See 2.4
$*$: See 2.5
K : Kautz graphs
K^+ :See 2.6
S : Sequence graphs (See 2.7)
d : operation described in [8]
$\times$:special product defined in [4] and [5]
r: Replacement of a vertex of degree Δ by a complete graph K_Δ
P : Petersen Graph
Coxeter : Coxeter Graph
P_q : Incidence graph associated to a projective plane of order q
P'_q : Factor graph of P_q by a polarity
HS : Hoffmann-Singleton Graph
Q_p : generalized quadrangle of order q.
C_m : cycle on m vertices.

$D' = D$

Δ \ D	2	3	4	5	6	7	8	9	10
2	C_4 4	C_5 5	C_6 6	C_7 7	C_8 8	C_9 9	C_{10} 10	C_{11} 11	C_{12} 12
3	$K_{3,5}$ 6	12	P_2 18	Coxeter 28	Coxeter_r 36	H 102	H_d 104	$K_2 * C_{61}$ 122	$K_2 * C_{85}$ 170
4	P 10	2*P 20	$C_3(9,2,2)$ 36	$5 \otimes P$ 50	$P[K_{11}]$ 110	K 192	$C_3(8,6,2)$ 512	$C_3(9,7,2)$ 1 152	K 1 536
5	16	3*P 30	Q_5 126		$15[K_{16}]$ 240	$15 \otimes Q_3$ 480	$40[K_{41}]$ 1 640	$5 \otimes H_3$ 4 368	$95[K_{96}]$ 9 120
6	$P(K_{2,2})$ 20	2*24 48		Q_6 462	K 972	K 2 916	K 8 748	K 26 224	K 78 732
7	$P(K_{2,2})d$ 22	3*24 72			Q_7 1 716	K^+ 3 240	$105[K_{106}]$ 11 130	$6 \otimes H_5$ 62 496	$320[K_{321}]$ 102 720
8	$15(K_{2,2})$ 30	2*HS 100	K 320	K 1 280	K 5 120	K 20 480	K 81 920	K 327 680	K 1 310 720

D' = D (continued)

D / Δ	2	3	4	5	6	7	8	9	10
	$15(K_{2,2})_d$	3*HS	K^+	K^+	K^+	K^+	K^+	$10 \otimes H_7$	K^+
9	32	150	360	1 440	5 760	23 040	92 160	392 160	1 474 560
	$24(K_{2,2})$	4*HS	K	K	K	K	K	K	K
10	48	200	750	3 750	18 750	93 750	468 750	2 343 750	11 718 750
	$24(K_{2,2})_d$	5*HS	K^+	K^+	K^+	K^+	$585[K_{1236}]$	K^+	K^+
11	50	250	810	4 050	20 250	101 250	723 060	2 531 250	12 656 250
	$32(K_{2,2})$	6*HS	K	K	K	K	K	K	K
12	64	300	1 512	9 072	54 432	326 592	1 959 552	11 757 312	70 543 872
	$32(K_{2,2})_d$	5*74	K^+	K^+	K^+	K^+	3* 723 060	K^+	K^+
13	66	370	1 638	9 828	58 968	353 808	2 169 180	12 737 088	76 422 528
	$HS(K_{2,2})$	5*91	K	K	K	K	K	K	K
14	100	455	2 744	19 208	134 456	941 192	6 588 344	46 118 408	322 828 856
	$HS(K_{2,2})_d$	6*91	K^+	K^+	K^+	K^+	K^+	K^+	K^+
15	102	546	2 912	20 384	142 688	998 816	6 991 712	48 941 984	342 593 888

$D' = D + 1$

D / Δ	2	3	4	5	6	7	8	9	10
2	C_5 5	C_6 6	C_7 7	C_8 8	C_9 9	C_{10} 10	C_{11} 11	C_{12} 12	C_{13} 13
3	P 10	PG 14	Coxeter 28	$Coxeter_r$ 36	$K_2 \times C_{25}$ 50	H 102	$K_2 * C_{61}$ 122	$K_2 * C_{85}$ 170	$K_2 * C_{113}$ 226
4	$K_3 \times 5$ 15	Q_4 35	$5 \otimes P$ 50	$P[K_{11}]$ 110	$C_3(4,5,2)$ 128	$20[K_{21}]$ 420	$5 \otimes H_3$ 630	$38[K_{39}]$ 1482	$C_3(5,8,2)$ 2 560
5	$K_3 \times 8$ 24	P_4 42	Q_3 126	$15[K_{16}]$ 240	$15 \otimes Q_5$ 480	$40[K_{41}]$ 1 640	$5 \otimes H_3$ 4 368	$95[K_{96}]$ 9 120	
6	$K_4 \times 8$ 32	P_5 62	$7 \otimes 24$ 168	$24[K_{25}]$ 600	$7 \otimes 174$ 1 418	$55[K_{57}]$ 4 422	$7 \otimes H_4$ 19 110	$175[K_{175}]$ 30 450	$C_3(5,8,3)$ 98 415
7	HS 50	3*24 72	$8 \otimes 32$ 256	$24[K_{49}]$ 1 174	$8 \otimes 317$ 2 536	$105[K_{106}]$ 11 130	$8 \otimes H_5$ 62 496	$320[K_{321}]$ 102 720	$8 \otimes 19110$ 152 880
8	P'_7 57	P_7 114	$9 \otimes 50$ 450	$HS[K_{51}]$ 2 550	K 5 120	$105[K_{211}]$ 22 155	K 81 920	K 327 680	$C_3(5,8,4)$ 1 310 720

D' = D + 1 (continued)

D Δ	2	3	4	5	6	7	8	9	10
	P'_8d	3*HS	Q'_5	$HS[K_{101}]$	$10 \otimes Q_7$	$200[K_{201}]$	$10 \otimes H_7$	$807[K_{808}]$	K^+
9	74	150	585	5 050	8 000	40 200	329 160	652 056	1 474 560
	P'_9	4*HS	2*585	$HS[K_{151}]$	K	$585[K_{651}]$	$11 \otimes H_8$	K	K
10	91	200	1 170	7 550	18 750	380 835	823 878	2 343 750	11 718 750
	P'_9d	5*HS	3*585	$P'_8[K_{156}]$	K^+	$585[K_{1236}]$	$12 \otimes H_9$	$1755[K_{1756}]$	K^+
11	94	250	1 755	11 388	22 250	723 060	1 594 320	3 081 780	12 656 250
	P'_{11}	6*HS	4*585	$P'_9[K_{193}]$	K	$585[K_{1821}]$	K	K	K
12	133	300	2 340	17 563	54 432	1 065 285	1 959 552	11 757 312	70 543 872
	$P'_{11}d$	5*74	5*585	$P'_9[K_{284}]$	K^+	$650[K_{2211}]$	$14 \otimes H_{11}$	$4680[K_{5210}]$	K^+
13	136	370	2 925	25 844	58 968	1 437 150	4 960 368	24 340 680	76 422 528
	P'_{13}	5*91	6*585	$P'_{11}[K_{279}]$	K	$910[K_{2341}]$	K	$4680[K_{9881}]$	K
14	183	455	3 510	37 107	134 450	2 130 310	6 588 348	46 243 080	322 828 428
	$P'_{13}d$	6*91	7*585	$P'_{11}[K_{412}]$	K^+	$910[K_{3251}]$	$15 \otimes H_{13}$	$4680[K_{14561}]$	K^+
15	186	546	4 095	54 796	142 688	2 958 410	12 871 488	68 145 480	342 593 888

$D' = D + 2$

D / Δ	2	3	4	5	6	7	8	9	10
2	C_5 5	C_7 7	C_8 8	C_9 9	C_{10} 10	C_{11} 11	C_{12} 12	C_{13} 13	C_{14} 14
3	P 10	20	Q_2 30	$K_2 \times C_{25}$ 50	$K_2 \times C_{41}$ 82	$K_2 \times C_{61}$ 122	$K_2 * C_{85}$ 170	$K_2 * C_{113}$ 226	$K_2 * C_{145}$ 290
4	$K_3 \times 5$ 15	O_4 35	Q_3 80	$P[K_{11}]$ 110	5×38 190	$20[K_{21}]$ 420	5×128 640	$38[K_{39}]$ 1482	$C_3(5,9,2)$ 2 560
5	$K_3 \times 8$ 24	P_4 42	Q_4 170	6×40 240	$6 \otimes 95$ 570	$40[K_{41}]$ 1 640	$6 \otimes 731$ 4 386	$95[K_{96}]$ 9 120	
6	$K_4 \times 8$ 32	P_5 62	Q_5 312	$24[K_{25}]$ 600	$7 \otimes 174$ 1 218	$66[K_{67}]$ 4 422	$7 \otimes 2\ 734$ 19 138	$174[K_{175}]$ 30 450	$C_3(5,9,3)$ 98 415
7	HS 50	3*24 72	$Q_5 d$ 346	$24[K_{49}]$ 1 176	$8 \otimes 317$ 2 536	$105[K_{106}]$ 11 130	$8 \otimes 7\ 817$ 62 536	$320[K_{321}]$ 102 720	$8 \otimes 19\ 138$ 153 104
8	P'_7 57	P_7 114	Q_7 800	$HS[K_{51}]$ 2 550	K 5 120	$105[K_{211}]$ 22 155	K 81 920	K 327 680	$C_3(5,9,4)$ 1 310 720

D' = D + 2 (continued)

D / Δ	2	3	4	5	6	7	8	9	10
	P'_8d	Q'_5	Q_8	$HS[K_{101}]$	$10 \otimes 807$	$200[K_{201}]$	$10 \otimes 59\ 223$	$807[K_{808}]$	K^+
9	74	585	1 170	5 050	8 070	40 200	592 230	652 056	1 409 024
	P'_9	Q'_5d	Q_9	$HS[K_{151}]$	K	$585[K_{651}]$	$11 \otimes 74\ 906$	K	K
10	91	650	1 640	7 550	18 750	380 835	823 966	2 343 750	11 718 750
	P'_9d	Q'_8d	Q_8d	$P'_8[K_{156}]$	S	$585[K_{1236}]$	$12 \otimes 132\ 869$	$1755[K_{1756}]$	S
11	94	715	1 734	11 388	27 000	723 060	1 594 428	3 081 780	24 300 000
	P'_{11}	Q'_8d	Q_{11}	$P'_9[K_{193}]$	K	$585[K_{1821}]$	K	K	K
12	133	780	2 928	17 563	54 432	1 065 285	1 959 552	11 757 312	70 543 872
	$P'_{11}d$	Q'_8d	$Q_{11}d$	$P'_9[K_{284}]$	S	$650[K_{2211}]$	$14 \otimes 354\ 323$	$4680[K_{5201}]$	S
13	136	845	3 064	25 844	74 088	1 437 150	4 960 522	24 340 680	130 691 232
	P'_{13}	Q'_8d	Q_{13}	$P'_{11}[K_{279}]$	K	$910[K_{2341}]$	K	$4680[K_{9881}]$	K
14	183	910	4 760	37 107	134 450	2 130 310	6 588 344	46 243 080	322 828 656
	$P'_{13}d$	Q'_8d	$Q_{13}d$	$P'_{11}[K_{412}]$	S	$910[K_{3251}]$	$16 \otimes 804\ 481$	$4680[K_{14561}]$	S
15	186	975	4 946	54 796	175 616	2 958 410	12 871 696	68 145 480	550 731 776

References

[1] C. Berge : "*Graphs and Hypergraphs*", North-Holland Publishing Co, New York, 1973.

[2] J-C Bermond, J.Bond, M. Paoli and C. Peyrat : "*Graphs and Interconnection Networks : Diameter and vulnerability*", Surveys in Combinatorics, Invited paper for the 9th British Combinatorial Conference, 1983, London Math. Soc. Lect. Note Ser. 82, Cambridge University Press, 1983, 1-30.

[3] J-C Bermond, C. Delorme and G. Farhi : "*Large graphs with given degree and diameter II*", J. Combinatorial Theory, Ser. B, Vol 36, N 1, Feb 1984, 32-48.

[4] J-C Bermond, C. Delorme and G. Farhi : "*Large graphs with given degree and diameter III*", in Proc. Coll. Cambridge 1981, Ann. Disc. Math., North-Holland., Vol 13, 1982, 23-32.

[5] J-C Bermond, C. Delorme and J-J Quisquater : "*Grands graphes de degre et diametre fixes*", in Proc. Coll. C.N.R.S., Marseille 1981, Ann. Disc. Math. 17, North Holland, 65-73.

[6] J. Bond : "*Constructions de grands reseaux d'interconnexion*", thesis, Universite Paris-Sud, Mars 1984.

[7] J. Bond and C. Peyrat : "*Diameter vulnerability of some large interconnection networks*", submitted.

[8] C. Delorme : "*Grands graphes de degre et diametre donnes*", to appear in Journal Europeen de Combinatoire.

[9] F.R.K. Chung and M.R. Garey : "*Diameter bounds for altered graphs*", to appear.

[10] M.A. Fiol, J.L.A. Yebra and J. Fabrega : "*Sequence graphs and interconnection networks*", to appear.

[11] C. Peyrat : "*Diameter Vulnerability of Graphs*", to appear in Disc. Applied Math.

[12] C. Peyrat : "*Vulnerabilite dans les reseaux d'interconnexion*", thesis, Universite Paris-Sud, Mars 1984.

[13] L. Uhr : "*Algorithms-structured computer arrays and networks*", Academic Press, 1984.

[14] G. Wegner : "*Graphs with given diameter and a coloring problem*", unpublished manuscript.

GENERALIZED COLORINGS OF OUTERPLANAR AND PLANAR GRAPHS

I. Broere[1]
Rand Afrikaans University

C.M. Mynhardt[2]
University of South Africa

ABSTRACT

Let F be a nontrivial graph. The F-*chromatic number* $F\langle G\rangle$ of a graph G is the minimum number of subsets into which V(G) can be partitioned in such a way that F is not an induced subgraph of the subgraph induced by each partition class.

If F has a nonempty edge set, then $F\langle G\rangle \leqslant \chi(G)$ for all G and hence $F\langle G\rangle \leqslant 3$ if G is outerplanar. We characterize, for $i = 1,2,3$, those graphs F for which $F\langle G\rangle \leqslant i$ for all outerplanar G.

---

[1]Research supported by the Rand Afrikaans University and the South African Council for Scientific and Industrial Research.

[2]Research supported in part by the South African Council for Scientific and Industrial Research.

The corresponding problems for planar graphs are investigated.

A companion paper [7] discusses other aspects of these chromatic numbers.

1. Introduction.

We use the notation and terminology of [4]. In particular, if G is a (finite) graph (without loops or multiple edges) and $\phi \neq U \subseteq V(G)$, then $\langle U \rangle$ denotes the subgraph of G induced by U while $F < G$ if $F = \langle U \rangle$ for some U.

If F and G are graphs, then a (k,F)-*coloring* of G (also called an F-*coloring* of G if the value of k is immaterial) is a partition $V_1,\ldots,V_k$ of V(G) such that $F \nless \langle V_i \rangle$ for each i. G admits an F-coloring iff F is nontrivial and hence *we assume henceforth that F is nontrivial*. The F-*chromatic number* of G, denoted by $F\langle G \rangle$, is the smallest k for which G admits a (k,F)-coloring. Clearly, $F\langle G \rangle$ exists (and is finite) for all G. Note that a (k,K_2)-coloring of G is nothing but a k-coloring of G and hence that $K_2\langle G \rangle = \chi(G)$ for all G.

Two remarks that will be used often are:

- If $H < F$, then $F\langle G \rangle \leqslant H\langle G \rangle$ for all G.
- If $G < H$, then $F\langle G \rangle \leqslant F\langle H \rangle$ for all F.

2. Outerplanar graphs.

It is well known that $\chi(G) \leqslant 3$ if G is an outerplanar graph (see for instance the (3,1)-THEOREM of [6]) and it follows that $F\langle G \rangle \leqslant 3$ if G is outerplanar and F has a nonempty edge set.

Our first result is trivial.

Proposition 1. $F\langle G\rangle \leqslant 1$ for all outerplanar G iff F is not outerplanar. ■

Our next result in this spirit is Theorem 5.

Lemma 2. If G is a maximal outerplanar graph with $p(G) \geqslant 4$ (embedded on the plane as a triangulation of a cycle), then G contains an edge $\{v_1,v_2\}$ on the exterior face with $\deg v_1 = 2$ and $\deg v_2 = 3$ or G contains two adjacent edges $\{w_1,w_2\}$ and $\{w_2,w_3\}$ on the exterior face with $\deg w_1 = \deg w_3 = 2$ and $\deg w_2 = 4$.

Proof (by induction on $p(G)$). The only maximal outerplanar graph (K_4-e) of order four clearly has such an edge.

Assuming the result if $p(G) \leqslant p$, let G be of order $p+1$ with v a vertex of degree two. Apply the induction hypothesis to G-v and rebuild G. It is easy to check that G has the desired edge(s). ■

Following [1], we call a forest *linear* if each of its components is a path.

Theorem 3. If G is an outerplanar graph, then there is a partition V_1,V_2 of $V(G)$ such that each $\langle V_i\rangle$ is a linear forest.

Proof. It is sufficient to prove the result for maximal outerplanar graphs and we do this by induction on $p(G)$; the cases in which $p(G) \leqslant 4$ being easy.

Hence assume the result if $p(G) \leqslant p$ and let G be of order $p+1$; embedded again as a triangulation of a cycle. If G has an edge $\{v_1,v_2\}$ on the exterior face with $\deg v_1 = 2$ and $\deg v_2 = 3$, apply the induction hypothesis to $G-v_1$ to obtain a partition V_1',V_2' of $V(G-v_1)$. If v_3

is the other neighbor of v_1 with (say) $v_3 \in V_1'$, let $V_1 = V_1'$ and $V_2 = V_2' \cup \{v_1\}$.

If, on the other hand by Lemma 2, G has two edges $\{w_1,w_2\}$ and $\{w_2,w_3\}$ on the exterior face with $\deg w_1 = \deg w_3 = 2$ and $\deg w_2 = 4$, apply the induction hypothesis to $G-w_1$ to obtain V_1' and V_2' with (say) $w_2 \in V_1'$. If $w_4 (w_5)$ is the other neighbor of w_1 (w_3 respectively) and $w_4 \in V_1'$, we let $V_1 = V_1'$ and $V_2 = V_2' \cup \{w_1\}$. If $w_4 \in V_2'$, then w_2 is an endvertex or an isolated vertex of $\langle V_1' \rangle$ and we let $V_1 = V_1' \cup \{w_1\}$ and $V_2 = V_2'$. ∎

Note that Theorem 3 sharpens the well known result that the vertex-arboricity of an outerplanar graph is at most two (see for instance the (3,2)-THEOREM of [6]). In the terminology of [1], Theorem 3 states that the point linear arboricity of every outerplanar graph is at most two.

Theorem 4. If F is a linear forest with nonempty edge set, then there exists an outerplanar graph G with $F\langle G \rangle = 3$.

Proof. Note that there exists a path P_n such that $F < P_n$. Hence it is sufficient to exhibit, for a given $n \geq 2$, an outerplanar graph X_n such that $P_n\langle X_n \rangle \geq 3$ (since then we have that $F\langle X_n \rangle \geq P_n\langle X_n \rangle \geq 3$ while $F\langle X_n \rangle \leq 3$ as remarked earlier).

We construct X_n as follows: $V(X_n) = \bigcup_{i=0}^{n-1} W_i$ with the W_i pairwise disjoint sets with $|W_i| = n^i$, $i = 0,1,\ldots,n-1$. For each $i < n - 1$ and each $w \in W_i$, choose a set $W \subseteq W_{i+1}$ with $|W| = n$ with disjoint sets chosen for different vertices. The edge set of X_n consists of those edges that join each $w \in W_i$ to each of the chosen vertices in W as well as those edges that make $\langle W \rangle \cong P_n$ for each chosen W. The graph X_3 is depicted in Figure 1.

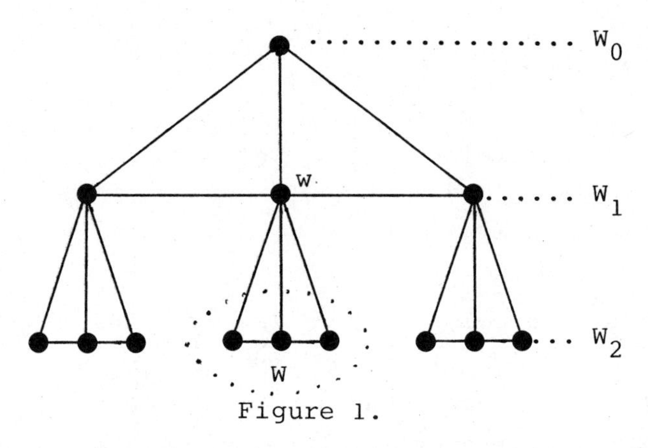

Figure 1.

Each X_n is outerplanar (since every block of X_n is isomorphic to $K_1 + P_n$) and we proceed to prove that $P_n\langle X_n \rangle \geqslant 3$: Assume V_1, V_2 is a $(2, P_n)$-coloring of X_n. Suppose that the only vertex w_o of W_o is in V_1. Since $\langle W_1 \rangle \cong P_n$, there is a vertex $w_1 \in W_1 \cap V_1$. By considering the set $W \subseteq W_2$ chosen for w_1, we find $w_2 \in W_2 \cap V_1$. Repeating this leads to $W' \subseteq V_1$ with $\langle W' \rangle \cong P_n$, a contradiction. ■

Note that X_n has a property slightly stronger than the graph G constructed in the proof of Theorem 3 of [6].

Theorem 5. $F\langle G \rangle \leqslant 2$ for all outerplanar G iff F is not a linear forest.

Proof. Theorem 3 and Theorem 4. ■

The last result in this spirit is again trivial.

Proposition 6. $F\langle G\rangle \leqslant 3$ for all outerplanar G iff F has a nonempty edge set. ∎

Note that one is able to deduce from the above results for which F there exists an outerplanar G with $F\langle G\rangle = 1$ (or 2 or 3). In particular we have: There exists an outerplanar G for which $F\langle G\rangle = 3$ iff F is a linear forest.

3. Planar graphs.

The four-color theorem (see [2] and [3]) tells us that $\chi(G) \leqslant 4$ for all planar graphs G — hence $F\langle G\rangle \leqslant 4$ if G is planar and F has a nonempty edge set. We now attempt to characterize, for i = 1,2,3 and 4, the graphs F for which $F\langle G\rangle \leqslant i$ for all planar G. The first case is again trivial.

Proposition 7. $F\langle G\rangle \leqslant 1$ for all planar G iff F is nonplanar. ∎

The next result in this spirit is Theorem 11.

Lemma 8. Every planar graph is an induced subgraph of a maximal planar graph.

Proof. Let G be planar; find an embedding of G in the plane. For each region of this plane graph, add a new vertex and join it to every vertex on the boundary of this region. Repeat this procedure at most two more times to find the required maximal plane graph. ∎

Theorem 9. If G is a planar graph and F contains a cycle, then $F\langle G\rangle \leqslant 2$.

Proof. If F contains a triangle, we partition V(G) into four sets W_1, W_2, W_3, W_4 using the four-color theorem (of [2] and [3]) and let $V_1 = W_1 \cup W_2$ and $V_2 = W_3 \cup W_4$. Clearly, $F \not< \langle V_i \rangle$ for i = 1,2. (In fact, $\langle V_1 \rangle$ and $\langle V_2 \rangle$ are bipartite and contain no odd cycles.)

Hence assume that (the girth of F) $g(F) \geqslant 4$. It is sufficient to prove that, for a given positive integer $m \geqslant 4$, $C_m\langle G \rangle \leqslant 2$ for all planar G (since for some such m we have $C_m < F$). By Lemma 8, it is sufficient to prove this for maximal planar graphs which we do now by induction on p(G).

If $p(G) \leqslant 6$ it is easy; hence assume that $C_m\langle G \rangle \leqslant 2$ if G is maximal planar and $p(G) \leqslant p$ and let G be a maximal plane graph of order p+1. Let v be a vertex of G with n = degv $\leqslant 5$; let $v_1,\ldots,v_n$ be the neighbors of v in G arranged cyclically around v. Let G' be a maximal plane graph obtained from G-v by adding v_i-v_j edges (if necessary) and apply the induction hypothesis to G' to obtain V_1' and V_2'. Note that v can be added to V_i' (to obtain a $(2,C_m)$-coloring of G) if $|\{v_1,\ldots,v_n\} \cap V_i'| \leqslant 1$ or if $|\{v_1,\ldots,v_n\} \cap V_i'| = 2$ and the two vertices in this intersection are adjacent in G.

This remark covers all but two cases:

Case 1. n = 4 and the nonadjacent pairs v_1,v_3 and v_2,v_4 are in V_1' and V_2' respectively. Note that either v_1 and v_3 are in different components of $\langle V_1' \rangle$ or v_2 and v_4 are in different components of $\langle V_2' \rangle$ — hence we can choose i and let $V_i = V_i' \cup \{v\}$ and $V_j = V_j'$ ($j \neq i$) such that v is a cut-vertex of $\langle V_i \rangle$ with degv = 2.

Case 2. n = 5 and v_1,v_3,v_4 in V_1' and the nonadjacent pair v_2, v_5 in V_2'. Again we can choose i and let $V_i = V_i' \cup \{v\}$ and $V_j = V_j'$ $(j \neq i)$ such that v is a cut-vertex of $\langle V_i \rangle$. Note that, in this case, $C_m \nless \langle V_i \rangle$ since $m \geq 4$. ■

Theorem 10. If F is a forest, then there exists a planar graph G such that $F\langle G \rangle \geq 3$.

Proof. Since $F < T$ for some tree T, it is sufficient to prove the result for trees with nonempty edge sets.

Hence let T be a tree with $p(T) = n \geq 2$ and $\Delta(T) = k$. We construct a graph G_T as follows: $V(G_T) = \bigcup_{i=0}^{n-1} W_i$ with the W_i pairwise disjoint sets with $|W_i| = (kn)^i$, $i = 0,1,\ldots,n-1$. For each $i < n-1$ and each $w \in W_i$, choose k sets $W \subseteq W_{i+1}$ with $|W| = n$ with every pair of chosen W's disjoint. The edge set of G_T consists of those edges that join each $w \in W_i$ to each of the kn vertices in the W's chosen for it as well as those edges that make $\langle W \rangle \cong T$ for each chosen W.

G_T is planar since each block of G_T is isomorphic to $K_1 + T$. A proof that $T\langle G_T \rangle \geq 3$, similar to the corresponding part of the proof of Theorem 4, can be given. ■

Theorem 11. $F\langle G \rangle \leq 2$ for all planar G iff F contains a cycle.

Proof. Theorem 9 and Theorem 10. ■

The last nontrivial case we were unable to solve completely; we formulate however

Conjecture 12. If G is planar, then $K(1,3)\langle G\rangle \leq 3$.

Theorem 13. If F is a linear forest with a nonempty edge set, then there exists a planar graph G with $F\langle G\rangle = 4$.

Proof. Again, since $F \prec P_n$ for some n, it is sufficient to exhibit for a given $n \geq 2$ a planar graph Y_n such that $P_n\langle Y_n\rangle \geq 4$ (since then $F\langle Y_n\rangle = 4$).

To construct Y_n, let X_n be as in the proof of Theorem 4 with $p(X_n) = p$. Then let $V(Y_n) = \bigcup_{i=0}^{n-1} W_i$ with the W_i pairwise disjoint sets with $|W_i| = p^i$, $i = 0,1,..,n-1$. Choose for each $i < n - 1$ and each $w \in W_i$ a set $W \subseteq W_{i+1}$ with $|W| = p$ with disjoint sets chosen for different vertices. The edge set of Y_n consists of those edges that join each w to each of the chosen vertices in W as well as those edges that make $\langle W\rangle \cong X_n$ for each chosen W.

In this case, each block of Y_n is isomorphic to $K_1 + X_n$ with X_n outerplanar; hence Y_n is planar. A proof that $P_n\langle Y_n\rangle \geq 4$ can again be given by imitating the corresponding part of the proof of Theorem 4 and using the essential property of X_n. ■

Note that Y_n has a property slightly stronger than the graph G in Theorem 5 of [5].

The next conjecture is equivalent to Conjecture 12.

Conjecture 14. $F\langle G\rangle \leq 3$ for all planar G iff F is not a linear forest.

Proof (of equivalence). It is clear that Conjecture 14 implies Conjecture 12. Conversely, Conjecture

14 is implied by Theorem 9, Conjecture 12 and Theorem 13. ∎

The last case to be considered is an easy deduction from the four-color theorem.

Proposition 15. $F\langle G\rangle \leqslant 4$ for all planar G iff F has a nonempty edge set. ∎

4. Discussion.

Conjecture 12 is the weakest form of a version of Theorem 3 for planar graphs we know of that is strong enough to give us Conjecture 14. We believe, however, that even the following stronger form is true.

Conjecture 16. If G is a planar graph, then there is a partition V_1, V_2, V_3 of V(G) such that each $\langle V_i\rangle$ is a linear forest.

Note that the statement of Conjecture 16 sharpens the well known result that the vertex-arboricity of a planar graph is at most three (see for instance the (4,2)-THEOREM of [6]).

REFERENCES

1. J. Akiyama, G. Exoo and F. Harary, Covering and packing in graphs III: Cyclic and acyclic invariants, *Math. Slovaca*, 30 (1980), 405-417.
2. K. Appel and W. Haken, Every planar map is four colorable (Part I: Discharging), *Illinois J. Math.*, 21 (1977), 429-490.
3. K. Appel, W. Haken and J. Koch, Every planar map is

four colorable (Part II: Reducibility), *Illinois J. Math.*, 21 (1977), 491-567.

4. M. Behzad, G. Chartrand and L. Lesniak-Foster, *Graphs & digraphs*, Prindle, Weber & Schmidt, Boston, 1979.
5. G. Chartrand, D.P. Geller and S. Hedetniemi, A generalization of the chromatic number, *Proc. Camb. Phil. Soc.*, 64 (1968), 265-271.
6. G. Chartrand, D.P. Geller and S. Hedetniemi, Graphs with forbidden subgraphs, *J. Combinatorial Theory Ser. B*, 10 (1971), 12-41.
7. C.M. Mynhardt and I. Broere, Generalized colorings of graphs, in this *Proceedings*.

The Ramsey Number for the Pair Complete Bipartite Graph-Graph of Limited Degree

by

S.A. Burr
City College, C.U.N.Y.
New York, N.Y.

P. Erdös
Hungarian Academy of Sciences
Budapest, Hungary

R.J. Faudree, C.C. Rousseau
and R.H. Schelp
Memphis State University
Memphis, Tennessee

ABSTRACT

A connected graph G is said to be F-good if the Ramsey number $r(F,G)$ has the value $(\chi(F)-1)(p(G)-1) + s(F)$, where $s(F)$ is the minimum number of vertices in some color class under all vertex colorings by $\chi(F)$ colors. It has been previously shown that certain "large" order graphs G with "few" edges are F-good when F is a fixed multipartite graph. We show when F is a complete bipartite graph that this edge condition can be relaxed.

Let F and G be (simple) graphs. The Ramsey number r(F,G) is the smallest positive integer p such that if each edge of the complete graph is colored one of the two colors red or blue, then either the red subgraph contains a copy of F or the blue subgraph contains a copy of G. Two surveys on this subject have been written by S. A. Burr [2,3]. Notation throughout the paper follows that given in [1].

Calculation of the Ramsey number for an arbitrary pair of graphs is known to be an extremely difficult problem. For a given pair of graphs, a starting place is knowing which graphical parameters affect the value of the Ramsey number.

Consider the pair (K_m, T_n) where T_n denotes a tree on n vertices and K_m (as usual) the complete graph on m vertices. V. Chvátal first observed that $r(K_m, T_n) = (m-1)(n-1) + 1$ [11]. The canonical example which determines the lower bound for this number is a two-colored $K_{(m-1)(n-1)}$ with the blue subgraph consisting of $m-1$ disjoint copies of K_{n-1} and with the red subgraph as its complementary graph. The example indicates that the important parameters for this pair of graphs are the chromatic number of K_m and the order of the connected graph T_n.

The lower bound implicit in Chvátal's result can be generalized. If F is a graph, let s(F) (the chromatic surplus of F) denote the smallest number of vertices in a color class under any χ coloring of F. The symbols s or s(F) will always denote this quantity.

Lemma 1 [4].

If F and G are graphs, with $p(G) \geq s(F)$, and

where G is connected, then

$$r(F,G) \geq (\chi(F)-1)(p(G)-1) + s(F).$$

With this lemma in mind, we say that a connected graph G is <u>F-good</u> if $r(F,G) = (\chi(F)-1)(p(G)-1) + s(F)$, that is, if Lemma 1 is sharp.

One would expect G to be F-good if G were "almost" a tree and F were "almost" a complete graph. Some families of graphs known to be K_m-good or F-good where F is almost a complete graph are given in [4,5,6,7,13].

Recently the following results have been obtained which show that large order trees T_n are F-good for certain fixed graphs F.

<u>Theorem.</u> [8]

$$r(K_\ell + \overline{K}_m, T_n) = \ell(n-1) + 1 \quad \text{for} \quad \ell \geq 2 \quad \text{and} \quad 3m \leq m$$

<u>Theorem.</u> [8]

Let $\ell_1 \leq \ell_2 \ldots \leq \ell_m$ be fixed positive integers. For n sufficiently large

$$r(K(1,1,\ell_1,\ell_2,\ldots,\ell_m),T_n) = (m+1)(n-1) + 1.$$

In each of the last two theorems a bit more is true. The same Ramsey number results if the graphs appearing in the first argument are replaced by subgraphs with the same chromatic number. In some sense these graphs are the most general F for which large order trees are F-good. This follows from the following theorems.

<u>Theorem</u>. [10]

Let $\ell_1 \leq \ell_2 \ldots \leq \ell_m$ be fixed positive integers. If n is sufficiently large, then

$$r(K(1,\ell_1,\ell_2,\ldots,\ell_m),K(1,n)) = \begin{cases} m(\ell_1 + n-2) + 1 & \text{for } \ell_1 \text{ and } n \text{ even} \\ m(\ell_1 + n-1) + 1 & \text{otherwise.} \end{cases}$$

Theorem. [9]

For n sufficiently large
$r(K(2,2),K(1,n)) > n + n^{1/2} - 5n^{3/10}$

These results suggest that large order "tree-like" graphs might be F-good for an arbitrary fixed graph F, if these tree-like graphs do not contain vertices of large degree. The following result confirms this.

Theorem. [12]

Let G be a connected graph on n vertices and F a fixed graph on p vertices with chromatic number χ and chromatic surplus s. There exist positive constants ε_1 and ε_2 such that if n sufficiently large and both $q(G) \leq n + \varepsilon_1 n^{1/(2p-1)}$ and $\Delta(G) \leq \varepsilon_2 n^{1/(2p-1)}$, then

$$r(F,G) = (\chi-1)(n-1) + s.$$

This result says that G is F-good when G has limited degree and "essentially" n edges. The focus of this paper will be to show this edge condition can be weakened when F is a bipartite graph. In particular we prove the following theorem.

Theorem 1.

Let $\ell \leq m$ be fixed positive integers and let G be a connected graph on n vertices. There exists a positive constant ε such that n sufficiently large and $\Delta(G) \leq \varepsilon n^{1/(\ell+2)}$ imply that $r(K(\ell,m),G) = n + \ell - 1$.

The proof of Theorem 1 requires three more lemmas. The first is very simple and we omit the proof.

Lemma 2.

Let G be a graph of order n and maximum degree $\Delta(G) \leqq d$. Then G contains at least $n/(d^2+1)$ vertices such that the distance between any two of them is at least three.

It is helpful at this point to decide on a uniform notation to be used in the following two lemmas and in the proof of Theorem 1. Throughout, $G(V,E)$ will denote a graph of order n. We shall use $[S]^k$ to denote the collection of k-element subsets of a set S. Let $U = \{1,2,...,p\}$ and consider a two-coloring of $[U]^2$ using colors red and blue. The resulting monochromatic graphs will be denoted R and B respectively. In the proof of Theorem 1, we need to show that, subject to an appropriate growth condition on $\Delta(G)$, when $p = n + \ell - 1$ there is either an embedding of $K(\ell,m)$ into R or else an embedding of G into B. The following lemma gives a start toward an embedding of G into B.

Lemma 3.

Suppose that G has s vertices $x_1,...,x_s$ such that the distance between any two of them is at least three. Further, suppose that with $U = \{1,2,...,n\}$ and R and B as described above, excluding $\{n-s+1,...,n\}$, every vertex has degree at most M in R. Let X consist of $x_1,...,x_s$ together with their additional neighbors $x_{s+1},...,x_k$ in G. If

$$\delta(B) > M(\Delta(G) - 1) + s - 1$$

then there is a map $\rho: X \to U$ which is an embedding of $\langle X \rangle$ into B and where $\rho(x_j) = n-s+j$, $j = 1,\ldots,s$.

<u>Proof</u>. Define ρ one vertex at a time. For some $s < j \leqq k$ we can fail to find an appropriate $\rho(x_j)$ only if for the unique x_i, $i \leqq s$, to which x_j is adjacent in G, every vertex in the neighborhood of $\rho(x_i)$ in B is either an x_k, $k \neq i$, or else adjacent in R to one of the at most $\Delta(G)-1$ vertices which are images of neighbors of x_j in G. But these images have degree at most M in R. Thus, the stated inequality assures us that we do not fail.

The next lemma is a version of a result used by Sauer and Spencer in [14] and the proof technique is exactly as in the proof of their Theorem 3. It is a key result in the proof of our Theorem 1.

<u>Lemma</u> <u>4</u>.

Let $G(V,E)$ be a graph of order n and let $U = \{1,..n\}$, R and B be as described above. If

$$n - k > 2\,\Delta(G)\,\Delta(R),$$

then given any $X \in [V]^k$ and any map $\rho: X \to U$ which is an embedding of $\langle X \rangle$ into B, ρ extends to a map $\sigma: V \to U$ which is an embedding of G into B.

<u>Proof</u>. Given any map $\sigma: V \to U$ which is an extension of ρ, let G_σ denote the corresponding image of G, i.e. $E(G_\sigma) = \{\sigma(uv): uv \in E(G)\}$. Of course, such a G_σ is not necessarily monochromatic. Let us denote by $E_R(G_\sigma)$ and $E_B(G_\sigma)$ the sets of red edges and blue edges respectively in G_σ. We claim that if the extension σ is chosen so that G_σ has as many blue

edges as possible, then, in fact, $E_R(G_\sigma)$ will be empty. Suppose not, i.e. suppose that there is an edge $uv \in E(G)$ for which $\sigma(uv)$ is red. As σ yields an embedding of $\langle X \rangle$ into B, we may assume that $v \notin X$. We would like to do an "exchange" by introducing a new map $\tau: V \to U$ given by $\tau(v) = \sigma(w)$, $\tau(w) = \sigma(v)$ and $\tau = \sigma$ otherwise. For this purpose, a vertex $w \in V \backslash X$ is <u>bad</u> if any one of the following four conditions is satisfied: (i) $w = v$, (ii) $\sigma(vw) \in E_R(G_\sigma)$, (iii) for some $z \in V$, $vz \in E(G)$ and $\sigma(zw)$ is red, (iv) for some $z \in V$, $zw \in E(G)$ and $\sigma(vz)$ is red. Suppose that $\sigma(v)$ is of degree d in the red subgraph of G_σ. Then d vertices w satisfy (ii) and at most $d(\Delta(R) - 1) + (\Delta(G) - d)\Delta(R) = \Delta(G)\Delta(R) - d$ vertices satisfy (iii). Similarly, at most $\Delta(G)\Delta(R) - d$ vertices satisfy (iv). Since $d \geqq 1$, there are at most $1 + d + 2(\Delta(G)\Delta(R)-d) \leqq 2\Delta(G)\Delta(R)$ bad vertices. In view of the assumed inequality, there is a vertex $w \in V\backslash X$ which is a good choice for the exchange. Using (i)-(iv) it is easily checked that all of the edges incident with either $\tau(v)$ or $\tau(w)$ in G_τ are blue. Edges which are not incident with either $\tau(v)$ or $\tau(w)$ are not affected. In particular, τ is still an extension of ρ but G_τ has more blue edges than does G_σ. This contradiction of our choice of completes the proof.

The following slight strengthening of Lemma 4 will be the version actually used in the proof of Theorem 1. It is a corollary to the proof of the lemma.

Corollary.

Lemma 4 remains valid if the inequality $n - k > 2\Delta(G)\Delta(R)$ continues to hold then $\Delta(R)$ is replaced by a bound M chosen so that no vertex x for which $\rho(x)$ is of degree $> M$ in R is adjacent to any vertex in $V\backslash X$.

Proof of Theorem 1. In view of Lemma 1, we need only show that if $U = \{1,2,\ldots,n + \ -1\}$ then in the two-coloring of $[U]^2$, either R will contain $K(\ell,m)$ or else B will contain G. Suppose that R contains no $K(\ell,m)$. Then an easy argument shows that in any collection of ℓ vertices at least one will have degree $\geqq \lceil (n-m)/\ell \rceil$ in B. Delete the $\ell - 1$ vertices of highest degree in R. For convenience, the vertices deleted are $n+1,\ldots,n+\ell-1$. Now let $U = \{1,2,\ldots,n\}$ and let R and B refer to the red and blue subgraphs of $[U]^2$. By the observation made earlier, we know that $\delta(B) > n/\ell - O(1)$. For an M yet to be chosen, let r denote the number of vertices which have degree at least M in R. Since there is no red $K(\ell,m)$ a standard argument yields the fact that

$$r\binom{M}{\ell} \leqq (m-1)\binom{n}{\ell}$$

and so

$$r < (m-1)(n/(M-\ell+1))$$

Set $a = 1/(\ell+2)$, $b = \ell/(\ell+2)$, $c = (\ell+1)/(\ell+2)$,

$$d(n) = c_1 n^a$$

and

$$M(n) = c_2 n^c.$$

Now the proof reduces to routine calculations. Setting

$$s(n) = \lceil n/(d^2 + 1) \rceil$$

and assuming that $\Delta(G) \leqq d(n)$, we can apply Lemma 3 if

$$s(n) \geqq r(n) \equiv (m-1)(n/(M-\ell+1))^{\ell}$$

and

$$n/\ell - O(1) > Md + s.$$

We can then apply the corollary to Lemma 4 if

$$n - s(d+1) > 2dM.$$

For $\ell \geqq 2$ the desired inequalities hold for all sufficiently large n if

$$C_2^{\ell}/C_1^{2} > m - 1$$

and

$$C_1 C_2 < 1/\ell.$$

Hence, Theorem 1 holds if we choose $\varepsilon < (m-1)^{-a} \;\; {}^{-b}$.
For $\ell = 1$ the theorem holds with $\varepsilon < (2(m-1))^{-1/3}$.

Theorem 1 holds in a slightly more general form when the complete bipartite graph is replaced by an arbitrary bipartite graph of order $\ell + m$ and with chromatic surplus ℓ.

<u>Theorem 2</u>.

Let F be a bipartite graph of order $\ell + m$ with chromatic surplus ℓ and let G be a connected graph of order n. There exists a positive constant ε such that $\Delta(G) \leqq \varepsilon\, n^{1/(\ell+2)}$ and n sufficiently large imply that G is F-good.

It should be noted that well-known Ramsey numbers

for special pairs of graphs appear as corollaries to Theorems 1 and 2 or to the theorem of [12], at least when n is large. Some examples, for n large, are $r(K_m,C_n)$, $r(C_m,C_n)$, $r(C_m,P_n)$, $r(T_m,P_n)$, $r(K(\ell_1,\ell_2,\ldots,\ell_m),P_n)$, $r(K(\ell_1,\ell_2,\ldots,\ell_m),C_n)$. In addition if G_n is a regular connected graph of fixed degree, then for n large Theorem 1 gives $r(K(\ell,m),G_n) = n+\ell-1$ $(\ell \leq m)$. This would not follow from the earlier results.

The most natural question left unanswered involves the improvement of the results given in Theorems 1 and 2 and the theorem of [12]. Specifically, can the edge condition or the maximum degree condition in any of these theorems be weakened? It is likely that there exists a constant $c < 1$ such that these results hold for $\Delta(G) \leq cn$ and G of bounded edge density. Here edge density is defined as $\max_{H \leq G} q(H)/p(H)$.

In another direction, what about $r(F,G)$ when F is not bipartite? In [4] it was conjectured that if F is any fixed graph, then any sufficiently large connected graph with bounded degree is F-good. This attractive conjecture seems difficult, but in view of the results proved here, it may yield to a determined attack.

REFERENCES

[1] M. Behzad, G. Chartrand, L. Lesniak-Foster, Graphs and Diagraphs. Wadsworth, Inc., Belmont, Calif. (1979).

[2] S. A. Burr, Generalized Ramsey Theory for Graphs-a Survey. Springer, Lecture Notes Math. 406 (1974) 52-75.

[3] S. A. Burr, What Can We Hope to Accomplish in Generalized Ramsey Theory for Graphs? (to appear).

[4] S. A. Burr, Ramsey Numbers Involving Graphs with Long Suspended Paths. J. London Math. Soc. 24 (1981) 03-413.

[5] S. A. Burr and P. Erdös, Generalizations of a Ramsey-Theoretic Result of Chvátal. J. Graph Theory 7(1983) 39-51.

[6] S. A. Burr and P. Erdös, Generalized Ramsey Numbers Involving Subdivision Graphs, and Related Problems in Graph Theory. Ann. Discrete Math. 9(1980) 37-42.

[7] S. A. Burr, P. Erdös, R. J. Faudree, C. C. Rousseau, and R. H. Schelp, Ramsey Numbers for the Pair Sparse Graph-Path or Cycle. Trans. Amer. Math. Soc. 269 (1982) 501-512.

[8] S. A. Burr, P. Erdös, R. J. Faudree, R. J. Gould, M. S. Jacobson, C. C. Rousseau, and R. H. Schelp, Trees are $K(1,1,s_1,s_2,\ldots,s_n)$- Good. (to appear).

[9] S. A. Burr, P. Erdös, R. J. Faudree, C. C. Rousseau, and R. H. Schelp, The Ramsey Number r(K(2,k),T). (to appear)

[10] S. A. Burr, R. J. Faudree, C. C. Rousseau, and R. H. Schelp, On Ramsey Numbers Involving Starlike Multipartite Graphs. J. Graph Theory 7 (1983) 395-409.

[11] V. Chvátal, Tree-Complete Graph Ramsey Numbers. J. Graph Theory 1 (1977) 93.

[12] P. Erdös, R. J. Faudree, C. C. Rousseau, and R. H. Schelp, Multipartite Graph-Sparse Graph Ramsey Numbers. (to appear).

[13] R. J. Gould and M. S. Jacobson, On the Ramsey Number of Trees Versus Graphs with Large Clique Number. J. Graph Theory 7(193) 71-78.

[14] N. Sauer and J. Spencer, Edge Disjoint Placement of Graphs. J. Combinatorial Theory 25 (1978), 295-302.

EMBEDDING GRAPHS IN BOOKS: A LAYOUT PROBLEM WITH APPLICATIONS TO VLSI DESIGN
Extended Abstract

Fan R. K. Chung

Bell Communications Research
Murray Hill, NJ 07974

Frank Thomson Leighton

Dept. of Mathematics and Laboratory for Computer Science
MIT
Cambridge, MA 02139

Arnold L. Rosenberg

Dept. of Computer Science
Duke University
Durham, NC 27706

ABSTRACT

We study the problem of embedding a graph in a book, with its vertices in a line along the spine of the book and its edges on the pages in such a way that edges residing on the same page do not cross. This problem abstracts layout problems arising in the routing of multilayer printed circuit boards and in the design of fault-tolerant processor arrays. In devising an embedding, one strives to minimize both the number of pages used and the "cutwidth" of the edges on each page. Our main results: (1) present optimal embeddings of a variety of families of graphs; (2) exhibit situations where one can achieve small pagenumber only at the expense of large cutwidth, and conversely; and (3) establish bounds on the minimum pagenumber of a graph based on various structural properties of the graph. Notable in this last category are proofs that (a) every n-vertex valence-d graph can be embedded using $O(dn^{1/2})$ pages, and (b) for every valence $d > 2$, for all large n, there are n-vertex valence-d graphs whose pagenumber is at least

$$\Omega\left(\frac{n^{1/2-1/d}}{\log^2 n}\right).$$

1. INTRODUCTION

1.1 The Problem

We study here a graph embedding problem that can be viewed in a variety of ways. We start with an undirected graph G.

Formulation 1. To embed G in a book, with its vertices in a line along the spine of the book and its edges on the pages, in such a way that edges residing on the same page do not cross.

We seek here embeddings that use pages that are few in number and small in width. (The *width* of a page is the maximum number of edges that are crossed by a line perpendicular to the spine of the book.) Our results are of four types:

(1) Characterizations of graphs that can be embedded in one- or two-page books. (E.g., the one-page graphs are the outerplanar graphs.)

(2) Upper and lower bounds on the numbers of pages required by graphs of valence (= maximum vertex-degree) d. (E.g., every n-vertex valence-($d>2$) graph can be embedded in a book with $\min(n/2, O(dn^{1/2}))$ pages, and there exist such graphs that require at least $\Omega\left(\frac{n^{1/2-1/d}}{\log^2 n}\right)$ pages.)

(3) Optimal or near-optimal embeddings for a variety of graphs, including trees, grids, X-trees, cyclic shifters, permutation networks, and complete graphs. (E.g., every n-vertex d-ary tree can be embedded in a one-page book of width $(d/2)\log n$.)

(4) Tradeoffs between pagenumber and pagewidth. (E.g., every one-page embedding of the depth-n "ladder graph" requires width $n/2$, but there are width-2 two-page embeddings for this graph.)

1.2 The Origins of the Problem

The problem has as many origins as formulations.

Sorting with Parallel Stacks. Even and Itai [6] and Tarjan [15] study the problem of how to realize fixed permutations of $\{1, \ldots, n\}$ with noncommunicating stacks. Initially, each number is PUSHed (in the order 1 to n) onto any of the stacks. After all of the numbers are on the stacks, the stacks are POPped to form the permutation. One can view this problem graph-theoretically: Let the graph G_n have vertices $\{a_1, \ldots, a_n, b_1, \ldots, b_n\}$ and edges connecting each a_i and b_i. Realizing the permutation π on $\{1, \ldots, n\}$ with k parallel stacks is equivalent to embedding G_n in a k-page book, with its vertices embedded in the order $a_1, \ldots, a_n, b_{\pi(1)}, \ldots, b_{\pi(n)}$.

The origins that led us to the problem come from VLSI design.

Single-Row Routing. So [14] simplified the problem of routing multilayer printed circuit boards (PCBs) in the following way. He arranged the circuit elements in a regular grid, with wiring channels separating rows and columns of elements, and then decomposed the circuit's net lists (possibly by adding dummy elements) so that every net connected elements in a single row or column. The PCB can now be routed by routing each row and column independently. The variant of this scenario that does not allow a net to run from the top of a row around to its bottom nor to change layers en route [12] corresponds to our layout problem applied to small-valence graphs.

Fault-Tolerant Processor Arrays. The DIOGENES approach to the design of fault-tolerant arrays of identical processing elements (PEs) [13,5] uses "stacks to wires" to configure around faulty PEs. The approach is illustrated by the following layout of a complete depth-d binary tree. One lays one's PEs out in a line, with a "bundle" of wires labeled $1, \ldots, d$ running above the line. After determining which PEs are good, one proceeds down the line from right to left. A good PE that is to be a leaf of the tree is connected to line 1 in the bundle, simultaneously having lines 1 through $d-1$ "shift up", to "become" lines 2 through d (switches are used to maintain correct connectivity); the bundle here behaves like a stack being PUSHed. A good PE that is to be a nonleaf of the tree is connected in stages. First it is connected to lines 1 and 2 of the bundle, simultaneously having lines 3 through d "shift down" to "become" lines 1 through $d-2$; the bundle here behaves like a stack being twice POPped. Second, the PE PUSHes a connection onto the stack. Thus, POPs have a PE adopt two children that lie to its right, while PUSHs have a PE request to be adopted by a higher level vertex to its left. This process lays the tree out in *preorder* and, hence, uses at most d lines.

Although not directly related to the research in this paper, the following relationship to Turing-machine graphs is also of interest.

Turing Machine Graphs. One can construct a T-vertex graph that "models" a given T-step Turing machine computation, as follows. Each vertex of the graph corresponds to a step of the computation; vertices t_1 and t_2 are adjacent in the graph just if one of the machine's tape heads visits the same tape square at times t_1 and t_2, but not in the interim. One shows easily that every k-tape Turing machine graph is embeddable in a $2k$-page book. Hence, a characterization of graphs that are embeddable in given numbers of pages might have applications to complexity theory. For example, a proof that such graphs have small bisection width would lead to several interesting complexity-theoretic results.

1.3 Additional Formulations

The problem's origins suggest several additional formulations.

Formulation 2. To place the vertices of G in a line and to assign its edges to stacks in such a way that the stacks can be used to lay out the edges.

Formulation 3. To embed the graph G so that its vertices lie on a circle and its edges are chords of the circle; to assign the chords to layers so that edges/chords on the same layer do not cross.

Formulation 3 suggests the following characterization and yet another formulation.

Theorem 1.1. [2] A graph can be embedded in a one-page book if it is outerplanar.

Formulation 4. To decompose G into outerplanar graphs all of whose outerplanarity is witnessed by the same embedding of G's vertices.

The many formulations of our problem suggest at least two variants: the first assumes that the layout of the vertices is fixed (as in sorting with parallel stacks and single-row routing); the second leaves the arrangement of the vertices as part of the problem (as in the construction of fault-tolerant processor arrays). We focus here on the harder, nonfixed vertices, version.

The many facets of our problem further allow us to draw on results obtained in a variety of contexts. Specifically, from [2] we learn of simple characterizations of 1- and 2-page graphs, from [15] we glean a technique for establishing lower bounds, and from [6] and [7] we infer the NP-completeness of the problem of minimizing pagenumber.

2. SPECIFIC EFFICIENT LAYOUTS

We look now at efficient layouts of a number of common graphs. Most have more modest pagenumber than do random graphs (cf. Theorems 3.7, 3.9). Here, we merely summarize our results in the following list. Several of these results were announced in [5].

n-vertex d-ary tree: 1 page of width $\leqslant (d/2)\log n$

(optimal in pagenumber and pagewidth)

$n \times n$ square grid: 2 pages, each of width n

(optimal in pagenumber; $\leqslant 2\times$optimal in cwidth [short for *cumulative pagewidth*])

depth-d X-tree: 2 pages, one of width $2d$, one of width $3d$

(optimal in pagenumber; $\leqslant 5\times$optimal in cwidth)

[An X-tree is obtained from a complete binary tree by connecting all nodes at each level into a line.]

n-input Benes network: 4 pages, each of width n

($\leqslant 4/3\times$optimal in pagenumber; $\leqslant 4\times$optimal in cwidth)

[This embedding is due to Buss and Shor, private communication.]

n-input ($\log n$)-stage barrel shifter: 5 pages, each of width n

($\leqslant 5/3\times$optimal in pagenumber; $\leqslant 5\times$optimal in cwidth)

Boolean n-cube: n pages, one of width 2^i for each $0 \leqslant i < n$

($\leqslant 2\times$optimal in pagenumber and cwidth)

complete graph K_n:$n/2$ pages, each of width n

(optimal in pagenumber and cwidth)

This last entry implies that any n-vertex graph can be realized in $n/2$ pages.

3. GRAPH STRUCTURE AND PAGENUMBER

Our main results concern structural features of a graph, that are related to the number of pages required to embed the graph in a book.

3.1 Planarity

Theorem 3.1. [2] A graph G admits a 2-page embedding if it is a subgraph of a planar hamiltonian graph.

Corollary 3.2. Every series-parallel graph is 2-page embeddable.

Corollary 3.3. [17] The problem of deciding 2-page embeddability is NP-complete, even for planar graphs.

Theorem 3.1 leaves unresolved the pagenumber of arbitrary planar graphs. Buss and Shor [4] have recently proved the surprising result that every planar graph can be embedded in 9 pages. Heath [9], the current champion, has lowered this number to 7.

Theorem 3.4. [9] Every planar graph is 7-page embeddable.

We still know of no planar graph that requires more than 3 pages.

3.2 Bisection Width

The ease of recursively cutting a graph into two equal size subgraphs (as measured, for example, by the notion of a graph *bifurcator* [3] or *separator* [10]) yields a nontrivial upper bound on pagenumber but no nontrivial lower bound. Using the former notion: the graph G has a *balanced* ρ bifurcator of size B (B an integer and $\rho > 1$) either if G has fewer than B edges or if G admits a *decomposition tree* of the following sort. The root of the tree (which is the sole level-0 node) is the graph G. Each graph H at level $k \geqslant 0$ of the tree that has more than one vertex gives rise to two disjoint graphs at level $k+1$, such that: (a) these graphs are equal in size (within one vertex); (b) their union is H; (c) each is connected to the other by at most $B\rho^{-k}$ edges.

Theorem 3.5. If a graph has a balanced ρ-bifurcator of size B, then it is embeddable in $\left(\frac{\rho}{\rho-1}\right)B$ pages.

In contrast to this nontrivial upper bound, we find:

Theorem 3.6. There exist n-vertex 3-page embeddable graphs whose smallest ρ-bifurcators have size $\Omega\left(\frac{n}{\log n}\right)$.

3.3 Valence

The valence of a graph affords nontrivial upper and lower bounds on pagenumber.

Theorem 3.7. Let G be an n-vertex graph of valence d. If $d \leqslant 2$, then G is 1-page embeddable. For any value of d, for all $\epsilon > 0$, G is $F(\epsilon,d,n)$-page embeddable, where

$$F(\epsilon,d,n) = \min[n/2, (1+\epsilon)(2+2^{1/2})(d+1)n^{1/2}].$$

Proof Sketch. If $d=1$, then G is a matching; if $d=2$, then G consists of disjoint paths and cycles.

For $d>2$: Say we are given an n-vertex valence-d graph G. We deduce that not all embeddings of G in books can use too many pages.

First use Vizing's algorithm [16] to decompose G into at most $d+1$ matchings G_i, $0 \leqslant i \leqslant d$, each having at most n vertices. Now consider all permutations of G's vertices (= all layouts of the vertices in the spine of a book).

Focus on an arbitrary permutation and its "behavior" on one of G's constituent matchings G_i. Say G_i has k edges that connect a vertex in the left half of the layout with a vertex in the right half. These edges can be viewed as specifying a random permutation on k integers. By applying a fundamental result of Hammersley [8, Theorem 6] about random permutations, we deduce that at most the fraction

$$\exp(-2\epsilon n^{1/2})$$

of all layouts will require as many as $(1+\epsilon)n^{1/2}$ pages for their realization. Now we remove these k edges and their incident vertices. We are left with two (roughly) half-size copies of the same problem. Moreover, the relative layout of the remaining vertices is independent of the removed vertices, so the permutations induced by the edges can again be viewed as random ones, hence within the purview of Hammersley's Theorem. We can, therefore, continue recursively to remove edges that have been considered, thereby creating at the i-th stage 2^i subproblems of size roughly $n/2^i$ each, each of which encounters "bad" layouts with probability less than

$$\exp(-2\epsilon(n/2^i)^{1/2}).$$

We stop generating half-size subproblems when $n/2^i \leqslant n^{1/2}$, for by that time (as we noted earlier) the probability of a "bad" layout is 0.

Carrying through this analysis, and noting that G is the disjoint union of its component matchings, we find that the fraction of layouts of G that require more than

$$(1+\epsilon)(2+2^{1/2})(d+1)n^{1/2}$$

pages cannot exceed

$$(const)(d+1)n^{1/2}\exp(-n^{1/4}),$$

which is significantly less than unity for large n. □

The (nonconstructive) upper bound on pagenumber of Theorem 3.7 holds for almost all orderings of the vertices. The following explicit construction works for all trivalent graphs. This result was proved in collaboration with Lenny Heath.

Theorem 3.8. Every n-vertex trivalent graph can be embedded in a $(\frac{3}{2}n^{1/2}+1)-$ page book, each page having width $O(n^{1/2})$.

Proof Sketch. The basic insight is that every trivalent graph is a tree of 2-connected components. We embed a trivalent graph by embedding its 2-connected components one by one and assembling their layouts using a tree layout as the glue. The strategy for laying out the 2-connected components is suggested by the following layout strategy for regular 2-connected trivalent graphs (whose correctness depends on Petersen's Theorem [11]). We embed a given 2-connected n-vertex regular trivalent graph G in a book via the following steps.

(1) Remove a matching from G, thereby decomposing G into disjoint paths and cycles. Tentatively lay G out, cycle by cycle. Call the edge removed in the matching the *extra* edges.

(2) Partition the linearized version of G into $n^{1/2}$ blocks of $n^{1/2}$ vertices each, from left to right.

(3) Rearrange the vertices in each block, from left to right, as follows. For block 1: sort the vertices in decreasing order of the block numbers to which their extra edges go. For block $k>1$: (a) place the vertices whose extra edges go to blocks $i<k$ to the left of those whose extra edges go to blocks $i>k$; (b) sort the leftgoing vertices in decreasing order of the block numbers to which their extra edges go; (c) analogously sort the rightgoing vertices; (d) within

each group of leftgoing vertices that are going to the same block, arrange the vertices in increasing order of the distance from the present block of their target vertex.

After the rearrangements in 3(a)-(d), each block needs just one page to realize all of its rightgoing extra edges. Since each vertex has only one extra edge, these pages have width at most $n^{1/2}$. To complete the layout, we need just realize the (now-scrambled) cycle edges. Since they all lie within disjoint blocks of size $n^{1/2}$ each, at most $\frac{1}{2}n^{1/2}$ additional pages suffice to realize all of these edges. □

We contrast these upper bounds with the following nontrivial lower bound.

Theorem 3.9. For all valences $d>2$, for all sufficiently large n, there are n-vertex graphs of valence d whose pagenumber is no less than

$$(const)\frac{n^{\frac{1}{2}-\frac{1}{d}}}{\log^2 n}.$$

Proof Idea. Fix on graphs of valence $d>2$. Each can be viewed as a set of d or $d+1$ matchings. Given an integer p, imagine that we have a table with each row labeled by one of the $n!$ layouts of n vertices and with each column labeled by an n-vertex matching: Table entry (i,j) is + if layout i uses p or fewer pages to realize matching j, and is – otherwise. Our proof strategy is to demonstrate that if p is no larger than indicated in the statement of the theorem, then some d-tuple of columns experiences at least one – in every row. Details appear in the paper. □

Our upper and lower bounds are rather close when d is around $\log n$ but are far apart for big or small valences.

4. COST TRADEOFFS

In this section we point out a rather interesting anomaly that could be important in the context of our study. We describe here two families of graphs that engender pagenumber-pagewidth *tradeoffs*. Each of the families can be laid out using some number p pages -- but only with pagewidths that are proportional to the size of the graph being laid out. If one is willing to use $p+1$ pages, then the widths of the pages can be kept bounded by a constant.

The *depth-k K_n-cylinder* $C(k,n)$ has as vertex-set the union of the k sets

$$V_{i,n} = \{v_{i,1} v_{i,2}, \ldots, v_{i,n}\}$$

and edges that (a) form each set $V_{i,n}$ into an n-clique, and (b) connect each vertex $v_{i,j}$ to vertex $v_{i+1,j}$, $1 \leqslant i < k$, $1 \leqslant j \leqslant n$.

The anomalies of interest appear in the first two parts of the next result. The third part indicates the failure of the obvious generalization of the first two parts.

Proposition 4.1.

(1a) Any 1-page layout of $C(k,2)$ has width at least $k/2$. (1b) There are 2-page layouts of $C(k,2)$ having width 2.

(2a) Any 2-page layout of $C(k,3)$ has width at least $k/2$. (2b) There are 2-page layouts of $C(k,3)$ having width 4.

(3) There are 3-page layouts of $C(k,4)$ having width 4.

ACKNOWLEDGEMENT

It is a pleasure to thank Lenny Heath, Ravi Kannan, and Gary Miller for helpful and stimulating conversations. A portion of this research was done while the second two authors were visiting AT&T Bell Laboratories, Murray Hill, NJ. The research of the second author

(FTL) was supported in part by a Bantrell Fellowship, by DARPA Contract N00014-80-C-0622, and by Air Force Contract OSR-82-0326. The research of the third author (ALR) was supported in part by NSF Grants MCS-81-16522 and MCS-83-01213.

REFERENCES

1. P. B. Arnold, Complexity results for single-row routing, Harvard Univ. Center for Research in Computing Technology Report TR-22-82 (1982).
2. F. Bernhart and P. C. Kainen, The book thickness of a graph, *J. Comb. Th. (B) 27* (1979) 320-331.
3. S. N. Bhatt and F. T. Leighton, A framework for solving VLSI graph layout problems, *J. Comp. Syst. Sci.*, (1984) to appear.
4. J. Buss and P. Shor, On the pagenumber of planar graphs, *16th ACM Symp. on Theory of Computing* (1984) 98-100.
5. F. R. K. Chung, F. T. Leighton, A. L. Rosenberg, DIOGENES -- A methodology for designing fault-tolerant processor arrays, *13th Intl. Conf. on Fault-Tolerant Computing* (1983) 26-32.
6. S. Even and A. Itai, Queues, stacks, and graphs, In *Theory of Machines and Computations* (Z. Kohavi and A. Paz eds.) Academic Press, NY (1971) 71-86.
7. M. R. Garey, D. S. Johnson, G. L. Miller, C. H. Papadimitriou, The complexity of coloring circular arcs and chords, *SIAM J. Algebr. Discr. Meth. 1* (1980) 216-227.
8. J. M. Hammersley, A few seedlings of research, *6th Berkeley Symposium on Math. Stat. and Prob.*, vol. 1 (1972) 345-394.
9. L. S. Heath, Embedding planar graphs in seven pages, (1984) typescript.
10. C. E. Leiserson, *Area-Efficient VLSI Computation* MIT Press, Cambridge, MA (1983).
11. J. Petersen, Die Theorie der regulaeren Graphen, *ACTA Math. 15* (1981) 193-220.
12. R. Raghavan and S. Sahni, Single row routing, *IEEE Trans. Comp., C-32* (1983) 209-220.
13. A. L. Rosenberg, The Diogenes approach to testable fault-tolerant arrays of processors, *IEEE Trans. Comp., C-32* (1983) 902-910.
14. H. C. So, Some theoretical results on the routing of multilayer printed-wiring boards, *IEEE Intl. Symp. on Circuits and Systems* (1974) 296-303.
15. R. E. Tarjan, Sorting using networks of queues and stacks, *J. ACM 19* (1972) 341-346.

16. V. G. Vizing, On an estimate of the chromatic class of a p-graph (in Russian), *Diskret. Analiz. 3* (1964) 25-30.

17. A. Wigderson, The complexity of the hamiltonian circuit problem for maximal planar graphs, Princeton Univ. EECS Dept. Report 298 (1982).

HAMILTON CYCLES AND QUOTIENTS OF BIPARTITE GRAPHS

I.J. Dejter
Universidade Federal de Santa Catarina, Brazil.

ABSTRACT

Paul Erdős conjectured that every graph in a certain family of bipartite graphs G_k where k runs over the positive integers is Hamiltonian. We consider a quotient pseudograph H_k of G_k such that a Hamilton path in H_k whose ends have one and two loops can be lifted to a Hamilton cycle in H_k. Edges in H_k are wholly characterized.

1. Introduction.

Let k be a positive integer and n=2k+1. For i=0,1 let V_i be the collection of all the subsets of $\{-k,..,k\}$ with cardinality k+i. Consider the bipartite graph G_k whose parts are V_0 and V_1 and whose adjacency is given by proper inclusion. P. Erdős conjectured that G_k is Hamiltonian for every k. (We use notation from [1],[2]). We consider a quotient graph H_k of G_k which is a (k+1)-regular graph where the degree of a vertex is given by the number of links and loops incident at it. The order of H_k is $\binom{n}{k}/n$, a Catalan number, see [3] pg. 265. The H_k have vertices with exactly i loops only for i=0,1,2. Theorem 10 reduces the search of Hamilton cycles in G_k to that of Hamilton paths in H_k whose ends are incident to one and two loops respectively.

This Research was supported by CNPq, Brazil and the Royal Society, United Kingdom.

2. Vectorial Definitions and Examples.

Let $C_r'=\{0,1\}^r$ with coordinates indexed in Z_r, $r>1$. An element c in $C'=C_n'$ is written $c=(c_{-k},..,c_k)=(c_i)_{i=-k}^k$, where i represents the class of i mod r. If $c \in C$ we write $|c|=\Sigma_{i=-k}^k|c_i|$. For i=0,1 we set $C_i=\{c\in C';|c|=k+i\}$ and $C=C_0 \cup C_1$. We say that $c\in C_0$, $c'\in C_1$ are adjacent iff $|c-c'|=1$. When this happens we write cEc' or $cc'\in E$. It is easy to see that (C,E) is graph isomorphic to G_k and we identify $G_k=(C,E)$. To define the quotient H_k we construct a quotient bipartite multigraph H_k'. For this we need to consider the group action of Z_n on C given by the map $\theta:Z_n\times C\to C$ given by $\bar{h}c=\theta(\bar{h},c)=(c_{i+h})_{i=-k}^k$, where $\bar{h}\in Z_n$ represents $h\in Z$ and $c=(c_i)_{i=-k}^k$. Let $D=C/\theta=C/Z_n$ be the orbit space of C under θ with orbit map $\pi:C\to D$. For i=0, 1 let $D_i=\pi(C_i)$ and $\pi_i=\pi|C_i:C_i\to D_i$. The graph H_k' with parts D_0 and D_1 is given by setting that $d\in D_0$, $d'\in D_1$ are adjacent in H_k' iff there are $c\in C_0$, $c'\in C_1$ with $d=\pi_0(c)$, $d'=\pi_1(c')$ and cEc'. When this holds we write $dd'\in F'$ or $dF'd'$. The multiplicity $\mu'(dd')$ of a $dd'\in F'$ is the number of different $c\in\pi_0^{-1}(d)$ with cEc', for a fixed $c'\in\pi_1^{-1}(d')$.

The following lemma will allow to define H_k as quotient graph of H_k'. Let $\rho:C\to C$ be given by $\rho(a)=(a_{-i})_{i=-k}^k$, where $a=(a_{-k},..,a_k)\in C$. It is easy to see that ρ induces a unique function $\bar{\rho}:D\to D$ such that $\bar{\rho}\pi=\pi\rho$. Let $\hat{0}=1$ and $\hat{1}=0$. Let $f:C_1\to C_0$ be given by $f(c)=\rho(\hat{c})$, where $c=(c_{-k},..,c_k)$ and $\hat{c}=(\hat{c}_{-k},..,\hat{c}_k)$.

Lemma 1. There exists a unique one-to-one correspondence $g:D_1\to D_0$ such that $g\pi_1=\pi_0 f$. Moreover, for every two elements $d,d'\in D_1$:
1. $dg(d')\in F'$ iff $g(d)d'\in F'$; 2. $\mu'(dg(d'))\in F'=\mu'(g(d)d')$.

Proof. The existence of g can be checked easily. Now observe that for every $c \in C_0$ and $c' \in C_1$ it holds $cf(c') \in E$ iff $f(c)c' \in E$, thus 1.,2.

We define H_k as the multigraph whose set of vertices is D_1 and whose adjacency F is given by $d'd \in F$ iff $g(d')F'd$, with $d,d' \in D_1$ and where the multiplicity $\mu(d'd)$ of an edge $d'd \in F$ is $\mu'(g(d')d)$. Since G_k is a (k+1)-regular graph we have the following.

Proposition 2. For every positive integer k, H_k is a (k+1)-regular pseudograph.

We will show by means of examples how a Hamilton path in H_k whose ends have loops can be lifted by means of the quotient map $H_k' \to H_k$ to a Hamilton cycle in G_k. When writing vectors we drop commas and parentheses are optional. The image under π of any of these vectors is indicated in the same way but in square brackets. For example if k=1, D_1,(resp.D_0), consists of the orbit class [110]= {110,101,011},(resp.[100]={100,010,001}). H_1' consists of [110] with two loops. It is easy to see that G_1 is Hamiltonian. If k=2, D_i={ a_{i0}, a_{i1}} for i=0,1, where a_{00}=[11000], a_{01}=[10100], a_{10}=[11100] and a_{11}=[11010]. H_2' consists of these four vertices such that $a_{0i}Fa_{1j}$, $\mu(a_{0i},a_{1i})=2$ and $g(a_{1i})=a_{0i}$ for $i,j \in \{0,1\}$ so that H_2' has two 2-links and two 1-links. H_2 consists of a_{10} and a_{11} with two loops each joined by a link. A Hamilton path in H_2 is (a_{10},a_{11}) lifting through the Hamilton cycle $(a_{10},a_{01},a_{11},a_{00})$ in H_2' to the Hamilton cycle in G_2

$$\begin{matrix}(11100,10100,10110,00110, & =(\tilde{a}_{10}, \tilde{a}_{01}, \tilde{a}_{11}, \tilde{a}_{00},\\ 01110,01010,01011,00011, & \tau\tilde{a}_{10}, \tau\tilde{a}_{01}, \tau\tilde{a}_{11}, \tau\tilde{a}_{00},\\ 00111,00101,10101,10001, & 2\tau\tilde{a}_{10}, 2\tau\tilde{a}_{01}, 2\tau\tilde{a}_{11}, 2\tau\tilde{a}_{00},\\ 10011,10010,11010,11000, & 3\tau\tilde{a}_{10}, 3\tau\tilde{a}_{01}, 3\tau\tilde{a}_{11}, 3\tau\tilde{a}_{00},\\ 11001,01001,01101,01100)= & 4\tau\tilde{a}_{10}, 4\tau\tilde{a}_{01}, 4\tau\tilde{a}_{11}, 4\tau\tilde{a}_{00}),\end{matrix}$$

where τ generates $Z_n=Z_5$, namely $\tau=\bar{1}$ and the $\tilde{a}_{ij}$ are convenient representatives of the a_{ij} in G_2. This procedure can be carried out for any k once we have a Hamilton path $(a_1,..,a_r)$, where $r=\binom{n}{k}/n$

and a_1, a_r are vertices with loops of H_k. A lifting of this to H'_k is

(*) $(a_1, g(a_2), a_3, .., g(a_r), a_r, g(a_{r-1}), .., g(a_1))$ or

$(a_1, g(a_2), a_3, .., a_r, g(a_r), a_{r-1}, .., g(a_1))$

according to whether r is even or odd, (that is $k \neq 2^m - 1$ or not, $m \in Z$), so that to get a further lifting to G_k as the one given above for k =2 giving a Hamilton cycle it is simply needed, when taking a path of representatives of (*) in G_k, say $(\tilde{a}_1, f(\tilde{a}_2), .., f(\tilde{a}_1))$, that $f(\tilde{a}_1)$ is adjacent to some $\tau\tilde{a}_1$, where τ generates Z_n. Theorem 10 assures this when a_1 is a vertex with two loops.

3. Characterization of Looped Vertices in H_k.

If G is a multigraph and x is a vertex of G we say that x is an i-looped vertex of G iff x is incident with exactly i loops. Let $C'_1 \subset C_1$ be the set of vectors $(c_{-k}, .., c_k) \in C_1$ with $c_j + c_{-j} = 1$ for every $j = 1, .., k$ and $c_0 = 1$. Consider $c = (c_{-k}, .., c_k)$ and $c' \in \pi_1^{-1}\pi(c)$, say $c' = (c'_{-k}, .., c'_k)$. Then there is a permutation σ of Z_n and $\sigma c = (c'_{\sigma(-k)}, .., c'_{\sigma(k)})$. We name $\sigma(0)$ a pivot coordinate of c'. In particular, representatives in C' of 1-looped vertices of H_k have exactly one pivot coordinate.

Theorem 3. <u>An element</u> $x \in D_1$ <u>is an</u> i-<u>looped vertex of</u> H_k <u>for</u> $i > 0$ <u>iff</u> $x = \pi(c)$, $c \in C'_1$.

<u>Proof.</u> If $c \in C'_1$ then $f(c)F'c$. Also if $c \in C_1$ and $f(c)F'c$ then $c \in C'_1$.

If $c \in D_1$ is an i-looped vertex of H_k for $i > 1$ then $x = \pi(c) = \pi(c')$ for $c, c' \in C'_1$ with $c \neq c'$ and $c = \rho(c')$. For example if k=2 it may happen c=11100, c'=00111; if k=3, c=1111000, c'=0001111. There exist elements $z, z' \in Z_n$ with $zc = z'c'$. In the examples above if k=2, zc=z'c'= 01110; if k=3, zc=z'c'=1100011. zc=z'c' is called a normal representation of $x \in C_1$. If $x \in D_1$ is a 2-looped vertex of H_k then there exists exactly one normal representation of x and if z,z' are as above then $z' = -z \neq 0$. We say that x is z- or z'-centered.

<u>Examples 4:</u> <u>Examples A.</u> <u>Let</u> $b^k = b^k(\emptyset, 1) = (b_{-k}, .., b_k) \in C$ <u>be given by</u>

$b_j=1$ iff $|j|<\lceil (k+1)/2 \rceil$. Then, for $i=0,1$, $k \not\equiv i \pmod 2$ implies $b^k \in C_i$. Moreover, $\pi(f^{1-i}(b)) \in D_1$ is a $\lceil k/2 \rceil$-centered 2-looped vertex of H_k.

This can be seen inductively. Define $c^k(\emptyset,1)=b^k(\emptyset,1)$ if $k \not\equiv 0 \pmod 2$ and $c^k(\emptyset,1)=\hat{b}^k(\emptyset,1)$ otherwise. If underlying (double) traces indicate (pivot) coordinates we have normal representations: $c^1(\emptyset,1)=101$, 1-centered; $c^2(\emptyset,1)=01110$, 1-centered; $c^3(\emptyset,1)=1100011$, 2-centered; $c^4(\emptyset,1)=001111100$, 2-centered; etc.

Examples B. (Extending A.) Let $u \in \{1,..,k\}$, $u|n$, $B=(b_1,..,b_{(u-1)}) \in \{0,1\}^{(u-1)}$ if $u>1$ or $B=\emptyset$ if $u=1$. Define $b^k(B)=b^k(B,u)=b_{-k}(B),..,b_k(B))$ as follows: A given index $j \in [-k,k]$ can be written in a unique way as $j=qu+p$ where p,q are integers and $0 \le p<u$. Then

i. If $p=0$, define $b_j(B)=1$ iff $|j|<\lceil (k+1)/2 \rceil$;

ii. If $0<p$, define $b_j(B)=b_p$.

Then, for $i=0,1$, $\lambda=((n/u)-1)/2 \not\equiv i \pmod 2$ implies $b^k(B) \in C_i$. Moreover $\pi(f^{1-i}(b^k(B))) \in D_1$ is incident to either a multilink or a multiloop this second case when $b_i=b_{u-i}$ for $i=1,..,(u-1)/2$, in which case it is a $\lceil \lambda/2 \rceil$-centered 2-looped vertex of H_k.

Multilinks in H_k are exemplified in example 11, multiloops below. Define $c^k(B,u)=b^k(B,u)$ if $\lambda \equiv 0 \pmod 2$ and $c^k(B,u)=\hat{b}^k(B,u)$ otherwise. If $u=1$ we are in A. Some normal representations now are:

$c^4((1),3)=110100110=101/0$,3-centered;
$c^4((0),3)=011001011=101/1$,3-centered;

These were obtained by interspersing 01, resp. 01 repeatedly between each two contiguous entries of $101=c^1(\emptyset,1)$. For $k=7$, $u=5$:

$c^7((00),5)=001110001100111=101/00$,5-centered;
$c^7((10),5)=101101001010110=101/10$,5-centered; etc.

Examples C. (Extending A.) For each $v \in \{1,..,k\}$ with $(v,n)=1$ let β_v: $Z_n \to Z_n$ be given by $\beta_v(\bar{z})=\overline{vz}$. Then β_v is a permutation of Z_n. Moreover, for $i=0,1$, $b^k(v)=b^k(\emptyset,v)=(b_{\beta_v(-\bar{k})},..,b_{\beta_v(\bar{k})})$ is in C_i whenever $k \not\equiv i \pmod 2$ and its image under f^{1-i} is a $\beta_v(\lceil k/2 \rceil)$-centered 2-looped vertex of H_k.

Define $c^k(\emptyset,v)=b^k(\emptyset,v)$ if $k\equiv 0 \pmod 2$ and $c^k(\emptyset,v)=b(\emptyset,v)$, otherwise. Then for example the normal representations now look like:
$c^5(\emptyset,2)=10101010101$, 5-centered; $c^5(\emptyset,3)=01011011010$, 2-centered;
$c^5(\emptyset,4)=10011011001$, 1-centered; $c^5(\emptyset,5)=01110001110$, 4-centered.

Examples D. (Extending B., C.) Let u,λ and B be as in B. and C. For each $uv\in\{1,..,k\}$ with $(uv,n)=u>0$ and $v>0$ let $\beta'_{uv}:uZ_n\to Z_n$ be given by $\beta'_{uv}(\overline{uz})=\overline{uvz}$ where $\overline{uz}\in Z_n$. Then β'_{uv} is a permutation of uZ_n extending to a permutation β_{uv} of Z_n with $\beta_{uv}(\bar{z})=\bar{z}$ for $\bar{z}\in Z_n-uZ_n$. Moreover, for $i=0,1$, $b^k(B,uv)=(b_{\beta_{uv}(-\bar{k})}(B),..,b_{\beta_{uv}(\bar{k})}(B))\in C_i$ whenever $\lambda\neq i \pmod 2$ and its image under f^{1-i} is incident to a multilink or multiloop, the latter as in B.

Define $c^k(B,uv)=b^k(B,uv)$ if $\lambda\equiv 0 \pmod 2$ and $c^k(B,uv)=\hat{b}^k(B,uv)$ otherwise. A normal representation for $k=7$, $u=3$ and $v=2$ is:
$c^7((1),6)=110100110100110=10101/1$, 6-centered.

4. Multiloop and Multilink Vertices of H_k.

The action of θ on $C\subset C'_n$ can be given in the same fashion in C'_r for any $r\in\{2,3,..\}$.

Lemma 5. Let $\gamma=(\gamma_1,..,\gamma_r)\in C'_r$. For $j=0,1$ let $\chi_j(\gamma)$ be the number of coordinates $i=1,..r$ with $\gamma_i=j$. Let $i,i'\in Z_r$ be two indices with $\gamma_i=1$, $\gamma_{i'}=0$. Let $\gamma'\in C'_r$ be such that $\gamma'_i=0$, $\gamma'_{i'}=1$ and $\gamma'_{i''}=\gamma_{i''}$ for every $i''\in Z_r$ with $i\neq i''\neq i'$. If: 1. γ' and γ differ in the action of θ on an element of Z_r and 2. $\chi_1(\gamma)-\chi_0(\gamma')=1$ then for every $s\in Z_r-\{0\}$, $\gamma_{i-s}\neq\gamma_{i+s}$.

Proof. Contradicting the lemma amounts to admit that there is an $s\in Z_r$ such that the only possible arrangements of the form

$$\begin{pmatrix} \gamma_{i-s} & \gamma_i & \gamma_{i+s} & \gamma_{i'-s} & \gamma_{i'} & \gamma_{i'+s} \\ \gamma'_{i-s} & \gamma'_i & \gamma'_{i+s} & \gamma'_{i'-s} & \gamma'_{i'} & \gamma'_{i'+s} \end{pmatrix}$$

are

$$I=\binom{010\ 100}{000\ 110},\quad II=\binom{010\ 001}{000\ 011},\quad III=\binom{111\ 100}{101\ 110},\quad IV=\binom{111\ 001}{101\ 011},$$
$$V=\binom{010\ 101}{000\ 111},\quad VI=\binom{111\ 000}{101\ 010},\quad VII=\binom{010\ 000}{000\ 010},\quad VIII=\binom{111\ 101}{101\ 111},$$

with possible identification of the third and fourth columns of II, III,VII and VIII and/or of the sixth and first columns of I,IV,VII and VIII. Cases I,II,III and IV are seen to be in contradiction to the hypothesis 1. Any of the resting cases implies $\chi_1(\gamma)-\chi_0(\gamma')\neq 1$.

Corollary 6. Under the hypothesis of lemma 5 there exists exactly one triple (s,ε,ℓ), where $s\in Z_r$, $\varepsilon=\pm 1$ and ℓ is a positive integer such that $\gamma_{i+\varepsilon sj}=1$ for $j=0,1,..,\ell$ and $\gamma_{i+\varepsilon sj}=0$ for $j=\ell+1,..,2\ell$.

Proof. Assume that there is no s as asked in the corollary. Then for each odd integer $t>0$ there exists $\varepsilon=\pm 1$ such that

$$\begin{pmatrix} \cdots & \gamma_{i-s\varepsilon 2^j} & \gamma_{i-s\varepsilon 2^{j-1}} & \cdots & \gamma_i & \cdots\gamma_{i+s\varepsilon 2^{j-1}} & \gamma_{i+s\varepsilon 2^j} & \cdots \\ \cdots & \gamma'_{i-s\varepsilon 2^j} & \gamma'_{i-s\varepsilon 2^{j-1}} & \cdots & \gamma'_i & \cdots\gamma'_{i+s\varepsilon 2^{j-1}} & \gamma'_{i+s\varepsilon 2^j} & \cdots \end{pmatrix}$$

equals

$$\binom{\ldots 101011010\ldots}{\ldots 101001010\ldots}$$

which cannot have a finite number of columns.

Corollary 7. Let $s\in Z_r$ be as in corollary 6. Then s is of the form $2\ell+1$, a divisor of r. Moreover, let $[s]\subset Z_r$ be the subgroup generated by s. Then for each nonzero class ζ of Z_r modulo $[s]$, γ_{i+j} remains constant when j varies in ζ.

Proof. The first assertion is a direct conclusion of corollary 6. The last assertion depends on the fact that the effective change of coordinate values from γ to γ' of lemma 5 happens, as assumed from the conclusion of corollary 6, in the zero orbit of Z_r modulo $[s]$. Thus, when passing from γ to γ' by changing the coordinate values of γ_i and γ'_i, which is equivalent to the action of θ on some $\bar{h}\in Z_r$, where h is of the form $\varepsilon s\ell$, the values of γ_j remain unchanged when j varies only in such a nonzero class.

Corollary 8. Under the hypothesis and notation of lemma 5 and corollaries let $\sigma \in Z$ be a representative of $s \in Z_r$ in the interval $(0,r)$. Let $(\sigma,r)=1$. Then $u(2\ell+1)=r$. Thus $v=\sigma/u$ and r are relatively prime. Moreover, if $[\bar{u}] \subset Z_r$ is the subgroup generated by $\bar{u} \in Z_r$, then for each nonzero class ζ' of Z_r modulo $[\bar{u}]$, γ_{i+j} remains constant when j varies in ζ', for every $i \in Z_r$.

Proof. The size of u is given in the statement, that is $u=r/(2\ell+1)$ because $2\ell+1$ is the size of the subgroup $[s]$. Corollary 7 implies the rest.

Assume that d,d' are the end vertices of a multiedge in H_k. Let $d=\pi(c)$, $d'=\pi(c')$ where $c=(c_{-k},..,c_k)$ and $c'=(c'_{-k},..,c'_k)$ are in C_1. To transform d into d' and back to d by different edges is equivalent to transform c into some $c'' \in C_1$ by changing two coordinates $c_i=1$ and $c'_i=0$ into $c_i=0$ and $c'_i=1$ in such a way that $\pi(c'')=d$. Then $c_{i-s} \neq c_{i+s}$ for every $s \in Z_n-\{s\}$ by lemma 5. All the corollaries of lemma 5 apply for $r=n$, $\gamma=c$ and $\gamma'=c''$. The arrangement

(**) $$(c \quad \rho(c') \quad c'')^t ,$$

where t stands for transpose, can be set in such a way that its jth column, for each $j \neq i,i'$, is either of the form $(101)^t$ or of the form $(010)^t$ while its ith column, (resp. i'^{th} column), is of the form $(110)^t$, (resp. $(011)^t$). From this and corollary 6 note that the class $J \subset Z_r$ modulo the subgroup $[\bar{u}] \subset Z_r$ that passes through i and i' is a set of subindices of columns of the arrangement (**), say $(c_J \quad \rho(c')_J \quad c''_J)^t$, with $\rho(c')_J$ and c_J differing by the action of θ on some $\bar{h} \in Z_r$. We also know by corollaries 7 and 8 that any other class $K \subset Z_r$ modulo $[\bar{u}]$ not passing through i, i.e. $K \neq J$ is a set of subindices of columns of the arrangement above, say $(c_K \quad \rho(c')_K \quad c''_K)^t$, and that in this case $c_K=c''_K$ equal either to 00..0 or to 11..1. while $\rho(c')_K$ is equal either to 11..1 or to 00..0. In order for d to have two loops it is necessary that $\rho(c')$ and c differ in the action of θ on a nontrivial $\bar{h}\epsilon Z_r$. This is equivalent to the following condition: For any class $K\subset Z_r$ modulo $[\bar{u}]$ with $K\neq J$ let $-K\subset Z_r$ be the class with $i+t\epsilon-K$ iff $i-t\epsilon K$ for any $t\epsilon Z_r$. Then $K\neq -K$ since u is odd. Then the condition we want is $(c_K\ \ \rho(c')_K\ \ c''_K)^t \neq (c_{-K}\ \ \rho(c')_{-K}\ \ c''_{-K})^t$ for some $K\neq J$. (Otherwise the middle row $\rho(c')$ of (**) would read back differently from c and c").

The discussion above shows that examples D are the totality of the 2-looped vertices of H_k provided that we show they are not i-looped vertices for i>2. In fact c is of the form $c=(..,c_{i+ju}, c_{i+j+u+1},..,c_{i+j+u+u-1},c_{i+(j+1)u},..)=(..,c_{i+ju},\ {}^u_1,..,\ {}^u_{u-1}, c_{i+(j+1)u},..)$, where the $i+ju+w^{th}$ entries, for each $w\neq 0$ fixed, are equal to a constant δ^u_w. If there were at least two different values of u, say u' and u", such that c fits in examples D both for u=u' and u", then $c=(..,c_{i'+ju'},\delta^{u'}_1,..,\delta^{u'}_{u'-1},..)=(..,c_{i''+ju},\delta^{u''}_1,..,\delta^{u''}_{u''-1},..)$. We can assume u'<u".

Case i. $(u',u'')=1$. Then $\delta^{u'}_1=\delta^{u'}_2=..=\delta^{u'}_{u'-1}$. This contradicts the fact that half of these δ's should take value 1, resp. 0.

Case ii. $(u',u'')=\phi>1$ and $u'\nmid u''$. Then for $j=1,..,\phi-1$ we have $\delta^{u'}_j=\delta^{u'}_{j+\phi}=..=\delta^{u'}_{j+((u'/\phi)-1}$ while $\delta^{u'}_1=\delta^{u'}_2=..=\delta^{u'}_{u'-\phi}$. This contradicts the fact that the number of 0's and 1's are equal among these δ's.

Case iii. $u'|u''$. Since all the numbers involved are odd we get again a contradiction to the fact that the $\delta^{u'}_j$'s are distributed into two equal parts according to the value taken, 0 or 1.

From the discussion above we get the following.

Theorem 9. For each triple (u,v,B) with $\{u,uv\}\subset\{1,..,k\}$, $(uv,n)=u$ and $B=(b_1,..,b_{(u-1)/2})\epsilon\{0,1\}^{(u-1)/2}$ if $u>1$ or $B=\emptyset$ if $u=1$, there exists a unique normal representation $c^k(u,v,B)=(c_{-k},..,c_0,\tilde{b}_1,..,\tilde{b}_{(u-1)/2},..,c_k)$ of a $uv\lceil\lambda/2\rceil$-centered 2-looped vertex of H_k where $\tilde{b}_i=\bar{b}_i$ if $\lambda\equiv 0 \pmod 2$ and $\tilde{b}_i=b_i$ otherwise. In fact, if $\lambda\equiv i \pmod 2$ then $c^k(u,v,B)=f^{1-i}(b^k(B,uv))$. The set of all the 2-looped vertices obtained this way comprises the totality of i-looped vertices of H_k, $i>1$.

Theorem 10. If for an integer $k>0$ there exists a Hamilton path $(a_1,..,a_r)$ of H_k, where $r=\binom{n}{k}/n$, a_1 is as in A. or C. and a_r is a looped vertex, then G_k is Hamiltonian in the following cases:

i. When n is a prime or a prime power.

ii. When there exists $j\in Z$ with $1<j<r-1$ such that there is a 2-link in H_k between a_j and a_{j+1}.

Proof. This is a conclusion of the final argument in §2 provided a_1 has a representative $\tilde{a}_1$ in G_k with $f(\tilde{a}_1)E\tau\tilde{a}_1$, τ generator of Z_n. The claim is clear when n is a prime. Since there are two loops of a_1 in H_k, therefore two links between a_1 and $g(a_1)$ in H'_k, we can modify a non Hamilton cycle to a Hamilton one in G_k for prime power n. If n is composite but ii. holds, a further modification is at hand:

Example 11. For k=7 it could happen $\tilde{a}_1$=111111110000000, $f(\tilde{a}_1)$= 000111111100. Both choices of a candidate for $\tau\tilde{a}_1$ happen with τ not generating Z_n: 000001111111100 and 000000111111110, lifting to cycles covering one third and one fifth respectively of C. To illustrate the use of ii. above consider $\tilde{a}_j$=011100111001010 so that $\tilde{a}_{j+1}$= either 101011000110101 or 100011010110101. If one of these $f(\tilde{a}_{j+1})$ leads to two non Hamilton cycles in G_k, the other choice of $f(\tilde{a}_{j+1})$ allows at least a Hamilton cycle.

We can extend theorem 9 to the case of multilinks as follows: A pair of vertices of H_k adjacent by means of exactly i links is an i-linked pair. If the vertices d,d' of a 2-linked pair have normal representations c,c' respectively, i.e. setting them as in D, we say that (d,d') is z-centered, where z is the integer radius of the pivot coordinates for the change from c to c' and backwards.

Theorem 9a. For each triple (u,v,B) with (uv,n)=u and $B=(b_1,..,b_{u-1})\in\{0,1\}^{u-1}$ with $b_i+b_j\neq 1$ at least for one pair of indices (i,j) *with* $i+j=u$, *there is a unique normal representation*

$$\begin{pmatrix} a^k \\ b^k \end{pmatrix} = \begin{pmatrix} a^k(u,v,B) \\ b^k(u,v,B) \end{pmatrix}$$

of a 2-*linked pair of* H_k *of the form*

$$\begin{pmatrix} c_{-k},\ldots,c_0,\bar{b}_1,\ldots,\bar{b}_u,\ldots,c_k \\ c_{-k},\ldots,c_0,\bar{b}_u,\ldots,\bar{b}_1,\ldots,c_k \end{pmatrix}$$

which is $uv\lceil\lambda/2\rceil$-*centered. In fact if* $\lambda\equiv i \pmod 2$ *then* $a^k(u,v,B)=f^{1-i}(b^k(B,uv))$. *The set of all* 2-*linked pairs obtained this way comprises the totality of* 2-*linked pairs of* H_k, *for* $i>1$.

REFERENCES

1. B.Bollobás, *Graph Theory, An Introductory Approach*, Springer-Verlag Graduate Courses in Mathematics, New York, 1979.
2. J.A.Bondy and U.S.R.Murty, *Graph Theory with Applications*, MacMillan, London, 1976.
3. L.E.Dickson, *History of the Theory of Numbers*, Vol. I, Chelsea Publ. Co., New York, 1966.

PROBLEMS AND RESULTS ON CHROMATIC NUMBERS IN FINITE AND INFINITE GRAPHS

P. Erdös
Hungarian Academy of Science

During my long life I wrote many papers of similar title. To avoid repetitions and to shorten the paper I will discuss almost entirely recent problems and will not give proofs.

First of all I discuss some problems which came up during a recent visit to Calgary. An old problem in graph theory states that if G_1 and G_2 both have chromatic number $\geq k$ then $G_1 \times G_2$ also has chromatic number $\geq k$. The product $G_1 \times G_2$ is defined as follows: If $x_1, \cdots, x_u$; $y_1, \cdots, y_v$ are the vertices of G_1 and G_2, then (x_i, y_j), $1 \leq i \leq u$; $1 \leq j \leq v$ are the vertices of $G_1 \times G_2$. Join (x_i, y_j) to (x_{i_1}, y_{j_1}) if and only if x_i is joined to x_{i_1} and y_1 to y_{j_1}. (Observe (x_i, y_j) and (x_{i_1}, y_{j_1}) are joined only if $i \neq i_1$ and $j \neq j_1$.)

This conjecture was known (and easy) for $k \leq 3$ and Sauer and El-Zahar proved it for $k = 4$ not long ago. The proof was surprisingly difficult and does not seem to generalize for $k > 4$. Hajnal proved that if G_1 and G_2 both have infinite chromatic number then their product also has infinite chromatic number.

Perhaps more surprisingly he showed that there are two graphs of chromatic number $\aleph_{k+1}$ whose product has chromatic number $\aleph_k$. The following two problems remain open: Are there two graphs of chromatic number $\aleph_{k+2}$ whose product has chromatic number $\leq \aleph_k$ are there two graphs of chromatic number $\aleph_\omega$ whose product has chromatic number $< \aleph_\omega$? These problems are analogous to some old problems of Hajnal and myself. We proved [4] that for every α there is a graph of power $(2^{\aleph_\alpha})^+$ and chromatic number $\geq \aleph_{\alpha+1}$ so that every subgraph of power $\aleph \leq 2^{\aleph_\alpha}$ has chromatic number $\aleph_\alpha$. We did not know (even assuming G.C.H.) if there is a graph of power and chromatic number $\aleph_{\alpha+2}$ so that each subgraph of power $\aleph_{\alpha+1}$ has chromatic number $\aleph_\alpha$. Recently Baumgartner proved that the existence of such a graph is consistent. In fact he proved that it is consistent with the generalised continuum hypothesis there is a graph of power and chromatic number ${}_2$ all of whose subgraphs of power $\leq \aleph_1$ have chromatic number $\leq \aleph_0$. At the moment it seems hopeless to find a graph of power and chromatic number $\aleph_3$ all of whose subgraphs of power $\leq \aleph_3$ have chromatic number $\leq \aleph_0$. Laver and Foreman showed that it is consistent (relative to the existence of a very large cardinal) that if every subgraph of power $\aleph_1$ of a graph of size has $\aleph_2$ chromatic number $\aleph_1$ then the whole graph has chromatic number $\leq \aleph_1$. Thus it is consistent that our example is best possible.

Shelah showed that in the constructible universe for every regular K that is not weakly compact, there is a graph of size K and chromatic number $\aleph_1$ all of whose subgraphs of size $< K$ have chromatic number $\leq \aleph_0$.

As far as we know, our old problem is still open: If G has power $\aleph_{\omega+1}$ and chromatic number $\aleph_1$, then it is consistent that it must hve a subgraph of power $\aleph_\omega$ and chromatic number $\aleph_1$.

An old theorem of Hajnal, Shelah and myself [5] states that if G has chromatic number $\aleph_1$, then there is an $n_0 = n_0(G)$

so that G contains a circuit C_n for every $n > n_0$. On the other hand, we know almost nothing of the 4-chromatic subgraphs that must be contained in G. In particular we do not know if G_1 and G_2 have chromatic number $\aleph_1$ whether there is an H of chromatic number 4 which is a subgraph of both G_1 and G_2. It seems certain that this is true and perhaps it remains true if 4 is replaced by any finite n and perhaps by $\aleph_0$. Hajnal, on the other hand, constructed $\aleph_1$ graphs G_α, $1 \le \alpha \le \omega_1$ of power $2^{\aleph_0}$ and chromatic number $\aleph_1$ no two of which contain a common subgraph H of chromatic number $\aleph_1$.

Now we have to state the fundamental conjecture of W. Taylor which, unfortunately, Hajnal and I missed (probably due to old age, stupidity and laziness): Let G be a graph of chromatic number $\aleph_1$. Is it then true that for every cardinal number m there is a graph G_m of chromatic number m all finite subgraphs of which are also subgraphs of G? No real progress has been made with this beautiful conjecture. Hajnal, Shelah and I investigated the following related problem: We call a family F of finite graphs good if there is an at least $\aleph_1$-chromatic graph G all whose finite subgraphs are in F. (We write at least $\aleph_1$-chromatic instead of $\aleph_1$-chromatic since Galvin [8] observed more than 15 years ago that it is not at all obvious that every graph of chromatic number greater than $\aleph_1$ contains a subgraph of chromatic number $\aleph_1$. In fact he proved that it is consistent that there is a graph of chromatic number $\aleph_2$ that does not contain an induced subgraph of chromatic number $\aleph_1$.) We call F very good if for every cardinal number m there is a graph G_m of chromatic number $\ge m$ all of whose finite subgraphs are in F. Hopefully good = very good. We observed that the set of all finite subgraphs of our [3] old r-shift graphs are very good for every r. The r-shift graph is defined as follows: Let $\{x_\alpha\}$ be a well ordered set. The vertices of the r-shift graph are the r-tuples

$\{x_{\alpha_1}, \cdots, x_{\alpha_r}\}$ $\alpha_1 < \alpha_2 < \cdots < \alpha_r$. Two such r-tuples $\{x_{\alpha_1}, x_{\alpha_2}, \cdots, x_{\alpha_r}\}, \{y_{\beta_1}, y_{\beta_2}, \cdots, y_{\beta_r}\}$ are joined if and only if $y_{\beta_1} = x_{\alpha_2}, \cdots, y_{\beta_{r-1}} = x_{\alpha_r}$.

We also stated the following problem: A family F_r of finite graphs is called r-good if there is a graph G_r of power $\leq \aleph_{r+1}$ and chromatic number $\geq \aleph_1$ all of whose finite subgraphs are in F_r. It is called r-very good if (for every cardinal $\aleph_\alpha$) there is a graph G of chromatic number $\leq \aleph_\alpha$ and power $\leq \aleph_{\alpha+r}$ all of whose finite subgraphs are in F_r. Hopefully r-good = r-very good. We proved that for $r < \omega$ $F_{r+1} \subset F_r$ and the inclusion is proper. We do not know what happens for $r > \omega$.

We proved that the number of vertices of an at least $\aleph_1$-chromatic graph all whose finite subgraphs are subgraphs of the r-th shift-graph must have power $\exp_r(\aleph_1)^+ = \aleph_{r+1}$. This last equation holds if the generalised continuum hypothesis is assumed.

We formulated as a problem that every good family must contain for some r the finite subgraphs of the r-th shift-graph. We expected that the answer to this question will be negative, but we could not show this. Recently A. Hajnal and P. Komjath [10] showed that the answer is negative. Hajnal conjecture that if F_n, $n = 1, 2, \cdots$ is a good family for all n then there is good family F satisfying $F \supset F_n$, $n = 1, 2, \cdots$. A much stronger (but also much more doubtful) conjecture is that there is a good family F which is almost contained in F_n for every n. Perhaps one should first try to disprove this. The answer is unknown even for the finite subgraphs of the r-th shift-graph.

The intersection of two good families is perhaps always good, but we cannot even exclude the possibility that there are c families of almost disjoint good families of finite graphs. We are, of course, interested only in finite graphs of chromatic number

≥ 4, since our old result with Shelah implies that every G of chromatic number $\geq \aleph_1$ contains all odd circuits for $n > n_0$.

Hajnal and I proved that every graph of chromatic number $\aleph_1$ contains a tree each vertex of which has degree $\aleph_0$, and we also proved that it contains for every n, a $K(n, \aleph_1)$ but it does not have to contain a $K(\aleph_0, \aleph_0)$. Hajnal [9] showed that if $c = \aleph_1$, it does not have to contain a $K(\aleph_0, \aleph_0)$ and a triangle. The problem is open (and is perhaps difficult) whether there is graph of chromatic number $\aleph_1$ which does not contain a $K(\aleph_0; \aleph_0)$ and has no triangle and no pentagon (and in fact no C_{2r+1} for $r \leq K$).

Hajnal and Komjath [10] recently proved the following result of astonishing accuracy: Every G of chromatic number $\aleph_1$ contains a half-graph (i.e. a bipartite graph whose white vertices are $x_1, x_2, \cdots$ and whose black vertices are $y_1, y_2, \cdots$, where x_i is joined to y_j for $j > i$) and another vertex which is joined to all the x_i. On the other hand, if $c = \aleph_1$ is assumed, it does not have to contain two such vertices.

To end this short excursion into transfinite problems, let me state an old problem of Hajnal and myself: Is it true that every G of chromatic number $\aleph_1$ contains a subgraph G' which also has chromatic number $\aleph_1$ and which cannot be disconnected by the omission of a finite number of vertices? We observed that, if true, this is best possible; we gave a simple example of a graph of chromatic number $\aleph_1$ every subgraph of which has vertices of degree $\aleph_0$.

P. Komjath recently proved that every graph G of chromatic number $\aleph_1$ contains for every n, a subgraph G_n of chromatic number $\aleph_1$ which cannot be disconnected by the omission of n vertices and he informed me that he can also insure that there is such a G_n all vertices of which have infinite degree.

As far as I know the following Taylor-like problem has not yet been investigated: Determine the smallest cardinal number m for which if G has chromatic number m, then there is a G' of arbitrarily large chromatic number all of whose denumerable subgraphs are also subgraphs of G. Hajnal observed that it is consistent that every G of chromatic number $\aleph_2$ contains a $K(\aleph_0, \aleph_0)$. Thus it is consistent that $m > \aleph_1$. He suggests that perhaps one can prove (assuming G.C.H.?) that every G of chromatic number $_2$ contains the Hajnal-Komjath graph as a subgraph. Thus the analog of Taylor's conjecture is perhaps $m = \aleph_2$.

Now I discuss some finite problems. El-Zahar and I considered the following problem: Is it true that for every k and ℓ there is an $n(k,\ell)$ so that if the chromatic number of G is $\geq n(k,\ell)$ and G contains no $K(\ell)$, then G contains two vertex-disjoint k-chromatic subgraphs G_1 and G_2 so that there is no edge between G_1 and G_2? We proved this for $k = 3$ and every ℓ, but great difficulties appeared for $k = 4$, and Rödl suggested that the probability method may give a counterexample. It seems to me that this method just fails.

For $k = 3$ the simplest unsolved problem is: Let G be a 5-chromatic graph not containing a $K(4)$. Is it then true that G contains two edges e_1 and e_2 so that the subgraph of G induced by the 4 vertices of e_1 and e_2 only contains these edges? The answer is certainly affirmative if we assume that the chromatic number of G is ≥ 9.

During a recent visit to Israel, Bruce Ruthschild was there and we posed the following problem:

Denote by $G(k;\ell)$ a graph of k vertices and ℓ edges. We say that the pair n, e forces k, ℓ, $(n,e) \to (k,\ell)$, if every $\mathcal{G}(n;e)$ contains a $\mathcal{G}(k;\ell)$ or a $\mathcal{G}(k;\binom{k}{2}-\ell)$ as an induced subgraph. It seems that the most interesting problems arise if $\ell = \frac{1}{2}\binom{k}{2}$. In this case we can of course assume that $c \leq \frac{1}{2}\binom{n}{2}$. We have unfortunately almost no positive results. We observed that if $e > \frac{2n}{3}$ then $(n,e) \to (4,3)$. This clearly does not hold for $e \leq \frac{2n}{3}$. This unfortunately is our only positive result. On the other hand, we observed that if $n > n_o$, then $(n,e) \not\to (5,5)$ for every e (and $n > n_o$). In other words, for every e there is a $\mathcal{G}(n;e)$ which does not contain a $\mathcal{G}(5;5)$ as an induced subgraph and the same holds for a $\mathcal{G}(8;14)$. Graham observed the same method gives that $(n,e) \not\to (12,33)$. We convinced ourselves that for $k > 12$ our method no longer will give a counterexample. The simplest unsolved problem is, unless we overlooked a trivial idea, perhaps interesting and non-trivial: Are there any values of n and e for which $(n,e_n) \to (9,18)$? Further and determine all these values of n and e_n.

Fan Chung and I spent (wasted?) lots of time on the following problem: Denote by $f(n;k,\ell)$ [1] the smallest integer for which every $\mathcal{G}(n,f(n;k,\ell))$ contains a $\mathcal{G}(k;\ell)$ as a subgraph. Here we of course do not insist that the subgraph should be induced. Also we do not prescribe the structure of our $\mathcal{G}(k;\ell)$. The first interesting and difficult case seems to be: Is it true that

$$\frac{f(n;\ 8,\ 13)}{n^{3/2}} \to \infty \qquad ? \tag{1}$$

We could not prove (1); the probability method seems to fail. Probably $f(n;8,13) > n^{3/2+\varepsilon}$ also holds. It is well known and easy to see that $f(n;8,12) < c\, n^{3/2}$ holds, since every $G(n;c_r\, n^{3/2})$ contains for sufficiently large c_r, a $K(r,2)$, and thus a $K(6,2)$ of 8 vertices and 12 edges. Completely new and interesting questions come up if we also consider the structure of $\mathcal{G}(k;\ell)$, e.g., Simonovits and I [7] proved that every $\mathcal{G}(n;c\, n^{8/5})$ contains a cube - the proof is quite difficult. We believe that our exponent 8/5 is best possible but could not even show that for every c and $n > n_o(c)$ there is a $\mathcal{G}(n;\ c\, n^{3/2})$ which contains no cube as a subgraph. A more general conjecture of Simonovits and myself states that if G is bipartite then the necessary and sufficient conditions of

$$\frac{f(n;\ \mathcal{G})}{n^{3/2}} \to \infty \tag{2}$$

is that $\mathcal{G}$ should have no induced subgraph each vertex of which has degree greater than 2. Perhaps this condition already implies

$$f(n;\ \mathcal{G}) > n^{3/2+\varepsilon}. \tag{3}$$

Conjectures (2) and (3), if true, will probably require some new ideas.

During a recent visit to Calgary, Sauer told me his conjecture: Let C be a sufficiently large constant. Is it true that for every k there is an $f_k(C)$ so that every $G(n; f_k(c)n)$ contains a subgraph each vertex of which has degree $v(x)$, $k < v(x) < Ck$. In other words, the subgraph is quasiregular. Related problems were also stated in our paper with Simonovits and we used the concept of quasiregularity to prove our $\mathcal{G}(n;\ c\ n^{8/5})$ theorem, but as far as I know the conjecture of Sauer is new and is very interesting.

During the 1984 international meeting on graph theory in Kalamazoo, Toft posed the following interesting question: Is there a 4-chromatic edge critical graph of $c_1\ n^2$ edges which can be made bipartite only by the omission of $c_2 n^2$ edges? It is not even known if for every c there is a 4-chromatic critical graph of $c\ n_1^2$ edges which can not be made 2-chromatic by the omission of $C\ n$ edges.

Perhaps I might be permitted to make a few historical remarks: A k-chromatic graph is called edge critical if the omission of every edge decreases the chromatic number to $k - 1$. This concept is due to G. Dirac. When I met him in London early in 1949 he told me this definition. I was already at that time interested in extremal problems and immediately asked: What is the largest integer $f(n;k)$ for which there is a $\mathcal{G}(n;\ f(n;k))$ that is k-chromatic and edge critical? In particular, can $f(n;k)$ be greater than $c\,n^2$? To my surprise Dirac showed very soon that for $k \geq 6$, $f(n;k) > c_k\,n^2$ and, in particular $f(n;6) > \frac{n^2}{4} + cn$. This result has not been improved for more than 35 years, and left the problem open for $k = 4$ and $k = 5$. In 1970 Toft [15] proved that $f(n;4) > \frac{n^2}{16} + cn$. Simonovits and I easily proved that $f(n;4) < \frac{n^2}{4} + cn$. It would be very desirable to determine $f(n;k)$, or, if this is too difficult, to determine

$$\lim \frac{f(n;k)}{n^2} = c_k.$$

The graph of Toft has many vertices of bounded degree. I asked: Is there a 4-chromatic critical graph $G(n)$ each vertex of which has degree $> cn$. (Dirac's 6-chromatic critical graph has this property.) Simonovits [14] and Toft [16] independently found a 4-chromatic critical graph each vertex of which has degree $> cn^{1/3}$. The following question occurred to me: Is there a 4-chromatic critical $G(n;\,c\,n^2)$ which does not contain a very large $K(t,t)$? All examples known to me contain a $K(t,t)$ for $t > c\,n$, but perhaps such an example exists with $t < C \log n$. (Rödl in fact recently constructed such an example).

To end this paper I want to mention some older problems which I find very attractive and which I have perhaps neglected somewhat and which have both a finite and an infinite version. First an old conjecture of Hajnal and myself:

Is it true that for every cardinal number m there is a graph G which contains no $K(4)$ and if one colors the edges of G by m colors there always is a monochromatic triangle. For $m = 2$ this was proved by Folkman and for every $m < \aleph_0$ it was proved by Nesetril and Rödl [11]. For $m \geq \aleph_0$ the problem is open. The strongest and simplest problem which is open is stated as follows (where we assume that the continuum-hypothesis holds): Is it then true that there is a G of power $\aleph_2$ without a K(4) so that if one colors the edges of G by $\aleph_0$ colors there always is a monochromatic triangle. If $c = \aleph_1$ is not assumed, then $\aleph_2$ must be replaced by c^+. I offer a reward of 250 dollars for a proof or disproof (perhaps this offer violates the minimum wage act).

An interesting finite problem remains. For m = 2 Folkman's graph is enormous, it has more than $10^{10^{10^{10^{10^{10^{10}}}}}}$ vertices and the graph of Nesetril and Rödl is also very large. This made me offer 100 dollars for such a graph of less than 10^{10} vertices (the truth in fact may be very much smaller, there very well could exist such a graph of less than 1000 vertices). Rödl and Szemerédi found such a graph which has perhaps $< 10^{12}$ vertices which does not fall very short of fulfilling my conditions and perhaps can be improved further.

Another old conjecture of Hajnal and myself states that for every k and r there is an f(k,r) so that if G has chromatic number $\geq f(k,r)$, then it contains a subgraph of girth $> k$ and chromatic number $> r$. For $k = 3$ this was answered affirmatively by Rödl [12]. The infinite version of our problem states: Is it true that every graph of chromatic number m contains a subgraph of chromatic number m the smallest odd circuit of which has size $> 2k + 1$? This problem is open even for $k = 1$.

Our triple paper with Hajnal and Szemerédi [6] contains many interesting unsolved finite and infinite problems. Is it true that every graph G of chromatic number $\aleph_1$ contains for every C a finite subgraph G(n) which cannot be made bipartite by the omission of C n edges? Perhaps one can further assume that our G(n) has chromatic number 4. The difficulty again is that so little is known about the critical 4-chromatic graphs.

Let $f(n)$ be a function that tends to infinity as slowly as we please. Is it true that for every k there is a k-chromatic graph so that for each n every subgraph of n vertices of G can be made bipartite by the omission of fewer than $f(n)$ edges. Lovász and Rödl [13] proved this for $f(n) = O(n^{(1/k)-2})$ and Rödl settled the conjecture for triple systems.

Let $F(n)$ tend to infinity as fast as we please. Is there an $\aleph_1$-chromatic G so that for each n every n-chromatic subgraph of G has more than $F(n)$ vertices?

Hajnal, Sauer and I asked in Calgary recently: Let G be n-chromatic and the smallest odd circuit of which is $2k + 1$. Is it then true that the number of vertices of G is greater than n^{c_k}, where c_k tends to infinity together with k? Perhaps we overlooked a trivial point, but we could not even show that the number of vertices of G must be greater than $n^{2+\varepsilon}$. It seems clear that this must hold if we only assume that G has no triangle and pentagon.

An old problem of mine which has been neglected [2] is stated as follows: Is it true that for every small $\varepsilon > 0$ and infinitely many n there is a regular $\mathcal{G}(n)$ with degree $v(x) = [n^{(1/2)+\varepsilon}]$ so that $\mathcal{G}(n)$ has no triangle and the largest stable set of which has size $v(x)$. I expect that the answer is negative and offer 100 dollars for a proof or disproof.

Here is a final question of mine which I had no time to think over carefully and which might turn out to be trivial. Let $\mathcal{G}(n)$ be a k-chromatic graph. Then clearly $\mathcal{G}(n)$ always has a subgraph of $\leq \frac{n+1}{2}$ vertices which has chromatic number $\geq \frac{k+1}{2}$. Can this be strengthened if we assume say that G has no triangle? (Without some assumption the complete graph shows that the original result is best possible.) As a matter of fact I now believe that no such strengthening is possible. The probability method seems to give that to every $\varepsilon > 0$ there is a $k_0(\varepsilon)$ so that for every $k > k_0(\varepsilon)$ and $n > n_0(\varepsilon,k)$ there is a k-chromatic $G(n)$ of girth ℓ so that every set of εn vertices of which spans a graph of chromatic number $(1+o(1))\ \varepsilon n$, but I may be wrong since I did not check the details.

References

1. P. Erdös, Extremal problems in graph theory. Theory of Graphs and its Applisations, Proc. Symp. Smolenice (1963) 29-36.
2. P. Erdös, On the construction of certain graphs, J. Combinatorial Theory 1(1966) 149-153.
3. P. Erdös and A. Hajnal, Some remarks on set theory IX. Combinatorial problems in measure theory and set theory. Michigan Math. Journal 11(1964) 107-127. See also [2].
4. P. Erdös and A. Hajnal, On chromatic number of infinite graphs. Theory of Graphs, Proc. Coll. Thinay Hungary Acad. Proc. (1968) 83-98.
5. P. Erdös, A. Hajnal and S. Shelah, On some general properties of chromatic numbers. Topics in Topology, Proc. Coll Kezszthely (1972), Coll. Math. Soc. T. Bolyai Vol. 8 North Holland, Amsterdam (1974) 243-255.
6. P. Erdös, A. Hajnal and E. Szemerédi, On almost bipartite large chromatic graphs. Annals Discrete Math. 12(1982) 117-123.
7. P. Erdös and M. Simonovits, Some extremal problems in graph theory. Combinatorial Theory and its Applications, Proc. Coll. Balatorfured 1969. North Holland, Amsterdam (1970) 377-390.
8. F. Galvin, Chromatic numbers of multigraphs. Periodica Math. Hungarica, 4(1973) 117-119.
9. A. Hajnal, A negative partition relation. Proc. Nat. Acad. Sci. U.S.A. 68(1971) 142-144.

10. A. Hajnal and P. Komjath, What must and what need not be contained in a graph of uncountable chromatic number, Combinatoria 4(1984) 47-52.

11. J. Nesetril and V. Rödl, Ramsey properties of graphs with forbidden complete subgraphs. J. Combinatorial Theory 20B (1976), 243-249.

12. V. Rödl, On the chromatic number of subgraphs of a given graph. Proc. Amer. Math. Soc. 64(1977) 370-371.

13. V. Rödl, Nearly bipartite graphs with large chromatic number. Combinatoria 2(1982) 377-383.

14. M. Simonovits, On colour initial graphs. Studia Sci. Math. Hungar. 7(1972) 67-81.

15. B. Toft, On the maximal number of edges of critical k-chromatic graphs. Studia Sci. Math. Hungar. 5(1970) 469-470.

16. B. Toft, Two theorems on critical 4-chromatic graphs. Studia Sci. Math. Hungar. 7(1972) 83-89.

SUPRACONVERGENCE AND FUNCTIONS THAT SUM TO ZERO ON CYCLES

V. Faber

Los Alamos National Laboratory*

Andrew B. White, Jr.

Los Alamos National Laboratory

ABSTRACT

The approximation of differential equations by difference equations defined on a given set of points $\{x_i\}$ separated sucessively by mesh spacings $\{\Delta_i\}$ is an important area of numerical mathematics. Classical analysis of these methods says that if the truncation error is proportional to Δ^n (where $\Delta = \max \Delta_i$), then the approximation error is also proportional to Δ^n. Recently, however, we have discovered that this is not necessarily the case for irregular grids; the approximation might be *supraconvergent* with convergence rate greater than n. We examine this anomaly by looking at periodic meshes. For supraconvergence to occur, it must be true that

$$\sum_{k=1}^{p} F(\Delta_k, \Delta_{k+1}, \cdots, \Delta_{k+t}) = 0 \ , \qquad (\dagger)$$

for all p where F is easily found by "usual" truncation error techniques. We use graph theory to reduce the amount of computation necessary to check $(\dagger)$.

*This work was performed under the auspices of the U.S. Department of Energy.

1. Supraconvergence

The numerical solution of two-point boundary-value problems is often accomplished using finite differences. Take, for example, the second-order equation

$$y'' = f(x) \tag{1a}$$

$$y(0) = A, \quad y(1) = B \ . \tag{1b}$$

We define a discrete mesh of points $\{x_i\}_{i=0}^{i=N}$ on $[0,1]$, where $x_0 = 0 < x_1 < \cdots < x_N = 1$. Associated with each mesh point, x_i, is a value v_i intended to be an approximation to $y(x_i)$. The differential equation may be approximated by a difference equation,

$$(L_h v)_i \equiv \frac{2}{\Delta_i(\Delta_{i+1} + \Delta_i)} v_{i-1} - \frac{2}{\Delta_i \, \Delta_{i+1}} v_i \tag{2a}$$

$$+ \frac{2}{\Delta_{i+1}(\Delta_{i+1} + \Delta_i)} v_{i+1} = f(x_i) \ ,$$

and the discrete boundary conditions are given by

$$v_0 = A, \quad v_N = B \ . \tag{2b}$$

In (2a) above, $\Delta_i = x_i - x_{i-1}$, the mesh spacing.

Classically, the error $(e_i = y(x_i) - v_i)$ of a stable difference scheme is determined by the truncation error. In this case, the truncation error, τ, is defined by

$$(L_h \, y)_i = f(x_i) + \tau_i \ .$$

That is, how close does the exact solution $y(x)$ come to solving the difference equation. In our example (2a),

$$\tau_i = -\frac{1}{3}\,[\Delta_{i+1} - \Delta_i]\; y^{(3)}\,(x_i) + \frac{1}{12}\,[\Delta_{i+1}^2 - \Delta_{i+1}\,\Delta_i + \Delta_i^2]\; y^{(4)}\,(x_i) + \cdots \quad .$$

The usual error estimates yield a relationship like

$$\|e\| \sim \|\tau\| \;.$$

Thus, there appear to be two cases. For a uniform mesh, $\Delta_i = h = \frac{1}{N}$, we have that the error is $O(h^2)$ or second order in mesh size; for an irregular mesh, we have that the error is $O(\Delta)$, $\Delta = \max \Delta_i$, *unless* the variation, $\Delta_{i+1} - \Delta_i$, in mesh sizes is small.

It has been discovered numerically that the smooth mesh restriction $|\Delta_{i+1} - \Delta_i| = O(\Delta^2)$ is not necessary. That is, even for mesh points randomly placed in $(0,1)$, we observe second order accuracy. This phenomenon is explained in detail in Manteuffel and White [4] and Kreiss et al. [3].

For this sort of thing to occur in general, it must hold for cubic functions $y(x)$. If $\bar{e}$ is the error in this case, then

$$(L_h\bar{e})_i = \frac{2}{\Delta_{i+1} + \Delta_i}\left(\frac{\bar{e}_{i+1} - \bar{e}_i}{\Delta_{i+1}} - \frac{\bar{e}_i - \bar{e}_{i-1}}{\Delta_i}\right) = -\frac{1}{3}\,(\Delta_{i+1} - \Delta_i) \;, \quad (3)$$

this last being the truncation error in this simple case. Solving equation (3) gives us a particular solution

$$\frac{\bar{e}_{i+1} - \bar{e}_i}{\Delta_{i+1}} = -\frac{1}{3}\sum_{k=1}^{i}\left(\Delta_{k+1}^2 - \Delta_k^2\right) .$$

Rearranging this sum yields

$$\frac{\overline{e_{i+1}} - \overline{e_i}}{\Delta_{i+1}} = -\frac{1}{3}\left(\Delta_{i+1}^2 - \Delta_1^2\right) = O(\Delta^2) \ .$$

From this point on, standard arguments yield the error estimate

$$\overline{e_i} = O(\Delta^2) \ ,$$

assuming appropriately accurate boundary data. Supraconvergence results of this sort require careful accounting of the error associated with $(\Delta_i + \Delta_{i+1})\tau_i$, which in this case is

$$-\frac{1}{3}\sum_{k=1}^{i}\left(\Delta_{k+1}^2 - \Delta_k^2\right) = \sum_{k=1}^{i} F(\Delta_k, \Delta_{k+1}) \ . \tag{4}$$

Another example (from [3]) of this sort of behavior is Numerov's method for solving (1a, b). In this difference scheme, the right-hand side of (2a) is replaced by an average

$$\alpha\ f(x_{i-1}) + (1 - \alpha - \beta)\ f(x_i) + \beta\ f(x_{i+1}) \ ,$$

where α and β are chosen locally (but independent of f) so that the order of the truncation error is as small as possible. In this case, the error term can be expressed as

$$\frac{\overline{e_{i+1}} - \overline{e_i}}{\Delta_{i+1}} = c_1 \sum_{k=1}^{i}\left(\Delta_{k+1}^4 + \Delta_{k+1}^3\,\Delta_k - \Delta_{k+1}\,\Delta_k^3 - \Delta_k^4\right)$$

$$\equiv c_1 \sum_{k=1}^{i} F(\ \Delta_k, \Delta_{k+1})$$

if $y^{(5)}(x)$ is constant. Here it is not clear whether we get cancellation in error as before or not. Numerical results on random meshes indicate fourth-order convergence. However, we can see that on the very special three-period mesh,

$$\Delta_1 = \frac{1}{2}h,\ \Delta_2 = h,\ \Delta_3 = \frac{3}{2}h,\ \Delta_4 = \frac{1}{2}h,\ \Delta_5 = h,\ \Delta_6 = \frac{3}{2}h,\ \cdots ,$$

the sum which determines the error,

$$\sum_{k=1}^{N-1} F(\Delta_k, \Delta_{k+1}) = \frac{5}{4}h^3 - \frac{15}{4}h^4 - \frac{3}{2}h^4 = O(\Delta^3) .$$

Since this is not $O(\Delta^4)$, the scheme is not supraconvergent.

For higher order equations or non-compact difference schemes, the stencil is larger, thus, we shall consider sums of the form

$$\sum_{k=1}^{i} F(\Delta_k, \Delta_{k+1}, \cdots, \Delta_{k+t}) ,$$

in order to examine this supraconvergence phenomena.

A mesh function $F(\Delta_1, \cdots, \Delta_{t+1})$ is defined to be *homogeneous* of order $n \geq 0$ if $F(\alpha_1 h, \cdots, \alpha_{t+1} h) = h^n F(\alpha_1, \cdots, \alpha_{t+1})$; and it is *bounded* if

$$F(\Delta_1, \Delta_2, \cdots, \Delta_{t+1}) \leq K$$

for all $\Delta_1, \Delta_2, \cdots, \Delta_{t+1}$ such that $0 < \Delta_l \leq 1$.

We shall always assume that our mesh functions are homogeneous and bounded (as is typically the case for the most significant part of τ when constant coefficient difference schemes are applied to the simplest of solutions of the approximated differential equations). We say F is n-supraconvergent if for all $i \leq N - t$,

$$\left| \sum_{k=1}^{i} F(\Delta_k, \cdots, \Delta_{k+t}) \right| \leq O(\Delta^n) . \tag{5}$$

Suppose that for some $p \geq 1$ there exists positive $s_1, s_2, \cdots, s_p$ such that

$$\sum_{k=1}^{p} F(s_k, s_{k+1}, \cdots, s_{k+t}) = C \neq 0 ,$$

where the indices are taken modulo p. Then we can construct a sequence of meshes for the whole problem for which supraconvergence fails. To accomplish this, we need only find one index, i, for which (5) is violated.

Let $N = pM$, $\sigma = \sum_{i=1}^{p} s_i$, $\Delta_{pm+l} = \frac{p}{N\sigma} s_l$. Thus $\sum_{l=1}^{p} \Delta_{pm+l} = \frac{p}{N}$ and $\sum_{l=1}^{N} \Delta_l = 1$ as required.

On this mesh

$$\sum_{l=1}^{(M-t)p} F(\Delta_l, \Delta_{l+1}, \cdots, \Delta_{l+t}) = \sum_{j=1}^{M-t} \sum_{k=1}^{p} F(\Delta_k, \cdots, \Delta_{k+t})$$

$$= \sum_{j=1}^{M-t} \sum_{k=1}^{p} \left(\frac{p}{N\sigma}\right)^n F(s_k, \cdots, s_{k+t})$$

$$= \left(\frac{p}{N\sigma}\right)^n C(M-t)$$

$$= h^n \left(\frac{p}{\sigma}\right)^n C\left(\frac{1}{hp} - t\right)$$

$$= C_1 \Delta^{n-1} + C_2 \Delta^n .$$

Thus (5) is violated and F is not n-supraconvergent.

Now suppose that for all p and $\Delta_1, \Delta_2, \cdots, \Delta_p$, the mesh function satisfies

$$\sum_{k=1}^{p} F(\Delta_k, \Delta_{k+1}, \cdots, \Delta_{k+t}) = 0 ,$$

where the indices are taken modulo p. In this case, we say that F *sums to zero on every periodic mesh.* Consider

$$\sum_{k=1}^{i} F(\Delta_k, \cdots, \Delta_{k+t}) = \sum_{k=1}^{i} F(\Delta_k, \cdots, \Delta_{k+t})$$

$$+ F(\Delta_{i+1}, \cdots, \Delta_{i+t}, \Delta_1) + \cdots + F(\Delta_{i+1+t}, \Delta_1, \cdots, \Delta_t)$$

$$- F(\Delta_{i+1}, \cdots, \Delta_{i+t}, \Delta_1) - \cdots - F(\Delta_{i+1+t}, \Delta_1, \cdots, \Delta_t)$$

$$= -\left[F(\Delta_{i+1}, \cdots, \Delta_{i+t}, \Delta_1) + \cdots + F(\Delta_{i+t}, \cdots, \Delta_t) \right]$$

$$= -\Delta^n \left[F(\frac{\Delta_{i+1}}{\Delta}, \cdots, \frac{\Delta_1}{\Delta}) + \cdots + F(\frac{\Delta_{i+t}}{\Delta}, \cdots, \frac{\Delta_t}{\Delta}) \right] .$$

Thus $\left| \sum_{k=1}^{i} F(\Delta_k, \cdots, \Delta_{k+t}) \right| \leq Kt\Delta^n$ and F is supraconvergent. We have proved the following theorem.

Theorem 1. Let the mesh function $F(\Delta_1, \cdots, \Delta_{t+1})$ be bounded and homogeneous of order n. Then F is n–supraconvergent if and only if F sums to zero on every periodic mesh.

In the next section we develop some graph theory which we shall use to find a more compact characterization of supraconvergent mesh functions.

2. The Cycle Space

In this section, we construct a directed graph whose directed cycles correspond to the periodic meshes. Then we use the theory of the cycle space to find the best (in some sense) characterization of supraconvergence. This is an important practical consideration, because it would be very difficult to directly verify that F sums to zero on every periodic mesh.

Suppose, for the moment, each of the $t+1$ arguments of F can take only values from a set M of m distinct positive numbers. In what follows x_i, y_i, $i = 1, 2, \cdots$, shall denote elements of M. We consider the directed graph $G = (V,E)$ with vertices in $V = M^t$ and arcs in E such that $((y_1, y_2, \cdots, y_t), (x_1, x_2, \cdots, x_t)) \in E$ if and only if $x_1 = y_2, x_2 = y_3, \cdots, x_{t-1} = y_t$. Let $y_{t+1} = x_t$; we denote this arc by $(y_1\ y_2\ \cdots\ y_t\ y_{t+1})$. The arcs correspond to the possible arguments of the mesh function $F(y_1, y_2, \cdots, y_{t+1})$.

Clearly the number of vertices in G is $p = m^t$ and the number of arcs is m^{t+1}. A *path* from a vertex v_0 to a vertex v_n is an alternating sequence of vertices and arcs $v_o, e_1, v_1, \cdots, v_{n-1}, e_n, v_n$ in which each arc e_i is either (v_{i-1}, v_i) (in this case, we say e_i is *oriented in the direction of the path*) or (v_i, v_{i-1}) (in this case, we say e_i is *oriented against the direction of the path.*) A path is *directed* if all of its arcs are oriented in the direction of the path. A graph is *connected* if every two vertices are connected by a path. Clearly our G is connected.

A path is *closed* if $v_o = v_n$. A closed path is a *cycle* if its vertices $v_1, v_2, \ldots, v_n$ are distinct. We denote a cycle by its list of arcs. The *cycle space* is the vector space over the two element field $F_2 = \{0,1\}$ spanned by the cycles. A *cycle vector* is any member of the cycle space and can be regarded as an arc-disjoint union of cycles. A *spanning tree* T of a connected graph G is a maximal subgraph of G, which contains no cycle.

Note that if an arc not in a spanning tree T of G is added to T, the resulting graph has exactly one cycle. We need the following fact.

Fact 1. (Harary [1, p.38]) The set of cycles formed by adding to T, one at a time, each arc not in T forms a basis for G. The dimension of the cycle space is

$$m^{t+1} - m^t + 1 \ .$$

Given a function $F : E \rightarrow R$ and a closed path C whose arcs are $e_1, e_2, \cdots, e_q$, the sum of F around the path is defined to be $F[C] = \epsilon_1 F(e_1) + \cdots + \epsilon_q F(e_q)$, where ϵ_i is $+1$ or -1 depending, respectively, on whether e_i is with or against the direction of the path.

The application we have in mind is as follows: suppose a periodic mesh, of period p, has successive mesh lengths $y_1, \cdots, y_p$. Then $\sum_{i=1}^{p} F(y_i, y_{i+1}, \cdots, y_{i+t})$ is $F[C]$ for the closed directed path $(v_0,v_1), (v_1,v_2), \cdots, (v_{p-1},v_p)$, where $v_i = (y_i, \cdots, y_{i+t})$. In this context, we may restate Theorem 1.

Theorem 1'. Let the mesh function $F(\Delta_1, \cdots, \Delta_{t+1})$ be bounded and homogeneous of order n. Then F is n-supraconvergent if and only if F sums to zero on every closed directed path of the directed graph G.

Now, we need a second easy-to-prove fact.

Fact 2. (Kirchhoff's Voltage Law [2]) There exists a function $f : G \to R$ such that for each $e = (u,v) \in E$, $F(e) = f(u) - f(v)$ if and only if $F[C] = 0$ for every closed path C.

For our graph G, we can prove more.

Theorem 2. Let y be a fixed element of M. The following are equivalent:

I) $F[C] = 0$ for every closed directed path C;

II) $F[C] = 0$ for every closed directed path C with $2t$ or $2t+1$ arcs;

III) $F[C] = 0$ for every closed path C of the form

$$(x_1\, x_2 \cdots x_t\, x_{t+1}),\ (x_2 \cdots x_{t+1}\, y)\ , \qquad (*)$$

$$(x_3 \cdots, x_{t+1}\, y\, y),\ \cdots,\ (x_{t+1}\, y\, y \cdots y),$$

$$(x_t\, y\, y \cdots y),\ \cdots,\ (x_3 \cdots x_t\, y\, y\, y),$$

$$(x_2 \cdots x_t\, y\, y), (x_1\, x_2 \cdots x_t\, y) ;$$

IV) $F[C] = 0$ for every closed path C.

Proof. Obviously, (I) implies (II) and (IV) implies (I). To show that (II) implies (III) consider the directed cycle C_1 with $2t+1$ arcs:

$$(x_1\, x_2 \cdots x_t\, x_{t+1}), (x_2 \cdots x_{t+1}\, y), (x_3 \cdots , x_{t+1}\, y\, y), \cdots ,$$

$$(y \cdots y\, y\, x_1), (y \cdots y\, x_1\, x_2), \cdots , (y\, x_1\, x_2 \cdots x_t) ,$$

and the directed cycle C_2 with $2t$ arcs:

$$(x_1\, x_2 \cdots x_t\, y), (x_2 \cdots x_t\, y\, y), (x_3 \cdots x_t\, y\, y\, y), \cdots ,$$

$$(y \cdots y\, y\, x_1), (y \cdots y\, x_1\, x_2), \cdots , (y\, x_1\, x_2 \cdots x_t) .$$

The statements $F[C_1] = 0$ and $F[C_2] = 0$ translate into the equations

$$F\,(x_1\, x_2 \cdots x_t\, x_{t+1}) + F\,(x_2 \cdots x_{t+1}\, y) + \cdots + F\,(x_{t+1}\, y\, y \cdots y)$$

$$+ F\,(y \cdots y\, y\, x_1) + F\,(y \cdots y\, x_1\, x_2) \cdots + F\,(y\, x_1\, x_2 \cdots x_t) = 0$$

and

$$F\,(x_1 \cdots x_t\, y) + F\,(x_2 \cdots x_t\, y\, y) \cdots + F\,(x_t\, y\, y \cdots y)$$

$$+ F\,(y \cdots y\, y\, x_1) + F\,(y \cdots y\, x_1\, x_2) \cdots + F\,(y\, x_1\, x_2 \cdots x_t) = 0 .$$

If we eliminate the common terms between these two equations, we have

$$F\,(x_1\,x_2\,\cdots x_t\,x_{t+1}) + F\,(x_2\,\cdots\,x_{t+1}\,y) + \cdots + F\,(x_{t+1}\,y\,y\cdots y)$$

$$- F\,(x_1\,\cdots x_t\,y) - F\,(x_2\,\cdots\,x_t\,y\,y)\,\cdots - F\,(x_t\,y\,y\cdots y) = 0\ ,$$

which is exactly the statement that $F[C] = 0$ for paths C of the form (*). Finally, to show that (III) implies (IV), we need only assume that the previous equation holds for an arbitrary closed path of the form (*) and let

$$f\,(x_1, x_2, \cdots, x_t) = F\,(x_1\,\cdots x_t\,y)$$

$$+ F\,(x_2\cdots x_t\,y\,y) + \cdots + F\,(x_t\,y\,y\cdots y)\ .$$

Then

$$F\,(x_1\,x_2\,\cdots\,x_t\,x_{t+1}) = f\,(x_1, x_2, \cdots, x_t) - f\,(x_2, x_3, \cdots, x_t, x_{t+1})\ ,$$

so (IV) holds by Kirchhoff's Law.

Theorem 3. A basis for the space of cycles consists of

i) Those closed paths of the form (*) having the property that $x_{t+1} \neq y$ and $(x_1, x_2, \cdots, x_t) \neq (y, \cdots, y)$;

ii) The cycles of the form $(y\,y\cdots y\,x_{t+1})$, $(y\cdots y\,x_{t+1}\,y)$, $\cdots$, $(x_{t+1}\,y\cdots y)$, with $x_{t+1} \neq y$;

iii) The cycle $(y\,y\cdots y)$.

Proof. The cycle vectors of the form (i) and (ii) are independent since each contains an arc, $(x_1\,x_2\,\cdots\,x_t\,x_{t+1})$, in case (i) and $(y\,y\,\cdots\,y\,x_{t+1})$ in case (ii) that is in no other. The cycle $(y\,y\cdots y)$ is independent of the others for the same reason. Finally, the number of such cycle vectors is $m^{t+1} - m^t + 1$, exactly the dimension of the cycle space.

3. Application to Supraconvergence

A basis for the space of cycles is a smallest collection of cycle vectors that we have to examine in order to find out if the mesh function $F(\Delta_1, \cdots, \Delta_{t+1})$ sums to zero on all periodic meshes. Thus Theorem 3 is a theorem about supraconvergence by Theorem 1. We can also use Theorem 2 to give another characterization of supraconvergence.

Theorem 4. Let $\delta > 0$ be fixed. Let F be bounded and homogeneous of order n. F is n-supraconvergent if and only if

i) $F(\Delta, \Delta, \cdots, \Delta) = 0$

and

ii) $H(x_1, \cdots, x_{t+1})$ is independent of x_{t+1}, where

$$H(x_1, \cdots, x_{t+1}) = F(x_1, \cdots, x_{t+1}) + F(x_2, \cdots, x_{t+1}, \Delta) + \cdots + F(x_{t+1}, \Delta, \cdots, \Delta)$$

for some fixed Δ.

Proof. If $H(x_1, \cdots, x_{t+1})$ is independent of x_{t+1}, then

$$H(x_1, \cdots, x_{t+1}) - H(x_1, \cdots, x_t, \Delta) = 0 \ .$$

This equation translates to

$$F(x_1, \cdots, x_{t+1}) + F(x_2, \cdots, x_{t+1}, \Delta) + \cdots + F(x_{t+1}, \Delta, \cdots, \Delta)$$
$$- F(x_1, \cdots, x_t, \Delta) - F(x_2, \cdots, x_t, \Delta, \Delta) - \cdots - F(\Delta, \cdots, \Delta) = 0 \ .$$

Since, by hypothesis, $F(\Delta, \cdots, \Delta) = 0$, we have exactly the condition of Theorem 2, part (III).

Conversely, if F sums to zero on every periodic mesh, then all of the equivalent conditions in Theorem 2 hold. Thus $F(\Delta, \cdots, \Delta) = 0$ for any Δ by (iii) of Theorem 3. Using this fact and Theorem 2, part (III), we must have

$H(x_1, \cdots, x_{t+1}) - H(x_1, \cdots, x_t, \Delta) = 0$ for any Δ which yields (ii).

4. Examples

We will examine the supraconvergence of two difference schemes. First, we look at a method for approximating

$$y^{(4)} = f(x) \ . \tag{4.1}$$

The scheme we are interested in is derived by writing (4.1) as a first-order system and then approximating those equations by the classical centered Euler method. The mesh function appropriate for that scheme is

$$F(a,b,c,d) = \frac{db - ca + b^2 - c^2}{(d+b)\,(c+a)} \ , \tag{4.2}$$

where we have taken the liberty of replacing the subscripted quantities used previously to define F with a, b, c, and d. This function is homogeneous of order 0, which means that the difference scheme is not even consistent with the differential equation. The simplest way to discover if this scheme is supraconvergent is to use Kirchhoff's Law; that is, find a function $f(a,b,c)$ such that

$$F(a,b,c,d) = f(a,b,c) - f(b,c,d) \ .$$

Combining several terms in (4.2) gives us

$$F = \frac{b}{a+c} - \frac{c}{b+d} \ ,$$

which is clearly in the proper form and we conclude that this F is supraconvergent.

In the second example, we wish to distinguish between those schemes which are supraconvergent and those which are not. The differential equation is

$$y''' = f(x) \ ,$$

and we approximate it using the following class of difference methods:

$$\frac{1}{\frac{1}{3}(\Delta_{i+2} + \Delta_{i+1} + \Delta_i)} \left[D_h^2 \, u_{i+1} - D_h^2 \, u_i\right]$$

$$= \frac{1}{\alpha+\beta+\gamma+\delta} \left[\alpha \, f_{i+2} + \beta \, f_{i+1} + \gamma \, f_i \, + \delta \, f_{i-1}\right] .$$

The problem is to decide for what values of α, β, γ, and δ is the difference scheme supraconvergent. The mesh function is (ignoring constants which do not matter)

$$F = (c^2 - a^2) + (a+b+c)\left[-\alpha a + \frac{1}{2}(\delta+\gamma-\beta-\alpha)b + \delta c\right] . \tag{4.3}$$

To prove supraconvergence, we have to show (1) that $F(a,a,a) = 0$ and (2) that $H(a,b,c)$ is independent of c. The first condition requires that

$$\frac{3}{2}(\alpha-\delta) + \frac{1}{2}(\beta-\gamma) = 0 \ .$$

Recalling the definition of H, we have

$$H(a,b,c) = c^2 \left(\frac{3}{2}(\delta-\alpha) + \frac{1}{2}(\gamma-\beta)\right) + c \left(a(\delta-\alpha)\right.$$

$$+ \, b(2(\delta-\alpha)+(\gamma-\beta)) + d(3(\delta-\alpha)+(\gamma-\beta))\Big) + \cdots ,$$

The second condition requires that both the quadratic and the linear terms in the above expression be identically zero. All of these requirements can be met if and only if a symmetric average is used in the difference equation,

$$\alpha = \delta \, , \;\; \beta = \gamma \; .$$

This result confirms that (for a third-order equation) the averages assumed in [3] are indeed the most general.

REFERENCES

[1] Harary, F., *Graph Theory*, Addison-Wesley, 1971.

[2] Kirchhoff, G., "Uber die Auflosung der Gleichungen, auf welche man bei der Untersuchung der linearen verteilung galvanischer Strome getuhrt wird," Annalen der Physik and Chemie 72 (1847) 497-508.

[3] Kreiss, H.-O., Manteuffel, T. A., Swartz, B., Wendroff, B., and White, A. B., "Supraconvergent schemes on irregular grids," Los Alamos National Laboratory document LA-UR 83-2818, submitted for publication.

[4] Manteuffel, T. A. and White A. B., Jr., "The numerical solution of second-order boundary value problems on nonuniform meshes," Los Alamos National Laboratory document LA-UR 84-196, submitted for publication.

EDGE-DISJOINT HAMILTONIAN CYCLES

R. J. Faudree
Memphis State University

C. C. Rousseau
Memphis State University

R. H. Schelp
Memphis State University

ABSTRACT

There are two well known conditions due to Ore which insure that a graph contains a Hamiltonian cycle. Generalizations of each of these conditions are shown to insure edge-disjoint Hamiltonian cycles. In particular, it is proved that if the sum of the degrees of any pair of nonadjacent vertices in a graph with n vertices is at least $n + 2k - 2$, then for n sufficiently large the graph has k edge-disjoint Hamiltonian cycles. Also, it is shown that a graph with n vertices and $\binom{n-1}{2} + 2k$ edges has k edge-disjoint Hamiltonian cycles when $n \geq 6k$. The later condition also insures that there are k edge-disjoint cycles of each length greater than or equal to three.

1. Introduction.

All graphs will be finite with no loops or multiple edges. As usual, the number of vertices and the number of edges in the graph will called the order and size of the graph respectively. A cycle (path) containing all of the vertices of the graph is called a Hamiltonian cycle (path), and a graph containing a Hamiltonian cycle is said to be Hamiltonian.

There have been numerous papers giving degree and edge conditions on graphs which are sufficient for the graph to have a Hamiltonian cycle (see for example [2] - [11]). Two of the first of this kind (one an edge condition and the other a degree condition) are due to O. Ore. They are stated below.

Theorem A: (Ore [8]) Let G be a graph of order $n \geq 3$. If the sum of the degrees of any pair of nonadjacent vertices is at least n, then G is Hamiltonian.

The following result is an immediate corollary of Theorem A.

Theorem B: (Ore [9]) If G is a graph or order $n \geq 3$ and size at least $\binom{n-1}{2} + 2$, then G is Hamiltonian.

If a graph G has a Hamiltonian cycle H, and the graph G - H (the graph obtained from G by deleting the edges in H) satisfies the condition of Theorem A, then G has two edge-disjoint Hamiltonian cycles. In fact, by using induction and a 'maximal counterexample' proof technique similar to that used in the standard proof of Theorem A (see e.g. [1], p. 137), one can show that if the sum of the degrees of any pair of nonadjacent vertices is at least $n + 4k - 4$ in a graph of order n, then the graph has k edge-disjoint Hamiltonian cycles. Note that this result does not follow immediately from a simple induction argument, since the deletion of edges changes the set of pairs of nonadjacent vertices. Also, the above condition is not a sharp sufficient condition to insure k edge-disjoint Hamiltonian cycles, as we shall see later.

We will prove the following theorems, which are generalizations of the above theorems of Ore. The degree and edge conditions in the following theorems are sharp, just as the conditions in Theorem A and Theorem B are sharp. However, the lower bounds for n are not sharp and are chosen for ease of calculation.

<u>Theorem 1</u>: Let G be a graph of order $n \geq 3$, and k a positive integer. If the sum of the degrees of any pair of nonadjacent vertices is at least $n + 2k - 2$, then for n sufficiently large ($n \geq 60k^2$ will suffice), G has k edge-disjoint Hamiltonian cycles.

<u>Theorem 2</u>: Let k be a positive integer. If G is a graph of order $n \geq 6k$, and has size at least $\binom{n-1}{2} + 2k$, then G has k edge-disjoint Hamiltonian cycles.

Consider the graph obtained from a complete graph of order n by deleting $n - 2k$ edges incident to one vertex. Since this graph has a vertex of degree $2k - 1$, it certainly does not contain k edge-disjoint Hamiltonian cycles. It verifies that the conditions in Theorem 1 and in Theorem 2 cannot be weakened.

A even ordered graph having k edge-disjoint Hamiltonian cycles also has $2k$ edge disjoint perfect matchings. The following result, which confirms a conjecture of S. Win for n sufficiently large (see [12]), is a direct consequence of Theorem 1 when s is odd. Also, the proof of Theorem 1 can be used to verify the result when s is even.

Theorem 3: Let s be a fixed non-negative integer and G a graph of even order n. If the sum of the degrees of any pair of nonadjacent vertices of G is $n + s - 1$ and n is sufficiently large, then G has $s + 1$ edge-disjoint perfect matchings.

Many conditions which imply that a graph is Hamiltonian, also imply that the graph is pancyclic (contains cycles of every possible length). In particular, the edge condition in Theorem B does this. The following result is a similar strengthening of Theorem 2.

Theorem 4: Let k be a fixed positive integer, and G a graph of order n and size at least $\binom{n-1}{2} + 2k$. If $n \geq 6k^2$, then G has k edge-disjoint cycles of any length from 3 to n.

2. Proofs

Before giving the proofs of these theorems, some additional notation will be needed. The vertex set of a graph G will be denoted by $V(G)$ and the edge set by $E(G)$. The edge determined by vertices x and y will be written as xy. However, a *path* and *cycle* with vertices $\{x_1,x_2,\ldots,x_n\}$ will be denoted by $(x_1,x_2,\ldots,x_n)$ and $(x_1,x_2,\ldots,x_n,x_1)$ respectively. The degree of a vertex x in a graph G will be denoted by $d_G(x)$. Also, if S is a subset of the vertices of a graph G, then the degree of x relative to S will be denoted by $d_S(x)$, and the subgraph of G induced by the set S will be denoted by $\langle S\rangle_G$. Any additional notation will be defined as needed, and notation not specifically mentioned will follow [1].

An additional result, which we state in the form that is most convenient for our use, will be needed. The case $k = 1$ of the following result merely states that the graph is *Hamiltonian connected* (there is a Hamiltonian path between each pair of vertices of the graph). The degree condition in this theorem is not sharp except when $k = 1$, but it is sufficient for our purposes.

Theorem C: (Ore [10]) Let k be a positive integer and G a graph of order n such that the sum of the degrees of any pair of vertices of G is at least $n + 4k - 3$. Then for any collection of k pairs of vertices of G, there exist k edge-disjoint Hamiltonian paths, one path between each pair of vertices.

Proof: (of Theorem 1) We will assume that the conclusion of Theorem 1 does not hold, and show that this leads to a contradiction. Let G be a counterexample with a maximal number of edges. Therefore, for any pair of nonadjacent vertices u and v of G, $G + uv$ contains k edge-disjoint Hamiltonian cycles.

Associated with each edge $uv \notin E(G)$, there are $k - 1$ edge-disjoint Hamiltonian cycles $H_1, \ldots, H_{k-1}$. Let H denote the subgraph of G generated by these cycles. Also, the graph $L = G - H$ has a Hamiltonian path $P = (u = x_1, x_2, \ldots, x_n = v)$, but no Hamiltonian cycle.

The remainder of the proof will be basically a series of lemmas, each assuming the results of the previous lemmas. Also, throughout the lemmas, we will assume that associated with a pair of nonadjacent vertices u and v, there are $k - 1$ Hamiltonian cycles and a Hamiltonian path from u to v. Although these are not necessarily unique, we will fix one such collection and use the notation of the previous paragraph to denote the Hamiltonian cycles and path. When it is clear which graphs G and L are being considered, we will shorten the notation for the degree of a vertex and express d_G by just d and d_L by d'. Thus, $d'(x) = d(x) - 2k + 2$ for all x in $V(G)$.

Lemma 5: For nonadjacent vertices u and v of G,

$$d(u) + d(v) \leq n + 4k - 5.$$

Proof: Since $d' = d - 2k + 2$, it is sufficient to show that

$$d'(u) + d'(v) \leq n - 1.$$

Note that if $ux_i \in E(L)$, then $vx_{i-1} \notin E(L)$, for otherwise there would be a Hamiltonian cycle in L. Thus $d'(v) \leq n - 1 - d'(u)$, and the lemma follows.

Lemma 6: If $d(x) \leq n/2$ for any $x \in V(G)$, then the vertices of G nonadjacent to x form a complete subgraph of G.

Proof: Let N be the vertices of G which are nonadjacent to x, and let y and z be two vertices in N. By assumption

$$d(y), d(z) \geq n + 2k - 2 - d(x).$$

Hence, $d(y) + d(z) \geq n + 4k - 4$, and Lemma 5 implies $yz \in E(G)$.

Lemma 7: For all $x \in V(G)$, $d(x) \geq 8k$.

Proof: Assume $d(x) = a < 8k$. Let A be the set of vertices adjacent to x and B the set of vertices nonadjacent to x in G, and let A' denote the set of vertices of A which are adjacent to each vertex of B. Denote the number of vertices in A' and B by a' and b respectively.

Claim: $a' \geq 2k$

Each vertex of B has degree at least $n + 2k - 2 - a$, and hence, by Lemma 5, each vertex of A of degree at least $a + 2k - 2$

is in A'. Therefore, if m is the number of edges in G between A and B, then m satisfies the inequalities

$$a'b + (a - a')(a + 2k - 3) \geq m \geq 2kb.$$

Even in the extreme case when $a' = 2k - 1$ and $a = 8k - 1$, this implies $n < 60k^2$, a contradiction.

By Lemma 5, the graph $\langle A' \rangle_G$ is a complete graph. Thus all of the vertices of $A' \cup B$ have degree at least $n + 2k - 2 - a$. Let $u = x$ and $v \in B$, and consider the Hamiltonian path P in L from u to v. Since $a' + b \geq n + 2k - 1 - a$, there is an x_i in P such that ux_i is an edge of L and x_{i-1} is in $A' \cup B$. If $x_{i-1}x_j \in E(L)$ for $i < j$, then $vx_{j-1} \notin E(L)$, for otherwise there would be a Hamiltonian cycle in L. Likewise, if $x_{i-1}x_j \in E(L)$ for $i > j$, then $vx_{j+1} \notin E(L)$. Hence $d'(x_{i-1}) + d'(v) \leq n$. This contradicts the fact that x_{i-1} and v are in $A' \cup B$, and completes the proof of Lemma 7.

Lemma 8: If ux_j and vx_i are both in $E(L)$ for $j > i$, then L contains a cycle of length at least $n - 2k + 2$.

Proof: Assume that L contains no such cycle, and that i and j are chosen such that $j - i$ is minimal. Therefore, ux_m and vx_m are not in $E(L)$ for $i < m < j$. Also, $j - i \geq 2k$, for otherwise L would contain the desired length cycle. For the same reason, ux_m in $E(L)$ implies vx_{m+1} is not in $E(L)$. Hence

$$d'(v) \leq n - 1 - d'(u) - (2k - 2).$$

This implies $d(u) + d(v) \leq n + 2k - 3$, a contradiction which completes the proof of Lemma 8.

Lemma 9: There is a cycle of length at least $n - 2k + 2$ in L.

Proof: Assume that L contains no such cycle. Therefore by Lemma 8 all of the adjacencies of u in L precede all of the adjacencies of v in L along the path P. Let r be the maximum integer such that ux_r is in $E(L)$ and s the minimum integer such that vx_s is in $E(L)$. We can assume that u and v have been chosen so that r is maximal and s is minimal over all possible choices of nonadjacent vertices u and v. Note that, since $d'(u) + d'(v) \geq n - 2k + 2$, $r \leq d'(u) + 2k - 2$ and $s \geq n - d'(v) - 2k + 3$.

Let R be the vertices of P which precede x_r, $\bar{R} = R \cup \{x_r\}$, S the vertices of P which succeed x_s, and $\bar{S} = S \cup \{x_s\}$. Let R' be the vertices x of R for which there is a Hamiltonian path in $\langle \bar{R} \rangle_L$ from x to x_r. For example, if ux_{i+1} is in $E(L)$, then x_i is in R'. Thus $|R'| \geq |R| - 2k + 2$, and any vertex in R' can play the role of u. Let S' be the corresponding vertices of S.

Claim 1: For x_a in R and x_b in S, $x_a x_b \notin E(L)$.

Assume that $x_a x_b$ is in $E(L)$. Select i minimal such that $i > a$ and ux_i is in $E(L)$, and j maximal such that $j < b$ and vx_j is in $E(L)$. Then the cycle

$$C = (x_1, x_i, x_{i+1}, \ldots, x_j, x_n, x_{n-1}, \ldots, x_b, x_a, x_{a-1}, \ldots, x_1)$$

uses all of the vertices of P except for those between x_a and x_i and between x_j and x_b. Since $d'(u) + d'(v) \geq n - 2k + 2$, at most $2k - 2$ vertices are avoided by C. This is a contradiction which proves the claim.

With no loss of generality, we can assume that $|R| \geq |S|$. The following claim states that neither R nor S can be "small".

<u>Claim 2</u>: $|S| \geq (n - 2k + 2)/(2k - 1)$.

If x is in R and $xv \notin E(H)$, then $d(x) + d'(v) \geq n$. Therefore there is an edge from H between S and any vertex x of R. The number of edges in H between R and S is a least $|R|$, but on the other hand at most $(2k - 2)|S|$ edges of H emanate from S. The inequalities $|R| \leq (2k - 2)|S|$ and $|R| + |S| \geq n - 2k + 2$ give the claim.

<u>Claim 3</u>: If x is in R, then $d_R(x) \geq r + 5 - 4k$.

If x is in S, then $d_S(x) \geq n - s + 6 - 4k$.

Each vertex in R has at most $2k - 2$ adjacencies in S, so there is a y in S such that xy and uy are not in $E(G)$. Thus we have the inequalities

$$d'(x) + d'(y) \geq n - 2k + 2 \quad \text{and} \quad d'(u) + d'(y) \leq n - 1.$$

This implies $d'(x) \geq d'(u) - 2k + 3 \geq r - 4k + 5$. The same argument gives the corresponding result for x in S.

<u>Claim 4</u>: $R' = R$ and $S' = S$.

Since each vertex in R has degree at least $r + 5 - 5k$ in $\langle R\rangle_L$, there is a Hamiltonian path in $\langle R\rangle_L$ between each pair of vertices of R by Theorem C. Thus each vertex in R can play the same role as u and $R' = R$. Likewise, $S' = S$.

Claim 5: Either $s = r$ or $s = r + 1$.

Assume $r < j < s$, and $x_i x_j$ is in $E(L)$ for some $i < r$. The maximality of r would be contradicted unless $x_i v$ is in $E(H)$. Hence, x_j is adjacent in L to at most $2k - 2$ vertices of R. The same is true for S, so x_j clearly has degree less than 8k. This contradicts Lemma 7 and proves the claim.

We are now prepared to complete the proof of Lemma 5. First consider the case when $s = r + 1$. The vertex x_r is adjacent in L to at most $2k - 2$ vertices of S, for otherwise v could be replaced by a vertex which would contradict the minimality of s. The proof of Claim 3 then gives that $d_R(x_r) \geq r + 7 - 6k$. By the same fashion, $d_S(x_s) \geq n - s + 8 - 6k$. If $s = r$, then either $d_R(x_r) \geq 8k - 8$ or $d_S(x_r) \geq 8k - 8$. If this were not true, then clearly x_r has degree less than $n/2$ in G. Since the vertices of G not adjacent to x_r form a complete graph, this would imply that there are edges in L between R and S, a contradiction.

Let $X = \bar{R}$ if $s = r + 1$ or $s = r$ and $d_R(x_r) \geq 8k - 8$, and let $X = R$ otherwise. Let Y be the complementary set $V(G) - X$. Both X and Y satisfy the hypothesis of Theorem C. Therefore for any set of k pairs of vertices of X (or Y), there exists edge-disjoint Hamiltonian paths in $\langle X\rangle_L$ (or $\langle Y\rangle_L$) between these k pairs of vertices.

We have already shown that each vertex of R is adjacent in H to at least one vertex of S. Since R has at least $4k^2$ vertices, there are 2k independent edges of H between R and S. Using these 2k edges and appropriate k Hamiltonian paths in each of X and Y, we can construct k edge-disjoint Hamiltonian paths in G. This gives a contradiction which completes the proof of Lemma 9.

We can now assume that L contains a maximal length cycle

$$C = (y_1, y_2, \ldots, y_{n-t}, y_1)$$

with $n - t$ vertices for $t \leq 2k - 2$. We will denote the remaining vertices of L by $Z = \{z_1, z_2, \ldots, z_t\}$. If $t = 0$, the theorem is proved, so assume $t \geq 1$.

<u>Lemma 10</u>: For $i = 1, 2, \ldots, t$, $d(z_i) > n/2$.

<u>Proof</u>: Assume $d(z_1) \leq n/2$. Since by Lemma 7 $d(z_1) \geq 8k$, z_1 is adjacent in L to at least 4k vertices of C. The maximality of the length of C implies that no two of these adjacencies are consecutive on C. Since z_1 has degree $2k - 2$ in H, z_1 is adjacent in L to at least 2k vertices of C (which we will denote by Y) and simultaneously nonadjacent in G to the successors along C of the vertices in Y. Denote the successors by Y'. The vertices in Y' form a complete graph in G by Lemma 6. Therefore there are y_i and y_j in Y' such that $y_{i+1}y_{j+1}$ is in E(L). The cycle C can be extended to a cycle of length $n - t + 1$ using the edges z_1y_i, z_1y_j, and $y_{i+1}y_{j+1}$, which gives a contradiction.

<u>Lemma 11</u>: $t = 1$ and $d(z_1) \geq (n/2) + k - 1$.

<u>Proof</u>: Assume $t \geq 2$. Let Y_i be the vertices of C which are adjacent in L to z_i and Y'_i the successors along C of the vertices in Y_i for $i = 1, 2$. Each of Y_1 and Y_2 has at least $(n/2) - t - 2k + 3$ vertices since $d(z_i) > n/2$. The maximality of the length of C implies $Y_i \cap Y'_i = \emptyset$ and $|Y'_i \cap Y_j| \leq 1$ for $\{i,j\} = \{1,2\}$.

Note that if y_s and y_{s+2} are in Y_1, then z_1 and y_{s+1} could be interchanged and there will be a cycle with $n - t$ vertices not containing y_{s+1}. Since Y_1 contains nearly half of the vertices of C, there will be at least one s such that $y_{s+1}z_2$ is not in $E(G)$. Hence, we can assume that z_1z_2 is not in $E(G)$.

Due to the restrictions on Y_1 and Y_2, $Y = Y_1 \cap Y_2$ has at least $|Y_1| - (n - t + 1 - 2|Y_2|) \geq (n/2) - 10k + 12$ vertices. Let $u = z_1$ and $v = z_2$, and consider the associated graphs H and L, and the Hamiltonian path P in L form u to v. If two vertices of Y are within a distance 2 of each other along the path P, there would be a cycle of length at least $n - 1$. This cannot occur, so we must have $3|Y| \leq n$. This gives a contradiction which verifies that $t = 1$.

Recall that z_1 can be interchanged with any vertex y_{s+1} if both z_1y_s and z_1y_{s+2} are in $E(L)$. We can select an s such that z_1y_{s+1} is not in $E(G)$. Since $d(z_1) + d(y_{s+1}) \geq n + 2k - 2$, there is no loss of generality in assuming that $d(z_1) \geq (n/2) + k - 1$.

The following lemma will complete the proof of Theorem 1.

<u>Lemma 12</u>: $t = 0$.

<u>Proof</u>: Let Y be the vertices of C which are adjacent in L to z_1, and let Y' contain z_1 and the successors of Y along the cycle C. Thus Y and Y' are disjoint sets with at least $(n/2) - k + 1$ and $(n/2) - k + 2$ vertices respectively. Partition Y' into two sets Y_1 and Y_2, where Y_1 is the union of $\{z_1\}$ and the set of vertices of Y' which are both predecessors and successors along C of vertices of Y. It is easy to verify that Y_1 has at least $(n/2) - 3k + 4$ vertices.

Note that there are no edges of L between any pair of vertices of Y', because any such edge would give a Hamiltonian cycle in L. Each vertex of Y_1 can be interchanged with z_1, and all but possibly $2k - 1$ of the vertices of Y' have high degree at least $(n/2) + k - 1$ in G. Each of these high degree vertices of Y' are incident to at least one edge of H in Y', so there is at least $((n/2) - 3k + 3)/2$ edges of H in Y'. Therefore, we can assume that Y' contains at least $((n/2) - 3k + 3)/(2(k - 1)) > n/4k$ edges (and hence $n/8k$ independent edges) from H_1.

Select an edge y_iy_j of H_1. In fact, we can assume that y_i and y_j both have high degree and are in Y_1. Then,

$$H'_1 = (z_1, y_{j-1}, \ldots, y_i, y_j, y_{j+1}, \ldots, y_{i-1}, z_1)$$

is a Hamiltonian cycle using one edge of H_1 and $n - 1$ edges of

L. Consider the $k - 1$ Hamiltonian cycles $H'_1, H_2, \ldots, H_{k-1}$ which generate the graph H' and the corresponding graph $L' = G - H'$. Since y_i and y_j are the endvertices of a Hamiltonian path P' in L' and they are commonly adjacent to nearly half of the vertices of P', the same counting argument used in Lemma 11 implies that L' contains a cycle of length $n - 1$.

There is a set X' with at least $(n/2) - k + 2$ vertices which is independent in L'. This corresponds to the set Y' which was independent in L. If X' and Y' are disjoint, then there are at least $2(|X'| - 1) + n/4k > n$ edges in H_1. Thus we can select a vertex x in $X' \cap Y'$ and a vertex y in Y' such that xy is not in $E(G)$. Since Y' contains at least $n/8k$ independent edges from H_1, there is at least $(n/8k) - 1$ vertices in Y' which are not in X'. Therefore,

$$d(x) + d(y) \leq 2n - |X'| - |Y'| - ((n/8k) - 1) + 2(2k - 1) < n.$$

This contradiction completes the proof of Lemma 12 and of Theorem 1.

We will not give the details of the proof of Theorem 3. For $s = 0$ the theorem is a consequence of the fact that G has a Hamiltonian path. As mentioned in the introduction, the case when s is odd, follows immediately from Theorem 1. When s is even, the same techniques used in the proof of Theorem 1 will verify that G has a perfect matching and $s/2$ edge-disjoint Hamiltonian cycles.

Theorem 2 is a consequence of Theorem 1, except that the lower bound on n is less restrictive. The proof is short, so we give the details.

Proof: (of Theorem 2) Assume the theorem is false and that G is a counterexample with a maximal number of edges. Therefore the addition of any edge to G will generate k edge-disjoint Hamiltonian cycles. If we use the notation used in the proof of Theorem 1, then Lemma 5 implies that $d_G(u) + d_G(v) \leq n + 4k - 5$ for any pair of nonadjacent vertices u and v in G. Thus $\overline{G}$, the complement of G, has at least $n - 4k + 2$ edges incident to at least one of any pair of nonadjacent vertices of G.

First consider the case when G has two independent edges. This implies $\overline{G}$ has at least $2(n - 4k + 2) - 4 = 2n - 8k$ edges. Since $\overline{G}$ has at most $n - 2k - 1$ edges, this implies $n < 6k$, a contradiction. Hence G has either three edges which form a triangle or at most $n - 2k - 1$ edges which form a star. It is well known that the complete graph K_n can be factored into $(n - 1)/2$ Hamiltonian cycles when n is odd and into $(n - 2)/2$ Hamiltonian cycles and a perfect matching when n is even. A triangle can be placed in two of the Hamiltonian cycles, and a star with m edges can be placed in $m/2$ Hamiltonian cycles when m is even and $(m - 1)/2$ Hamiltonian cycles and a matching when m is odd. In either situation, there would be at least k of the Hamiltonian cycles in the factorization which are edge-disjoint from the star or the triangle. This completes the proof of Theorem 2.

Proof: (of Theorem 4) We know that G has k edge-disjoint Hamiltonian cycles from Theorem 2. An induction on the length and number of the shorter cycles will be used. We start by showing that G has k edge-disjoint triangles.

Observe that G has a vertex of degree at least $n - 2$, for otherwise there would be at most $n(n-3)/2 < 2k + (n-1)(n-2)/2$ edges in G. Let N be the neighborhood of a vertex of degree at least $n-2$. If there are k independent edges in N, then there will be k edge-disjoint triangles. If this is not true, then there are at least $\binom{n-2k}{2}$ edges of $\overline{G}$ in N. However, $\binom{n-2k}{2} \leq n - 2k - 2$ implies that $n \leq 2k + 1$, a contradiction. Thus, G has k edge-disjoint triangles.

Assume that for some i $(3 \leq i < n-1)$ and for some $t < k$, that G contains edge-disjoint cycles

$$L_1, L_2, \ldots, L_t, M_{t+1}, \ldots, M_k$$

where each cycle L has length $i + 1$ and each M has length i. We also assume that none of the cycles M can be extended to a cycle of length $i + 1$ which is edge-disjoint from the remaining cycles. Let H be the subgraph of G generated by the $k - 1$ cycles above excluding M_{t+1}. Thus no vertex of H has degree more than $2(k-1)$.

We will first consider the case when $i < n - 2$. If x is a vertex not in $V(M_{t+1})$, then x is not adjacent in $G - H$ to two consecutive vertices of M_{t+1}. Hence there are at most $(n-i)i/2$ edges of $G - H$ and at most $2(k-1)\min\{n-i, i\}$ edges of H between $V(M_{t+1})$ and the remaining vertices of G. Therefore, $\overline{G}$ has at least $(n-i)i/2 - 2(k-1)\cdot\min\{n-i, i\}$ edges, and it is straightforward to confirm that this number is greater than $n - 2k - 1$. This contradiction confirms that we need only consider the case $i = n - 2$.

Let x_1 and x_2 be the vertices of G not in M_{t+1}, N_i the vertices of M_{t+1} adjacent to x_i, and N'_i the successors of N_i along the cycle M_{t+1} for $i = 1$ or 2. There are no edges of $G - H$ between vertices of $N'_i \cup \{x_i\}$ for $i = 1$ or 2, for this would imply the cycle could be extended to a cycle of length $n - 1$. If n_i is the number of vertices in N'_i for $i = 1$ or 2, then $\overline{G}$ has at least

$$(n - n_1 - 2) + (n - n_2 - 2) + \binom{n_1}{2} + \binom{n_2}{2} - (n_1 + n_2 + 2)(2k - 2)$$

edges. On the other hand $\overline{G}$ has at most $n - 2k - 1$ edges, which is impossible for $n \geq 6k^2$. This completes the proof of Theorem 4.

REFERENCES

[1] M. Behzad, G. Chratrand, and L. Lesniak-Foster, Graphs and Digraphs, Prindle, Weber and Schmidt, (1979).

[2] J. A. Bondy, Properties of Graphs with Constraints on Degrees, Studia Ac. Math. Hung., 4 (1969), 473-475.

[3] J. A. Bondy, Cycles in Graphs, Combinatorial Structure and Applications, Gordon and Breach, New York, 1970, 15-18.

[4] V. Chvátal, On Hamiltonian Ideals, J. Comb. Theory, 12(B), (1972), 163-168.

[5] G. A. Dirac, Some Theorems on Abstract Graphs, Proc. London Math. Soc., 2 (1952), 69-81.

[6] H. V. Kronk, A Note on k-path Hamiltonian Graphs, J. Comb. Theory, 7, (1969), 104-106.

[7] M. Las Vergnas, Sur l'Existence des Cycles Hamiltonian dans un Graphe, C. R. Acad. Sc. Paris, 270, (1960), A-1361-1364.

[8] O. Ore, Note on Hamiltonian Circuits, Amer. Math. Monthly, 67, (1960), 55-56.

[9] O. Ore, Arc Coverings of Graphs, Ann. Mat. Pura. Appl., 55, (1961), 315-321.

[10] O. Ore, Hamiltonian Connected Graphs, J. de Math. Pures et Appl., 42, (1963), 21-27.

[11] L. Posa, A Theorem Concerning Hamilton Lines, Magyar Tud. Akad. Mat. Kutato Int. Kozl., 7 (1962), 225-226.

[12] S. Win, A Sufficient Condition for a Graph to Contain Three Disjoint 1-Factors, J. Graph Theory, 6, (1982), 489-492.

STUDIES RELATED TO THE RAMSEY NUMBER $r(K_5-e)$

R. J. Faudree
Memphis State University

C. C. Rousseau
Memphis State University

R. H. Schelp
Memphis State University

ABSTRACT

As part of an effort to determine $r(K_5-e)$, we find all critical colorings for the pair (K_4-e, K_5-e).

1. Introduction

Modifying a slogan of the U.S. Army Service Forces in World War II, Erdős and Ulam have remarked "the infinite we do immediately, the finite takes a little longer." The Erdős-Ulam comment is nowhere more appropriate than in Ramsey theory. In fact, in the realm of finite Ramsey theory there is some justification for the slogan "the large we do immediately, the small takes a little longer." For, however much progress is made in understanding Ramsey problems involving sparse graphs and however many theorems have been proved to hold when "n is sufficiently large," relatively small examples are sharp reminders of the limitations of our knowledge and technique. Success in this area will doubtless require the best mixture of combinatorial technique and computing power. There are not likely to be many shortcuts.

The Ramsey number $r(K_5-e)$ is a case in point. Further progress toward determining this Ramsey number apparently requires a knowledge of all critical colorings for the pair (K_4-e, K_5-e). In [2], Clancy found that $r(K_4-e, K_5-e) = 13$. Thus we seek all graphs of order 12 which contain no K_4-e and the complements of which contain no K_5-e.

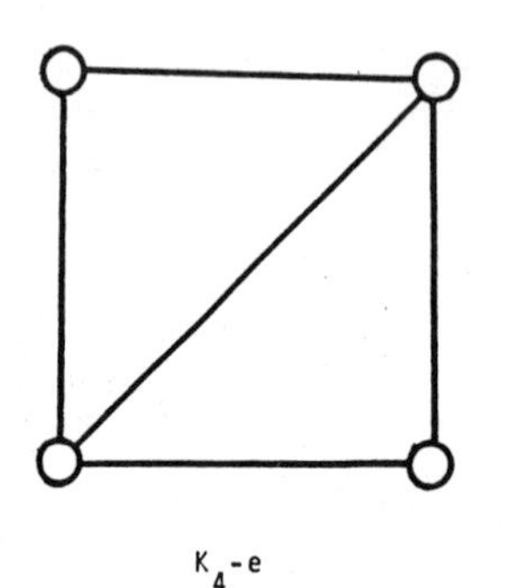

K_5-e

Figure 1.

As has become customary, we shall formulate the Ramsey problem in terms of a partition or coloring (R,B) = (red, blue) of the edges of a complete graph. The red and blue subgraphs so induced will be denoted $\langle R \rangle$ and $\langle B \rangle$ respectively. The Ramsey number $r(G,H)$ is the smallest p such that in every such coloring of the edges of K_p there is either a red copy of G or else a blue copy of H. A coloring of the edges of the complete graph of order $m < r(G,H)$ in which there is neither a red G nor a blue H is called (G,H)-good. In the case of $m = r(G,H)-1$ such a good coloring is called critical. If (R,B) is a coloring of the edges of the complete graph on the vertex set V and $X \subset V$, the red and blue subgraphs induced on X will be denoted $\langle X \rangle_R$ and $\langle X \rangle_B$ respectively. In the way of general graph theoretic terminology and notation, we shall follow the textbook of Behzad, Chartrand and Lesniak-Foster [1].

2. (K_4-e)-Good Colorings

As a preliminary to the study of the critical colorings for the pair (K_4-e, K_5-e), we need to know some things about good colorings for the diagonal case (K_4-e). It is well-known that $r(K_4-e) = 10$ and that $\langle R \rangle \simeq \langle B \rangle \simeq L(K_{3,3})$ is a critical coloring. The graph $L(K_{3,3})$ (equivalently, the Paley graph of order nine) is self-complementary and strongly regular. The following lemma summarizes the needed information concerning (K_4-e)-good colorings.

Lemma. For $n = 7$, 8 or 9 let (R,B) be a coloring of the edges if K_n which is $(K_4\text{-}e)$-good. Then

$$\begin{aligned}
&\text{(a) } n = 9 \Rightarrow \langle R\rangle \simeq L(K_{3,3}),\\
&\text{(b) } n = 8 \Rightarrow \langle R\rangle \simeq G_0, \text{ and}\\
&\text{(c) } n = 7 \Rightarrow \langle R\rangle \simeq G_1, G_2, G_3 \text{ or } G_4
\end{aligned}$$

where $L(K_{3,3})$ and $G_0 - G_4$ are the graphs shown in Figure 2.

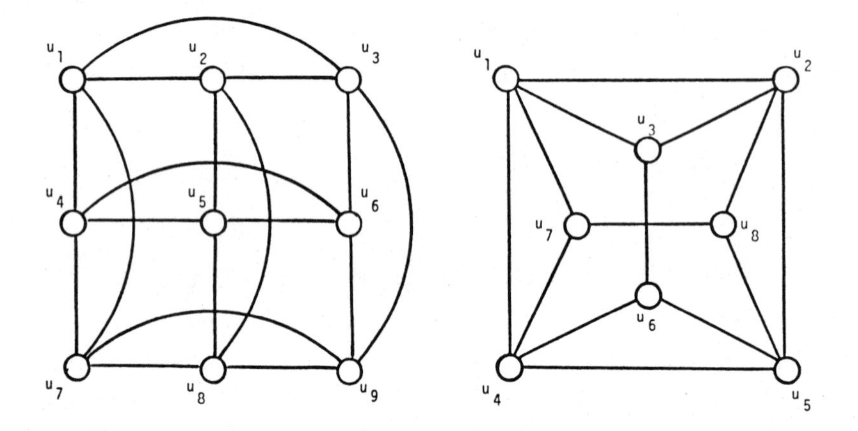

(a) $L(K_{3,3})$ (b) G_0

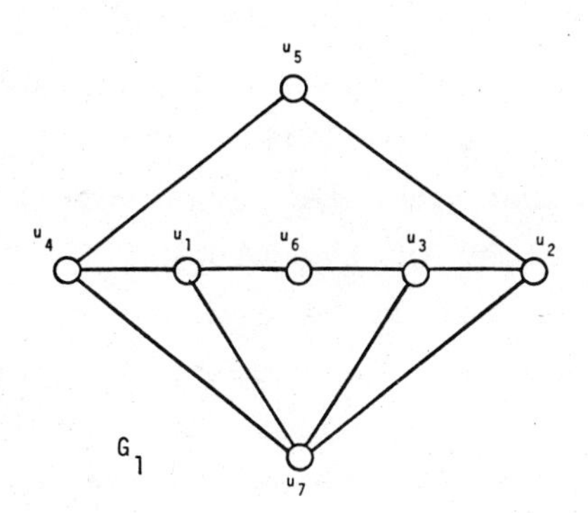
u5
u4
u1
u6
u3
u2
G1
u7

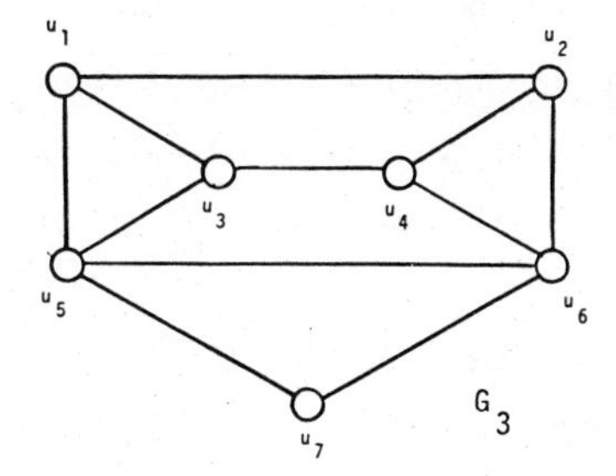
u1
u2
u3
u4
u5
u6
u7
G3

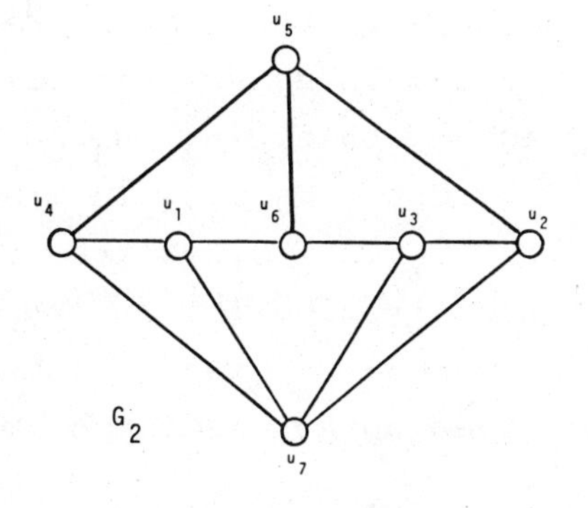
u5
u4
u1
u6
u3
u2
G2
u7

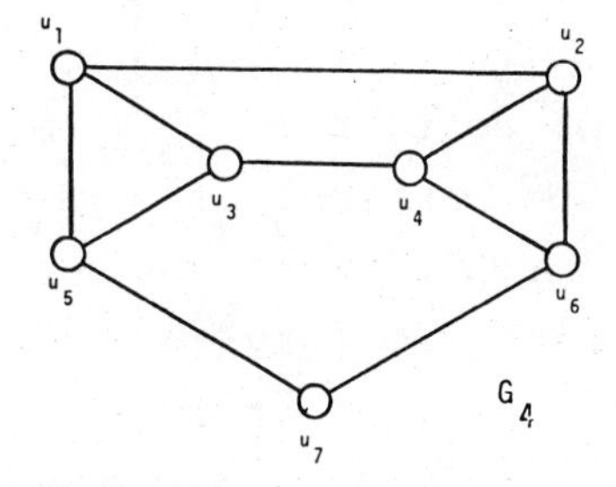
u1
u2
u3
u4
u5
u6
G4
u7

(c)

Figure 2.

Proof. The following fact concerning (K_4-e)-good colorings is useful. There does not exist a monochromatic C_3 and, vertex disjointly, a C_m, $m<6$, in the opposite color. For, suppose that there is a red C_3 and, disjointly, a blue C_m for $m = 3$, 4 or 5. Since there is no red K_4-e, there are at least $2m$ blue edges from the blue C_m to the red C_3. Hence one of the three vertices must be adjacent in $\langle B\rangle$ to at least $\lceil 2m/3 \rceil$ vertices of the blue C_m. It then follows that a blue K_4-e cannot be avoided.

(a) The uniqueness of the critical coloring for K_4-e is known [4]. However, the proof of this result serves as a good illustration of the usefulness of the fact cited above. Without loss of generality, we may assume that $\{u_1, u_2, u_3\}$ are the vertices of a red C_3. Since the blue graph on the remaining six vertices cannot contain a C_3, we may assume that $\{u_4, u_5, u_6\}$ are the vertices of a second red C_3. Keeping in mind the prohibition against a blue C_m ($m = 3$, 4 or 5) disjoint from any red C_3, we readily find (i) $\{u_7, u_8, u_9\}$ are the vertices of a third red C_3 and (ii) the remaining red edges are determined up to isomorphism. We thus obtain $\langle R\rangle \simeq L(K_{3,3})$.

(b) As in the preceding argument, we may assume that $\{u_1, u_2, u_3\}$ are the vertices of a red C_3. Since the blue graph on the remaining five vertices contains neither C_3 nor C_5, again we may assume that $\{u_4, u_5, u_6\}$ are the vertices of a second red C_3. The lack of a blue C_m ($m = 3$, 4 or 5) disjoint from either of the red C_3s now can be used to imply that (i) $u_7u_8 \in R$ and (ii) there is no loss of generality in assuming that each of the following edges is red: u_1u_7, u_2u_8, u_4u_7, u_5u_8. The remaining red edges, namely u_1u_4, u_3u_6 and u_2u_5, are then completely determined and so we arrive at $\langle R\rangle \simeq G_0$. It should be noted that G_0 is obtained from $L(K_{3,3})$ by deleting one vertex (u_9 in Figure 2). Also, it is worth pointing out that G_0 provides us with a rarely observed phenomenon. It is not unusual for there to be a unique critical coloring; it is striking that in this case there is a unique good coloring of the complete graph with two fewer vertices than the Ramsey number.

(c) Noting that $G_3 \simeq G_1$ and $G_4 \simeq G_2$, it suffices to show that if a majority of the edges are red then $\langle R\rangle \simeq G_2$ or $\langle R\rangle \simeq G_3$. If $|R| > 10$ then there is a vertex of degree four in $\langle R\rangle$. (It is obvious that there is no vertex of degree five or more.) Since there is no monochromatic K_4-e, there is no choice except for the red graph on the four neighborhood vertices to consist of two independent edges. Neither of the two remaining vertices can be adjacent in $\langle R\rangle$ to more than two of the neighborhood vertices. Since there are at least 11 red edges, this means that the two remaining vertices are joined by a red edge and that each is adjacent to precisely two of the neighborhood vertices. There are two non-isomorphic ways to complete the coloring, and each is successful. We thus obtain $\langle R\rangle \simeq G_2$ or $\langle R\rangle \simeq G_3$ when there are a majority of red edges and $\langle R\rangle \simeq G_1$

or $\langle R\rangle \simeq G_4$ when there are a majority of blue edges.

3. $(K_4\text{-}e,\ K_5\text{-}e)$-Critical Colorings

Theorem. Let (R,B) be a critical coloring for the pair $(K_4\text{-}e,\ K_5\text{-}e)$. Then either $\langle R\rangle \simeq H_1$ or else $\langle R\rangle \simeq H_2$, where H_1 and H_2 are the two non-isomorphic four-regular graphs shown in Figure 3. In case $\langle R\rangle \simeq H_1$ there may be as many as six additional red edges. In case $\langle R\rangle \simeq H_2$ two additional red edges are possible. There is just one case in which $\langle R\rangle$ is five-regular.

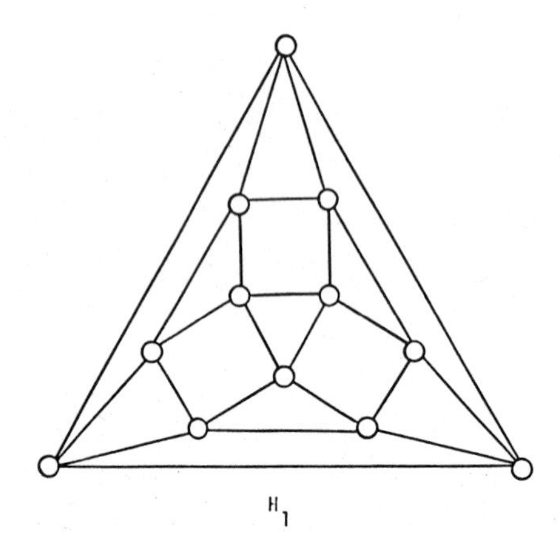

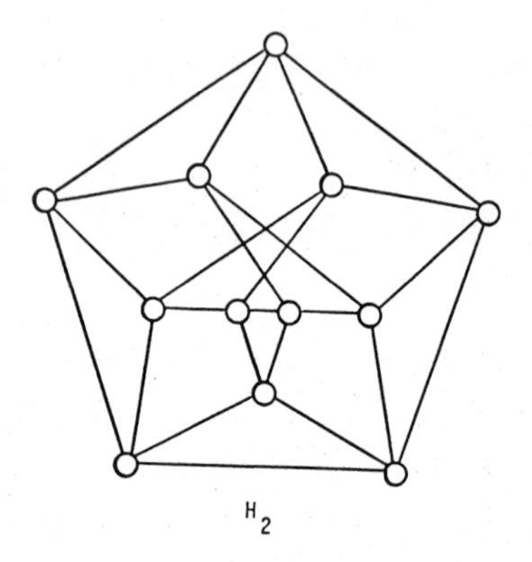

Figure 3.

Proof. The proof is facilitated by identifying an appropriate set of forbidden structures in addition to basically forbidden red K_4-e and blue K_5-e.

For example, it is helpful to note that there can be no red K_3 and blue K_4 which are vertex disjoint. For, denying a red K_4-e and blue K_5-e, we can account for only $(4)(1) + (3)(2) = 10$ of the 12 edges which join the vertices of the K_3 to those of the K_4.

Secondly, there can be no set X of five vertices for which $\langle X \rangle_R \simeq K_{1,4}$. For, consider the center w of such an alleged star and let U and V be its neighborhoods in $\langle B \rangle$ and $\langle R \rangle$ respectively. If $|V| = 4$ then the fact that $r(K_3, k_4-e) = 7$ means that on U we must find either a red K_3 or else a blue K_4-e. In the first case we have a disjoint red K_3 and blue K_4 and in the second there is a blue K_5-e. If $|V| > 4$, then the fact that V already supports a blue K_4 means that successful completion of $\langle V \rangle_R$ is impossible.

No vertex is of degree six or more in $\langle R \rangle$. Since $r(K_{1,2}, K_4) = 7$, it is obvious that in $\langle R \rangle$ no vertex is of degree seven or more, but the case of six requires more detailed consideration. Suppose that w is of degree six in $\langle R \rangle$ and let U and V be its neighborhoods in $\langle B \rangle$ and $\langle R \rangle$ respectively. There is no choice except for $\langle V \rangle_R \simeq 3K_2$. Let V_1, V_2, V_3 be the vertex sets of the three K_2s. Each vertex u U must be adjacent in $\langle B \rangle$ to a vertex in each of the V_is. But then u cannot be adjacent in $\langle B \rangle$ to a second vertex in one of the V_is, for then u and the four vertices of V to which it is adjacent in <B> yield a blue K_5-e. Since $|U| > 4$ a pair of vertices in U are commonly adjacent in <B> to a pair of vertices in V. Without loss of generality, we may assume that u_1 and u_2 are both adjacent in <B> to v_1 and v_3, where the three red edges on V are v_1v_2, v_3v_4 and v_5v_6. By previous remarks, u_1 and u_2 are adjacent in <R> to v_2 and v_4. Finally, if $u_1u_2 \in R$ there is a red K_4-e and if $u_1u_2 \in B$ there is a red K_3 on $\{w, v_5, v_6\}$ and a blue K_4 on $\{u_1, u_2, v_1, v_3\}$. As each of these alternatives is forbidden, so is our vertex of degree six in <R>.

We now begin a process of considering the coloring from the point of view of a minimum degree vertex in <R>. Let w be such a vertex and suppose that its degree is δ. Following the previously used format, its neighborhoods in <R> and <B> will be V and U respectively. A basic constraint is that the induced coloring on U must be $(K_4$-e)-good, so this coloring is available from the Lemma. In particular, we must have $|U| < 10$ and so $\delta > 1$. In the next two arguments, we eliminate the cases $\delta = 2$ and $\delta = 3$.

$\delta = 2$. In view of the Lemma, $\langle U\rangle_R \simeq L(K_{3,3})$. Let $v \in V$ and refer to the drawing of $L(K_{3,3})$ in Figure 2(a). The coloring of the edges u_iv, $i = 1,..,9$, induces (in a trivial way) a vertex coloring of U. The requirements of the coloring are (i) no two red vertices are in the same row or column, and (ii) any three blue vertices must have two in the same row or column. Now any one red vertex implies - by (i) - four blue vertices and these - by (ii) - imply four more red vertices and a contradiction of (i).

$\delta = 3$. In this case $\langle U\rangle_R \simeq G_0$, the graph shown in Figure 2(b). As above, let $v \in V$ and, with reference to Figure 2(b), consider the vertex coloring of U induced by the edge coloring of $u_i v$, $i = 1,..,8$. Note the index triples which give rise to K_3s in $\langle U\rangle_R$ - 123, 456, 147 and 258 - and those which yield K_3s in $\langle U\rangle_B$ - 168, 267, 357 and 348. The requirements for a good coloring are (i) for each K_3 in $\langle U\rangle_R$ at least two vertices are blue, and (ii) for each K_3 in $\langle U\rangle_B$ at least one vertex is red. We now claim that either u_3 and u_6 are red or else u_7 and u_8 are red. Otherwise, we may assume that u_3 is blue. By symmetry and in view of (i), we take u_4 to be blue. By (ii), u_8 must be red. Then (i) u_5 must be blue and this forces u_7 to be red. Thus, u_7 and u_8 have been forced to be red and our claim is justified. Since $|V| = 3$ the fact just established implies a red K_4-e and so eliminates $\delta = 3$ from contention.

$\delta = 4$. As might be expected, the bulk of the proof concerns this case. From the perspective of a vertex w of degree four, $\langle R\rangle$ must be as represented schematically in (a) or (b) of Figure 4, where G_1-G_4 are the graphs shown in Figure 2(c). We shall prove that there are no good colorings of the type depicted in Figure 4(a). Thus the perspective from every vertex of degree four is as represented in Figure 4(b). This alternative leads to the critical colorings described in the Theorem.

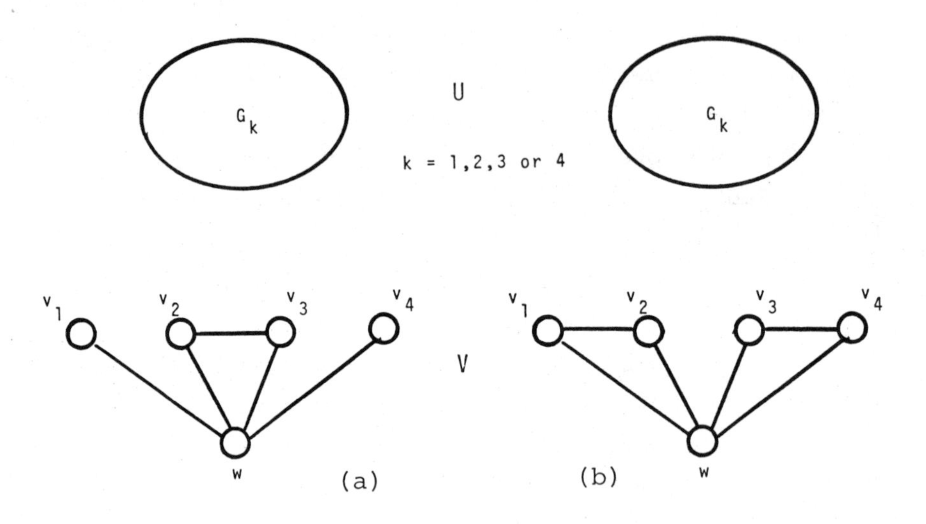

Figure 4.

Although completely straightforward, the negative proofs for the four cases of Figure 4(a) tend to be somewhat tedious. For this reason, it is appropriate to abandon leisurely discussion in favor of a terse representation of the facts. To this end, we introduce five Boolean variables, the truth of each being tied to the corresponding condition shown below. Each of the first four is associated with a forbidden structure and the last reflects the fact that two vertices play symmetric roles (at the appropriate stage in the coloring.)

RED(X):	$\langle X\rangle_R \supset K_4 - e$
BLUE(X):	$\langle X\rangle_B \supset K_5 - e$
STAR(X):	$\langle X\rangle_R \simeq K_{1,4}$
DEG(v):	deg(v) $\neq$ 4 or 5 in $\langle R\rangle$
SYM(u,v):	u and v play symmetric roles

The proofs will be given as blocks of statements, typically of the form

edge $\in$ color - else FORBIDDEN

or

edge $\in$ color - SYM, else FORBIDDEN

where FORBIDDEN is RED, BLUE, STAR or DEG. In words, the latter statement is: edge _______ must be of color ___________ because, in view of the symmetry of _________ and ___________, the contrary choice leads to the forbidden structure ____________ . Having set up the appropriate code, we now proceed with the next four cases, each leading to the conclusion "no graph."

$\langle U\rangle_R = G_1$

$u_7v_i \in R$ for some i - else BLUE(u_7, v_1, v_2, v_3, v_4)

$\langle U \rangle_R = G_1$ (Cont.)

$u_j v_i \in B$ for $j = 1,2,3,4$ - else $RED(u_1, u_4, u_7, v_i)$

$u_5 v_i \in R$ - else $BLUE(u_1, u_3, u_5, v_i, w)$

$u_6 v_i \in R$ - else $BLUE(u_2, u_4, u_6, v_i, w)$

$STAR(u_5, u_6, u_7, v_i, w) \Rightarrow$ no graph

Note that only the last statement made use of the fact that $u_5 u_6 \in B$. Thus, all but the last statement holds equally well for the case $\langle U \rangle_R = G_2$. Now there are two distinguished cases, $i = 1$ and $i = 2$.

$\langle U \rangle_R = G_2$ $i = 1$

$u_j v_4 \in B$ for $j = 1$ or 4 - else $RED(u_1, u_4, u_7, v_4)$

$u_k v_4 \in B$ for $k = 2$ or 3 - else $RED(u_2, u_3, u_7, v_4)$

$u_j v_2 \in R$ and $u_k v_2 \in R$ - else $BLUE(u_j, u_k, v_1, v_2, v_4)$

$u_j v_3 \in R$ and $u_k v_3 \in R$ - else $BLUE(u_j, u_k, v_1, v_3, v_4)$

$RED(u_j, v_2, v_3, w) \Rightarrow$ no graph

$\langle U \rangle_R = G_2$ $i = 2$

$u_7 v_j \in B$ for $j = 1,3,4$ - else $DEG(u_7)$

$\langle U\rangle_R = G_2$ $i = 2$ (Cont.)

$u_5v_3 \in B$ - else RED(u_5,v_2,v_3,w)

$u_6v_3 \in B$ - else RED(u_6,v_2,v_3,w)

u_5v_1, $u_5v_4 \in R$ - else BLUE(u_5,u_7,v_1,v_3,v_4)

u_6v_1, $u_6v_4 \in R$ - else BLUE(u_6,u_7,v_1,v_3,v_4)

RED$(u_5,u_6,v_1,v_4) \Rightarrow$ no graph

$\langle U\rangle_R = G_3$

$u_7v_1 \in R$ - else DEG(v_1), RED(u_1,u_3,u_5,v_1) or

RED(U_2,u_4,u_6,v_1)

$u_7v_4 \in R$ - SYM(v_1,v_4)

u_5v_1, u_6v_1, u_5v_4, $u_6v_4 \in B$ - else RED(u_5,u_6,u_7,v_1)

or RED(u_5,u_6,u_7,v_4)

$u_1v_1 \in R$ - SYM(v_1,v_3), else DEG(v_1) or

RED(u_2,u_4,u_6,v_1)

$u_4v_1 \in B$ - else STAR (u_1,u_4,u_7,v_1,w)

$u_2v_1 \in R$ - else DEG(v_1) or RED(u_1,u_3,u_5,v_1)

$\langle U \rangle_R = G_3$ (Cont.)

u_3v_4, $u_4v_4 \in R$ - $SYM(v_1,v_4)$, else $RED(u_1,u_2,v_1,v_4)$

$u_7v_2 \in B$ - $SYM(v_2,v_3)$, else $RED(u_7,v_2,v_3,w)$

$u_1v_2 \in B$ - $SYM(u_1,u_3)$, else $RED(u_1,u_3,u_5,v_2)$

$u_4v_2 \in R$ - else $BLUE(u_1,u_4,u_7,v_2,w)$

$u_2v_2 \in B$ - else $RED(u_2v_4,u_6,v_2)$

$u_3v_2 \in R$ - else $BLUE(u_2,u_3,u_7,v_2,w)$

$RED(u_3,u_4,v_2,v_4) \Rightarrow$ no graph

The reader should note that certain parts of the last argument apply as well to the case $\langle U \rangle_R = G_4$. In particular, u_7 is adjacent to v_1 and v_4 in $\langle R \rangle$ and we may assume that v_2 is adjacent to u_3 and u_4 in $\langle R \rangle$ and to u_1, u_5, u_2, u_6 and u_7 in $\langle B \rangle$. The following argument starts at this point.

$\langle U \rangle_R = G_4$

$u_5v_1 \in R$ - $SYM(u_5,u_6)$, $SYM(v_1,v_4)$, else

$STAR(u_5,u_6,u_7,v_1,v_4)$

u_1v_1, $u_3v_1 \in B$ - else $RED(u_1,u_3,u_5,v_1)$

$\langle U \rangle_R = G_4$ (Cont.)

$u_6v_1 \in B$ - else $RED(u_5, u_6, u_7, v_1)$

u_1v_4, $u_6v_4 \in R$ - else $BLUE(u_1, u_6, v_1, v_2, v_4)$

u_3v_4, $u_5v_4 \in B$ - else $RED(u_1, u_3, u_5, v_4)$

u_2v_4, $u_4v_4 \in B$ - else $RED(u_2, u_4, u_6, v_4)$

$u_2v_1 \in R$ - else $BLUE(u_2, u_5, v_1, v_2, v_4)$

u_4v_1, $u_6v_1 \in B$ - else $RED(u_2, u_4, u_6, v_1)$

u_3v_3, $u_4v_3 \in B$ - else $RED (u_3, v_2, v_3, w)$ or $RED(u_4v_2, v_3, w)$

$BLUE(u_3, u_4, v_1, v_3, v_4) \Rightarrow$ no graph

We are now ready to consider the four cases as depicted in Figure 4(b). Having found that none of the cases of Figure 4(a) leads to a good coloring, we have the advantage of being able to add one more forbidden structure to our list. We define NOMATCH(w) to be true if $|V| = 4$ and $\langle V \rangle_R \neq 2K_2$, where V is the neighborhood of w in $\langle R \rangle$.

$\langle U \rangle_R = G_1$

$u_5v_1,\ u_5v_3 \in R$ - SYM(v_1,v_2), SYM(v_3,v_4), else DEG(u_5), RED(u_5,v_1,v_2,w) or RED(u_5,v_3,v_4,w)

$u_5v_2,\ u_5v_4 \in B$ - else RED(u_5,v_1,v_2,w) or RED(u_5,v_3,v_4,w)

$u_6v_2 \in R$ - else DEG(v_2) or BLUE(u_5,u_6,u_7,v_2,w)

$u_6v_4 \in R$ - SYM(v_2,v_4)

$u_6v_1, u_6v_4 \in B$ - else RED(u_6,v_1,v_2,w) or RED(u_6,v_3,v_4,w)

$u_2v_1 \in R$ - SYM(u_2,u_4), else BLUE(u_2,u_4,u_6,v_1,w)

$u_1v_2 \in R$ or $u_3v_2 \in R$ - else BLUE(u_1,u_3,u_5,v_2,w)

At this point, the reader should have no trouble following the implication of each of the two alternatives in the last line. The argument is completed as follows:

$u_1v_2 \in R \Rightarrow \langle R \rangle \supset H_2$
$u_3v_2 \in R \Rightarrow \langle R \rangle \supset H_1.$

$\langle U\rangle_R = G_2$

u_5v_1, $u_5v_3 \in R$ - SYM(v_1,v_2), SYM(v_3,v_4), else
NOMATCH(u_5), RED(u_5,v_1,v_2,w)
or RED(u_5,v_3,v_4,w)

$u_6v_1 \in B$ - DEG(v_2), NOMATCH(v_2) or RED(u_6,v_1,v_2,w),
RED(u_1,u_4,u_7,v_2) etc.

u_6v_2, $u_6v_4 \in R$ - SYM(u_5,u_6)

At this point, the argument follows that of the last case. The conclusion is the same, either $\langle R\rangle \supset H_1$ or else $\langle R\rangle \supset H_2$.

$\langle U\rangle_R = G_3$

u_7v_1, $u_7v_3 \in R$ - SYM(v_1,v_2), SYM(v_3,v_4), else
DEG(u_7), RED(u_7,v_1,v_2,w) or
RED(u_7,v_3,v_4,w)

NOMATCH(u_7) $\Rightarrow$ no graph

Reference to the proof which dealt with case(a) of Figure 4 reveals the following general fact: if $v \in V$ is adjacent to u_7 in $\langle B\rangle$, then either u_1v, $u_2v \in R$ or else u_3v, $u_4v \in R$. In view of the last argument, for $\langle U\rangle_R = G_4$ we may again assume that
u_7v_1, $u_7v_3 \in R$.

$\langle U \rangle_R = G_4$

u_5v_1, $u_6v_3 \in R$ - SYM(v_1,v_3), else NOMATCH(u_7)

$u_7v_2 \in B$ - else RED(u_7,v_1,v_2,w)

$u_7v_2 \in B \Rightarrow$ SYM(u_1,u_3), u_1v_2, $u_2v_2 \in R$

u_3v_4, $u_4v_4 \in R$ - SYM(v_2,v_4), else RED(u_1,u_2,v_2,v_4)

$\langle R \rangle \supset H_2$.

The reader may recognize that H_1 is the graph of the cube octahedron. Six edges may be added to H_1 without producing a K_4-e. These are the edges which join pairs of antipodal vertices of the polyhedron. In H_2, four vertices have the property that their non-neighborhood induces G_1 and these four vertices are paired. Thus in H_2 only two edges can be added. The reader is encouraged to find the unique five-regular graph - it practically draws itself - and to verify that this is the graph obtained from H_1 by adding all six possible edges.

4. Final Remarks

This investigation began with a question raised by John Sheehan: is it true that $r(K_5$-e$) = 26$? The answer to this question is no, and it is found by comparatively simple arguments based on edge counts and the uniqueness of the five-regular $(K_4$-e, K_5-e$)$-critical coloring. From an example in [4], it was known that $r(K_5$-e$)$ is at least 21. It was decided that final determination of $r(K_5-e)$ would require all critical colorings for the pair (K_4-e), $K_5-e)$. Since these results were originally presented, Harborth and Mengersen [3] have shown that $r(K_5-e)$ is at most 23. In their proof, use is made of the of the facts given in our main theorem. These facts were apparently known to Harborth and Mengersen earlier, but no published proof existed. At this time, it remains open whether $r(K_5-e)$ is 21, 22 or 23.

REFERENCES

1. M. Behzad, G. Chartrand and L. Lesniak-Foster, Graphs and Digraphs. Wadsworth, Belmont, CA (1979).

2. M. Clancy, Some small Ramsey numbers, J. Graph Theory, 1 (1977), 89-91.

3. H. Harborth and I. Mengersen, An upper bound for for the Ramsey number $r(K_5-e)$, to appear.

4. C. C. Rousseau and J. Sheehan, On Ramsey numbers for books, J. Graph Theory, 2 (1978), 77-87.

THE STRUCTURAL COMPLEXITY OF FLOWGRAPHS

Norman E. Fenton
Oxford University

ABSTRACT

We describe an important measure of the structural complexity of flowgraphs, and show that for labeled flowgraphs there is no 'computationally equivalent' flowgraph of lower complexity. The ramifications of this result for structured programming in general are discussed.

We begin by recalling some definitions given in [2] (to which the reader is also referred for a fuller account of all the underlying concepts).

<u>Definition 1</u> A <u>flowgraph</u> F is a triple (G,a,z) consisting of a finite digraph G together with distinguished nodes a,z of G satisfying :-

i) All nodes in G except z (which has outdegree 0) either have outdegree 1 (we shall call such nodes <u>procedure nodes</u>) or outdegree 2 (<u>predicate nodes</u>).

ii) The distinguished node a (the <u>start node</u>) is the

source of some directed spanning tree of G (i.e. a can reach all other nodes of G). The distinguished node z (the stop node) is the sink of some directed spanning tree of G (i.e. all nodes can reach z).

Suppose that F' = (G',a',z') is a flowgraph for which G' is a subgraph of G (in the usual sense for digraphs). A node x ($\neq$ z') of G' is an entry node of G' if indegree(x) is smaller in G' than in G. Similarly x is an exit node of G' if outdegree(x) is smaller in G' than in G. In addition, if a or z are in G' then these are always defined to be an entry and an exit node respectively of G'. We say that F' is a subflowgraph of F if a' is an entry node of G' and z' is the only exit node of G' (so that predicate nodes of F do not become procedure nodes of F'). A flowgraph if irreducible if it contains no non-trivial proper subflowgraphs (the trivial flowgraphs are shown in the figure).

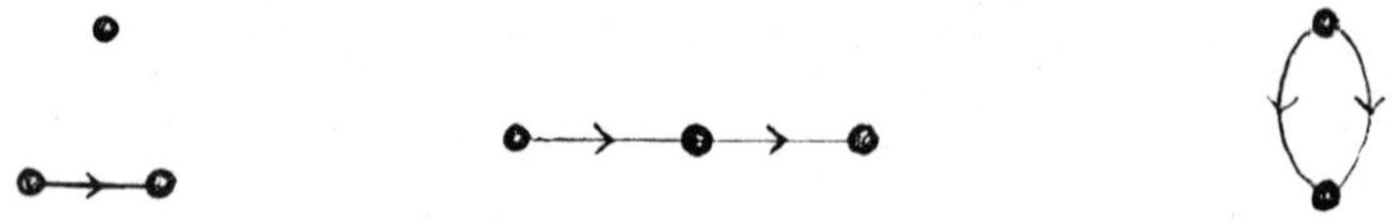

the two trivial flowgraphs

the flowgraph F_0

the flowgraph D_1

A flowgraph having no procedure nodes is called a CGK-graph; given any flowgraph F we associate a uniquely defined CGK-graph C(F) with F by contracing out the procedure nodes. Most important properties of flowgraphs (for example irreducibility) are preserved under this mapping to CGK-graphs, so that we are usually able to restrict our attention to the study of CGK-irreducibles (i.e. CGK-graphs which are irreducible flowgraphs). The main result in this paper is no exception to this.

For any family S of flowgraphs, we define the class of <u>S-graphs</u> to be the smallest family of flowgraphs containing each member of S and which is closed under the operation of composition in S (this operation in which procedure nodes are 'replaced' by flowgraphs is defined precisely in [2]). Thus an S-graph is 'structured' with respect to S in the usual hierarchical sense. Although this generalised notion of structuredness is valid for an arbitrary family S we are particularly interested in families S which satisfy :-

i) every member of S is irreducible,

ii) the flowgraph F_0 (see figure above) is in S, and

iii) for any irreducible F, $F \in S$ iff $C(F) \in S$.

Such a family S is said to be <u>complete</u>. In particular, for each $n \geq 0$ let S_n denote the family of irreducible flowgraphs having $\leq n$ predicate nodes. Then S_n is complete. Note that the S_1-graphs are precisely the traditionally accepted 'structured' flowgraphs (i.e. D-graphs, GOTO-less graphs etc.). We shall require the following result (lemma 5.6. of [2]):-

<u>Lemma 1</u> Suppose the family S is complete. Then for any flowgraph F, F is an S-graph if and only if C(F) is an S-graph.

In this paper we are especially concerned with flowgraphs whose nodes have labels (which could for example correspond to encoded procedures/predicates of some programming language). Formally :-

<u>Definition 2</u> Let Ω, Π be disjoint alphabets. A <u>labeled flowgraph</u> (F, Ω_0, Π_0, f) consists of

i) A flowgraph $F = (G,a,z)$ and finite sets $\Omega_0 \subset \Omega, \Pi_0 \subset \Pi$.

ii) A surjective labeling function f which assigns a letter in Ω_0 (resp. Π_0) to each procedure (resp. predicate) node of F.

If in addition f is injective then we say that F is distinctly labeled.

The definition above is slightly different from that given in [2], since we have ignored the TRUE, FALSE labelings on arcs leaving predicate nodes. This has been done to avoid unnecessary complications -the results given here can be extended in a natural way to this case.

Suppose then that (F, Ω_0, Π_0, f) (which we will subsequently refer to as just F) is a labeled flowgraph where $F = (G,a,z)$. The language of F, denoted L(F), is defined to be the set of all finite strings of node labels got by concatenating the labels on all paths from a to z (where a 'path' is allowed to pass nodes more than once). If F' is another labeled flowgraph then F' is L-equivalent to F if $L(F) = L(F')$. Thus the language of F can be regarded as the set of all possible computational histories of F (as in [3]) or as the set of language strings accepted by the natural finite automata associated with F (see [1,2]). From a practical viewpoint it could be argued that L(F) is the single most important invariant of F, and it is natural to look for flowgraphs which are L-equivalent to F and which are in some sense simpler than F. In [2] we found a systematic method for constructing such flowgraphs, namely by 'unfolding' subflowgraphs having more than one possible entry node. This unfolding leads us to define F to be a folded S-graph if every subflowgraph of F can be reduced to an S-graph by contracting each of its maximal subflowgraphs to a single node. Thus a folded S-graph may not

not necessarily be an S-graph itself, but it will be L-equivalent to an S-graph (namely the 'unfolded' form of F). This suggests a natural measure of the structural complexity of a flowgraph F :-

Definition 3 The structural complexity n of F is the least integer n for which F is a folded S_n-graph.

For example, the structural complexity of any GOTO-less flowgraph (with at least one predicate) is equal to 1.

We now ask the obvious question : 'Given a labeled flowgraph of structural complexity n, is there any flowgraph F' which is L-equivalent to F for which the structural complexity of F' is strictly less than n ?' For distinctly labeled flowgraphs the answer to this question is always no . Our main theorem proves this for the most important case, namely when F is irreducible (the full result follows by a routine but laborious argument), i.e. we shall prove :-

Theorem 1 Let F be a distinctly labeled irreducible flowgraph of n predicate nodes. Then every flowgraph L-equivalent to F has structural complexity $\geq$ n.

Before proving the theorem, let us consider the ramifications of the result. It means that no 'ad-hoc' method could ever yield a flowgraph which is L-equivalent to F that is intrinsically any simpler in structure than the flowgraph obtained by the systematic method of unfolding. This is a powerful and useful result which shows that structural complexity (of definition 3) is an important (and moreover valid) complexity measure, comparing most favourably with the rather trivial measure given in [4] . Moreover the result also infers that F is a folded S_n-graph if and only if it is L-equivalent to an S_n-graph.

The proof of Theorem 1 hinges on the following rather technical lemma :-

Lemma 2 Let F be a distinctly labeled CGK-graph with no subflowgraphs of type D_1 (see figure above). Suppose that F is L-equivalent to F'. If π_1 is a node label in F with arcs to nodes labeled π_2 and π_3 (which must be distinct by hypothesis), then every occurence of the node label π_1 in F' has arcs to nodes labeled π_2 and π_3.

Proof For each node x of $F = (G,a,z)$ define m(x) by

$$m(x) = \min\{m;\ m \text{ is the length of a simple path from a to } x\}$$

For any integer $m \geqslant 0$ we shall show that if $m(x) = m$ then the node x (or more specifically the label of x) satisfies the required condition. Since every node lies on a finite path from a this will prove the lemma. We use induction on m. For $m = 0$ there is just one node (namely $x = a$) for which $m(x) = 0$. So we have to show that the result holds for the node a. Suppose then that a is labeled π_1 and that it has arcs to nodes labeled π_2, π_3. Since every string in L(F) is prefixed by π_1 it follows that the start node a' of F' must be labeled π_1; also, since there are strings $\pi_1\pi_2\ldots$ and $\pi_1\pi_3\ldots$ in L(F) (= L(F')) this particular occurence of π_1 in F' must be followed by nodes labeled π_2 and π_3. If this is the only occurence of π_1 in F' we are done. If not then we will be able to trace the next occurence of a node labeled π_1 along some path from a', i.e. we will trace a path in F' whose language string in L(F') is $\pi_1 P \pi_1 \ldots$

(where P is a string of π's not containing π_1). Morover by considering the first two labels in this string it is clear that no other path in F' has this string. Now since $\pi_1 P \pi_1 \in L(F)$ and F has no duplicated node labels, we must have a path C in F from a back to itself with string L(C) = $\pi_1 P \pi_1$. But with such a path in F it follows that F also contains the paths whose labeled strings are $\pi_1 P \pi_1 \pi_2$ and $\pi_1 P \pi_1 \pi_3$. By the above remarks these strings can only occur in F' if the second occurence of π_1 has arcs to nodes labeled π_2 and π_3 . We may continue to pick out nodes labeled π_1 in this way - each one must lie on a path from some previously considered node labeled π_1 . Since F' is finite we must eventually exhaust all possibilities. This proves the result for $m = 0$. Next assume $m > 0$ and that the result holds for all nodes y with $m(y) < m$. We have to show that if $m(x) = m$ then x satisfies the required condition. Suppose then that x (which cannot be equal to a) is labeled π_1' and has arcs to nodes labeled π_2', π_3' . There are only a finite number of simple paths from a to x in F. Consider any one such path, say it has string P (so that P must be of the form $\pi_1 \pi_2 \ldots \pi_1'$ or $\pi_1 \pi_3 \ldots \pi_1'$). Since L(F)= L(F'), there must be a path from a' to a node labeled π_1' in F' whose string is precisely P. We claim that there is only one such path in F'; for two distinct paths with this property would have to 'split' at some node labeled by π' say in F'. So if y is the unique node in F labeled by π' then we have $m(y) < m$ and so the inductive hypothesis contradicts the fact that both arcs leaving π' in F' go to nodes with the same label. Now in L(F) we have strings $P\pi_2'$ and $P\pi_3'$ which means by the uniqueness of P in F' that the occurence of π_1' just considered must be followed by arcs to nodes labeled π_2' and π_3'. We can repeat this argument for each of the simple paths from a to x. If these are the only occurences of π_1' in F' then we are done.

If not, then any other occurence can be reached by continuing along one of these paths and we may just repeat the argument used in the proof of $m = 0$, i.e. by considering the relevant 'loop' in F.

Lemma 3 (For a complete family of flowgraphs S). Let F be a distinctly labeled flowgraph for which C(F) has no subflowgraphs of type D_1. Then F is L-equivalent to an S-graph if and only if C(F) is L-equivalent to an S-graph.

Proof ($\Longrightarrow$) Suppose that F is L-equivalent to the S-graph F'. Then certainly C(F) is L-equivalent to C(F') which is an S-graph by lemma 1.

($\Longleftarrow$) Suppose C(F) is L-equivalent to the S-graph F'. If $\langle \pi_1, \pi_2 \rangle$ is any (labeled) arc in F' it follows that the (unique nodes labeled π_1, π_2 in F are connected by a path punctuated only by procedure nodes (if any at all), say labeled in order $\omega_1, \ldots, \omega_K$. We form a new flowgraph from F' by inserting procedure nodes labeled $\omega_1, \ldots, \omega_K$ between each arc labeled $\langle \pi_1, \pi_2 \rangle$ in F'. If we do this for each arc in F' we end up with a flowgraph $\tilde{F}$ which is L-equivalent to F (this follows from lemma 2). But $C(\tilde{F}) = F'$, whence by lemma 1 F is L-equivalent to an S-graph.

Proof of Theorem 1 By lemma 3, it suffices to consider the case when F is a CGK-irreducible. Suppose the theorem is false. Then we can find an S_m-graph F' which is L-equivalent to F and for which $m < n$. In this case F' must have an irreducible subflowgraph F_1' with $\leq m$ predicates. Let F_1 be the subgraph of F consisting of precisely those nodes whose labels appear at least once in F_1' together with all the arcs joining them (except for the arcs from the stop node of F_1'). It is a direct consequence of lemma 2 that F_1' must actually be a subflowgraph of F. But then $F_1 = F$ and hence

$$n = \# \text{ predicates of } F_1 \leq \# \text{ predicates of } F_1' \leq m < n$$

and this contradiction completes the proof of the theorem.

Acknowledgements This research was funded by the SERC who also financed the visit to Kalamazoo. I would also like to thank THFC - a constant source of inspiration.

REFERENCES

1. V.Aho & D.Ullman 'Principles of Compiler Design', Addison-Wesley, 1977.

2. N.Fenton, R.Whitty & A.Kaposi, A generalised mathematical theory of structured programming, to appear Thoer. Comp.Sci.

3. S.Kosaraju, Analysis of structured programming, JCSS 9 (1974) 232-255

4. T.McCabe, A complexity measure, IEEE SE 2 (1976) 308-320.

n-DOMINATION IN GRAPHS

John Frederick Fink
University of Louisville

Michael S. Jacobson
University of Louisville

ABSTRACT

The purpose of this work is to generalize the classical notion of domination in graphs to include a prescribed degree of redundance in domination. We call a set D of vertices in a graph an n-dominating set if each vertex not in D is adjacent with at least n members of D. The n-domination number $\gamma_n(G)$ is defined to be the cardinality of a minimum n-dominating set in G. In this paper we study the behavior of $\gamma_n(G)$, obtaining both bounds and exact values for this parameter.

1. Introduction.

A nonempty subset D of the vertex set V(G) of a graph G is a dominating set if every vertex in V(G)-D is adjacent with a member of D. If $u \in D$

and v is a member of $V(G)-D$ that is adjacent with u, we say that u *dominates* v and v *is dominated by* u. A dominating set having smallest cardinality among all dominating sets in a given graph is called a *minimum dominating set*. The cardinality of a minimum dominating set in a graph G is termed the *domination number* of G and is denoted $\gamma(G)$. (Terminology and notation not defined herein follows that of Behzad, Chartrand and Lesniak-Foster [2].)

The study of domination in graphs was initiated by Ore [13], who observed that for every graph G, the relation $\gamma(G) \leq \beta(G)$ holds, where $\beta(G)$, the *independence number* of G, is the cardinality of a largest independent set of vertices in G. Since Ore's initial presentation, a limited amount of work has been done with dominating sets and the domination number (see [5] and [7] for surveys). Hypothetical applications of minimum dominating sets have been suggested by Berge [4] and Liu [11] who discuss the use of the notion in devising optimal methods of radar surveillance and transmitter placement in communications networks. That the implementation of these methods to problems modelled by large graphs may be difficult is suggested by the fact that the determination of the domination number of an arbitrary graph is an NP-complete problem (see Garey and Johnson [9]). It should be noted that bounds on $\gamma(G)$ do exist (see e.g. Ore [13], Vizing [14], and Berge [4]), though the parameter values on which these bounds depend may also be difficult to determine. Nieminen [12] showed that if $\varepsilon(G)$ denotes the maximum number of end-edges in a spanning forest of a graph G, then the sum $\gamma(G) + \varepsilon(G)$ equals the number of vertices in G. Other results concerning the domination number and related parameters can be found in papers by Jaegar and Payan [10], Allan and Laskar [1], and Cockayne, Favaron, Payan and Thomason [6].

In [8] Fink, Jacobson, Kinch and Roberts defined the <u>bondage number</u> of a nonempty graph G to be the cardinality of a smallest subset F of the edge set E(G) of G such that $\gamma(G-F) > \gamma(G)$. In a certain sense, the bondage number measures the integrity of the domination <u>number</u> under edge-removal (or, in the context of a communications network, under link failure). In this paper, we generalize the concepts of dominating sets and the domination number to consider the impact of edge (or vertex) removal on individual dominating sets. This study is motivated by the following theorem which indicates that the dominating property of a minimum dominating set can be destroyed by the removal of only one or two edges (or vertices) from the graph.

<u>Theorem 1</u>. If D is a minimum dominating set in a nonempty graph G, then at least one vertex in V(G)-D is dominated by at most two members of D.

<u>Proof</u>. Assume to the contrary that each vertex in V(G)-D is adjacent with at least three members of D. Let $u \in V(G)-D$, and let v and w be members of D that dominate u. By assumption, each vertex in V(G)-D is adjacent with at least one member of $D-\{v,w\}$. Thus, since v and w are adjacent with u, the set $D' = (D-\{v,w\}) \cup \{u\}$ is a dominating set in D. However, $|D'| < |D|$, a contradiction of the hypothesis that D is a minimum dominating set. The theorem follows. □

As mentioned before, Theorem 1 shows that the dominating property of a minimum dominating set can be destroyed by removing at most two edges or vertices from the graph. In many cases, the removal of only one edge or vertex from a graph will leave some vertex undominated by what had been a minimum dominating set in the unaltered graph. For example, if D is any minimum dominating set in the complete bipartite graph $K(m,n)$, with $3 \le m \le n$, then each vertex not in D is dominated by only one member of D. If the graph under consideration in some domination problem models a system susceptible to the failure of components corresponding to edges or vertices, it might be desirable to require a certain degree of redundance in domination. With this in mind, we introduce some new definitions.

If D is a subset of $V(G)$ and $u \in V(G)-D$ that is adjacent with at least n members of D, we say that u is n-*dominated* by D. If every vertex in $V(G)-D$ is n-dominated by D, then D is called an n-*dominating set*. If D has smallest cardinality among all n-dominating sets of the graph G, then D is a *minimum* n-*dominating set* and its cardinality is the n-*domination number* $\gamma_n(G)$. Note that every n-dominating set $(n \ge 1)$ is a dominating set in the usual sense; thus, for every graph G we have $\gamma(G) \le \gamma_n(G)$ for each $n \ge 1$. In particular, a minimum 1-dominating set is a minimum dominating set and $\gamma(G) = \gamma_1(G)$. More generally, if $1 \le m \le n$, then every n-dominating set in G is also an m-dominating set and thus $\gamma_m(G) \le \gamma_n(G)$.

The Behavior of $\gamma_n(G)$

Since every minimum n-dominating set is a dominating set, one may be led to ask about the magnitude of the n-domination number $\gamma_n(G)$ in relation to the usual domination number $\gamma(G)$. We first remark that if the maximum degree, denoted $\Delta(G)$, of a graph G is at least 3, and if $n \geq 3$, then by Theorem 1 no n-dominating set in G is a minimum dominating set. Hence we have the following.

Corollary 1.1. If G is a graph with maximum degree $\Delta(G) \geq 3$, then $\gamma_n G) > \gamma(G)$ for all $n \geq 3$.

By considering $G = W_p$, the wheel of order p, we see that even $\gamma_3(G)$ and $\gamma(G)$ can differ by a substantial amount, for $\gamma(W_p) = 1$ while $\gamma_3(W_p) = \lceil (p + 1)/2 \rceil$. (Note: $\lceil x \rceil$ denotes the least integer greater than or equal to x).

The following theorem gives a bound on $\gamma_n(G)$, for $n \geq 2$, that is more informative than Corollary 1.1 and yields Corollary 1.1 as a consequence.

Theorem 2. If G is a graph with $\Delta(G) \geq n \geq 2$, then

$$\gamma_n(G) \geq \gamma(G) + n - 2.$$

Proof. Let D be a minimum n-dominating set in G, let $u \in V(G)-D$, and let $v_1, v_2, \ldots, v_n$ be distinct members of D that dominate u. (Note: the condition $\Delta(G) \geq n$ implies that $V(G)-D \neq \emptyset$). Since D is an n-dominating set, each vertex in $V(G)-D$ is dominated by at least one member of $D-\{v_2, v_3, \ldots, v_n\}$. Therefore, since u is adjacent with each of $v_2, v_3, \ldots, v_n$, the set

$$D^* = (D - \{v_2, v_3, \ldots v_n\}) \cup \{u\}$$

is a dominating set in G. Hence,

$$\gamma(G) \leq |D^*| = \gamma_n(G) - (n-1) + 1,$$

so that

$$\gamma_n(G) \geq \gamma(G) + n - 2. \qquad \square$$

Although Theorem 2 yields a lower bound on the difference between $\gamma_n(G)$ and $\gamma(G)$ for $n \geq 2$, it often does not provide an easily computed lower bound on $\gamma_n(G)$ -- for in many cases the determination of $\gamma(G)$ is difficult. The next two theorems, however, give lower bounds based on parameters whose values can be quickly determined.

Theorem 3. If G is a graph with p vertices and maximum degree Δ, then

$$\gamma_n(G) \geq np/(\Delta + n).$$

Proof. Let D be a minimum n-dominating set in G, let $S = V(G)-D$, and let t denote the number of edges in G having one endpoint in D and the other in S. Then, since the degree of each vertex in D is at most Δ, we have

$$t \leq \Delta \cdot |D| = \Delta \cdot \gamma_n(G) . \qquad (3.1)$$

Also, since each vertex in D is adjacent with at least n members of D, we have

$$t \geq n \cdot |S| = n\,[p - \gamma_n(G)]. \qquad (3.2)$$

From (3.1) and (3.2) we get

$$n\,[p - \gamma_n(G)] \leq \Delta \cdot \gamma_n(G).$$

whence

$$\gamma_n(G) \geq np\,/\,(\Delta + n). \qquad \square$$

If G is the n-regular complete bipartite graph K(n,n), the bound on $\gamma_n(G)$ given in Theorem 3 is attained. The bound given by our next theorem is sharp for those bipartite graphs, called n-<u>semiregular</u>, whose vertex sets can be bipartitioned in such a way that every vertex in one of the partite sets has degree n (e.g. see Figure 1). The partite set each of whose vertices has degree n is called the n-<u>regular partite set</u>.

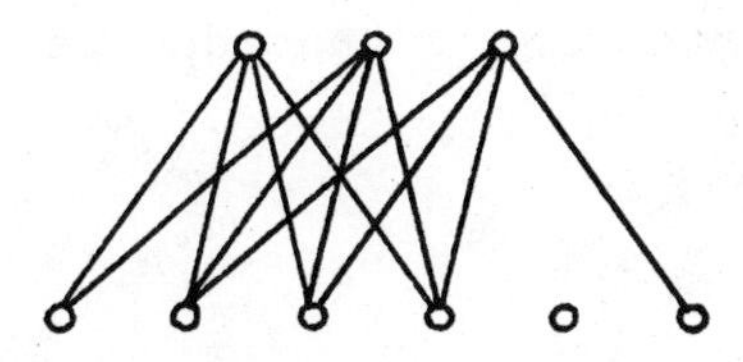

a 4-semiregular bipartite graph

Figure 1

<u>Theorem 4</u>. If G has p vertices and q edges, then

$$\gamma_n(G) \geq p - (q/n)$$

for each $n \geq 1$. Furthermore, if $q \neq 0$, then $\gamma_n(G) = p - (q/n)$ if and only if G is an n-semiregular bipartite graph.

Proof. Let D be a minimum n-dominating set in G, and let $S = V(G)-D$. Since each vertex in S is adjacent with n or more vertices of D, it follows that

$$q \geq n \cdot |S| = n[p - \gamma_n(G)]. \tag{4.1}$$

and thus

$$\gamma_n(G) \geq p - (q/n). \tag{4.2}$$

If $\gamma_n(G) = p - (q/n)$ and $q \neq 0$, then from the counting argument that yielded (4.1), we see that each edge of G joins a vertex of D with a vertex in S. Thus D and S are independent sets and each vertex of S has degree n; that is, G is an n-semiregular bipartite graph.

Conversely, suppose that G is an n-semiregular bipartite graph $(n \geq 1)$ with partite sets A and B such that each vertex in B has degree n. Then A is an n-dominating set in G. Since

$$q = n \cdot |B| = n(p - |A|)$$

we have $|A| = p - (q/n)$. Thus it follows from (4.2) that $\gamma_n(G) = p - (q/n)$. □

If T is a tree of order p, then $q = p-1$ and Theorem 4 yields the following bound on the 2-domination number of T.

Corollary 4.1. If T is a tree of order $p \geq 2$, then

$$\gamma_2(T) \geq (p + 1)/2$$

By observing that a 2-semiregular bipartite graph is the subdivision graph of a multigraph M (i.e. the graph obtained by subdividing each edge of M), we see that Theorem 4 also yields the following two results concerning the 2-domination number.

Corollary 4.2. If G is a nonempty graph with p vertices and q edges, then $\gamma_2(G) = p - (q/2)$ if and only if G is the subdivision graph of some multigraph.

Corollary 4.3. If T is a tree of order $p \geq 2$, then $\gamma_2(T) = (p + 1)/2$ if and only if T is the subdivision graph of another tree.

Theorem 4 also allows us to obtain at least one formulation of the exact value of the n-domination number for any nonempty graph; this is given in the next theorem.

Theorem 5. If G is a graph with maximum degree $\Delta \geq n$, then

$$\gamma_n(G) = \min \{\gamma_n(H)\} \tag{5.1}$$

where this minimum is taken over all spanning n-semiregular bipartite subgraphs H of G.

Proof. If H is any spanning subgraph of G, then it is readily seen that $\gamma_n(G) \leq \gamma_n(H)$. Thus, the right-hand side of (5.1) is an upper bound on $\gamma_n(G)$.

Let D be a minimum n-dominating set in G, and let $S = V(G)-D$. Then each vertex in S is adjacent with n or more members of D. Hence we can construct a spanning subgraph H of G as follows: let $V(H) = V(G)$, and for each vertex v of S, select exactly n edges of G that join v to vertices of D. Then H is a spanning n-semiregular bipartite subgraph of G, and D is an n-dominating set in H. Thus $\gamma_n(G) \geq \gamma_n(H)$, and we see that the right-hand side of (5.1) is also a lower bound for $\gamma_n(G)$. By trichotomy, (5.1) is true. □

By observing that the n-domination number of an n-semiregular bipartite graph G of order p equals $p - |S|$, where S is the n-regular partite set of G, we have the following direct consequence of Theorem 5.

Corollary 5.1. If G is a graph of order $p \geq 2$, and if, over all spanning n-semiregular bipartite subgraphs of G, η denotes the maximum number of vertices in an n-regular partite set, then

$$\gamma_n(G) = p - \eta$$

As a consequence of Theorem 5 and Corollary 5.1, we may obtain the following well-known theorem of Nieminen [12] concerning the usual domination number.

Theorem 6. (Nieminen - 1974). If ε denotes the maximum possible number of end-edges in a spanning forest of a graph G of order p, then

$$\gamma(G) + \varepsilon = p .$$

Proof. Let F be a spanning forest of G having ε end-edges, let H be a spanning 1-semiregular bipartite of G having a largest 1-regular partite set S, and let $\eta = |S|$. Then H is a spanning forest with η edges, each of which is an end-edge. Thus $\eta \le \varepsilon$.

Let F' be the subgraph of F derived by deleting from F all nonend-edges. Then F' is the union of stars and (possibly) isolated vertices. Hence F' is a spanning 1-semiregular bipartite subgraph of G whose 1-regular partite set has ε members; thus $\eta \ge \varepsilon$.

By trichotomy, $\varepsilon = \eta$. Hence, by Corollary 5.1, we have $\gamma(G) + \varepsilon = p$. □

n-Domination and n-Dependence

Perhaps the first result obtained concerning the usual domination number was the following theorem of Ore [13]; its proof rests on the observation that every maximal independent set is a dominating set.

Theorem 7. (Ore - 1962). If $\beta(G)$ denotes the independence number of a graph G, then $\gamma(G) \le \beta(G)$.

We begin this section by studying similar relationships between n-domination and a generalized notion of independence.

As usual, $\Delta(H)$ will denote the maximum degree of a graph H. Also, if S is a nonempty subset of the vertex set of a graph G, then $\langle S \rangle$ denotes the subgraph of G induced by S. A subset S of $V(G)$ is an n-*dependent set* if and only if $\Delta(\langle S \rangle) \le n$. An n-dependent set of largest possible cardinality in a graph G is a *maximum n-dependent set*, and its cardinality, denoted $\beta_n(G)$, is called the *n-dependence number* of G. We remark that if $k \leq m$, then every k-dependent set is an m-dependent set and $\beta_k(G) \leq \beta_m(G)$. Also, an independent set is precisely a 0-dependent set and $\beta(G) = \beta_0(G)$. Therefore, Theorem 7 expresses the relationship $\gamma_1(G) \leq \beta_0(G)$. The following theorem provides an analogous relationship between $\gamma_2(G)$ and $\beta_1(G)$.

Theorem 8. For every graph G,

$$\gamma_2(G) \leq \beta_1(G)$$

Proof. The result is obvious if $\Delta(G) \leq 1$, so we suppose that $\Delta(G) \geq 2$. Among all maximum 1-dependent sets in G, let D be selected so that $\langle D \rangle$ has as few edges as possible. We shall show that D is a 2-dominating set in G.

Assume to the contrary that there is a vertex $u \in V(G)-D$ that is not 2-dominated by D. Since $D \cup \{u\}$ cannot be a 1-dependent set, there are two vertices v and w in D such that v is adjacent with both u and w. But then, since v is the only vertex of D that is adjacent with u, the set $D = (D-\{v\}) \cup \{u\}$ is a 1-dependent set of cardinality $\beta_1(G)$ and $\langle D' \rangle$ has one fewer edge than $\langle D \rangle$ has. With this contradiction to our choice of D, we conclude that D is a 2-dominating set. Thus $\gamma_2(G) \leq |D| = \beta_1(G)$. □

Having seen that $\gamma_1(G) \leq \beta_0(G)$ and $\gamma_2(G) \leq \beta_1(G)$ the authors attempted (unsuccessfully) to prove the following.

Conjecture 1. For any graph G and $n \geq 0$,

$$\gamma_{n+1}(G) \leq \beta_n(G).$$

As an example of a graph G for which $\gamma_{n+1}(G) = \beta_n(G)$, take G to be the complete graph K_p. To find examples of graphs for which $\gamma_{n+1}(G) < \beta_n(G)$, one might find a graph satisfying the hypothesis of the following theorem.

Theorem 9. If $\gamma_n(G) = \gamma_{n+1}(G)$ and D is a minimum $(n+1)$-dominating set in the graph G, then D is a maximal $(n-1)$-dependent set.

Proof. Since each vertex of $V(G)-D$ is $(n+1)$-dominated by D, it suffices to show that D is an $(n-1)$-dependent set, for then the maximality condition will follow.

Assume to the contrary that there is a vertex u in the minimum $(n+1)$-dominating set D that is adjacent with at least n other members of D. Then u and every vertex in $V(G)-D$ are n-dominated by $D-\{u\}$; that is, $D-\{u\}$ is an n-dominating set in G. This contradicts the hypothesis that $\gamma_n(G) = \gamma_{n+1}(G)$ and completes the proof. □

As an immediate corollary to Theorem 9, we have the following.

Corollary 9.1. If G is a graph for which $\gamma_n(G) = \gamma_{n+1}(G)$, then $\gamma_{n+1}(G) \leq \beta_{n-1}(G)$.

Thus, in order to prove Conjecture 1, it remains to consider those graphs G for which $\gamma_n(G) < \gamma_{n+1}(G)$.

Since every maximal independent set of vertices is a dominating set, Theorem 9 also yields the following result concerning the _independent dominating number_ $i(G)$ which may be defined as the cardinality of a smallest maximal independent set in the graph G.

Corollary 9.2. If G is a graph for which $\gamma(G) = \gamma_2(G)$, then every minimum 2-dominating set in G is a maximal independent set and $\gamma(G) = i(G) = \gamma_2(G)$.

Another result relating our generalized versions of the independence and domination numbers is the following.

Theorem 10. If G is a graph with p vertices and maximum degree $\Delta \geq n \geq 1$, then

$$\gamma_n(G) \geq p - \beta_{\Delta-n}(G).$$

Proof. If D is a minimum n-dominating set in G and $S = V(G)-D$, then $\Delta(\langle S \rangle) \leq \Delta-n$. Hence S is a $(\Delta-n)$-dependent set and $|S| \leq \beta_{\Delta-n}(G)$. Since $\gamma_n(G) = |D| = p - |S|$, we have $\gamma_n(G) \geq p - \beta_{\Delta-n}(G)$. □

With the aid of Theorem 10, we can prove several results concerning regular graphs.

Theorem 11. If G is an r-regular graph of order p, and $1 \leq n \leq r$, then

$$\gamma_n(G) = p - \beta_{r-n}(G).$$

Proof. Let S be a maximum $(r-n)$-dependent set in G. Then, if $D = V(G)-S$, each vertex in S is adjacent with at least $r - (r-n) = n$ vertices of D. Thus D is an n-dominating set and $\gamma_n(G) \le |D| = p - \beta_{r-n}(G)$. By Theorem 10, $\gamma_n(G) \ge p - \beta_{r-n}(G)$. Hence $\gamma_n(G) = p - \beta_{r-n}(G)$. □

As a result of Theorem 11, we can develop a relationship between $\gamma_n(G)$ and the chromatic number $\chi(G)$ for regular graphs G.

Corollary 11.1. If G is an r-regular graph of order p and $1 \le n \le r$, then

$$\gamma_n(G) \le p \cdot [1-(1/\chi(G))].$$

Proof. Since $\gamma_n(G) \le \gamma_r(G)$, it suffices to suppose that $n = r$.

By Theorem 11 we have $\gamma_r(G) = p - \beta(G)$. Since $\beta(G) \ge p/\chi(G)$, we have

$$\gamma_r(G) = p - \beta(G) \le p - p/\chi(G).$$

which gives the desired inequality. □

Since the chromatic number is often difficult to compute, we offer the following weakened version of Corollary 11.1.

Corollary 11.2. If G is an r-regular graph of order p, and no component of G is a complete graph or an odd cycle, then

$$\gamma_n(G) \le p - p/r$$

for $1 \le m \le r$.

Proof. Since no component of G is a complete graph or an odd cycle, Brooks' Theorem gives $\chi(G) \le \Delta(G) = r$. Thus $[1-(1/\chi(G))] \le [1-(1/r)]$ and the result follows from Corollary 11.1. □

As an application of Theorem 11 and Corollary 11.1 we compute the n-domination numbers for a large class of n-regular graphs.

Theorem 12. If $1 \le m \le (n+2)/2$ and G is the Cartesian product $K_m \times K_{n+2-m}$, then

$$\gamma_n(G) = p - m,$$

where $p = m(n+2-m)$ is the order of G.

Proof. From a result of Behzad and Mahmoodian [3], $\chi(G) = n+2-m = p/m$. Thus, since G is n-regular, Corollary 11.1 gives

$$\gamma_n(G) \le p - p(m/p) = p - m \tag{12.1}$$

Since $V(G)$ can be partitioned into m subsets each of which induces a subgraph isomorphic to K_{n+2-m}, we see that $\beta(G) \le m$. Hence Theorem 11 gives

$$\gamma_n(G) = p - \beta(G) \ge p - m. \tag{12.2}$$

From (12.1) and (12.2) we have $\gamma_n(G) = p - m$. □

In concluding this section, we remark that Theorem 12 shows that the bounds on $\gamma_n(G)$ given in Theorem 11 and Corollary 11.1 are sharp. Also, with $m = 2$, it shows that the bound in Corollary 11.2 is attainable.

2. Conclusion.

Throughout this paper, we have studied the relationships between the n-domination number and various other parameters associated with a graph. We have also ventured to guess an n-domination analogue (Conjecture 1) to Ore's classic result relating domination and independence (Theorem 7). We close with an open problem concerning the rate at which the n-domination number increases as n increases.

Problem. Find a sharp bounding function $f(n)$ such that if G is a graph with $\delta(G) \geq n$ and $m \geq f(n)$ then $\gamma_n(G) < \gamma_m(G)$.

REFERENCES

1. R. B. Allan and R. Laskar, On domination and independent domination numbers in graphs. Discrete Math. 23 (1978) 73-76.

2. M. Behzad, G. Chartrand, and L. Lesniak-Foster, Graphs & Digraphs. Wadsworth International, Belmont, CA (1979).

3. M. Behzad and S. E. Mahmoodian, On topological invariants of the products of graphs. Canad. Math. Bul. 12 (1969) 157-166.

4. C. Berge, Graphs and Hypergraphs, North-Holland, Amsterdam (1973).

5. E. J. Cockayne, Domination of undirected graphs--a survey. Theory and Applications of Graphs (Proc. Internat. Conf., Western Mich. Univ.,

Kalamazoo, Mich., 1976), pp. 141-147, Lecture Notes in Math., 642, Springer, Berlin, 1978.

6. E. J. Cockayne, O. Favaron, C. Payan and A. G. Thomason, Contributions to the Theory of domination, independence, and irredundance in graphs, Discrete Math. 33 (1981) 249-258.

7. E. J. Cockayne and S. T. Hedetniemi, Towards a theory of domination in graphs. Networks 7 (1977) no. 3, 247-261.

8. J. F. Fink, M. S. Jacobson, L. F. Kinch and J. Roberts, The bondage number of a graph. Submitted for publication.

9. M. R. Garey and D. S. Johnson, Computers and Intractability: A Guide to the Theory of NP-Completeness. W. H. Freeman, San Francisco, CA (1979).

10. F. Jaegar and C. Payan, Relations du type Nordhaus-Gaddum pour le nombre d'absorption d'un graphe simple. C. R. Acad. Sc. Paris Series A, t. 274 (1972) 728-730.

11. C. L. Liu, Introduction to Combinatorial Mathematics, McGraw-Hill, New York. (1968).

12. J. Nieminen, Two bounds for the domination number of a graph. J. Inst. Maths. Applics. 14 (1974) 175-177.

13. O. Ore, Theory of Graphs, Amer. Math. Soc. Colloq. Publ. 38. Providence, R. I. (1962).

14. V. G. Vizing, A bound on the external stability number of a graph, Doklady A. N. 164 (1965) 729-731.

ON n-DOMINATION, n-DEPENDENCE AND FORBIDDEN SUBGRAPHS

John Frederick Fink
University of Louisville

Michael S. Jacobson
University of Louisville

ABSTRACT

A set of vertices S of a graph G is an n-dominating set whenever every vertex of $V(G)-S$ is adjacent to at least n vertices of S. An n-dependent set T is a subset of the vertex set such that $\Delta(\langle T\rangle)\leq n$. The n-domination number, $\gamma_n(G)$, and the n-dependence number, $\beta_n(G)$, of a graph G is the order of the smallest n-dominating set and the largest n-dependent set respectively. It is the purpose of this paper to present various relationships between $\gamma_n(G)$ and $\beta_m(G)$ for graphs G not containing specified sets of induced subgraphs. One class of graphs in particular that is considered are the claw-free graphs.

1. Introduction.

In this paper we will consider finite undirected graphs with no multiple edges, and with no loops. All definitions not presented here can be found in [2]. For a graph G, we will refer to $V(G)$ and $E(G)$ as the vertex and edge set respectively. A subgraph $H \subseteq G$ is a graph with $V(H) \subseteq V(G)$ and with $E(H)$ a subset of the edges of G whose end vertices are both elements of $V(H)$. For a subset $D \subseteq V(G)$, the induced subgraph on D, denoted $\langle D \rangle$, is the subgraph of G with $V(\langle D \rangle) = D$ and $E(\langle D \rangle)$ precisely the set of all the edges in G whose end vertices are both elements of D. Also, for $D \subseteq V(G)$, denoted by $N(D)$, the neighborhood of D, the set of vertices in $V(G)-D$ adjacent to at least one vertex of D. When $D=\{x\}$, $N(D)$ will be abbreviated by $N(x)$.

A set $D \subseteq V(G)$ is a *dominating set* of G if $N(D)=V(G)-D$. The *domination number* of a graph G, $\gamma(G)$, is the minimum order of any dominating set. The theory of domination in graphs is a field that has attracted a great deal of interest in the last ten years, see [5,6] for surveys. One reason for this interest is because of the large number of applications of dominating sets in graphs, see [3,9]. The parameter $\gamma(G)$ has been studied in relation to a number of other parameters; one in particular is the independence of a graph, $\beta(G)$, the order of the largest independent set of vertices in G. Clearly, every maximal independent set is a minimal dominating set, hence $\gamma(G) \leq \beta(G)$. Also, if $i(G)$ is the order of the smallest maximal independent set, then it follows that $\gamma(G) \leq i(G)$. A question that has received a good deal of consideration [1] is the problem of when $\gamma(G)=i(G)$. In such graphs, the problem of finding a minimum dominating set reduces to the problem of finding a smallest maximal independent set of vertices. In [1], Allan and Laskar consider only a special class of graphs, claw-free graphs. They showed:

<u>Theorem A</u>. If G is a graph which does not have an induced subgraph isomorphic to $K_{1,3}$, then $\gamma(G)=i(G)$.

In [8], the authors introduced the following generalizations. Let G be a graph and $S \subseteq V(G)$ then S is <u>k-dependent</u> if $\Delta(\langle S \rangle) \leq k$. The <u>k-dependence number</u> of a graph G, denoted $\beta_k(G)$, is the order of the largest k-dependent set in G. Note, $\beta(G)=\beta_0(G)$. Let $D \subseteq V(G)$ then D is an <u>n-dominating set</u> if for all $x \in V(G)-D$, $|N(x) \cap D| \geq n$, and the n-domination number, $\gamma_n(G)$, is the order of the smallest n-dominating set. Note, $\gamma(G)=\gamma_1(G)$.

It is the purpose of this paper to study various relationships between these parameters for various classes of graphs which do not contain a particular induced subgraph or set of subgraphs. In particular, Theorem A is generalized for the notions of n-domination and k-dependence. Let the j-dependent-n-domination number $i(j,n;G)$ be the order of the smallest j-dependent, n-dominating set. Note, $i(j,n;G)$ does not exist for all pairs, j and n, but if $i(j,n;G)$ exists for some j and n then $\gamma_n(G) \leq i(j,n;G)$. We prove in the next section:

<u>Theorem</u>. If G is a graph which does not contain an induced subgraph isomorphic to $K_{1,3}$, then

$$i(2n-2,n;G) = \gamma_n(G) .$$

Observe, for the case $n=1$, this is precisely Theorem A. We consider various other sets of forbidden induced subgraphs.

<u>Main Results</u>. For convenience, we define the following notation. Let G, $H_1, H_2, \ldots, H_m$ be graphs; G is $(H_1, H_2, \ldots, H_m)$-free whenever G contains no induced subgraph isomorphic to any of $H_1, H_2, \ldots,$ or H_m. In [8], the authors make the following conjecture:

<u>Conjecture 1</u>. If G is a graph with $\delta(G) \geq n$ then

$$\gamma_n(G) < \gamma_{2n+1}(G) .$$

We now present a result which verifies this conjecture in the case of $K_{1,3}$-free graphs.

<u>Theorem 1</u>. If G is a $K_{1,3}$-free graph with $\Delta(G) \geq n$ then

$$\gamma_n(G) < \gamma_{2n+1}(G) .$$

<u>Proof</u>. If $n \leq \Delta(G) \leq 2n$, then it follows that

$$\gamma_n(G) < |V(G)| = \gamma_{2n+1} .$$

Hence, we may assume $\Delta(G) \geq 2n+1$. Clearly then, $\gamma_{2n+1}(G) < |V(G)|$. Let D be a $2n+1$-dominating set. We proceed by showing that D is not a minimal n-dominating set. If there is a vertex of degree n in $\langle D \rangle$, by removing it an n-dominating set still remains, and thus D is not a minimal n-dominating set. Hence $\Delta(\langle D \rangle) \leq n-1$. Let $x \in V(G)-D$ and consider $H = \langle N(x) \cap D \rangle$ which has order at least $2n+1$. Since G is $K_{1,3}$-free, H contains at most two independent vertices. Also, since $H \subset \langle D \rangle$, if $v \in V(H)$ there are at least $n+1$ vertices in H non-adjacent to v, no pair of which can be non-adjacent. This implies that $K_{n+1} \subseteq H$, but then it follows that $\Delta(\langle D \rangle) \geq n$. Therefore, no 2n+1-dominating set can be a minimal n-dominating set and result follows.

This result can be improved slightly. The proof will be omitted since it is essentially the same as the previous proof with an extra case to consider when looking at the graph H as defined above.

<u>Theorem 2</u>. If G is $K_{1,3}$-free with $\Delta(G) \geq n$ then

$$\gamma_n(G) < \gamma_{2n}(G)$$

This result is the best possible; for example, let G be the graph obtained by identifying a vertex in a K_{n+1} with a vertex in a K_n. The graph G is $K_{1,3}$-free with $\Delta(G) \geq n$ and $\gamma_n(G) = \gamma_{2n-1}(G) = 2n-1$. We also note that there are graphs which contain induced $K_{1,3}$ with $\gamma_n(G) = \gamma_{2n}(G)$. For example, let $G = K_{2n,2n}$; $\gamma_n(G) = \gamma_{2n}(G) = 2n$.

For the sake of convenience, we prove the following result.

<u>Proposition 3</u>. An n-dominating set D is minimal if and only if for each $v \in D$ either $|N(v) \cap D| < n$ or there exists $x \in V(G)-D$ such that $|N(x) \cap D| = n$ and $x \in N(v)$.

<u>Proof</u>. Let D be a minimal n-dominating set. Suppose there exists $v \in D$ such that $|N(v) \cap D| \geq n$ and for every $x \in V(G)-D$ either $|N(x) \cap D| > n$ or $x \in N(v)$. Then consider $D' = D - v$. Since v is adjacent to at least n vertices of D' and each vertex in $V(G)-D$ is adjacent

to at least n vertices of D, D is an n-dominating set. But this contradicts the assumption that D was minimal.

Now suppose D is an n-dominating set such that for every $v \in D$ either $|N(v) \cap D| < n$ or there exists $x \in V(G)-D$ such that $|N(x) \cap D| = n$ and $x \in N(v)$. Consider $D' = D-v$. If $|N(v) \cap D| < n$ then D' could not be an n-dominating set. If there exists $x \in V(G)-D$ such that $|N(x) \cap D| = n$ and $x \in N(v)$ then x would be adjacent to only $n-1$ vertices of D'. Hence D is a minimal n-dominating set.

Throughout the remainder of this section, we present results relating the concepts of n-domination and m-dependence. Recall, the j-dependent-n-domination number, $i(j,n;G)$ is the order of the smallest j-dependent, n-dominating set of G.

<u>Theorem 4</u>. If G is $K_{1,3}$-free then

$$i(2n-2,n;G) = \gamma_n(G) .$$

<u>Proof</u>. To prove this result, we must first show there exists a $2n-2$ dependent, n-dominating set; thus $i(2n-2,n;G)$ exists and hence $\gamma_n(G) \le i(2n-2,n;G)$. Next we must show $i(2n-2,n;G) \le \gamma_n(G)$ assuring equality. We accomplish both by showing that there exist minimum n-dominating sets that are 2n-2-dependent.

Over all minimum n-dominating sets D, choose one such that $\langle D \rangle$ has as few edges as possible. If D is 2n-2-dependent, we would be done, so we will suppose there exists $v \in D$ such that $|N(v) \cap D| \ge 2n-2$. By Proposition 3, since D is a minimal n-dominating set, there must exist $x \in V(G)-D$ such that $|N(x) \cap D| = n$ and

$x \in N(v)$. Let $T = \{x \in V(G)-D$ such that $|N(x) \cap D| = n$ and $x \in N(v)\}$. As noted, $T \neq \phi$. Choose any two vertices $x,y, \in T$. Since $|N(x) \cap D| = n$, $|N(y) \cap D| = n$, $x,y \in N(v)$ and $|N(v) \cap D| \geq 2n-1$, it follows that there must be a vertex $z \in N(v) \cap D$ such that $xz, yz \notin E(G)$. Now since $\langle\{v,x,y,z\}\rangle \neq K_{1,3}$, it follows that $xy \in E(G)$. Hence $\langle T\rangle$ is complete. Let $D' = D - v \cup \{x\}$ where $x \in T$. Clearly $\langle D'\rangle$ contains fewer edges than $\langle D\rangle$, thus there must exist a vertex p in $V(G)-D'$ adjacent to fewer than n vertices of D'. Clearly $p \neq v$ since v is adjacent to at least $2n-2$ vertices of D'. If $p \in T-x$, it would be adjacent to exactly n vertices of D' since it is adjacent to x and $n-1$ vertices of $D-v$. But clearly if $p \notin T$, then it would be adjacent to at least n vertices of D' contradicting the assumption that D' is not an n-dominating set. Consequently, this would contradict the minimal number of edges in $\langle D\rangle$ hence it must be the case that D is a $2n-2$ dependent n-dominating set and thus $i(2n-2,n;G) = \gamma_n(G)$.

Corollary 5. (Allan and Laskar [1]). If G is $K_{1,3}$-free, then

$$i(0,1;G) = \gamma_1(G)$$

(i.e. independent domination number = domination number).

As in [1], appropriate results for $L(G)$ and $M(G)$, the line graph and middle graph respectively, follow as corollaries to Theorem 4.

Corollary 6. For any graph G,

$$\gamma_n(L(G)) = i(2n-2,n;L(G))$$

where $L(G)$ denotes the line graph of G.

The middle graph of a graph G, denoted by $M(G)$, is the intersection graph on $\Omega(F)$ in $V(G) = \{v_1, v_2, \ldots, v_p\}$ where $F = E(G) \cup \{\{v_1\}, \{v_2\}, \ldots, \{v_p\}\}$.

Corollary 7. For any graph G,

$$\gamma_n(M(G)) = i(2n-2, n; M(G)) ,$$

where $M(G)$ denotes the middle graph of G.

Theorem 8. If G is $(K_{1,3}, K_{1,3} + e)$-free, then

$$i(n-1, n; G) = \gamma_n(G) .$$

Proof. Again we proceed by showing that there exists a minimum n-dominating set which is n-1 dependent. Over all minimum n-dominating sets D, choose one such that $\langle D \rangle$ has as few edges as possible. If D is n-1 dependent, we would be done, so we will assume that there exists $v \in D$ such that $|N(v) \cap D| \geq n$. By Proposition 3, since D is a minimal n-dominating set, there must exist $x \in V(G) - D$ such that $|N(x) \cap D| = n$ and $x \in N(v)$. Let T be the set of all such vertices x. Choose any two vertices $x, y \in T$. Since $|N(x) \cap D| = n$, $x \in N(v)$ and $|N(v) \cap D| \geq n$, it follows that there must exist a vertex $z \in N(v) \cap D$ such that $xz \notin E(G)$. Since $\langle x, y, v, z \rangle = K_{1,3}$ or $K_{1,3} + e$, it follows that $xy \in E(G)$. Hence $\langle T \rangle$ is complete. Let $D' = D - v \cup \{x\}$. As in the previous result, D' is an n-dominating set with $\langle D' \rangle$ containing fewer edges than $\langle D \rangle$. But this contradicts the choice of D. Hence it must be the case that D is an n-1 dependent, n-dominating set. Therefore $i(n-1, n; G) \leq \gamma_n(G)$ and the result follows.

For the next result, we define the following graphs:

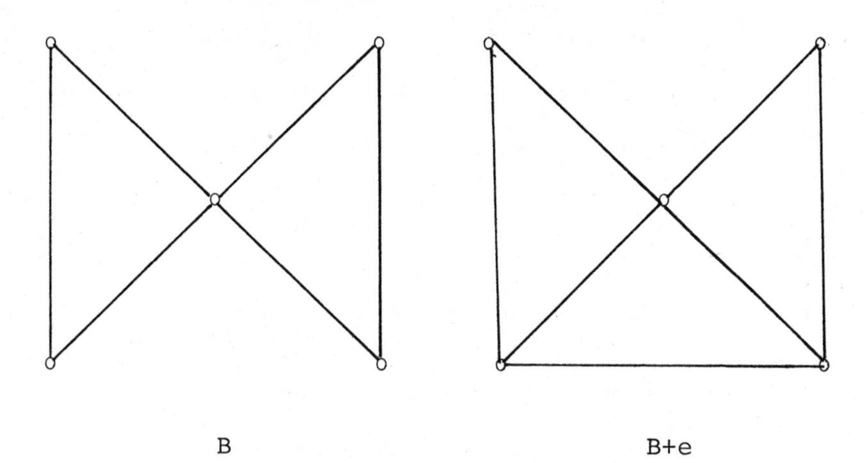

Figure 1

<u>Theorem 9</u>. If G is $(K_{1,3}, B, B+e)$-free, then

$$i(n-1,n;G) = \gamma_n(G).$$

<u>Proof</u>. We show $i(n-1,n;G) \leq \gamma_n(G)$. Over all minimum n-dominating sets D, choose one such that $\langle D \rangle$ has as few edges as possible. If D is n-1-dependent, we would be done, so we will assume that there exists $v \in D$ such that $|N(v) \cap D| \geq n$. By Proposition 3, there must exist $x \in V(G)-D$ such that $|N(x) \cap D| = n$ and $x \in N(v)$. Again, let T be the set of all such vertices. Choose any two vertices $x, y \in T$. Since $|N(x) \cap D| = n$, $|N(v) \cap D| \geq n$ and $x \in N(v)$, there exists $x_1 \in N(v) \cap D$ such that $xx_1 \notin E(G)$. Similarly, there exist $y_1 \in N(v) \cap D$ such that $yy_1 \in (G)$. If $x_1 = y_1$, then consider $\langle \{v, x, y, x_1\} \rangle$. Since G is $K_{1,3}$-free, it follows that $xy \in E(G)$. If $x_1 \neq y_1$, then it must be the case that either $xy \in E(G)$ or else both yx_1 and $xy_1 \in E(G)$. Now consider $\langle \{v, x, y, x_1, y_1\} \rangle$. Since this subgraph $\not\cong$ B or B+e, it

must be the case that $xy \in E(G)$. Therefore T is complete and $D' = D - v \cup \{x\}$ contradicts the choice of D being a minimum n-1-dominating set containing a minimum number of edges. Consequently, D is an n-1 dependent, n-dominating set. Therefore, $i(n-1,n;G) \le \gamma_n(G)$ and the theorem follows.

Conclusion. There are a number of interesting problems related to the parameters $\gamma_n(G)$ and $\beta_m(G)$ as well as $i(j,n;G)$. Some possible directions are listed below.

Conjecture 1. If G is a graph with $\delta(G) \ge n$, then

$$\gamma_n(G) < \gamma_{2n+1}(G).$$

Conjecture 2. For any graph G,

$$\gamma_n(G) \le \beta_{n-1}(G).$$

These conjectures are discussed in [8]. Other related questions include:

3. Are there other classes of graphs which emit various relationships between $\gamma_n(G), i(k,n;G)$ and $\beta_m(G)$?

4. Are there other parameters which naturally relate to $\gamma_n(G), i(k,n;G)$ or $\beta_m(G)$?

Addendum: Since the conference, both Conjecture 1 and Conjecture 2 have been settled. Dick Schelp, private communication, has constructed a class of examples for which the conjecture is false and, in fact, finds a G with $\gamma_n(G) = \gamma_{\frac{n^2+n}{4}}(G)$. Odile Favaron, private communication, has given a very nice proof of Conjecture 2. In fact, she has shown that every graph contains n-dominating sets which are n-1 dependent.

Acknowledgement. Both authors would like to thank the University of Louisville whose support partially funded this research.

References

1. R. B. Allan and R. Laskar, On domination and independent domination numbers of a graph, Disc. Math. 23(1978), 73-76.

2. M. Behzad, G. Chartrand and L. Lesniak-Foster, Graphs and Diagraphs, Prindle, Weber and Schmidt, Boston (1979).

3. C. Berge, Graphs and Hypergraphs, North-Holland, Amsterdam (1973).

4. B. Bollabas and E. J. Cockayne, Graph Theoretic parameters concerning domination, independence and irredundance. Journal of Graph Theory 3 (1979), 241-249.

5. E. J. Cockayne, Domination in undirected graphs - a survey in: Y. Alavi and D. R. Lick, eds., Theory and Applications of Graphs in America's Bicentennial Year, Springer-Verlag, Berlin, 1978, 141-147.

6. E. J. Cockayne and S. Hedetniemi, Towards a Theory of domination in graphs, Networks, Fall (1977), 247-261.

7. E. J. Cockayne, O. Favoron, C. Payan and A. G. Thomason, Contributions to the theory of Domination, Independence and Irredundance in Graphs, Disc. Math. 33 (1981), 249-258.

8. J. F. Fink and M. S. Jacobson, On the n-domination number of a graph, to appear.

9. C. L. Liu, Introduction to Combinatorial Mathematics, McGraw Hill, New York, (1968).

DIGRAPHS WITH WALKS OF EQUAL LENGTH BETWEEN VERTICES

M.A. Fiol
Universidad Politécnica de Catalunya

I. Alegre
Universidad Politécnica de Catalunya

J.L.A. Yebra
Universidad Politécnica de Catalunya

J. Fábrega
Universidad Politécnica de Catalunya

ABSTRACT

This paper studies digraphs that have walks of equal length ℓ between vertices. When such a digraph models a communication network, this means that any message can be sent from its origin to its destination with precisely ℓ delay time units. It is shown that a digraph D has this property unless it is a generalized cycle.

When D has the maximum possible order there is just one such walk between vertices. Among such digraphs are the Good-de Bruijn's. Other families are constructed by adequately modifying these digraphs and using some well-known constructions: line digraphs and conjunction of digraphs.

1. Equi-reachable digraphs

Let $D = (V,A)$ be a strongly connected digraph, and for any vertex x call $\Gamma(x)$ the set of vertices adjacent from x. Analogously call $\Gamma(W)$ the set of vertices adjacent from the vertices of $W \subset V$, and $\Gamma^n(W) = \Gamma(\Gamma^{n-1}(W))$. The digraph D has walks of equal length m between vertices iff for all $x \in V$

$$\Gamma^m(x) = V \qquad (1)$$

If ℓ is the smallest such m we say that D is ℓ-reachable. For convenience, we shall use the term equi-reachable for digraphs that are ℓ-reachable for some ℓ, that is for digraphs with walks of equal length between vertices.

In fact, for D to be equi-reachable it suffices that (1) holds for some $x \in V$, since then

$$\Gamma^{m+1}(x) = \Gamma(\Gamma^m(x)) = \Gamma(V) = V$$

because D is strongly connected. Thus $\Gamma^n(x) = V$ for all $n \geqslant m$ and then, if D has diameter k, there are walks through x of length $m + k$ between any two vertices of D. Therefore D is ℓ-reachable for some $\ell \leqslant m + k$.

Trivially, (1) can not hold when D is a cycle or a bipartite digraph. More generally, if D is a generalized cycle in the sense that V is the disjoint union of $r > 1$ subsets,

$$V = \{ \bigcup V_i,\ 0 \leqslant i \leqslant r-1 \}, \quad V_i \cap V_j = \emptyset \quad \text{for } i \neq j \qquad (2)$$

and, for i modulo r

$$\Gamma(V_i) = V_{i+1}, \quad 0 \leqslant i \leqslant r-1, \qquad (3)$$

it can not be equi-reachable. The following result shows that this is necessarily the structure of such digraphs.

Theorem 1.- A strongly connected digraph D is equi-reachable unless it is a generalized cycle.

Proof.- Suppose that D is not equi-reachable and consider for $x \varepsilon V$ the sequence

$$x,\ \Gamma(x),\ \Gamma^2(x),\ \ldots,\ \Gamma^n(x),\ \ldots \tag{4}$$

of nontrivial subsets of 2^V. Since necessarily repetitions will occur, let $\Gamma^{m+r}(x) = \Gamma^m(x)$ be the first one. Then

$$\Gamma^{m+r+t}(x) = \Gamma^t(\Gamma^{m+r}(x)) = \Gamma^t(\Gamma^m(x)) = \Gamma^{m+t}(x)$$

so that for $n \geqslant m$ the sets

$$V_i = \Gamma^{m+i}(x) \qquad 0 \leqslant i \leqslant r-1 \tag{5}$$

recur periodically in the above sequence. As D is strongly connected any $y \varepsilon V$ must appear in the periodic part of the sequence. Therefore

$$V = \{\bigcup V_i,\quad 0 \leqslant i \leqslant r-1\}$$

and in particular $r > 1$, since $V_i = \Gamma^{m+i}(x) \neq V$.

From its construction and the periodicity the sets V_i satisfy (3). Therefore to prove that D is a generalized cycle it suffices to show that they are disjoint. Suppose on the contrary that there exists $y \varepsilon V_i \cap V_j$, $i \neq j$, and let $h = d(y,x)$, so that $x \varepsilon \Gamma^h(y)$. Then

$$x \varepsilon \Gamma^h(y) \Longrightarrow \Gamma^m(x) \subset \Gamma^{m+h}(y)$$

$$y \in V_i \Rightarrow \Gamma^{m+h}(y) \subset \Gamma^{m+h+m+i}(x) = \Gamma^{m+t_1}(x)$$

$$y \in V_j \Rightarrow \Gamma^{m+h}(y) \subset \Gamma^{m+h+m+j}(x) = \Gamma^{m+t_2}(x)$$

where $0 \leqslant t_1,\ t_2 \leqslant r-1$ and $t_1 \neq t_2$ because $i \neq j$. So $\Gamma^m(x) \subset \Gamma^{m+t}(x)$ for some $1 \leqslant t \leqslant r-1$. But this implies $\Gamma^{m+t}(x) \subset \Gamma^{m+2t}(x)$ and then

$$\Gamma^m(x) \subset \Gamma^{m+t}(x) \subset \ldots \subset \Gamma^{m+rt}(x) = \Gamma^m(x)$$

so that $\Gamma^m(x) = \Gamma^{m+t}(x)$ with $1 \leqslant t \leqslant r-1$ against the choice for r.

Remarks

1.- When the digraph D is the Cayley diagram of a (finite) group G, the above decomposition of V corresponds to the partitioning of G into cosets given by the normal subgroup H of those elements that can be expressed in terms of the generators in such a way that the sum of exponents equals zero. Then r equals the greatest common divisor of the sums of exponents in the defining relators.

2.- We can characterize ℓ-reachable digraphs as those whose adjacency matrix A is such that A^ℓ is positive (that is, $(A^\ell)_{ij}>0$ for all i, j), and therefore by the fact that its spectral radius $\rho(A)$ is —a simple eigenvalue— greater in magnitude than any other eigenvalue, see, for instance, |1|.

2. Some constructions

We examine in this section the behaviour of equi-reachable and non-equi-reachable digraphs under some well-known graph constructions. It is evident that a digraph D is ℓ-reachable if and only if its converse digraph (i.e., the digraph obtained by reversing the orientation of the arcs in D) is ℓ-reachable. More important are the results on the line digraph of D and the conjunction of two digraphs D_1 and D_2.

2.1. The line digraph

In the line digraph L(D) of a digraph D = (V,A) each vertex represents an arc of D, that is

$$V(L(D)) = \{uv \mid [u,v] \in A(D)\} \,, \qquad (6)$$

and two vertices are adjacent when the corresponding arcs are adjacent in D. Its order is the size (i.e., number of arcs) of D. Then if D is strongly regular of degree d, the order of L(D) is d times the order of D.

The main result in our context is

Theorem 2.- A digraph is ℓ-reachable if and only if its line digraph is $(\ell+1)$-reachable.

Proof.- If D is ℓ-reachable there is a walk of length $\ell+1$ between any two vertices uv, wz of L(D) that uses the walk of length ℓ from v to w in D. And the argument can be reversed.

Using Theorem 1 it follows that D is a generalized cycle if and only if L(D) is a generalized cycle. In fact, in this case both decompose into the same number r of subsets.

2.2. Conjunction of digraphs

The conjunction $D_1 * D_2$ of two digraphs $D_1 = (V_1, A_1)$ and $D_2 = (V_2, A_2)$ is the digraph with set of vertices $V = V_1 x V_2$ and adjacency rule

$$[(x,y),(z,t)] \in A \Leftrightarrow [x,z] \in A_1 \text{ and } [y,t] \in A_2 \qquad (7)$$

It follows that its order is the product of the orders of D_1 and D_2 and its maximum out-degree the corresponding product of out-degrees, and that $D_1 * D_2$ is strongly regular if and only if both D_1 and D_2 are strongly regular.

When D_1 is ℓ_1-reachable and D_2 is ℓ_2-reachable their conjunction is ℓ-reachable with $\ell = \max\{\ell_1, \ell_2\}$, (use ℓ length walks from x to z in D_1 and from y to t in D_2 to construct a length ℓ walk from (x,y) to (z,t) in $D_1 * D_2$). Analogously it is easily seen that if one or both digraphs are not equi-reachable nor is their conjunction.

3. Largest digraphs

If D is ℓ-reachable and has maximum out-degree d its order is at most $N = d^{\ell}$, since this is the maximum number of different walks of length ℓ. To attain this bound there should be just one walk of length ℓ between any two vertices. It follows that the adjacency matrix A of D must verify the matrix equation $A^{\ell} = J$, and therefore D ought to be strongly regular of degree d (i.e., with in- and out-degree of all vertices equal to d), see |4|. Note also that these digraphs must be geodetic (i.e., with just one shortest path between any two vertices).

The ℓ-reachable digraphs with d^{ℓ} vertices have been studied by N.S. Mendelsohn in |6| as "UPP digraphs" (digraphs with the unique path property of order ℓ), and by Conway and Guy, unaware of the work of Mendelsohn, in |2| as "tight precisely ℓ-steps digraphs", using them to construct large transitive digraphs of given diameter. We use here Mendelsohn's terminology.

Among the UPP digraphs are the well-known Good-de Bruijn digraphs, whose set of vertices consist of all length ℓ words from an alphabet of d letters, and with a vertex x adjacent to y if the last ℓ-1 letters of x coincide with the first ℓ-1 letters of y.

But they are not the only UPP digraphs. For instance, for d=3 and ℓ=2 Mendelsohn presents in |6| five other non-isomorphic such digraphs that can be seen as models of grupoids. More generally, UPP digraphs can be seen as models of a universal algebra, see |7|.

We describe below two direct methods of constructing digraphs of this kind for any $d \geq 3$ and $\ell \geq 2$ by adequately modifying the Good-de Bruijn digraphs, that when d=3 and ℓ=2 produce the five above-mentioned digraphs.

Since, from Theorem 2, the line digraph of an ℓ-reachable digraph of order $N = d^{\ell}$ is a $(\ell+1)$-reachable digraph of order $dN = d^{\ell+1}$, it suffices to construct 2-reachable digraphs with order $N = d^2$ and then their iterated line-digraphs.

Consider then the Good-de Bruijn digraph of all length 2 sequences of d digits, $d \geq 3$, that is

$$V = \{x_0x_1,\ \ x_i \varepsilon X\}\ , \qquad X = \{0,1,\ldots,d-1\}$$

and x_ox_1 is adjacent to x_1x_2 for all choices of $x_2 \varepsilon X$.

For the first method, let $\alpha_0, \alpha_1, \ldots, \alpha_{d-1}$ be d (not necessarily different) permutations of X such that $\alpha_j(0) = 0$, and modify the Good-de Bruijn digraph by replacing each arc $[0i,ij]$ with the arc $[0i,\alpha_j(i)j]$ for $i,j = 0,1,\ldots,d-1$ (in fact for i=0 there is no alteration). To see that the resulting digraph is still 2-reachable (hence strongly regular) it suffices to consider the following walks from x_0x_1 to y_0y_1:

(1) If $x_0 \neq 0,\ x_1 \neq 0$ $\quad x_0x_1 \to x_1y_0 \to y_oy_1$.

(2) If $x_0 = 0,\ x_1 \neq 0$, setting $j = y_0$

$$0x_1 \to \alpha_j(x_1)j \to jy_1 = y_0y_1$$

since $x_1 \neq 0$ implies $\alpha_j(x_1) \neq 0$.

(3) If $x_1 = 0$, setting $j = y_1$, $i = \alpha_j^{-1}(y_0)$

$$x_0 0 \to 0i \to \alpha_j(i)\, j = y_0 y_1.$$

Alternatively, consider 2 permutations of X, α and β, such that $\alpha(1)=1$ and $\beta(0)=0$, and modify the Good-de Bruijn digraph by replacing each arc $[0i,i1]$ with the arc $[0i,\alpha(i)1]$ for all i, and also each arc $[1i,i0]$ with the arc $[1i,\beta(i)0]$ for all i. The new digraph is still 2-reachable as it can be seen by adequately adapting the previous reasoning. For example the length 2 walk from $x_0 0$ to $y_0 1$, $x_0 \neq 0$, go through $0y_0$ if $y_1 \neq 1$ and through $0\alpha^{-1}(y_0)$ if $y_1 = 1$.

When d=3 the choices of α_0, α_1 and α_2 as

$$\alpha_0 = (1,2) \qquad \alpha_1 = \iota \qquad \alpha_2 = \iota \qquad (D_1)$$

$$\alpha_0 = \iota \qquad \alpha_1 = (1,2) \qquad \alpha_2 = \iota \qquad (D_2)$$

for the first method, and the choices of α and β as

$$\alpha = (0,2) \qquad \beta = (1,2) \qquad (D_3)$$

for the second, yield three non-isomorphic 2-reachable digraphs. The other two, besides Good-de Bruijn's, are the converse digraphs of D_2 and D_3.

The digraphs D_2 and D_3 are non-planar, so they are counterexamples to the conjecture of Mendelsohn in |5| and |6|.

The above constructions requiere $d \geq 3$. For d=2 and $k \geq 3$ analogous techniques yield non-isomorphic UPP digraphs. For instance if $\ell=3$, besides the Good-de Bruijn digraph, there are two other 3-reachable digraphs. One is shown in Figure 1 and the other one is its converse digraph.

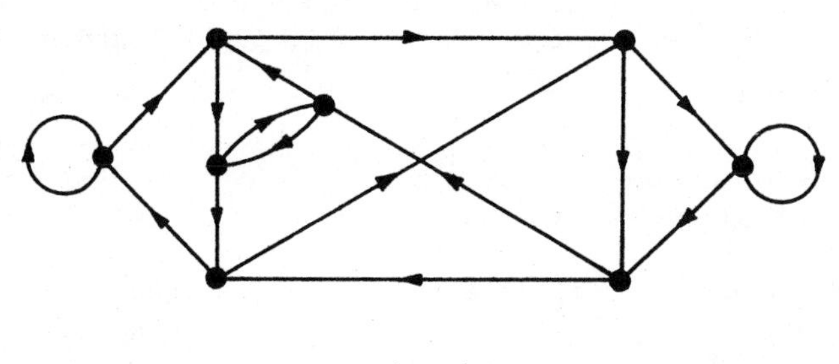

Fig. 1

This digraph is useful to verify a question raised by Mendelsohn in |6|. For given values of ℓ and d, he calculates the number of elementary circuits of length $n \leq \ell+1$ in the Good-de Bruijn digraphs. He shows that all UPP digraphs have the same number of elementary circuits of length $n \leq \ell$ and wonders if this result still holds for $n > \ell (n \leq d^{\ell})$. However, already for d=2, $\ell=3$ and n=5 the Good-de Bruijn digraph has 2 elementary circuits while the digraph of Figure 1 has 3. Moreover, their line digraphs may be used to verify that even for $n=\ell+1$ the result does not hold.

REFERENCES

1. A. Berman & R.J. Plemons, *Nonnegative matrices in the mathematical sciences*, Ac. Press, 1979.

2. J.H. Conway & M.J.T. Guy, Message Graphs, *Annals of Discrete Mathematics*, 13 (Proc. of the Conf. on Graph Theory, Cambridge, 1981), North Holland, 1982, 61-64.

3. M.A. Fiol, J.L.A. Yebra & I. Alegre, Line digraph iterations and the (d,k) digraph problem, *IEEE Trans. Comp.* C-33 (1984) 400-403.

4. A.J. Hofmann & M.H. McAndrew, The polynomial of a directed graph, *Proc. Amer. Math. Soc.* 16, (1965) 303-309.

5. D.M. Johnson & N.S. Mendelsohn, Planarity properties of the Good-de Bruijn graphs, *Combinatorial Structures and their Applications*, (Proc. Calgary Int. Conf. 1969), Gordon and Breach, 1970, 177-183.

6. N.S. Mendelsohn, Directed graph with the unique path property, *Combinatorial Theory and its Applications* II (Proc. Colloq. Balatonfürer, 1969), North Holland, 1970, 783-799.

7. N.S. Mendelsohn, An application of matrix theory to a problem in universal algebra, *Linear Algebra and its Applications*, 1 (1968) 471-478.

APPLICATION OF NUMBERED GRAPHS IN THE DESIGN OF MULTI-STAGE TELECOMMAND CODES

S. Ganesan
Department of Electrical Engineering
Western Michigan University
Kalamazoo, Michigan 49008 U.S.A.

M. O. Ahmad
Department of Electrical Engineering
Concordia University
Montreal, Canada

ABSTRACT

Many types of telecontrol systems are guided by commands consisting of time sequences of RF pulses. The design of certain non-periodic telecommand codes resistant to random interference pulses, is equivalent to numbering the complete graph in such a way that all the edge numbers are different. The vertex numbers represent the time positions at which pulses are transmitted. This design problem corresponds to numbering as gracefully as possible K number of disconnected undirected graphs, each with n vertices. It is also shown that design of such codes with reduced protection corresponds to numbering K disconnected directed graphs with the required conditions on the vertex (or edge) numbers. Similarly the security function of the multistage code can be considered as a single undirected graph numbered as gracefully as possible and the command function can be chosen similar to the single stage code. Example of graph models and design conditions for the above codes are presented in this paper. A computer program which can be used for choosing the vertex numbers is also explained.

I. INTRODUCTION

Graph theory has establised itself as an important mathematical tool in a wide variety of subjects. Numbered graphs have found usage in various coding theory problems. The design of certain non-periodic telecommand codes is equivalent to numbering the complete graph in such a way that all the edge numbers are different. The node numbers represent the time positions at which pulses are transmitted. The telecommand code receiver passes all incoming signal pulses through a shift register. If it is tapped at several places which correspond to the actual time interval between incoming pulses then the coincedence detection of these pulses by a multi input AND gate, can initiate some control action. In order to make the code insensitive to random interference pulses, all the delays between the pulses for one command must totally differ from those of every other command. A brief description of this type of code is given in section 2. After a brief description of the definitions in section 3, the graph models and design methods for these codes are described in sections 4 and 5.

II. TELECOMMAND CODES

Many types of telecontrol systems are guided by commands consisting of time sequences of RF pulses. Eckler [1] has described missile guidance codes i.e. 1) Single stage n pulse codes with maximum and reduced protection against random additive type interference. 2) Two stage codes where the security and command functions of the code are separated. The following section briefly reviews

Eckler's single stage and two stage codes. Detailed analysis and design can be found in [1].

Single Stage Code

A command is represented by a set of n pulses; the command information is contained in the (n-1) time spacings between pairs of successive pulses. There are K distinct commands. These K commands are encoded such that i false pulses cannot combine in any way with (n-1) pulses from any command to form either the same command (shifted in time) or one of the other (K-1) commands. The maximum value of i is equal to (n-2). Since it is desirable to keep all the commands as short as possible, the time spacings are selected such that the length of the longest command is minimum. Eckler has given the equation to be satisfied while choosing the time spacings for codes with maximum protection and reduced protection. Tables I, II, and III show the pulse single stage code spacings for 2 and 3 commands, secure against 3, 2 and 1 interference pulses. Only a few values are shown here as an example. Fig. 1 shows the pulse waveform for the 3 types of 2 command 5 pulse codes. Fig. 2a shows the encoder and decoder circuit for one type of 5 pulse single stage code. In a multistage code the security function of the code and the command function of the code are separeated. In the decoder the output from one (e.g. security) decoding operation becomes as input to the next (e.g. command) decoding operation. The symbol i/j denotes two stage code containing i/j pulses. The methods for the construction of these codes are given in Section IV. Fig. 3 shows 3/3 two-stage code waveforms. Fig. 2b shows the decoder circuit for one type of two-stage code.

III. NUMBERED GRAPHS

A graph with n vertices is numbered if each vertex is assigned an integer value and each edge is assigned a value which depends

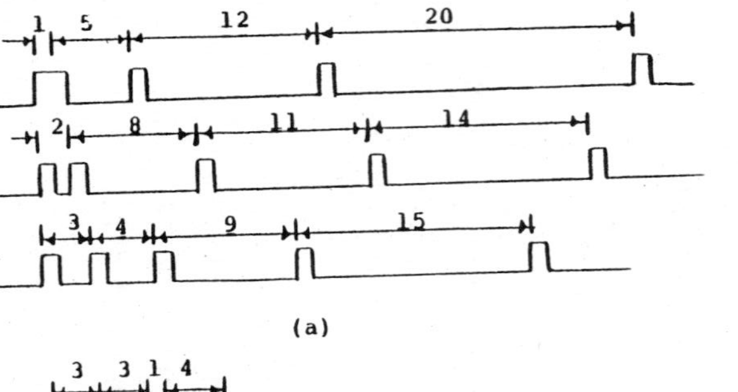

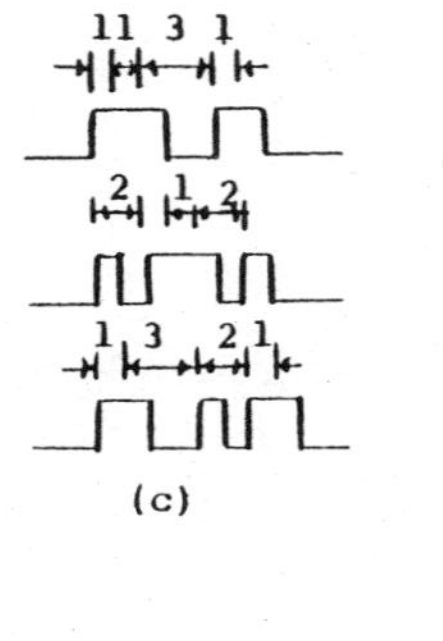

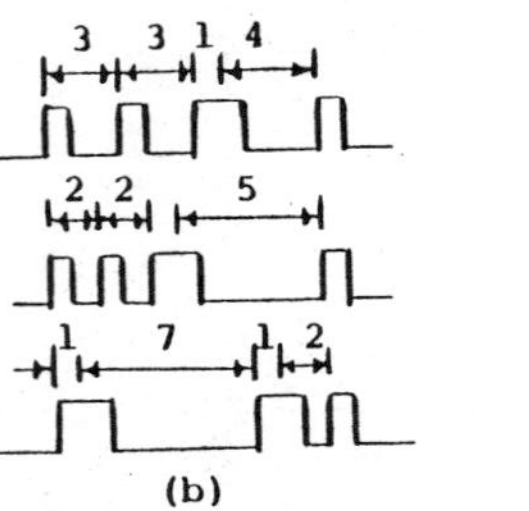

Fig.1. 5-pulse single-stage codes, number of commands equals three (a) Secure against 3 false pulses (b) secure against 2 false pulses (3) secure against 1 false pulse.

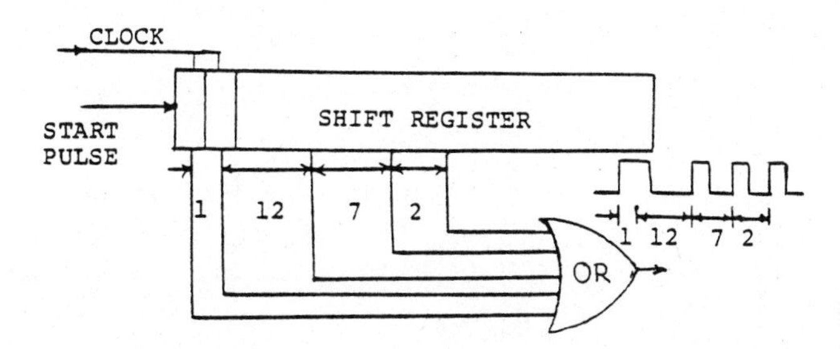

(i)

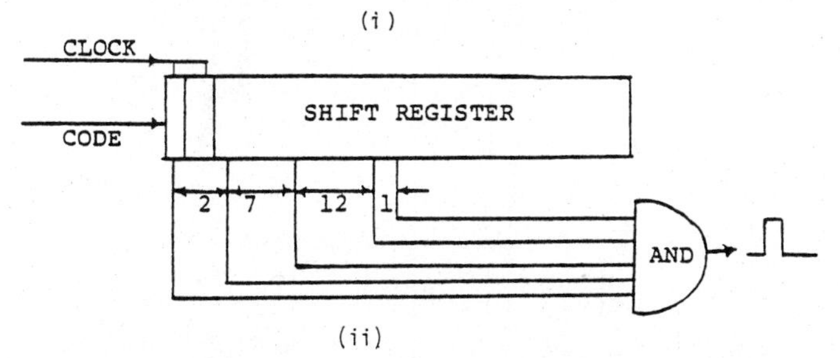

(ii)

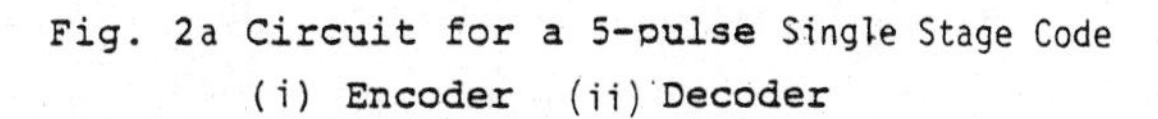
Fig. 2a Circuit for a 5-pulse Single Stage Code
(i) Encoder (ii) Decoder

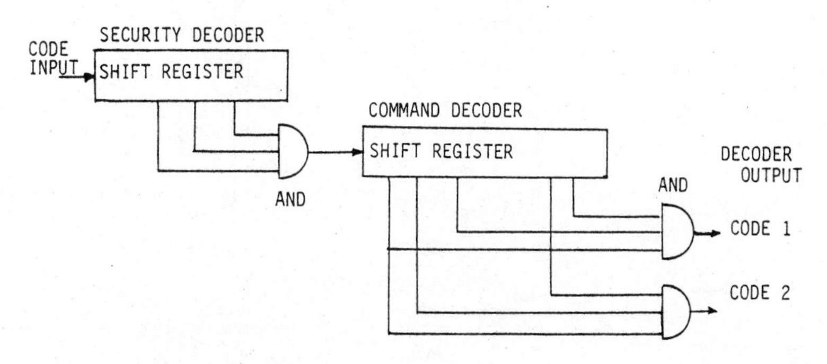

Fig. 2b. TWO STAGE CODE DECODER

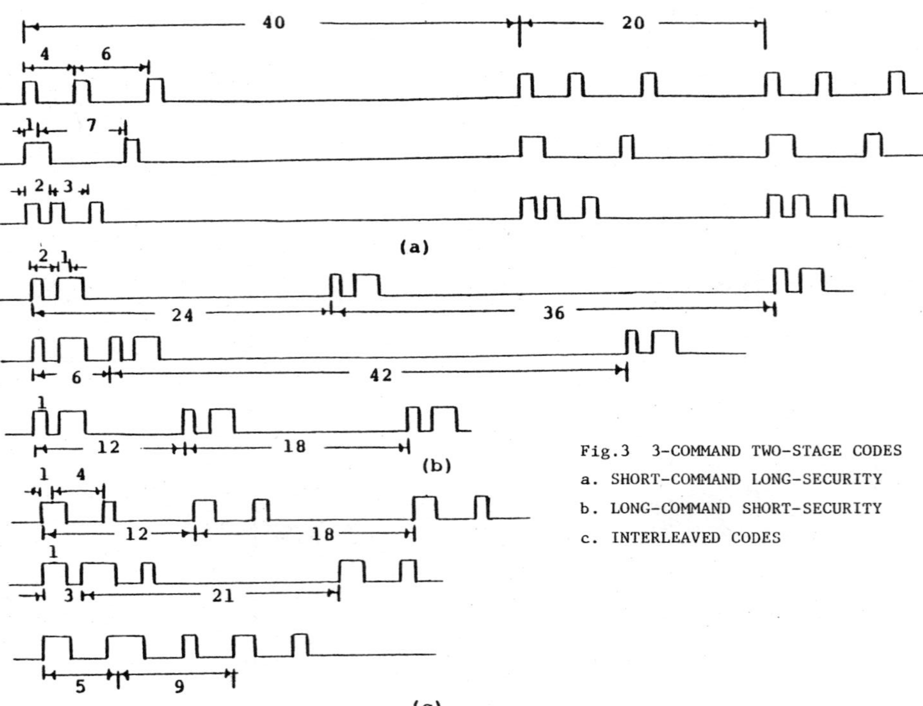

Fig.3 3-COMMAND TWO-STAGE CODES

a. SHORT-COMMAND LONG-SECURITY

b. LONG-COMMAND SHORT-SECURITY

c. INTERLEAVED CODES

on the end vertices numbers of that edge. The complete graph of n vertices has $(n^2-n)/2$ lines joining all pairs of points. A graph with E edges is called graceful if its vertices can be numbered with distinct non-negative integers no larger than E. Figure 4 shows the graphs with graceful numberings. Golomb [3] has shown that for $n \geq 5$ the complete graph cannot be labelled gracefully. The existence of such numberings and their properties have been investigated by Golomb [3], Rosa [4] , Kotzig [5] and are surveyed by Bloom [6]. A semigraceful graph is one where the largest edge number has the optimal-minimal value. A semigraceful graph for n = 5 is shown in fig. 5. Bloom and Golomb [6] have presented the application of numbered undirected graphs in various coding theory problems, X-ray crystallography, communication network addressing system and others. In the following section the application of undirected and directed graph in the design of telecommand codes mentioned in Section II, are described.

IV. GRAPHS MODELS FOR TELECOMMAND CODES

Eckler [1] designed the interval time spacings for many feasible codes. In 1967, Robinson and Bernstein [2] improved some of his results. Bloom and Golomb [8] have described that the design of Eckler's code is identical to finding a set of n rulers or numbering as gracefully as possible a disconnected or connected undirected graph. In following sections the usage of undirected graph in the design of single stage codes and multi stage codes with maximum protection, and directed graph in the design of codes with reduced protection, are described.

Undirected Graph Model for Single Stage Codes with Maximum Protecttection

The design of single stage codes with maximum protection corresponds to numbering the Kn vertices of K copies of the complete

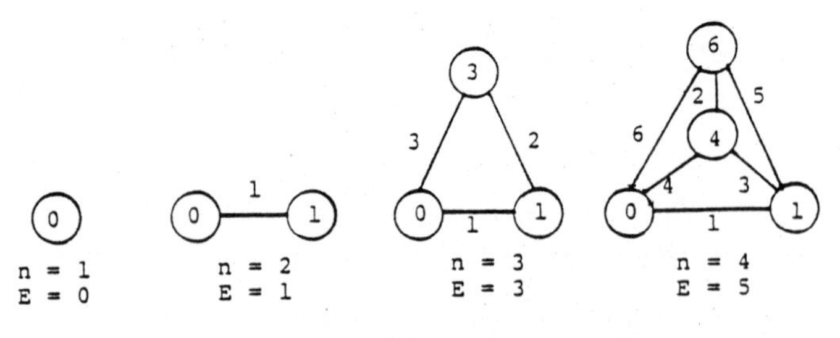

Fig. 4 Graph with graceful numbering

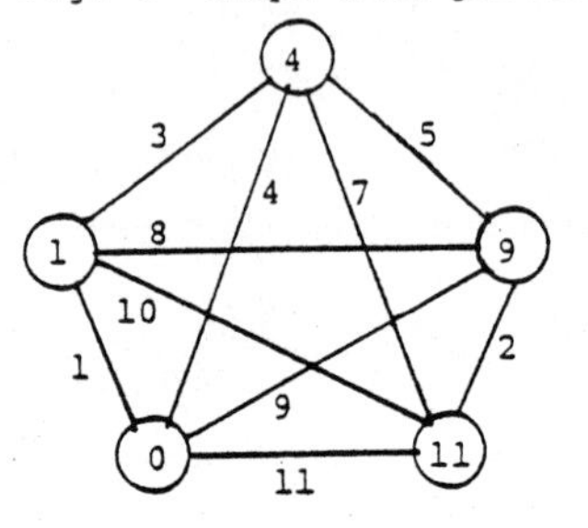

Fig. 5 Graph with semigraceful numbering, n=5, E=10

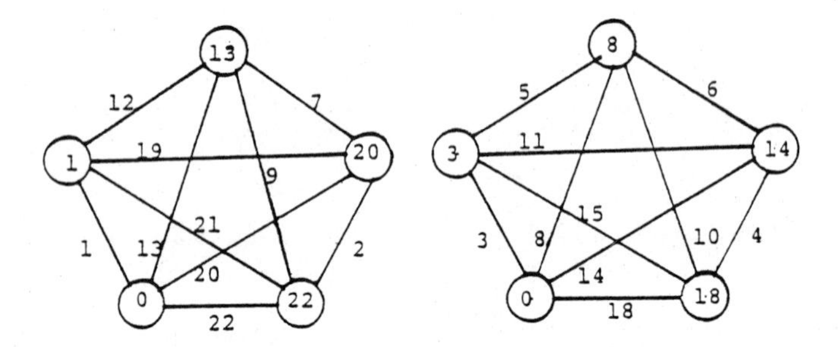

Fig. 6 Numbered undirected graph model for 5-pulse 2-command codes secure against 3 false pulse

graph. If t_i^j represents the i th edge number j th graph

(i.e the time interval spacing between i th pulse and (i + 1)th pulse),

then $V_i^j - V_{i-1}^j = t_i^j$ where $i = 1$ to $n - 1$, $j = 1$ to K

n = number of pulses/command

K = number of command

The numbers t_i^j are chosen such that [1] the following elements are all different.

$$t_i^j \ ,$$

$$t_i^j + t_{i+1}^j \ ,$$

$$\vdots$$

$$\sum_{i=1}^{n-1} t_i^j \ . \qquad (1)$$

i.e., the $K \binom{n}{2}$ differences $V_i^j - V_m^j$ must be distinct for all j and all i > m.

Fig. 6 shows the graph model for 5 pulse 3 command codes, secure against 3 false pulses, whose time spacings are shwon in Table I.

Directed Graph Model For Single Stage Codes With Reduced Protection

This problem corresponds to designing a directed graph such that [1] all the edge directions from vertex (V_1) are going away from V_1. The last vertex V_n is a sink i.e. no edge is directed away from V_n. All the edge numbers are chosen such that, the possible sets of consecutive edge numbers, chosen along the directed paths, are different. The number of elements in the sets depends on the protection against the number of false pulses. For example, if we represent $V_{i+j} - V_i = t_i$ and $V_{i+2} - V_i = t_i + t_{i+1}$ then for a 5 pulse code secure against 2 false pulses the following sets should be different.

$$(t_1^j, t_2^j), (t_2^j, t_3^j), (t_3^j, t_4^j), (t_1^j + t_2^j, t_3^j),$$

$$(t_2^j + t_3^j, t_4^j), (t_1^j, t_2^j + t_3^j), (t_2^j, t_3^j + t_4^j),$$

$$(t_1^j, t_2^j + t_3^j + t_4^j), (t_1^j + t_2^j, t_3^j + t_4^j),$$

$$(t_1^j + t_2^j + t_3^j, t_4^j) . \qquad (2)$$

for $j = 1, 2 \cdots K$

To construct 5 pulse code secure against one false pulse, the following sets must be different.

$$(t_1^j, t_2^j, t_3^j), (t_2^j, t_3^j, t_4^j), (t_1^j + t_2^j, t_3^j, t_4^j),$$

$$(t_1^j, t_2^j + t_3^j, t_4^j), (t_1^j, t_2^j, t_3^j + t_4^j)$$

for $i = 1, 2 \cdots K$ (3)

The graph model for 5 pulse codes with reduced protection are shown in Fig. 7. 5 pulse single stage code spacings secure against 2 and 1 false pulses are shown in Table II and III.

Graph Model for Multi-Stage Codes

In two stage codes there are three methods of combining security and command codes [1],

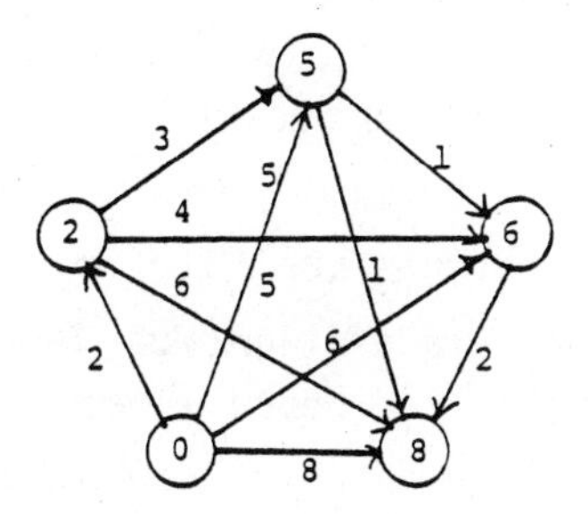

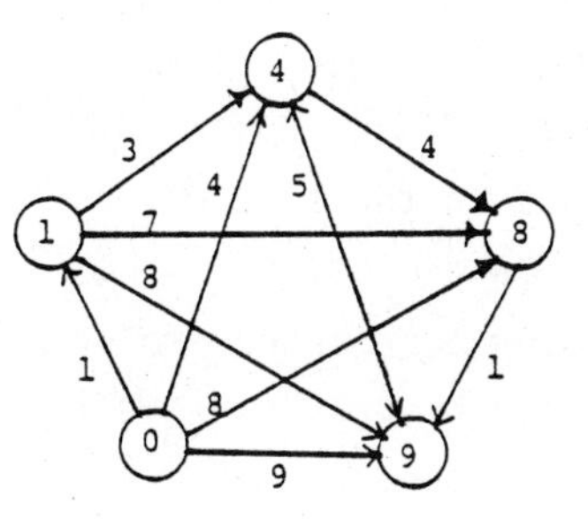

Fig. 7(a) Numbered directed graph model for 5-pulse 2-command codes secure against 2 false pulses

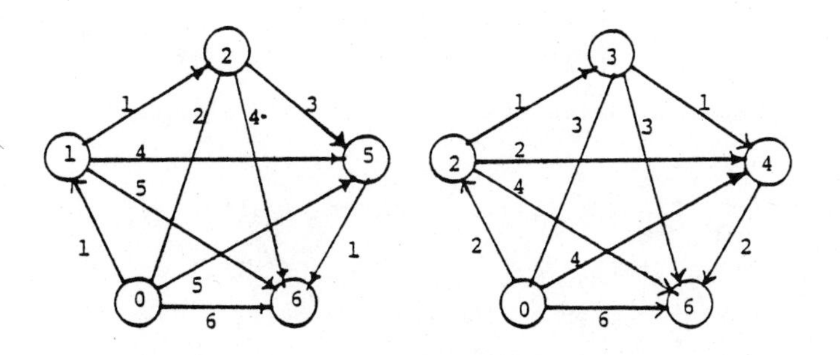

Fig. 7(b) Numbered directed graph model for 5-pulse 2-command codes secure against 1 false pulse.

(i) Short command codes, long security codes

(ii) Long command codes, short security codes

(iii) Interleaved security and command codes

The first two methods require that the security code spacings be designed in the same way as for a 1 command single stage code with maximum protection. The time spacings are shown in Table IV. This problem corresponds to designing graceful graphs. When $n \leq 5$, graceful graphs are not possible and hence a semi-graceful graph has to be designed. For interleaved code, the i-pulse security code spacings are chosen, i.e., the i-vertex numbered undirected graph is designed such that

(i) the edge numbers satisfy all the conditions for the i-pulse single-stage code with maximum protection, and

(ii) none of the $i(i-1)$ integers $t_p, (t_p+t_{p+1})$, ..., $(t_1+t_2+\ldots t_{i-1})$, where V_p is the vertex number of the pth vertex, $t_p=|V_{p+1}-V_p|$, and p=1, ..., (i-1), is multiple of i.

Table V shows the time spacings for interleaved security code. Fig. 3 depicts the three types of 3/3 two-stage codes.

V. COMPUTER PROGRAM TO SELECT THE VERTEX NUMBERS

For a n pulse, K command codes one has to fix $K(n-1)$ elements such that $Kn(n-1)/2$ partial sums are all different and longest code length is as short as possible. A straight forward algorithm will require searching $K(n-1)!$ combinations and a few of the combinations will produce the minimum length time spacings.

The algorithm developed here requires far less search, though this search itself may be large for higher values of K and n. The program has been divided into two parts: (i) Initialization and (ii) Refinement.

The initialization program provides a nearby solution quickly. The principle of this part is explained by an example. For example, when K = 3 and n = 4, the elements can be fixed in the following order (the upper script denotes command or graph number, and the lower script the time spacing or edge number).

$$t_1^1,\ t_1^2,\ t_1^3,\ t_3^3,\ t_3^2,\ t_3^1,\ t_2^3,\ t_2^2,\ t_2^1$$

This produces the time spacings for 3 commands as:

(1, 16, 6), (2, 13, 5), (3, 7, 4)

The length of the longest command is 23. The minimum achieved after refinement program is 19.

The flow chart for the initialization program is shown in Fig. 8. The order in which the elements have to be fixed is chosen by the programmer. The action of initialization program involves

1. Selection of an element
2. Generating the corresponding partial sums
3. Comparison with other elements and partial sums
4. If the condition is not satisfied repeat steps 1 to 3.

The refinement algorithm tries to refine the output of initialization program by systematic search. Fig. 9 shows the flow chart for the refinement program. The combinations which will produce higher length codes than the minimum so far obtained, are not considered. The order in which the elements must be refined is specified. The

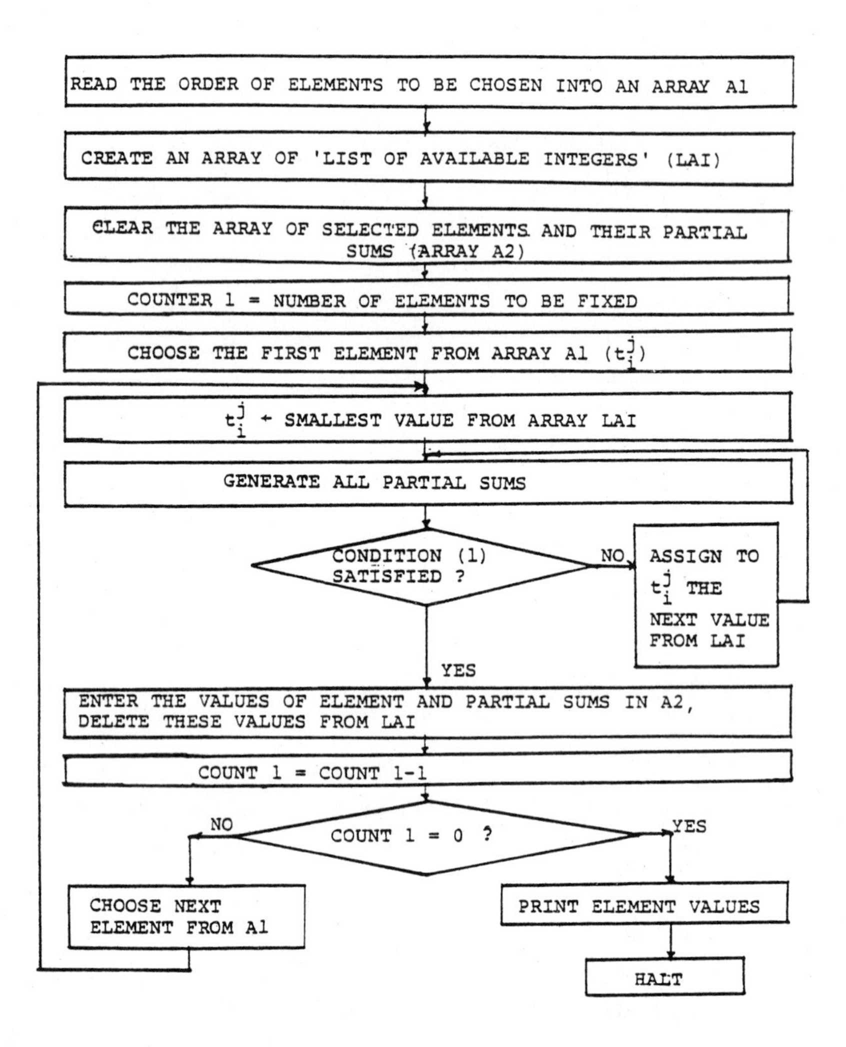

Fig. 8 Flowchart for initialization program

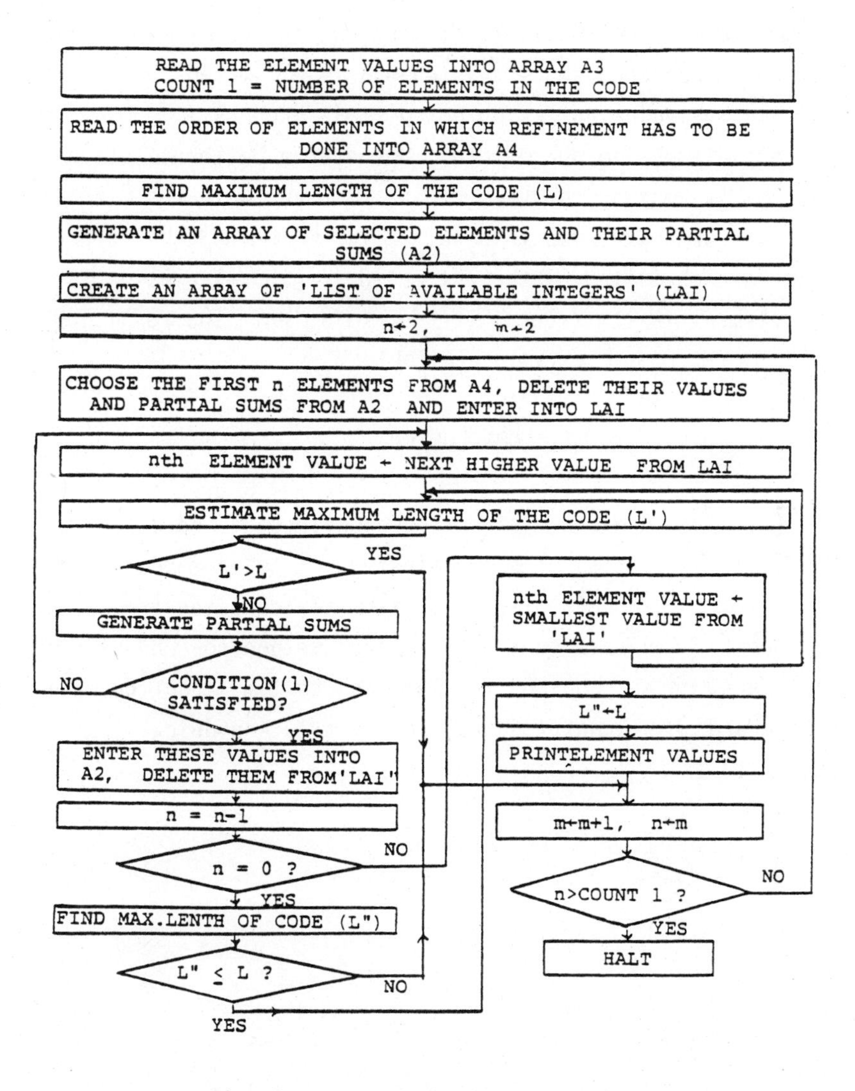

Fig. 9 Flowchart for refinement program

value of the element chosen for refinement, is given a new higher value from the list of available elements. The approximate maximum length of the code because of this change is estimated. If it is smaller than the value so far obtained, this process is repeated for the next element to be refined. Whenever a better solution is obtained, it is kept as the value to be compared with the estimated maximum. The program comes to a halt after trying all the combinations that are permitted by the estimated length calculation for the last element in the order given for refining. The result obtained for K = 2, n = 5 is shown in Table VI.

VI. CONCLUSION

In this paper, the usage of numbered undirected and directed graph model in the design of a certain type of telecommand codes, is shown. There is no efficient algorithm for choosing the optimal vertex numbers. A simple systematic search method for choosing the vertex numbers, which requires far less search time than the full search time and which can produce a nearby solution quickly has been suggested.

TABLE I

5-PULSE SINGLE-STAGE CODE SPACINGS SECURE AGAINST 3 FALSE PULSES

K	Max. length	Time spacings
2	22	(1, 12, 7, 2) (3, 5, 6, 4)
3	38	(1, 5, 12, 20) (2, 8, 11, 14) (3, 4, 9, 15)

TABLE II

5-PULSE SINGLE-STAGE CODE SPACINGS SECURE AGAINST 2 FALSE PULSES

K	Max. length	Time spacings
2	22	(2, 3, 1, 2) (1, 3, 4, 1)
3	11	(3, 3, 1, 4) (2, 2, 1, 6) (1, 7, 1, 2)

TABLE III

5-PULSE SINGLE-STAGE CODE SPACINGS SECURE AGAINST 1 FALSE PULSE

K	Max. length	Time spacings
2	6	(1, 1, 3, 1) (2, 1, 1, 2)
3	7	(1, 1, 3, 1) (2, 1, 1, 2) (1, 3, 2, 1)

TABLE IV

MINIMUM LENGTH SINGLE-STAGE CODE SPACING FOR ONE COMMAND

Number of pulses	Minimum length	Code spacings
2	1	1
3	3	(2, 1)
4	6	(2, 3, 1)
5	11	(2, 5, 3, 1)
6	17	(4, 2, 3, 7, 1)
9	45	(3, 6, 7, 4, 1, 14, 8, 2)

TABLE V

THE SECURITY CODE SPACINGS FOR 2-STAGE INTERLEAVED CODES

Number of secuirty pulses (i)	Minimum length	Time spacings
2	1	1
3	5	(1, 4)
4	9	(2, 1, 6)
5	17	(6, 7, 1, 3)
6	20	(4, 5, 8, 2, 1)

TABLE VI

TIME SPACINGS FOR 3-PULSE CODES SECURE AGAINST ONE FALSE PULSE
Number of commands = 2

Number of possible combinations	Time spacings	Minimum length of longest code
1	(1, 3) (2, 5)	7
2	(1, 5) (3, 4)	7
3	(2, 3) (1, 6)	7

REFERENCES

[1] A. R. Eckler, "The construction of missile guidance codes resistant to random interference," BELL SYST.-TECH. J., Vol. 39, pp. 973-974, July 1960.

[2] J. P. Robinson and A. J. Bernstein, "A class of binary recurrent codes with limited error propagation," IEEE Trans. Inform. Theory, Vol. IT-13, pp. 106-113, January 1967.

[3] S. W. Golomb, "How to number a graph," Graph theory and computing , New York, Academic Press, 1972.

[4] A. Rosa, "On certain valuations of the vertices of a graph," in theory of graphs: Int. Symp. Rome, July, 1966, pp. 349-355, Paris, France :Dunod, 1967.

[5] A. Kotzig, "On certain vertex-valuations of finite graphs," Utilitas Mathematica, Vol. 4, pp. 261-290, March 1973.

[6] G. S. Bloom,"Numbered undirected graphs and their uses," Ph.D. dissertation, Univ. Southern California, Los Angeles, 1975.

[7] G. S. Bloom and S. W. Golomb, "Numbered complete graphs, unusual rulers, and assorted applications," in theory and applications of graphs - Ed Y. Alavi and D. R. Lick, Springer Verlag, 1978.

[8] G. S. Bloom and S. W. Golomb, "Applications of numbered undirected graphs," Proc. IEEE, Vol. 65, No. 4, pp. 562-570, April 1977.

THE COCHROMATIC NUMBER OF GRAPHS IN A SWITCHING SEQUENCE

John Gimbel
Colby College

ABSTRACT

By $S_H(G)$ we mean the graph G switched on H, where H is a subset of the vertex set of G. Given a labeling $v_1, v_2, \ldots, v_p$ of the vertex set of G, let $G_0 = G$ and $G_i = S_{\{v_1, v_2, \ldots, v_i\}}(G)$, for $1 \leq i \leq p$. We refer to the sequence of graphs $G_0, G_1, \ldots, G_p$ as a switching sequence. In this paper we discuss the cochromatic number of graphs in switching sequences.

1. Introduction.

For undefined concepts the reader is referred to [1]. Let G denote a graph and H a nonempty proper subset of V(G). By G switched on H, denoted $S_H(G)$, we mean the graph with vertex set V(G) and edge set $E(G-H) \cup \{uv \in E(\bar{G}) \mid u \in H\}$. If $H = \emptyset$ let $S_H(G) = G$. If $H = V(G)$ let $S_H(G) = \bar{G}$. This concept was first introduced in [2]. We will only be concerned with graphs which have nonempty vertex sets.

Given a labeling $v_1, v_2, \ldots, v_p$ on the vertices of G, let $G_0 = G$ and $G_i = S_{\{v_1, v_2, \ldots, v_i\}}(G)$ for $1 \leq i \leq p$. This provides us with a sequence of graphs $G_0, G_1, \ldots, G_p$. Let us refer to this sequence as the switching sequence on G associated with the labeling $v_1, v_2, \ldots, v_p$. A sequence of graphs is a switching sequence if it is a switching sequence on some graph with some vertex labeling. A family F of graphs is a spectrum if there is a switching sequence $G_0, G_1, \ldots, G_p$ such that $F = \{G_0, G_1, \ldots, G_p\}$.

2. Cochromatic Number.

Given G, a graph, a vertex partition of G is a collection of nonempty, pairwise disjoint sets $V_1, V_2, \ldots, V_n$ whose union is V(G). The order of a vertex partition is the number of sets in the partition. That is, the order of $V_1, V_2, \ldots, V_n$ is n. A vertex partition of G is a cocoloring of G if each set in the partition induces a complete or an empty graph in G. A cocoloring of order n will be called an n-cocoloring. The cochromatic number of G, denoted Z(G), is the minimum order of all cocolorings of G. Cochromatic theory has been studied in [3], [4], [5]. We are interested here in studying the cochromatic numbers of graphs in spectra. We will examine sets of the form $\{Z(G) | G \in F\}$ where F is a spectrum.

By $K_{n(n)}$ we mean the complete n-partite graph with n vertices in each partite set. As a corollary to a theorem of Lesniak and Straight [3], it is known that $Z(K_{n(n)}) = n$.

Our first lemma is presented without proof. It tells us that the cochromatic number increases or decreases by at most one when moving from a graph to its successor in a switching sequence.

Lemma 1. Let $G_0, G_1, \ldots, G_p$ be a switching sequence. Then, for $i = 1, 2, \ldots, p$ we have $Z(G_{i-1}) - 1 \leq Z(G_i) \leq Z(G_{i-1}) + 1$. Furthermore, $Z(G_1) - 1 \leq Z(G_0) \leq Z(G_1) + 1$.

Lemma 2. Let $A = \bigcup_{i=1}^{n} S_i$ where $S_1, S_2, \ldots, S_n$ is a sequence of pairwise disjoint sets with $|S_i| = n$ for each i. Let $B_1, B_2, \ldots, B_m$ be a partition of A, where $m<n$. Then, there is a set B_K containing distinct elements w, x, y, and z such that

$\{w, x\} \subseteq S_i$,
$\{y, z\} \subseteq S_j$,
and $i \neq j$.

Proof:

Since $n>m$ each S_i must have at least two elements in some B_j. Define a function f with the property that S_i has at least two elements in $B_{f(i)}$, for $i = 1, 2, \ldots, n$. By the pigeonhole principle, there exist i and j such that $1 \leq i \neq j \leq n$ and $f(i) = f(j)$.

Now, S_i has two elements in $B_{f(i)}$ as does S_j. ■

Lemma 3. For each positive integer N, there is a graph G_N with the property that for any $H \subseteq V(G_N)$ the cochromatic number of $S_H(G_N)$ is N.

Proof:

Let $G_N = K_{N(N)}$. Let H be a subset of the vertex set of G_N. We first show that $Z(S_H(G_N)) \leq N$. Let $S_1, S_2, \ldots, S_N$ denote the partite sets of G_N. Now, if $S_i \cap H$ is nonempty label the vertices of $S_i \cap H$ as $u_2^i, u_3^i, \ldots, u_{p_i}^i$ if $i = 1$; as $u_1^i, u_2^i, \ldots, u_p^i$ if $i>1$ and $|S_i \cap H|<i$ and as $u_1^i, u_2^i, \ldots, u_{i-1}^i, u_{i+1}^i, \ldots, u_{p_i}^i$ in all other cases. Note, in any case, $p_i \leq N+1$.

We now describe a three step process for partitioning $V(S_H(G_N))$ into N cocolor classes $V_1, V_2, \ldots, V_N$.

Step 1

For $i = 1, 2, \ldots, N$ let V_i initially be defined to equal $S_i - H$.

Step 2

If V_i is the empty set, add $u_{p_i}^i$ to V_i. Note that in such a case $S_i \subseteq H$. Thus, $p_i = N + 1$.

Also note that at this point, each V_i is nonempty and induces an empty graph in G_N as well as $S_H(G_N)$.

Step 3

For each $i = 1, 2, \ldots, N$ and $j = 1, 2, \ldots, N$, if u_j^i exists, add this vertex to V_j. Since u_j^i is in H, it is not adjacent in $S_H(G_N)$ to any elements in S_k, $k \neq i$. In particular, u_j^i is not adjacent to any u_j^k nor to any vertex in S_j. Thus, each V_j induces an empty graph in $S_H(G_N)$.

Since each vertex is a member of some V_i, it follows that $Z(S_H(G_N)) \leq N$.

We now use the preceding lemma to show that $N \leq Z(S_H(G_N))$. Suppose, to the contrary, that $Z(S_H(G_N)) < N$. Let $C_1, C_2, \ldots, C_m$ be a cocoloring of $S_H(G_N)$, where $m<N$. By the lemma, there are distinct vertices u, v, w, x and some integer k such that $\{u, v, w, x\} \subseteq C_k$, and there are sets S_i and S_j, $i \neq j$, such that $\{u, v\} \subseteq S_i$ and $\{w,x\} \subseteq S_j$.

We now produce a contradiction by showing that C_k induces neither a complete nor an empty graph in $S_H(G_N)$. Without loss of generality, there are two cases to be considered. We will denote $V(G_N)$ by V.

Case 1

Suppose $\{u, v\} \subseteq V-H$. Since $\{u, v\} \subseteq S_i-H$, it follows that u and v are not adjacent in $S_H(G_N)$. Hence, C_k must induce an empty graph in $S_H(G_N)$. If w or x is in H, then w is adjacent with x in $S_H(G_N)$, contradicting the fact that C_k induces an empty graph in $S_H(G_N)$. So neither w nor x is in H. Since v and w are adjacent in G_N they are adjacent in $S_H(G)$. Thus C_k does not induce an empty graph in $S_H(G_N)$ and a contradiction is reached.

Case 2

Suppose that $u \in H$. Since u and v are non-adjacent in G_N they are adjacent in $S_H(G_N)$. But u and w are adjacent in G_N so they are not adjacent in $S_H(G_N)$. So C_k does not induce either a complete or an empty graph in $S_H(G)$. Again, a contradiction is reached.

We conclude that $N \leq Z(S_H(G_N))$. ■

In the following, we will use the fact that the cochromatic number of a graph is bounded below by the cochromatic number of any of its induced subgraphs.

A set S of integers is cochromatic if there is a spectrum F such that $S = \{Z(G) | G \in F\}$. A set of integers is convex if it is finite and contains each integer between its minimum and its maximum.

Theorem 4. A set S of integers is cochromatic if and only if it is convex and Max S $\leq$ 2 Min S.

Proof. Suppose S is cochromatic. Let $G_0, G_1, \ldots, G_p$ be a switching sequence where $S = \{Z(G_0), Z(G_1), \ldots, Z(G_p)\}$. From lemma 1, we see that S must be convex. Now, let G be a graph and $v_1, v_2, \ldots, v_p$ be a vertex labeling on G so that $G_0 = G$ and $G_i = S_{\{v_1, v_2, \ldots, v_i\}}(G)$ for $i = 1, 2, \ldots, p$. Select k so that $Z(G_k) = \text{Min } S$. Select l so that $Z(G_l) = \text{Max } S$. We shall consider the special case where $0 \leq k \leq l$. The other case follows similarly. Let $V_1, V_2, \ldots, V_m$ be a cocoloring of G_k where $m = Z(G_k)$. We shall show that $Z(G_l) \leq 2m$. Hence, Max S $\leq$ 2 Min S. If k = l this is clearly true. So, suppose $k < l$. Let $L = \{v_{k+1}, v_{k+2}, \ldots, v_l\}$. Note, if V_i induces a complete graph in G_k then $V_i - L$ induces a complete graph in G_l and $V_i \cap L$ induces an empty graph in G_l. Likewise, if V_i induces an empty graph in G_k then $V_i - L$ induces an empty graph in G_l and $V_i \cap L$ induces a complete graph in G_l. Hence, after removing the empty sets from $V_1 - L, V_2 - L, \ldots, V_m - L, V_1 \cap L, V_2 \cap L, \ldots, V_m \cap L$ a cocoloring of G_l is formed. Thus, $Z(G_l) \leq 2m$.

Now, suppose $S = \{m, m+1, \ldots, n\}$, where m and n are integers with $1 \leq n \leq 2m$. We wish to show that such a set is cochromatic. We do this by constructing a graph G and a labeling $v_1, v_2, \ldots, v_p$ of V(G) so that the corresponding set $\{Z(G_0), Z(G_1), \ldots, Z(G_p)\}$ is in fact equal to the set S.

If S is a singleton then by Lemma 3 the set S is cochromatic. So, assume $m < n$. Thus $n \geq 2$.

Let $k = n - m$. Also, let A_i be a set of n vertices for each $i = 1, 2, \ldots, k$. Define each A_i so that $A_1, A_2, \ldots, A_k$ are mutually disjoint. Also, let $B_1, B_2, \ldots, B_m$ be pairwise disjoint sets of new vertices, each set with cardinality n. Now define S_i so that

$$S_i = \begin{cases} A_i \cup B_i, & 1 \leq i \leq k \\ B_i, & k < i \end{cases}$$

for each $i = 1, 2, \ldots, m$. Let G be the complete m-partite graph with partite sets $S_1, S_2, \ldots, S_m$. Note, $p(G) = n^2$.

Label the elements of $A_1 \cup A_2 \cup \ldots \cup A_k$ with $v_1, v_2, \ldots, v_{kn}$ so that

$$A_1 = \{v_1, v_2, \ldots, v_n\}$$
$$A_2 = \{v_{n+1}, v_{n+2}, \ldots, v_{2n}\}$$

and so forth. Label the elements of $B_1 \cup B_2 \ldots \cup B_m$ with v_{kn+1}, $v_{kn+2}, \ldots, v_{n^2}$, where

$$B_1 = \{v_{kn+1}, v_{kn+2}, \ldots, v_{kn+n}\}$$
$$B_2 = \{v_{kn+n+1}, v_{kn+n+2}, \ldots, v_{kn+2n}\} \text{ and so forth.}$$

We show (i) $Z(G)=m$; (ii) $Z(S_{\{v_1, v_2, \ldots, v_{kn}\}}(G)) = n$; and (iii) for any $r = 1, 2, \ldots, n^2$, $m \leq Z(S_{\{v_1, v_2, \ldots, v_r\}}(G)) \leq n$.

To verify (i) we note that G contains $K_{m(m)}$ as an induced subgraph; thus $m = Z(K_{m(m)}) \leq Z(G)$. However, G is an m-partite graph; thus $Z(G) \leq m$. We conclude that $Z(G) = m$.

To prove (ii), let $H = A_1 \cup A_2 \cup \ldots \cup A_k$. Now note, $A_1, A_2, \ldots, A_k, B_1, B_2, \ldots, B_m$ is a cocoloring of $S_H(G)$. Thus, $Z(S_H(G)) \leq m+k = n$. Suppose then that $Z(S_H(G) < n$. Note that $A_1, A_2, \ldots, A_k, B_1, B_2, \ldots, B_m$ is a collection of n pairwise disjoint sets of cardinality n. By Lemma 2 any cocoloring of $S_H(G)$ with less than n colors produces a cocolor class containing four distinct vertices, say u, v, w, and x such that for some i and j one of the following three cases occurs:

(1) $\{u, v\} \subseteq A_i$
$\{w, x\} \subseteq A_j$, $i \neq j$

(2) $\{u, v\} \subseteq A_i$
$\{w, x\} \subseteq B_j$

(3) $\{u, v\} \subseteq B_i$
$\{w, x\} \subseteq B_j$, $i \neq j$.

Each case, however, leads to a contradiction. To see this, we will examine Case 1; the other two cases follow similarly.

Note that $\{u, v, w, x\} \subseteq H$. Since u and v are not adjacent in G, they are adjacent in $S_H(G)$. Since u and w are adjacent in

G, they are not in $S_H(G)$. So {u, v, w, x} does not induce either a complete or an empty graph in $S_H(G)$. Hence, {u, v, w, x} does not lie in a single cocolor class and a contradiction is reached. Thus, $n \leq Z(S_H(G))$. We have now verified (ii).

To see that $m \leq Z(S_H(G))$, where $H = \{v_1, v_2, \ldots, v_r\}$ for any $r = 1, 2, \ldots, n^2$ let $B = B_1 \cup B_2 \cup \ldots \cup B_m$ and let $H' = H \cap B$. We note that $K_{m(m)} \triangleleft \langle B \rangle_G$ and $\langle B \rangle_{S_H(G)} = S_{H'}(\langle B \rangle_G)$. From the proof of Lemma 3 we know that this graph has a cochromatic number of m. Hence $Z(S_H(G)) \geq m$.

We now conclude the proof by showing that for any $H = \{v_1, v_2, \ldots, v_r\}$ with $r \leq n^2$, the cochromatic number of $S_H(G)$ is no more than n. There are three cases to consider.

Case 1

Suppose $r \leq kn$, that is, $H \subseteq A_1 \cup A_2 \cup \ldots \cup A_k$. Choose i so that $v_r \in A_i$. Let $T = A_i \cap H$ and $T' = B_i \cup (A_i - H)$. Then $A_1, A_2, \ldots, A_{i-1}, T, T', B_1, B_2, \ldots, B_{i-1}, S_{i+1}, S_{i+2}, \ldots, S_m$ defines a t-cocoloring of $S_H(G)$, where $t = i + m$. Since $i \leq k$, we have $Z(S_H(G)) \leq k+m = n$.

Case 2

Suppose $v_r \in B_1 \cup B_2 \cup \ldots \cup B_k$. Choose i so that $v_r \in B_i$. In this case, let $T = A_i \cup (B_i \cap H)$ and $T' = B_i - H$. Then a t-cocoloring of $S_H(G)$ is defined by $S_1, S_2, \ldots, S_{i-1}, T, T', A_{i+1}, A_{i+2}, \ldots, A_k, B_{i+1}, B_{i+2}, \ldots, B_m$, where $t = 1+k-i+m$. Now since $i \geq 1$, we see that $1+k-i+m \leq k+m = n$. Thus, in this case, $Z(S_H(G)) \leq n$.

Case 3

Suppose $v_r \in B_{k+1} \cup B_{k+2} \cup \ldots \cup B_m$. Choose i so that $v_r \in B_i$. Let $T = B_i \cap H$ and $T' = B_i - H$. Now, we can define a cocoloring on $S_H(G)$ with the decomposition $S_1, S_2, \ldots, S_{i-1}, T, T', S_{i+1}, \ldots, S_m$. Thus, $Z(S_H(G)) \leq m+1$. Since $m < n$, in this final case $Z(S_H(G)) \leq n$. ■

It has been shown [5] that if $S = \{m, m+1, \ldots, n\}$ is a set of natural numbers with $n \leq 2m$ then for any $k \geq n^2$ there is a graph G on k vertices with a labeling on V(G) such that the corresponding set $\{Z(G_0), Z(G_1), \ldots, Z(G_p)\} = S$, provided $S \neq \{1\}$.

BIBLIOGRAPHY

[1] M. Behzad, G. Chartrand, and L. Lesniak, Graphs and Digraphs, Prindle, Weber and Schmidt, Boston, 1979.

[2] G. Chartrand, S. Kapoor, D. Lick and S. Schuster, The Partial Complement of Graphs, To Appear.

[3] L. Lesniak and H. Straight, The Cochromatic Number of a Graph, Ars Comb. 3 (1977) 34-45.

[4] H. Straight, Cochromatic Number and the Genus of a Graph, J. Graph Theory 3 (1979) 43-51.

[5] J. Gimbel, The Chromatic and Cochromatic Number of a Graph, Doctoral Dissertation, Western Michigan University (1984).

A RECURSIVE ALGORITHM FOR HAMILTONIAN CYCLES IN THE (1,j,n)-CAYLEY GRAPH OF THE ALTERNATING GROUP

Ronald J. Gould*
Emory University

Robert L. Roth
Emory University

ABSTRACT

A recursive algorithm is presented which accepts as input a permutation length $(n \geq 5)$ and a permutation $W \varepsilon A_n$ (where W sends $1 \to n$). As output, the algorithm produces a directed Hamiltonian path in the (1,j,n)-Cayley Graph D_n of A_n that begins with the vertex representing the identity element of A_n and ends with the vertex representing W. The algorithm makes use of a collection of twelve distinct Hamiltonian paths in D_5.

*Research supported by Emory University Research Grant No. 8399/02

1. Notation.

We use the standard cycle notation for permutations; and the composition, $\Pi_1 \circ \Pi_2$, will always mean Π_1 followed by Π_2. We denote the identity permutation by ι. We often abbreviate $h^{\Pi} = k$ by writing "Π sends $h \rightarrow k$". Let A_n denote the alternating group on n letters. We define

$H_n(h_1, h_2, \ldots, h_m; k_1, k_2, \ldots, k_m) =$
$\{\Pi \in A_n : h_r^{\Pi} = k_r, r \in \{1, 2, \ldots, m\}\}$. Clearly $H_n(k;k)$ is the stabilizer of k in A_n. The coset $H_n(j;k)$ will often be called the j coset when the value of k is clear from the context. We denote by τ_j the element $(1,j,n)$ and we let $B_n = \{\tau_j : j \in \{2, 3, \ldots, n-1\}\}$. For $n \geq 3$, we let D_n denote the Cayley Graph of A_n with respect to B_n. That is D_n is the directed graph with $V(D_n) = A_n$ and
$E(D_n) = \bigcup_{j=u}^{n-1} E_j$ where $E_j =$
$\{(\Pi,\Psi) : \Pi,\Psi \in A_n$ and $\Pi \circ \tau_j = \Psi\}$. We refer to the elements of E_j as j-edges.

Because $V(D_n) = A_n$, we will feel free to use the terms permutation and vertex interchangeably.

2. Introduction

Recently interest has arisen (see [4], [5], [6], [7] and [2]) in generating a sequencing of the elements of a permutation group subject to various constraints. Of special interest is the problem of generating a sequencing $\Pi_1, \Pi_2, \ldots, \Pi_{|G|}$ of the elements of a permutation group G so that the total cost

$$C = \sum_{i=1}^{|G|-1} c(\Pi_i^{-1} \circ \Pi_{i+1})$$

is minimized, where $c : G \rightarrow R^+$ is a cost function. Of course

$\Pi_i \circ \Pi_i^{-1} \circ \Pi_{i+1}) = \Pi_{i+1}$, so that $c(\Pi_i^{-1} \circ \Pi_{i+1})$ is the cost of "proceeding by multiplication" from Π_i to Π_{i+1} in the sequencing.

Tannenbaum, in [6], raised the problem of finding such a sequencing when $G = A_n (n \geq 3)$ with its natural action on $\{1, 2, \ldots, n\}$, $c(\Pi) = |\{j : j^{\Pi} \neq j\}|$ so that $c(\Pi) \geq 3$ for each non-identity element $\Pi \in A_n$, and the set of allowable multipliers for use in the sequencing is $B_n = \{(1,j,n) : j \in \{2, 3, \ldots, n-1\}\}$ which is a minimal generating set of A_n. This question was answered in [2]. In this paper we present an algorithm for determining many such sequencings for each $n \geq 5$. Terms not defined in this article can be found in [1] or [3].

3. Overview of the Algorithm

In constructing sequencings for A_n, several results from [2] will be useful. We state them below.

Proposition 1 ([2]). For $k \in \{1, 2, \ldots, n\}$, the left cosets of the stabilizer, $H_n(k;k)$, of k in A_n are precisely the sets $H_n(1;k), H_n(2;k), \ldots, H_n(n;k)$.

When $n \geq 4$ and $k \in \{2, 3, \ldots, n-1\}$, the induced subgraph $\langle H_n(h;k) \rangle$ of D_n is isomorphic to D_{n-1} for each $h \in \{1, 2, \ldots, n\}$. Further, every j-edge of D_n $(j \neq k)$ is an edge of exactly one of these n induced subgraphs, and every k-edge of D_n has its end vertices in two of these induced subgraphs. More precisely, if $\Pi \in H_n(h_1, h_2, h_3; k, 1, n)$ and $\Pi \circ \tau_k = \Psi$ then, $\Psi \in H_n(h_2, h_3, h_1; k, 1, n)$.

Proposition 2 ([2]). For $n = 3$ and $n \geq 5$, given any $W \in A_n = V(D_n)$ such that W sends $1 \to n$, there exists a directed $\iota - W$ Hamiltonian path. For $n = 4$, there exists a directed i - (1, 4)(2, 3) Hamiltonian path.

We remark that if a directed i - W Hamiltonian path exists then a directed $\Pi - \Pi \circ W$ Hamiltonian path exists since premultiplication by any $\Pi \in A_n$ is an automorphism of D_n.

<u>Proposition 3 ([2])</u>. If $n \geq 5$, $\{\Psi \in A_n$: there exists a directed $\Pi - \Psi$ Hamiltonian path in $D_n\} \supseteq \{\Psi \in A_n : n^{\Psi^{-1}} = 1^{\Pi^{-1}}\}$.

Our goal is to explicitly construct a Hamiltonian path in D_n from the vertex representing the identity element to a specified vertex W, where W sends $1 \to n$ (by Proposition 2 such a path exists).

The algorithm PATHFINDER accepts as input a permutation length $N (N \geq 5)$, the permutation W we wish to be the end vertex of the directed i - W path and twelve basepaths in A_5 used in our recursive construction. We require that W be a permutation that sends $1 \to N$. Each of the basepaths is a directed Hamiltonian path in A_5 having initial vertex i and terminal vertex one of the twelve elements of A_5 which send $1 \to 5$. The output is a directed Hamiltonian path in the Cayley Graph D_N of A_N that begins with the identity element of A_N and ends with W.

The main idea of the algorithm is to use the cosets $H_N(j;k)$ (for a fixed value of k which is moved by W and $j = 1, 2,\ldots,N$) as much as possible in the construction of the path. Some care must be taken to ensure we do not enter the coset containing W too early in our coset process. It is also important to maintain control of the vertices we traverse first and last in each subgraph representing these cosets. It is fairly easy to convince oneself that it is not possible to traverse all N of the cosets one by one in some fixed order. Thus, we must break out of at least one coset at some time and return later to traverse the remaining vertices in that coset. We choose to begin our path with the identity vertex (which of course is in $H_N(k;k)$) and immediately move across to (1,k,N) (and hence to the coset $H_N(k;N)$). We now traverse all of $H_N(k;N)$ and return to traverse the remainder of $H_N(k;k)$ (see Figure 1).

At this point we must determine if a problem exists with the next coset. Since our target vertex W represents a permutation in $H_N(HSTAR;k)$, where HSTAR is the inverse image of K under W, we must be sure at this point that we do not enter this coset next. If on multiplying the terminal vertex of $H_N(k;k)$ by (1,k,N) we determine that our next vertex lies in $H_N(HSTAR;k)$, we modify the path constructed so far to avoid this problem. This is done by conjugating the elements of the path by the involution CYC2 = (HSTAR, GOODCOS3). The values of HSTAR and GOODCOS3 are chosen to avoid entering the HSTAR coset upon leaving $H_N(k;k)$.

Having thus modified our path, we are now able to traverse the remaining cosets, ending with $H_N(HSTAR;k)$, and construct the Hamiltonian path beginning with the identity of A_N and ending with W.

4. The PATHFINDER Algorithm

We now describe the PATHFINDER Algorithm (see Appendix 1) in detail. The PATHFINDER Algorithm relies on the straightforward routines that perform permutation composition (COMPOSE), permutation conjugation (CONJ), upshifting (UPSHFT), downshifting (DWNSHFT), and permutation inversion (INVERSE). Algorithms for these routines are listed in Appendix 2.

If the permutation length N equals 5, then we have recursed to a base case. The path is taken to be the unique one of the twelve basepaths which has W (that is a W suitably

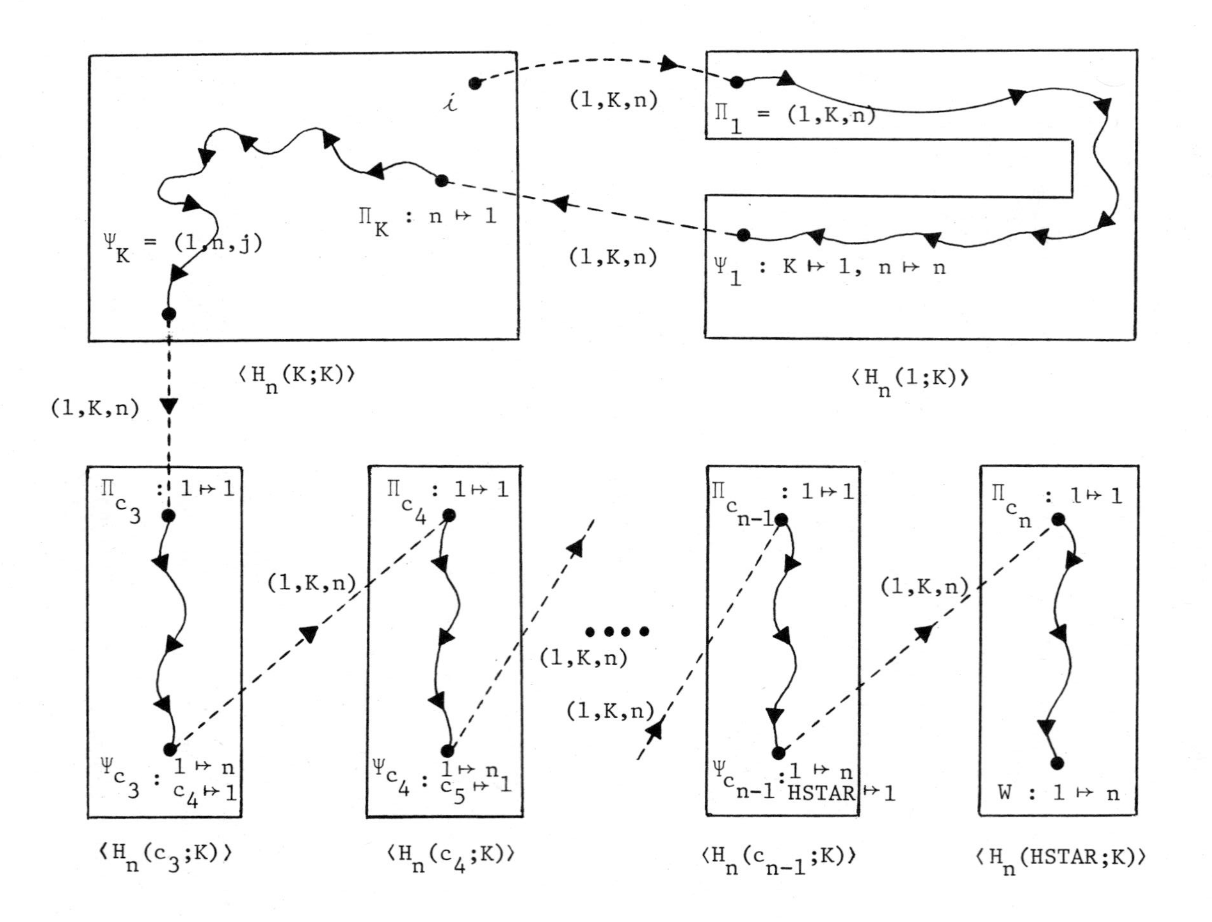

i
(1,K,n)
Π1 = (1,K,n)
ΠK : n ↦ 1
ΨK = (1,n,j)
(1,K,n)
Ψ1 : K ↦ 1, n ↦ n
⟨Hn(K;K)⟩
⟨Hn(1;K)⟩
(1,K,n)
Πc3 : 1 ↦ 1
Πc4 : 1 ↦ 1
Πcn-1 : 1 ↦ 1
Πcn : 1 ↦ 1
(1,K,n)
(1,K,n)
(1,K,n)
(1,K,n)
Ψc3 : 1 ↦ n, c4 ↦ 1
Ψc4 : 1 ↦ n1, c5 ↦ 1
Ψcn-1 : 1 ↦ n, HSTAR ↦ 1
W : 1 ↦ n
⟨Hn(c3;K)⟩
⟨Hn(c4;K)⟩
⟨Hn(cn-1;K)⟩
⟨Hn(HSTAR;K)⟩

modified by the downshifting described below) as its terminal vertex. The terminal vertex W is compared to the terminal vertices of each of the twelve basepaths in order to find the proper basepath to use for PATH.

If $N > 5$ we begin the coset process. We first compute K, the smallest letter between 2 and $N - 1$ (inclusive) which is moved by W. The value of K is used to determine the coset decomposition of A_N. We next compute HSTAR which is the inverse image of K under W. The HSTAR coset will be the last coset traversed in our coset process since $W \in H_N(\text{HSTAR};K)$.

We are now ready to begin to construct PATH. First the identity vertex is placed in TMPLST (a temporary storage array for the first two cosets. As noted later, this much of the path sometimes must be modified.) Next the values T1 and T2 are computed. These are the two smallest values in the set $\{2,3,\ldots,N - 1\}$ other than K. The values T1 and T2 are used in computing the first permutation in PSI and the second permutation in PI. The permutations in PI represent the first vertices to be traversed in each of the cosets, while those of PSI represent the last vertices to be traversed in each of the cosets. The first and Kth rows of PI are the permutations (1,K,N) and (1,N)(T1, T2) respectively; while the first and Kth rows of PSI are (1,K)(T1, T2) and the identity respectively.

The permutation WW1 is computed to be the composition of the inverse of the first permutation in PI with the first permutation of PSI. That is, $WW1 = (1,K,N)^{-1} \circ (1,K)(T1, T2) = (1,N)(T1, T2) \in H_N(K;K)$. The permutation WW2 is the composition of the inverse of the Kth permutation of PI with the Kth permutation of PSI. So $WW2 = (1,N)(T1, T2)$. The permutation WW1 (the terminal vertex of the automorphic copy of the 1 coset) is downshifted to W1, that is, W1 is the permutation of A_{n-1} acting as $S_K = \{1, 2,\ldots,N\}\setminus\{K\}$ which agrees with

WW1 on S_K. (We observe that K is fixed by WW1.) The routine is then recursively called to obtain a directed Hamiltonian path in D_{n-1} from the identity to W1. This path is returned as SMLPATH. The vertices of SMLPATH are now successively upshifted to A_N (being stored in UU), premultiplied by the first row of PI (to obtain a path beginning at (1,K,N) and ending at (1, K)(T1, T2)) and placed in TMPLST. We now repeat the actions taken on WW1 for WW2 and the K coset except that we omit the last vertex of the path obtained (since it is the identity, which has already appeared).

We have now traversed the first two cosets of A_N in our process (see Figure 1). At this point some care must be taken. We must check the last vertex of TMPLST and if N is sent to HSTAR under this permutation, then the path must be modified to free the HSTAR coset to be the last one traversed. (We recall that $W \in H_N$ (HSTAR;K).) Thus, in this case we conjugate all of the vertices in TMPLST using the involution (HSTAR, GOODCOS3). The value of GOODCOS3 is set to be the smallest value in {2, 3,...,N - 1} other than K and HSTAR. The involution CYC2 is set equal to (HSTAR, GOODCOS3) to perform the conjugation. The vertices of TMPLST are then conjugated (if necessary) and stored in PATH. In any case, GOODCOS3 is computed to be the image of N under the permutation which represents the last vertex presently in PATH.

The distinct values K, GOODCOS3 and HSTAR are sorted in ascending order and stored in BAD1, BAD2 and BAD3. These represent the distinguished cosets that must be treated first, third and last in the coset process. We now place these and the remaining N - 3 cosets into COSET, which is a listing of the order in which the cosets will be traversed.

The COSET(3)-th row of PI is computed to be the permutation

(K,N,COSET(3)) = (1,N,COSET(3)) ∘ (1,K,N) which is correct since (1,N,COSET(3)) is the last permutation presently in PATH (whether or not conjugation by CYC2 was necessary) because it (or its pre-image under the conjugation by CYC2) was the penultimate vertex in the traversal of the K coset which terminated with the identity.

Next, for I = 3, 4,...,N - 1, the COSET(I)-th row of PSI and the COSET(I+1)-th row of PI are computed. The COSET(I)-th row of PSI is computed to be (1,N)(COSET(I),K) if N = COSET(I+1), (1,N,K,T4,COSET(I+1)) (where T4 is the smallest letter different from I,N,K, and COSET (I+1)) if N = COSET(I) and (1,N,COSET(I),K,COSET(I+1)) otherwise. The COSET(I+1)-th row of PI is computed to be (K,COSET(I),N) if N = COSET(I+1), (K,T4,COSET(I+1)) if N = COSET(I), and (K,COSET(I+1))(COSET(1),N) otherwise.

Finally, the COSET(N)-th row of PSI is computed to be W.

We are now ready to generate the paths in the remaining cosets. Their order, beginning vertices and ending vertices have been computed to allow us to move from a completed coset to the next coset via multiplication in A_N by (1,K,N). Writing c_i for COSET(I) we see that for each $H_N(c_i;\, K)$, $i = 3, 4, \ldots, N$, the directed Hamiltonian path constructed has initial vertex Π_{c_i} (the c_ith entry of PI) which sends $1 \to 1$ and for $i = 3, 4, \ldots, N - 1$ this path has terminal vertex Ψ_{c_i} (the c_ith entry of PSI) which sends $1 \to N$ and $c_i \to 1$. Thus $\Psi_{c_i} \circ (1,K,N) = \Pi_{c_{i+1}}$ as is required.

For the remaining cosets, WWL is computed to be the composition of the inverse of the c_Lth entry of PI with the c_Lth entry of PSI. That is, WWL is the terminal vertex for the automorphic image of the c_Lth coset. As before, WWL is downshifted to WL in A_{N-1} and the PATHFINDER routine is recursively called to obtain a directed Hamiltonian path from the identity to WL in A_{N-1}. This path is returned as SMLPATH and the vertices of SMLPATH are successively upshifted to A_N (and stored in UU), premultiplied by the c_Lth entry of PI and placed into PATH.

APPENDIX 1

/* This is the PATHFINDER Algorithm described in section 4. Input to the algorithm is the permutation length N, the final permutation W and the basepaths in the group A(5). Output from the algorithm is a Hamiltonian path in PATH in the group A(N) that begins with the identity and ends with W. The Algorithm is written in a C-like pseudocode. */

```
PATHFINDER (N, W, PATH)
/* Compare the downshifted W to the end vertices of the base-
paths to choose the correct path. */
if (N=5) {   I = 1;
        while (I<= 12) {  IT = I;
            if(W does not equal BP[60][I]) { I += 1; }
            if ( I=IT )  {  for (II = 1; II<=60; ++II) {
            PATH[ II ] = BP[II][I]; }
            I=13
        }
}
return;
}

/* We begin the coset process by finding K and HSTAR */
else {  I=2;
        while (2<=I and I<=N-1) {
          if (W does not send I -> I) { K=I; I=N; }
          else I = I + 1;
        }

        I=2;
        while( 2<=I and I<= N ){
          if (W sends I -> K)    { HSTAR = I; I = N+1; }
          else I = I + 1;
  }
```

```
/* Place the identity first, then find  T1  and  T2; the smallest
values other than  K. */
  TMPLST[1] = identity;
        if ( K = 2 )   { T1=3; T2=4; }
        else {   T1=2;
                if (K=3) T2=4;
                else T2=3;
        }
/*  Store PI and PSI for the 1 and K cosets. */
  PI[1]=(1,K,N); PI[K]=(1,N)(T1,T2);
  PSI[1]=(1,K)(T1,T2); PSI[K]=identity;

                           -1
/*  Compute WW1= PI(1)  o PSI(1)    */
  inverse (N,PI[1],PI1INV) ;
  compose (N,PI1INV, PSI[1], WW1) ;

/*  Compute WW2  in a similar manner. */
  inverse (N,PI[K],PIKINV) ;
  compose (N, PIKINV, PSI[K], WW2) ;

/*  Downshift to A(N-1) and recurse. */
  dwnshft (N, K, WW1, W1) ;
  PATHFINDER (N-1, W1, SMLPATH) ;

/*  Rebuild the vertices by upshifting to A(N), premultiplying by
PI(1) to adjust the path, and store the vertices. */
  for (I=1; I<= FACT[N-1]/2; ++I) {
        upshft (N-1, K, SMLPATH[I], VV);
        compose (N, PI[1], VV, PI1VV) ;
        TMPLST[I+1] = PI1VV ;
  }
```

```
/* Repeat the process on WW2. */
  dwnshft (N, K, WW2, W2) ;
  PATHFINDER (N-1, W2, SMLPATH) ;
  for (I=1; I<=FACT[N-1]/2 -1; ++I) {
        upshft (N-1, K, SMLPATH[I], VV) ;
        compose (N, PSI[K], VV, PIKVV) ;
        TMPLST[  FACT[N-1]/2 + I+1] = PIKVV;
  }
/* Determine if the third coset must be adjusted to avoid the HSTAR
coset.  Then set  CYC2 for conjugating and adjust the temporary
path.
*/
  if (TMPLST[ FACT[N-1]] sends N -> HSTAR )  {  I=2;
        while (2<=I and I<= N {
                if (I does not equal K and I does not equal HSTAR)
                   {GOODCOS3 = I; I = N+1;}
                   else I = I + 1;
  CYC2 = (GOODCOS3,HSTAR);
    for ( I=1; I<= FACT[N-1]; ++I) {
        conj (N, CYC2, TMPLST[I], PATH[I] ;
    }
  }
/*  No adjusting, just copy to PATH, then get GOODCOS3.  */
  else  { for (I=1; I<= FACT[N-1]; ++I) {  PATH[I] = TMPLST[I]; }
  GOODCOS3 = the image of N  in TPMLST[FACT[N-1]];
  }
/*  Build the coset order for processing.  */
  BAD1 = K;  BAD2 = GOODCOS3;  BAD3 = HSTAR;
  SORT(BAD1,BAD2,BAD3);
  COSET[1] = 1;  COSET[2] = K;  COSET[3] = GOODCOS3;
  COSET[N] = HSTAR;
```

```
    J = 2;    JJ = 3;
    while (2<=J and J<= N) {
          if (J = BAD1) J += 1;
          if (J = BAD2) J += 1;
          if (J = BAD3) J += 1;
          JJ += 1;
          if(J<=N) COSET[JJ] = J;
          J += 1;
    }
/*  Construct the PI and PSI permutations for the remaining
    cosets. */
    PI[COSET[3]] = (K,N,COSET[3]);

    for (I=3; I<= N-1; ++I) {
      if(COSET[I+1]=N {  PSI[COSET[I]] = (1,N)(COSET[I],K);
                         PI[COSET[I+1]] = (K,COSET[I],N);
      }
    else {
      if ( COSET[I] = N {
          for(J=2; J<=N-1; ++J) {
             if(J does not equal K and J does not equal COSET[I+1])
                 { T4 = J; J =N; } }
          PSI[COSET[I]] = (1,N,K,T4,COSET[I+1]);
          PI[COSET[I+1]] = (K,T4,COSET[I+1])
      }
      else { PSI[COSET[I]] = (1,N,COSET[I],K,COSET[I+1]);
             PI[COSET[I+1]] = (K,COSET[I+1])(COSET[I],N); }
    }
    }
    PSI[COSET [N]] = W;
```

```
/*  Now construct paths in the remaining cosets in a manner
similar to that used on the special cosets.  COSET determines the
order.  */
        for (L=3; L<=N; ++L){
                inverse (N, PI[COSET[L]], PICLINV);
                compose (N, PICLINV, PSI[COSET[L]], WWL) ;
                dwnshft (N, K, WWL, WL);
                PATHFINDER (N-1, WL, SMLPATH);

                for (I=1; I<= FACT[N-1]/2; ++I) {
                upshft (N-1, K, SMLPATH[I], VV);
                compose (N, PI[COSET[L]], VV, PICLVV);
                PATH[((L-1)*FACT[N-1]/2 +I)] = PICLVV:
                }
        }
return;
}
```

APPENDIX 2

```
/*  ROUTINE INVERSE to compute the inverse of a permutation of
length N.  The routine accepts a permutation length N, a permu-
tation alfa, and returns alfa's inverse in invalfa.  Both alfa
and invalfa are thought of as arrays of length N.  */

inverse ( N , alfa , invalfa )
{
for (I=1; I =N; ++I)   invalfa[ alfa[I] ] = I;
return;
}

/*  ROUTINE COMPOSE to multiply permutations of length N.  The
routine accepts a permutation length N and permutations alfa and
beta (thought of as arrays of length N).  The composition is
returned in athenb.  That is, athenb = alfa o beta.  */

compose ( N, alfa, beta, athenb)
{
for (i=1; i =N; ++i)   athenb[i] = beta[ alfa[i] ];
}

/*  ROUTINE CONJ to conjugate permutations.  This routine accept
a permutation length N and two permutations of length N, namely
alfa and beta.  The routine computes the conjugation of beta by
alfa.  That is, iaba = invalfa o beta o alfa.  Note the temporar
variables iab and iaba to pass returned values.  */

conj (N,alfa,beta,iaba)
{
inverse(N,alfa,invalfa);
compose(N,invalfa,beta,iab);
compose(N,iab,alfa,iaba);
}
```

```
/*  ROUTINE DWNSHFT to downshift a permutation to one in A(n-1).
The routine accepts a permutation length N, a specified value k,
and a permutation of length N in VV.  The permutation to be
downshifted is VV and the returned permutation of length N-1 is
V.  */

dwnshft(N,k,VV,V)
{

for( i=1; i<= k-1; ++i)  {
        if (VV[i] < k)      V[i] = VV[i];
        else   V[i] = VV[i] - 1;
}

for(i=k+1; i<=N; ++i)     {
        if (VV[i] < k)      V[i-1] = VV[i];
        else    V[i-1] = VV[i]-1;
}
return;
}

/*  ROUTINE UPSHFT to reverse the process of dwnshft and insert
the k -> k element of a permutation of length N-1 to build it to
a permutation of length N.  The routine accepts a permutation
length N, a value k, and a permutation to rebuild V (thought of
as an array of length N-1).  The routine returns the rebuilt
permutation in VV, a permutation of length N.  */

upshft(N,k,V,VV)
{
for (i=1; i<= k-1; ++i)   {
       if (V[i] < k)     VV[i] = V[i];
       else    VV[i] = V[i]+1;
}
```

```
VV[k] = k;
for(i=k+1; i<= N+1; ++i)   {
        if(V[i-1] < k)     VV[i] = V[i-1];
        else    VV[i] = V[i-1] +1;
}
return;
}

/*  ROUTINE SORT, a routine to sort three values into ascending
order.  Any sorting technique will work here.  */
```

REFERENCES

1. M. Behzad, G. Chartrand, and L. Lesniak-Foster, "Graphs and Digraphs", Wadsworth International Mathematics Series, 1979.

2. R.J. Gould and R. Roth, Cayley Graphs and (1,j,n)-Sequencings of the Alternating Group, preprint.

3. J.J. Rotman, "The Theory of Groups (Second Edition)", Allyn and Bacon Series in Advanced Mathematics, 1973.

4. R. Sedgewick, Permutation Generation Methods, Computer Surveys 9(1977), 137-164.

5. P.J. Slater, Generating all Permutations by Graphical Transportations, Ars Combinatoria 5(1978), 219-225.

6. P. Tannenbaum, Minimal Cost Permutation Generating Algorithms, Proceedings of the Fourteenth Southeastern Conference on Combinatorics, Graph Theory, and Computing, to appear.

7. D.S. Witte, On Hamiltonian Circuits in Cayley Diagrams, Discrete Mathematics 38 (1982), 99-108.

FURTHER RESULTS ON A GENERALIZATION OF EDGE-COLORING

S. Louis Hakimi
Department of Electrical Engineering
and Computer Science
Northwestern University
Evanston, Illinois 60201

Abstract

Hakimi and Kariv have recently introduced the following generalization of the edge-coloring problem: color the edges of a multigraph $G(V,E)$ with the least number, $\chi'_m(G)$, of colors such that at vertex $v \in V$ there are at most $m(v)$ representatives of each color. Hakimi and Kariv present results which are generalizations of theorems of Vizing, Ore, Shannon and König. In this paper, Goldberg's technique is used to provide unified proofs for some of the above results and some variations of these results.

1. Introduction

Let $G(V,E)$ be a multigraph without loops with finite vertex set V and non-empty edge set E. Let us associate positive integers $1,2,\ldots$ with colors and call C a k-edge-coloring of G if $C: E \rightarrow \{1,2,\ldots, k\}$. Let $C_v^{-1}(i)$ be the number of edges of G incident at vertex v which receive color i by the coloring C. Suppose a positive integer $m(v)$ is associated with each vertex $v \in V$. We call C a proper k-edge-coloring of G if for each vertex $v \in V$ and each $i \in \{1,2,\ldots, k\}$, $C_v^{-1}(i) \leq m(v)$. The chromatic index, $\chi'_m(G)$ is the smallest integer k for which a proper k-edge-coloring of G exists. Let d(v) denote the number of edges of G incident at $v \in V$, and let $\Delta_m = \max \{ \lceil d(v)/m(v) \rceil \mid v \in V\}$. Let $\mu(u,v)$ denote the number of edges of G joining u and $v \in V$ and $\mu(u) = \max \{\mu(u,v) \mid v \in V\}$. Recently, Hakimi and Kariv [1] have shown that:

(a) $\chi'_m(G) \leq \max \{ \lceil (d(u) + \mu(u))/m(u) \rceil \mid u \in V\}$;

(b) if G is bipartite, or if m(v) is even for all $v \in V$, $\chi'_m(G) = \Delta_m$; and

(c) $\chi'_m(G) \leq \max \{\Delta_m, \sigma_m\}$, where $\sigma_m = \max\left\{\left\lfloor \frac{1}{2}\left(\left\lfloor \frac{d(v)}{m(v)} \right\rfloor + \left\lfloor \frac{d(u)}{m(u)} \right\rfloor + \left\lfloor \frac{d(w)}{m(w)} \right\rfloor\right)\right\rfloor \right\}$ and the maximum is taken over all simple paths v-u-w of length two in G with $m(v) \leq m(u)$.

It should be noted that these results represent generalizations of the results of Vizing [2,3], König [4], and Ore-Shannon [5,6] for the classical edge-coloring.

Goldberg [7] and Anderson [8] have developed a unified proof technique for obtaining Vizing and Ore-Shannon bounds for the classical chromatic index. In the next section, we will use an adaptation of Goldberg's proof technique to obtain bounds for $\chi'_m(G)$. These bounds are sometimes superior to the bounds described above in (a) and (c); however, when m(v) = 1 for all $v \in V$, these bounds are not reduced to Goldberg's.

Two other interesting papers on the generalization of edge-coloring or its applications have recently come to our attention. Hilton and de Werra [9] are primarily concerned about "equitable edge-coloring" and obtain results which indirectly imply some of the results of Hakimi and Kariv [1]. Coffman, Garey, Johnson and La Paugh [10] obtain certain algorithmic and complexity oriented results intended for scheduling applicators of edge-coloring type problems.

2. A Unified Approach to Proving Bounds for the Generalized Chromatic Index

Goldberg [7] uses the notion of a critical graph to obtain his bounds. Here instead, we assume integer $k \geq \Delta_m$ is such that $G - e_0$ has a proper k-edge-coloring with $e_0 = (b,a_0)$. Our goal, then, is to obtain lower bounds for k such that e_0 can also be (properly) colored. Toward this goal, let C be a proper k-edge-coloring of $G - e_0$. For $x \in V$, let $\Omega(x) = \{i \mid 1 \leq i \leq k$ and $C_x^{-1}(i) < m(x)\}$ and denote $|\Omega(x)|$ by $\omega(x)$. If $\gamma \in \Omega(x)$, we say γ is an available color at vertex x.

A sequence $e_0e_1e_2 \ldots e_\ell$ of distinct edges of G starting with the uncolored edge $e_0 = (b,a_0)$ is called a <u>fan sequence</u> at b (FS(b)) if $e_i = (b,a_i)$ for $i = 0,1,\ldots,\ell$, the vertices $a_0,a_1,\ldots,a_\ell$ are distinct, $C(e_i) \in \Omega(a_{i-1})$ for $i = 1,2,\ldots,\ell$, the colors $C(e_1), C(e_2), \ldots, C(e_\ell)$ are distinct, and $C(e_i) \notin \Omega(b)$ for $i = 1,2,\ldots\ell$. We note that the fan sequence provides a mechanism for obtaining a proper k-edge-coloring $G - e_\ell$ by recoloring e_{i-1} by the color $C(e_i)$ for $i = 1,2,\ldots,\ell$ and uncoloring edge e_ℓ and leaving the colors for all other edges unchanged. Furthermore, we observe that if at this stage a certain color $\gamma \in \{1,2,\ldots,k\}$ is available at both a_ℓ and b, then we can complete our coloring by coloring edge e_ℓ by γ.

Suppose $G - e_0$ has a proper k-edge-coloring and let $\alpha, \beta \in \{1,2,\ldots,k\}$ and $\alpha \neq \beta$. Let $u \in V$ and $C_u^{-1}(\alpha) = m(u)$ and $C_u^{-1}(\beta) < m(u)$. An (α,β)-path starting with vertex u is a "path," denoted by p(u), which starts at u with an edge of color α and proceeds alternately with edges of colors α and β without traversing any edge twice while vertices are allowed to be visited more than once. Furthermore, if p(u) is to leave vertex v with an edge of color $\gamma \in \{\alpha,\beta\}$, then it may do so only if $\gamma \notin \Omega(v)$, unless $v = u$ and also p(u) arrives at the vertex u with an edge of color α, then p(u) may leave u with an edge of color β only if $C_u^{-1}(\beta) = m(u) - 1$. A maximal (α,β)-path p(u) is a path whose length cannot be extended by an addition of an edge at its end. Suppose now that p(u) is a maximal (α,β)-path, observe that if we switch the colors of the edges of $p(u)$, we obtain another proper k-edge-coloring of $G - e_0$. Finally a <u>maximal (α,β)-path $p(u)$ in the present of FS(b)</u> is a maximal (α,β)-path with the following additional properties: **(1)** if $u \neq b$, then the path uses an edge of the fan only if necessary, and **(2)** if $u = b$, the path uses an edge of the fan at its first opportunity. We are now prepared to prove the following basic lemma:

Lemma 1. Let C be a proper k-edge-coloring of $G - e_0$ and $e_0e_1\ldots e_\ell$ be FS(b) with $e_i = (b,a_i)$ for $i = 0,1,\ldots,\ell$. If $k \geq \Delta_m$ and at least one of the following conditions fails, then G has a proper k-edge-coloring:

1. $\Omega(a_\ell) \cap \Omega(b) = \phi$.

2. Let $\alpha \in \Omega(a_\ell)$ and $\beta \in \Omega(b)$. Then we have: **(i)** any maximal (β,α)-path $p(a_\ell)$ in the presence of FS(b) terminates at b, and **(ii)** any maximal (α,β)-path $p(b)$ in the presence of FS(b) terminates at a_ℓ.

3. Let $V_b(C)$ be the set of all vertices $a_j \neq b$ and a_j is an end vertex of an edge in any FS(b) (which begins with e_0). Then we have: **(i)** if x and $y \in V_b(C) \cup \{b\}$ and $x \neq y$, then $\Omega(x) \cap \Omega(y) = \phi$, and **(ii)** if $m(a_0) \leq m(b)$, then $|V_b(C)| \geq 2$.

4. For each $x \in V_b(C) \cup \{b\}$ and each $\gamma \in \Omega(x)$, we have $C_x^{-1}(\gamma) = m(x) - 1$.

Proof. We will prove each part separately; however, in proving each part we will assume that the previous parts do hold.

1. If $\gamma \in \Omega(a_\ell) \cap \Omega(b)$ $(\neq \phi)$, then we can obtain a proper k-edge-coloring of G by recoloring the fan $e_0e_1\ldots e_\ell$ and coloring edge e_ℓ by color γ.

2. **(i)** Let $\alpha \in \Omega(a_\ell)$ and $\beta \in \Omega(b)$; by Part 1, we have that $\beta \notin \Omega(a_\ell)$ and $\alpha \notin \Omega(b)$. Let $p(a_\ell)$ be a maximal (β,α)-path in the presence of FS(b). Suppose $p(a_\ell)$ terminates at v_t with $v_t \neq b$. Note that there are no edges of color β (by Part 1) and at most one edge of color α in the fan. Suppose, for the moment, that (b,a_j) is an edge of color α in the fan. Note that (b,a_j) could not have been traversed from b to a_j in $p(a_\ell)$, because $\alpha \notin \Omega(b)$ and by the fact that $C_b^{-1}(\beta) < C_b^{-1}(\alpha)$ there must be at least

one other edge of color α at b. On the other hand if (b,a_j) was traversed from a_j to b in $p(a_\ell)$, then $p(a_\ell)$ would have terminated at b which is contrary to our assumption. Therefore $p(a_\ell)$ does not contain any edge of the fan. We now consider three cases: **(1)** If $v_t \notin \{a_0,\ldots,a_\ell\}$, then switching the colors of the edges of $p(a_\ell)$ provides a proper k-edge-coloring C' of $G - e_0$ for which color β is available at both a_ℓ and b. As the fan sequence is still a fan sequence with respect to C', we can recolor the fan and then color $e_\ell = (b,a_\ell)$ by color β. **(2)** If $v_t = a_r$ for some r, $0 \le r < \ell$, then $p(a_\ell)$ must terminate with an edge of color β (because $\beta \notin \Omega(a_i)$ by Part 1). Now if we switch the colors of the edges of $p(a_\ell)$, color β becomes available at both b and a_r and the fan sequence $(e_0e_1\ldots e_r)$ is a fan-sequence with respect to this new coloring. We can then complete the coloring of G by recoloring fan $(e_0\ldots e_r)$ and color e_r by color β. **(3)** If $p(a_\ell)$ terminates at a_ℓ, then it must terminate with an edge of color β and consequently by the definition of (β,α)-path, $C^{-1}_{a_\ell}(\alpha) < m(a_\ell) - 1$. Then, as before, we can switch the colors of the edges of $p(a_\ell)$ and obtain a proper coloring which permits recoloring of the fan and the coloring of edge e_ℓ by color β.

2. (ii) Suppose p(b) is a maximal (α,β)-path starting at vertex b in the presence of the fan sequence $e_0e_1\ldots e_\ell$ which terminates at v_t with $v_t \neq a_\ell$. Let (b,a_j) be the edge of color α in FS(b); and if no such an edge exists assume $j = \ell + 1$ and observe that vertex $a_{\ell+1}$ can be selected such that $(b,a_{\ell+1})$ has color α. Note that if the edge (b,a_j) of color α is in the fan sequence, then by definition of (α,β)-path in the presence of FS(b), (b,a_j) will be the first edge in the path. If $v_t \notin \{a_0,a_1,\ldots,a_{j-1}\}$, then switching the colors of the edges of p(b) would provide us with a proper coloring for which $(e_0e_1\ldots e_{j-1})$ is a fan sequence and for which color α is available at both b and a_{j-1}. Therefore, we can complete the coloring by first recoloring the fan $e_0e_1\ldots e_{j-1}$ and the coloring $e_{j-1} = (b,a_{j-1})$ by color α. If $v_t \in \{a_0,a_1,\ldots,a_{j-1}\}$ but $v_t \neq a_{j-1}$, then again we can first switch the colors of the edges of p(b) and consequently have color

α available at both b and a_{j-1}, then recolor the fan and color edge e_{j-1} by color α. Now suppose $v_t = a_{j-1}$. Observe that now $j \neq \ell + 1$, because otherwise $v_t = a_\ell$ which is contrary to our hypothesis. If we switch the colors of the edges of p(b), then the edge (b,a_j) would have color β, but β now becomes available at a_{j-1}. This means that $e_0e_1e_2\ldots e_\ell$ is still a fan-sequence and color α is available at both b and a_ℓ. Therefore, we can complete our coloring by recoloring the fan $e_0e_1\ldots e_\ell$ and then coloring e_ℓ by color α.

3. (i) If $x \in V_b(C)$ and $y = b$, then Part (1) implies the desired result. Thus we need to show that if x and $y \in V_b(C)$, $x \neq y$, and $\Omega(x) \cap \Omega(y) \neq \phi$, then G is a k-edge-colorable. Let $\alpha \in \Omega(x) \cap \Omega(y)$ and let $\beta \in \Omega(b)$, then by Part (2) of this lemma the (α,β)-path p(b) in the presence of the fan sequence must terminate at both x and y which is a contradiction. This means that Part (2) would fail and thus we arrive at the desired conclusion.

3. (ii) Suppose $m(a_0) \leq m(b)$. As edge $e_0 = (b,a_0)$ is uncolored and $k \geq \Delta_m$, $\Omega(a_0) \neq \phi$. Let $\gamma \in \Omega(a_0)$. If $\gamma \in \Omega(b)$, we can complete the coloring by coloring e_0 by color γ. Thus, we may assume $\gamma \notin \Omega(b)$. This means that there are m(b) edges of color γ at b and at most $m(a_0) - 1$ edges of color γ at a_0. This implies that there exists at least one edge of color γ at vertex b which does not join b and a_0. By the definition of fan-sequence, this would mean that $|V_b(C)| \geq 2$.

4. Suppose to the contrary that for some $x \in V_b(C) \cup \{b\}$ and some color $\gamma \in \Omega(x)$, $C_x^{-1}(\gamma) < m(x) - 1$. Assume, for the moment, that $x \neq b$, then $x = a_\ell$ for some FS(b), $e_0e_1\ldots e_\ell$. Let $\delta \in \Omega(b)$. Then by Part (2) of this lemma, a maximal (γ,δ)-path p(b) must terminate at $a_\ell = x$. This path must end with an edge of color δ. As $C_x^{-1}(\gamma) < m(x) - 1$, if we switch the colors of the edges of p(b) color γ would become available at both b and x. Furthermore, it can be proved that the fan-sequence would remain a fan-sequence as follows. Because if the fan-sequence contains any edge of p(b) it must be an edge of color γ, but if edge (b,a_j) of the fan-sequence has color γ, then $\gamma \in \Omega(a_{j-1})$. This would imply that $\gamma \in \Omega(a_{j-1}) \cap \Omega(a_\ell)$. This, however, is a contradiction to Part 3(i). Therefore, the fan-sequence $e_0 e_1 \ldots e_\ell$ contains no edges of p(b) and consequently switching colors of p(b) would leave it intact. Now, it can be seen that we can recolor the fan sequence and then color e_ℓ by color γ. It remains to consider the case when x = b. Let $\alpha \in \Omega(a_0)$ and $p(a_0)$ be a (γ,α)-path. By Part 2(i), this path terminates at b. But as $C_b^{-1}(\gamma) < m(b) - 1$, switching the colors of the edges of $p(a_0)$ would make γ available at both a_0 and b and thus enables us to color e_0 by color γ.

Observe that by Part 4 of Lemma 1 at each vertex $x \in V_b(C) \cup \{b\}$, we have $m(x) - 1$ edges of each color $\gamma \in \Omega(x)$ and m(x) edges of each color $\gamma \notin \Omega(x)$. Thus we have $d(x) - 1 = \omega(x)(m(x) - 1) + (k-\omega(x))\, m(x)$ for $x \in \{a_0, b\}$ and $d(x) = \omega(x)(m(x) - 1) + (k - \omega(x))\, m(x)$ for $x \in V_b(C) - \{a_0,b\}$. The above relations immediately imply that

$$(1)\quad \omega(x) = \begin{cases} km(x) - d(x) + 1\,, & \text{if } x \in \{a_0,b\} \\ km(x) - d(x)\,, & \text{if } x \in V_b(C) - \{a_0,b\}\,. \end{cases}$$

Lemma 2. Let $V_b(C)$, $\omega(x)$, k, m(v), d(v) and $\mu(u,v)$ be as defined before. If $k \geq \Delta_m$ and any of the following three inequalities fails, then G has a proper k-edge-coloring.

1. $\sum_{x \in V_b(C)} \omega(x) \leq \left(\sum_{x \in V_b(C)} \mu(b,x)\right) - 1.$

2. $\sum_{x \in V_b(C)} \omega(x) \leq \left\lfloor \frac{d(b)}{m(b)} \right\rfloor - 1.$

3. $\sum_{x \in V_b(C) \cup \{b\}} \omega(x) \leq k.$

Proof. 1. By the definition of $V_b(C)$, for each $x \in V_b(C)$ and each color $\gamma \in \Omega(x)$, there exists an edge (b,y) of color γ with $y \in V_b(C)$. Therefore,

$$\left| \bigcup_{x \in V_b(C)} \Omega(x) \right| \leq \left(\sum_{x \in V_b(C)} \mu(b,x) \right) - 1\,.$$

But by Lemma 1 Part 3(i), we have

$$\left| \bigcup_{x \in V_b(C)} \Omega(x) \right| = \sum_{x \in V_b(C)} \omega(x)$$

which yields the desired result.

2. As the sets of colors $\Omega(x)$ for $x \in V_b(C)$ are disjoint and each of these colors must appear exactly m(b) times at b. The number of colors appearing exactly m(b) times at b is exactly $\big(d(b)-1-\omega(b)\,(m(b)-1)\big)/m(b)$. Thus, we have

$$\sum_{x \in V_b(C)} \omega(x) \le \frac{d(b)-1-\omega(b)\big(m(b)-1\big)}{m(b)}$$

or, as $\omega(b) \ge 1$,

$$\sum_{x \in V_b(C)} \omega(x) \le \frac{d(b)}{m(b)} - 1.$$

As the left side is an integer, the above fails if and only if part 2 of this lemma fails.

3. The proof of part 3 follows from the proof of part 2 and the observation that the number colors which appear exactly m(b) times at b is $k-\omega(b)$.

Theorem 1. Let $V_b(C)$, $\omega(x)$, k, m(v), d(v), $\mu(u,v)$ be as defined before. Then G(V,E) has a proper k-edge-coloring if $k \ge \Delta_m$ and at least one of the following four inequalities is satisfied:

1. $$k \ge \frac{\sum_{x \in V_b(C)} \big(\mu(b,x) + d(x)\big) - 1}{\sum_{x \in V_b(C)} m(x)}$$

2. $$k \ge \frac{\left\lfloor \frac{d(b)}{m(b)} \right\rfloor - 1 + \sum_{x \in V_b(C)} d(x)}{\sum_{x \in V_b(C)} m(x)}$$

3. $$k \geq \frac{\left(\sum_{x \in V_b(C) \cup \{b\}} d(x)\right) - 1}{\left(\sum_{x \in V_b(C) \cup \{b\}} m(x)\right) - 1}$$

4. $$k \geq \frac{\left\lfloor \frac{d(b)}{m(b)} \right\rfloor + \left\lfloor \frac{d(a_0)}{m(b_0)} \right\rfloor + \sum_{x \in V_b(C) - \{a_0\}} \left\lfloor \frac{d(x) + 1}{m(x)} \right\rfloor}{|V_b(C)|} - 1 .$$

Proof. Using Eq. (1), we have:

$$\sum_{x \in V_b(C)} \omega(x) = 1 + k \sum_{x \in V_b(C)} m(x) - \sum_{x \in V_b} d(x)$$

Substituting this value in parts 1 and 2 of Lemma 2, the inequalities of parts 1 and 2 of Theorem 1 immediately follow. The proof of part 3 is entirely similar and is omitted.

To prove part 4, we begin by assuming that

$$(2) \quad \sum_{x \in V_b(C)} \omega(x) \geq \left\lfloor \frac{d(b)}{m(b)} \right\rfloor$$

which by Lemma 2 implies that a proper k-edge-coloring would then exist. We note that $\omega(x) \geq \lceil (\omega(x)-1)/m(x) \rceil + 1$, therefore, if Eq. (3) below is satisfied, then so would be Eq. (2).

$$(3) \quad \sum_{x \in V_b(C)} \left(\left\lceil \frac{\omega(x) - 1}{m(x)} \right\rceil + 1 \right) \geq \left\lfloor \frac{d(b)}{m(b)} \right\rfloor .$$

By Eq. (1), we may rewrite Eq. (3) as follows

$$\left\lceil \frac{km(a_0) - d(a_0)}{m(a_0)} \right\rceil + 1 + \sum_{x \in V_b(C) - \{a_0\}} \left(\left\lceil \frac{km(x) - d(x) - 1}{m(x)} \right\rceil + 1 \right) \geq \left\lfloor \frac{d(b)}{m(b)} \right\rfloor .$$

Or equivalently,

$$(k+1) - \left\lfloor \frac{d(a_0)}{m(a_0)} \right\rfloor + \sum_{x \in V_b(C) - \{a_0\}} \left((k+1) - \left\lfloor \frac{d(x) + 1}{m(x)} \right\rfloor \right) \geq \left\lfloor \frac{d(b)}{m(b)} \right\rfloor$$

which implies the desired result.

We now wish to use the results of Theorem 1 to provide more meaningful upper bounds for $\chi'_m(G)$. Before we do that, we need the following lemma.

Lemma 3. Consider real positive numbers $m_1, m_2, \ldots, m_t$ and real non-negative numbers $\delta_1, \delta_2, \ldots, \delta_t$ such that

$$\frac{\delta_1}{m_1} \geq \frac{\delta_2}{m_2} \geq \cdots \geq \frac{\delta_t}{m_t}$$

If s is an integer with $1 \leq s \leq t$, then

$$\frac{\sum_{i=1}^{t} \delta_i}{\sum_{i=1}^{t} m_i} \leq \frac{\sum_{i=1}^{s} \delta_i}{\sum_{i=1}^{s} m_i}$$

Proof. The above inequality may be rewritten as

$$\frac{\sum_{i=1}^{s} \delta_i + \sum_{i=s+1}^{t} \delta_i}{\sum_{i=1}^{s} m_i + \sum_{i=s+1}^{t} m_i} \leq \frac{\sum_{i=1}^{s} \delta_i}{\sum_{i=1}^{s} m_i}$$

By cross multiplication, we have equivalently

$$\sum_{i=s+1}^{t} \delta_i \sum_{i=1}^{s} m_i \leq \sum_{i=s+1}^{t} m_i \sum_{i=1}^{s} \delta_i .$$

By hypothesis, for each pair of integers i and j with $i > j$, we have $\delta_i \, m_j \leq m_i \, \delta_j$. Thus each term in the left hand side of the above inequality can be paired up with a unique term on the right side which is no smaller than itself. This proves the validity of the above inequality and consequently the lemma.

Theorem 2. Let $G(V,E)$ be a multigraph and for $v \in V$, let Γv be the set of vertices adjacent to v. Let

$$A(u) = \begin{cases} \max\limits_{\substack{x_1 \text{ and } x_2 \,\varepsilon\, \Gamma u \\ x_1 \neq x_2}} \left\{ \left\lceil \dfrac{\mu(u,x_1) + d(x_1) + \mu(u,x_2) + d(x_2)}{m(x_1) + m(x_2)} \right\rceil \right\}, & \text{if } |\Gamma u| > 1 \\ 0, \text{ if } |\Gamma u| \leqq 1, & \end{cases}$$

$$B(u) = \begin{cases} \max\limits_{\substack{x_1 \text{ and } x_2 \,\varepsilon\, \Gamma u \\ x_1 \neq x_2}} \left\{ \left\lceil \dfrac{d(x_1) + d(x_2) + \left\lfloor \frac{d(u)}{m(u)} \right\rfloor - 1}{m(x_1) + m(x_2)} \right\rceil \right\}, & \text{if } |\Gamma u| > 1 \\ 0, \text{ if } |\Gamma u| \leqq 1, & \end{cases}$$

$$C(u) = \begin{cases} \max\limits_{\substack{x_1, x_2, x_3 \,\varepsilon\, \Gamma u \cup \{u\} \\ x_1 \neq x_2 \neq x_3}} \left\{ \left\lceil \dfrac{d(x_1) + d(x_2) + d(x_3) - 1}{m(x_1) + m(x_2) + m(x_3) - 1} \right\rceil \right\}, & \text{if } |\Gamma u| > 1 \\ 0, \text{ if } |\Gamma u| \leqq 1, & \end{cases}$$

and

$$D(u) = \begin{cases} \max\limits_{\substack{x_1 \text{ and } x_2 \,\varepsilon\, \Gamma u \\ x_1 \neq x_2}} \left\{ \left\lfloor \dfrac{\left\lfloor \frac{d(x_1)}{m(x_1)} \right\rfloor + \left\lfloor \frac{d(x_2)}{m(x_2)} \right\rfloor + \left\lfloor \frac{d(u)}{m(u)} \right\rfloor}{2} \right\rfloor \right\}, & \text{if } |\Gamma u| > 1 \\ 0, \text{ if } |\Gamma u| \leqq 1. & \end{cases}$$

Let $h(u) = \min\{A(u), B(u), C(u), D(u)\}$. Let $e = (u,v)$ be an edge of G and define

$$R(e) = \begin{cases} h(u), & \text{if } m(u) > m(v) \\ h(v), & \text{if } m(u) < m(v) \\ \min\{h(u), h(v)\}, & \text{if } m(u) = m(v). \end{cases}$$

Finally, let $R_m = \max \{R(e) | e \in E\}$. Then

$$\chi'_m(G) \leq \max \{\Delta_m, R_m\}.$$

Remarks. In spite of the length of the statement of this theorem its proof is not difficult and will be given next. After that, we will state without proof a slightly weaker but more transparent version of this theorem as Theorem 2'. Then we proceed to show that Theorem 2' implies the known results.

Proof. We prove this theorem by showing that if $k = \max\{\Delta_m, R_m\}$, then G has a proper k-edge-coloring. By induction, we assume that G - e has a proper k-edge-coloring C, then we will prove (using Theorem 1 and Lemma 3) that so does G. Let $e = (b,a_0)$ and assume, without loss of generality, that $R(e) = h(b)$ and note that this implies that $m(b) \geq m(a_0)$. We observe that by hypothesis, $k \geq \Delta_m$ and $k \geq R_m \geq R(e) \geq h(b)$. By Lemma 1, Part 3 (ii), this implies that $|V_b(C)| \triangleq t \geq 2$ which in turn implies that $|\Gamma b| \geq 2$. We now consider four cases:

Case (1). $h(b) = A(b)$. This means that

$$k \geq A(b) = \max_{\substack{x_1 \text{ and } x_2 \in \Gamma b \\ x_1 \neq x_2}} \left\{ \left\lceil \frac{\mu(b,x_1) + d(x_1) + \mu(b,x_2) + d(x_2)}{m(x_1) + m(x_2)} \right\rceil \right\}$$

If we could prove that

$$A(b) \geq \frac{\sum_{x \in V_b(C)} \big(\mu(b,x) + d(x)\big) - 1}{\sum_{x \in V_b(C)} m(x)}$$

then, by Theorem 1, we can complete the coloring G in this case. Toward this goal, let $x'_1, x'_2, \ldots, x'_t \in V_b(C)$ be such that

$$\frac{\mu(b,x'_1) + d(x'_1) - \frac{1}{t}}{m(x'_1)} \geq \frac{\mu(b,x'_2) + d(x'_2) - \frac{1}{t}}{m(x'_2)} \geq \cdots \geq$$

$$\frac{\mu(b,x'_t) + d(x'_t) - \frac{1}{t}}{m(x'_t)}$$

Then by Lemma 3,

$$\frac{\sum_{x \in V_b(C)} (\mu(b,x) + d(x)) - 1}{\sum_{x \in V_b(C)} m(x)} = \frac{\sum_{i=1}^{t} (\mu(b,x'_i) + d(x'_i) - \frac{1}{t})}{\sum_{i=1}^{t} m(x'_i)} \leq$$

$$\frac{\sum_{i=1}^{2} (\mu(b,x'_i) + d(x'_i)) - \frac{2}{t}}{\sum_{i=1}^{2} m(x'_i)} \leq \frac{\mu(b,x'_1) + d(x'_1) + \mu(b,x'_2) + d(x'_2)}{m(x'_1) + m(x'_2)}$$

$$\leq A(b).$$

This completes the proof in this case.

Case (2). $h(b) = B(b)$. This means that

$$k \geq B(b) = \max_{\substack{x_1 \text{ and } x_2 \in \Gamma b \\ x_1 \neq x_2}} \left\{ \left\lceil \frac{d(x_1) + d(x_2) + \left\lfloor \frac{d(b)}{m(b)} \right\rfloor - 1}{m(x_1) + m(x_2)} \right\rceil \right\}.$$

If we could prove that

$$B(b) \geq \frac{\left(\sum_{x \in V_b(C)} d(x)\right) + \left\lfloor \frac{d(b)}{m(b)} \right\rfloor - 1}{\sum_{x \in V_b(C)} m(x)}$$

then, by Theorem 1, we can complete the coloring of G in this case. Let $x'_1, x'_2, \ldots, x'_t \in V_b(C)$ be such that

$$\frac{d(x'_1) + \frac{1}{t}\left(\left\lfloor \frac{d(b)}{m(b)} \right\rfloor - 1\right)}{m(x'_1)} \geq \frac{d(x'_2) + \frac{1}{t}\left(\left\lfloor \frac{d(b)}{m(b)} \right\rfloor - 1\right)}{m(x'_2)} \geq \cdots \geq$$

$$\frac{d(x'_t) + \frac{1}{t}\left(\left\lfloor \frac{d(b)}{m(b)} \right\rfloor - 1\right)}{m(x'_t)} .$$

Then by Lemma 3,

$$\frac{\left(\sum\limits_{x \in V_b(C)} d(x)\right) + \left\lfloor \frac{d(b)}{m(b)} \right\rfloor - 1}{\sum\limits_{x \in V_b(C)} m(x)} = \frac{\sum\limits_{i=1}^{t}\left(d(x'_i) + \frac{1}{t}\left(\left\lfloor \frac{d(b)}{m(b)} \right\rfloor - 1\right)\right)}{\sum\limits_{i=1}^{t} m(x'_i)} \leq$$

$$\frac{d(x'_1) + d(x'_2) + \frac{2}{t}\left(\left\lfloor \frac{d(b)}{m(b)} \right\rfloor - 1\right)}{m(x'_1) + m(x'_2)} \leq$$

$$\frac{d(x'_1) + m(x'_2) + \left(\left\lfloor \frac{d(b)}{m(b)} \right\rfloor - 1\right)}{m(x'_1) + m(x'_2)} \leq B(b) .$$

This completes the proof of Case 2.

Case (3). $h(b) = C(b)$. This means that

$$k \geq C(b) = \max_{x_1, x_2, x_3 \in \Gamma b \cup \{b\}} \left\{ \left\lceil \frac{d(x_1) + d(x_2) + d(x_3) - 1}{m(x_1) + m(x_2) + m(x_3) - 1} \right\rceil \right\} .$$

If we could prove that

$$C(B) \geq \frac{\left(\sum\limits_{x \in V_b(C) \cup \{b\}} d(x)\right) - 1}{\left(\sum\limits_{x \in V_b(C) \cup b} m(x)\right) - 1}$$

then, by Theorem 1 part 3, we can complete our coloring of G. Let $x'_1, x'_2, \ldots, x'_{t+1} \in V_b(C) \cup \{b\}$ be such that

$$\frac{d(x'_1)}{m(x'_1)} \geq \frac{d(x'_2)}{m(x'_2)} \geq \cdots \geq \frac{d(x'_{t+1})}{m(x'_{t+1})} ,$$

then by Lemma 3, we have

$$\frac{\sum_{x \in V_b(C) \cup \{b\}} d(x)}{\sum_{x \in V_b(C) \cup \{b\}} m(x)} = \frac{\sum_{i=1}^{t+1} d(x_i')}{\sum_{i=1}^{t+1} m(x_i')} \leq \frac{\sum_{i=1}^{3} d(x_i')}{\sum_{i=1}^{3} m(x_i')} .$$

However, because, without loss of generality, one may assume that $d(v) \geq m(v)$ for all $v \in V$,

$$\frac{\left(\sum_{x \in V_b(C) \cup \{b\}} d(x)\right) - 1}{\left(\sum_{x \in V_b(C) \cup \{b\}} m(x)\right) - 1} \leq \frac{\left(\sum_{i=1}^{3} d(x_i')\right) - 1}{\left(\sum_{i=1}^{3} m(x_i')\right) - 1} \leq C(b) .$$

This completes the proof of Case (3).

Case (4). $h(b) = D(b)$. This means that

$$k \geq D(b) = \max_{\substack{x_1 \text{ and } x_2 \in \Gamma b \\ x_1 \neq x_2}} \left\{ \left\lfloor \frac{\left\lfloor \frac{d(x_1)}{m(x_1)} \right\rfloor + \left\lfloor \frac{d(x_2)}{m(x_2)} \right\rfloor + \left\lfloor \frac{d(b)}{m(b)} \right\rfloor}{2} \right\rfloor \right\} .$$

If we could prove that

$$D(b) \geq \frac{\left\lfloor \frac{d(b)}{m(b)} \right\rfloor + \left\lfloor \frac{d(a_0)}{m(a_0)} \right\rfloor + \sum_{x \in V_b(C) - \{a_0\}} \left\lfloor \frac{d(x) + 1}{m(x)} \right\rfloor}{|V_b(C)|} - 1$$

then, by Theorem 1, part 4, we can complete the coloring of G in this case. We first observe that

$$\frac{\left\lfloor \frac{d(b)}{m(b)} \right\rfloor + \left\lfloor \frac{d(a_0)}{m(a_0)} \right\rfloor + \sum\limits_{x \in V_b(C)-\{a_0\}} \left\lfloor \frac{d(x)+1}{m(x)} \right\rfloor}{V_b(C)} - 1 \le$$

$$\frac{\left\lfloor \frac{d(b)}{m(b)} \right\rfloor - 1 + \sum\limits_{x \in V_b(C)} \left\lfloor \frac{d(x)}{m(x)} \right\rfloor}{|V_b(C)|} .$$

We now proceed as before, by letting $x'_1, x'_2, \ldots, x'_t \in V_b(C)$ be such that $d(x'_1) + \frac{1}{t}(\left\lfloor \frac{d(b)}{m(b)} \right\rfloor - 1) \ge d(x'_2) + \frac{1}{t}(\left\lfloor \frac{d(b)}{m(b)} \right\rfloor - 1) \ge \ldots \ge d(x'_t) + \frac{1}{t}(\left\lfloor \frac{d(b)}{m(b)} \right\rfloor - 1)$.

Then by Lemma 3,

$$\frac{(\sum\limits_{x \in V_b(C)} \left\lfloor \frac{d(x)}{m(x)} \right\rfloor) + (\left\lfloor \frac{d(b)}{m(b)} \right\rfloor - 1)}{|V_b(C)|} = \frac{\sum\limits_{i=1}^{t} (\left\lfloor \frac{d(x'_i)}{m(x'_i)} \right\rfloor + \frac{1}{t}(\left\lfloor \frac{d(b)}{m(b)} \right\rfloor - 1))}{t} \le$$

$$\frac{\left\lfloor \frac{d(x'_1)}{m(x'_1)} \right\rfloor + \left\lfloor \frac{d(x'_2)}{m(x'_2)} \right\rfloor + \frac{2}{t}(\left\lfloor \frac{d(b)}{m(b)} \right\rfloor - 1)}{2} \le$$

$$\frac{\left\lfloor \frac{d(x'_1)}{m(x'_1)} \right\rfloor + \left\lfloor \frac{d(x'_2)}{m(x'_2)} \right\rfloor + \left\lfloor \frac{d(b)}{m(b)} \right\rfloor - 1}{2} \le D(b) .$$

This completes the proof of the Theorem.

Theorem 2'. Let $G(V,E)$ be a graph and $A(u)$, $B(u)$, $C(u)$, and $D(u)$ be defined as in Theorem 1. Let $A_m = \max\{A(u) \mid u \in V\}$, $B_m = \max\{B(u) \mid u \in V\}$, similarly define C_m and D_m. Then,

1. $\chi'_m(G) \leqq \max\{\Delta_m, A_m\}$,
2. $\chi'_m(G) \leqq \max\{\Delta_m, B_m\}$,
3. $\chi'_m(G) \leq \max\{\Delta_m, C_m\}$, and
4. $\chi'_m(G) \leq \max\{\Delta_m, D_m\}$.

Corollary 1. Let $G(V,E)$ be a graph and let $\mu(b) = \max\{\mu(b,v) \mid v \in \Gamma b\}$. Then, we have

1. $\chi'_m(G) \leq \max\left\{\left\lceil \frac{d(b) + \mu(b)}{m(b)} \right\rceil \;\middle|\; b \in V\right\}$

2. $\chi'_m(G) \leq \max\{\Delta_m, \sigma_m\}$

   where $\sigma_m = \max\left\{ \left\lceil \frac{d(x_1) + d(x_2) + \left\lfloor \frac{d(b)}{m(b)} \right\rfloor - 1}{m(x_1) + m(x_2)} \right\rceil \right\}$ and the maximum is taken over all paths $x_1 - b - x_2$ of length 2 in G.

3. $\chi'_m(G) \leq \max\{\Delta_m, \sigma'_m\}$

   where $\sigma'_m = \max\left\{ \left\lceil \frac{\left\lfloor \frac{d(x_1)}{m(x_1)} \right\rfloor + \left\lfloor \frac{d(b)}{m(b)} \right\rfloor + \left\lfloor \frac{d(x_2)}{m(x_2)} \right\rfloor}{2} \right\rceil \right\}$ and the maximum is taken over all paths $x_1 - b - x_2$ of length 2 in G.

4. $\chi'_m(G) \leq \max\{\Delta_m, \sigma''_m\}$

   where $\sigma''_m = \max\left\{ \left\lceil \frac{d(x_1) + d(x_2) + d(x_3) - 1}{m(x_1) + m(x_2) + m(x_3) - 1} \right\rceil \right\}$ and the maximum is taken over all $x_1, x_2, x_3 \in \Gamma(b) \cup \{b\}$ and over all $b \in V$.

Proof. 1. It is easy to see that $\max\left\{ \left\lceil \frac{d(b) + \mu(b)}{m(b)} \right\rceil \;\middle|\; b \in V\right\} \geq \Delta_m$ and also that

$$\max_{b \in V} \left\{ \left\lceil \frac{d(b) + \mu(b)}{m(b)} \right\rceil \right\} \geq \max_{b \in V} \left\{ \max_{x_1 \in \Gamma b} \left\lceil \frac{\mu(b,x_1) + d(x_1)}{m(x_1)} \right\rceil \right\}.$$

However, using Lemma 3, we can see that

$$\max_{x_1 \in \Gamma b} \left\lceil \frac{\mu(b,x_1) + d(x_1)}{m(b)} \right\rceil \geq \max_{\substack{x_1, x_2 \in \Gamma b \\ x_1 \neq x_2}} \left\lceil \frac{\mu(b,x_1) + d(x_1) + \mu(b,x_2) + d(x_2)}{m(x_1) + m(x_2)} \right\rceil$$

As the right side of the above inequality is equal to A(b), the desired result follows immediately.

The proofs of parts 2, 3 and 4 of the above corollary routinely follow from parts 2, 3 and 4 of Theorem 2'.

Part (1) of the the above corollary is the generalization of the Vizing bound due to Hakimi and Kariv [1]. Part (2) represents a new generalization of the Ore bound. It is easy to see that parts (2)-(4) of this corollary imply the Shannon bound.

Finally, one may be able to use the results of Theorem 1 to obtain other interestisng bounds. Examing the Goldberg's bound [7], one would have expected that the "ceiling" operator in the definition of A(u) in Theorem 2 could be possibly replaced by a "floor" operator. This, however, is not possible.

References

[1] S. L. Hakimi, O. Kariv, "On a Generalization of Edge-Coloring in Graphs," to appear in J. of Graph Theory, preliminary draft completed Sept. 1983.

[2] V. G. Vizing, "On an estimate of the chromatic class of a p-graph," Discrete Analiz., 3(1964), pp. 25-30 (Russian).

[3] V. G. Vizing, "The chromatic class of a multigraph," Kibernetica (Kiev), 3(1965), pp. 29-39 / Cybernatics 3(1965), pp. 32-41.

[4] S. Fiorini and R. J. Wilson, Edge-Colouring of Graphs, Pitman, London, 1977.

[5] C. E. Shannon, "A theorem on coloring the lines of a network," J. Math. Phys., 28(1949), pp. 148-151.

[6] O. Ore, The Four Color Problem, Academic Press, New York, 1967.

[7] M. K. Goldberg, "Edge-coloring of multigraphs: recoloring techniques," Journal of Graph Theory, Vol. 8, No. 1, 1984, pp. 122-136.

[8] L. D. Anderson, "On Edge-colourings of graphs," Math. Scand 40(1977), pp. 161-175.

[9] A. J. W. Hilton and D. de Werra, "Sufficient conditions for balanced and for equitable edge-coloring of graphs," Department of Mathematics, University of Reading, Whiteknights, Reading, UK; or Department of Mathematiques, Ecole Polytechnique Federal de Lausanne, Av. de Cour 61, DH-1007-Lausanne, Switzerland.

[10] E. G. Coffman, Jr., M. R. Gary, D. S. Johnson, and A. S. La Paugh, "Scheduling File Transfers in a Distributed System," Proc. of 2nd Annual ALM Symp. on Principle of Distributed Computing, 1983, pp. 254-266.

THE DIRECTED SHANNON SWITCHING GAME
AND THE ONE-WAY GAME

Yahya Ould Hamidoune*
Michel Las Vergnas*
Université Pierre et Marie Curie
U.E.R. 48 - Mathématiques
4 place Jussieu, 75005 Paris (France)

Abstract. We present directed versions of the well-known Shannon Switching Game. The main results give classifications and winning strategies for the Directed Switching Game and the One-Way Game.

1. THE SHANNON SWITCHING GAME

The Shannon Switching Game is a well-known two-player game with complete information introduced by C. Shannon circa 1960 [1]. The board is an undirected graph G with two distinguished vertices x_0, x_1. The two players are called Cut and Short. At each move Cut deletes an unplayed edge of G, Short makes an unplayed edge invulnerable to deletion. The objectives of Short is to connect x_0 and x_1 by an invulnerable path.

The outcome of the game may depend on the identity of the first player. An easy argument shows that if there is a winning strategy for Short resp. Cut playing second then there is also a winning strategy for Short resp. Cut playing first. Hence there are 3 types of games: either Short playing second has a winning strategy (the game is called a short game in [1]), or Cut playing second has a winning strategy (cut game), or the first player has a winning strategy (neutral game).

* C.N.R.S.

Theorem (A. Lehman [1]): The Shannon Switching Game on a graph G with respect to two vertices x_0, x_1 is winning for Short playing second if and only if there exist two edge-disjoint trees of G on a same subset of vertices containing x_0 and x_1.

The above theorem characterizes short games. A complete classification of the Shannon Switching Game can be easily derived, using duality considerations.

We observe that (1) if Short can win on subgraph of G than he can also win on G, (2) given a first move $b \in E(G)$ of Cut playing first there is a winning strategy for Short with $w \in E(G)$ as a response to b if and only if there is a winning for Short playing second on $G \setminus b / w$ (Deletion/Contraction Lemma).

Hence to prove the sufficiency in Lehamn theorem it suffices in the case when G is a union of two spanning trees, $E(G) = T_1 + T_2$, to indicate a response $w \in E(G)$ for Short to a first move $b \in E(G)$ of Cut such that $G \setminus b / w$ is again a union of two disjoint spanning trees. The "if" part follows by induction.

Lehman Strategy: Let $b \in T_1$ be the first move of Cut. Then Short should play any $w \in T_2$ (the cocycle of G determined by $T_1 \setminus \{b\}$). In general if $b_1, w_1, b_2, w_2, \cdots, b_i, w_i$ have been played and b_{i+1} is the $(i + 1)$-th move of Cut, then Short forms $G \setminus \{b_1, b_2, \cdots, b_i\}/\{w_1, w_2, \cdots, w_i\}$ and apply the first move strategy with $b = b_{i+1}$.

Actually in his paper Lehman generalized the Shannon Switching Game to a game played on a matroid with respect to a distinguished element. The Shannon Switching Game corresponds to the case of a cycle matroid of a graph. The solution of the Shannon Switching Game is given directly in terms of matroids in [1] (see also [3] for a survey).

2. DIRECTED SWITCHING GAMES

We have introduced in [2] and [3] several directed versions

of the Shannon Switching Game and their generalizations to oriented matroids. The main ones, which we will describe here, are the Directed Shannon Switching Game and the One-Way Game.

The Directed Shannon Switching Game

The board is an undirected graph G with two distinguished vertices x_o, x_1. Short and Cut play alternatively. At each move Cut deletes an unplayed edge, Short makes an unplayed edge invulnerable to deletion and in addition *directs* it. The objective of Short is to join x_0 to x_1 by a *directed* path.

The One-Way Game

The board is an undirected graph G with two distinguished vertices x_0, x_1. In this game both Short and Cut directs edges. The objective of Short is to join x_0 to x_1 by a directed path, which may contain edges directed by Cut.

The reader is referred to [3] for a solution of these games. We state here without proofs the main results:

Theorem 1

The classification of the Directed Shannon Switching Game on a graph G *with respect to two vertices* x_0, x_1 *is identical to the classification of the (undirected) Shannon Switching Game on* G *with respect to* x_0, x_1.

In particular the Directed Switching Game on G *with respect to* x_0, x_1 *is winning for Short playing second if and only if there exist two edge-disjoint trees of* G *on a same subset of vertices containing* x_0, x_1.

The proof of Theorem 1 is more elaborate than that of the undirected case. A simple example shows that Lehman strategy cannot be directed. Furthermore, as easily seen, the Deletion/Contraction lemma does not hold in the directed case. However, as in the undirected case but for different reasons, the desired strategy can be reduced to an inductive step in the case when G is a union of two-edge-disjoint spanning trees (for breviety we say that G is a _block_).

Strategy A

Suppose G _is a block_, $E(G) = T_1 + T_2$. _Let_ $b \in T_1$ _be the first move of Cut._ _Construct_ P _the minimal connected block of_ G _incident to_ x_0 _and containing_ b. _Construct_ Q _the maximal connected block of_ G _incident to_ x_0 _and containing_ b. _A winning response for Short is an edge_ $w \in P\setminus\{b\}\setminus Q$ _incident to_ $\{x_0\} \cup V(Q)$ _directed outgoing from_ $\{x_0\} \cup V(Q)$.

Inductively suppose $b_1, w_1, b_2, w_2, \cdots, b_i, w_i$ _have been played._ _Let_ b_{i+1} _be the_ (i+1)-th _move of Cut._ _Then Short forms_ $G\setminus\{b_1, b_2, \cdots, b_i\}/\{w_1, w_2, \cdots, w_i\}$ _and apply the first move strategy with_ $b = b_{i+1}$.

We illustrate Strategy A in Figures 1, 2, 3 by the 3 first moves of a game. We point out that constrasting with Lehman Strategy, in Strategy A after deletion of b and contraction of w the two trees T_1, T_2 of G do not necessarily give two trees of G b / w (of the tree with heavy lines in Figure 3).

Theorem 2

The classification of the One-Way Game on a graph G _with respect to two vertices_ x_0, x_1 _is identical to the classification of the Shannon Switching Game on_ G _with respect to_ x_0, x_1.

In particular the One-Way Game on G _with respect to_ x_0, x_1 _is winning for Short playing second if and only if there exist two edge-disjoint trees of_ G _on a same subset of vertices containing_ x_0, x_1.

Theorem 1 provides half of the proof of Theorem 2. The second half is given by a Theorem 1*, counterpart of Theorem 1 for the Cographic Directed Switching Game. In this game Cut contracts edges, Short directs edges, his objective being to form

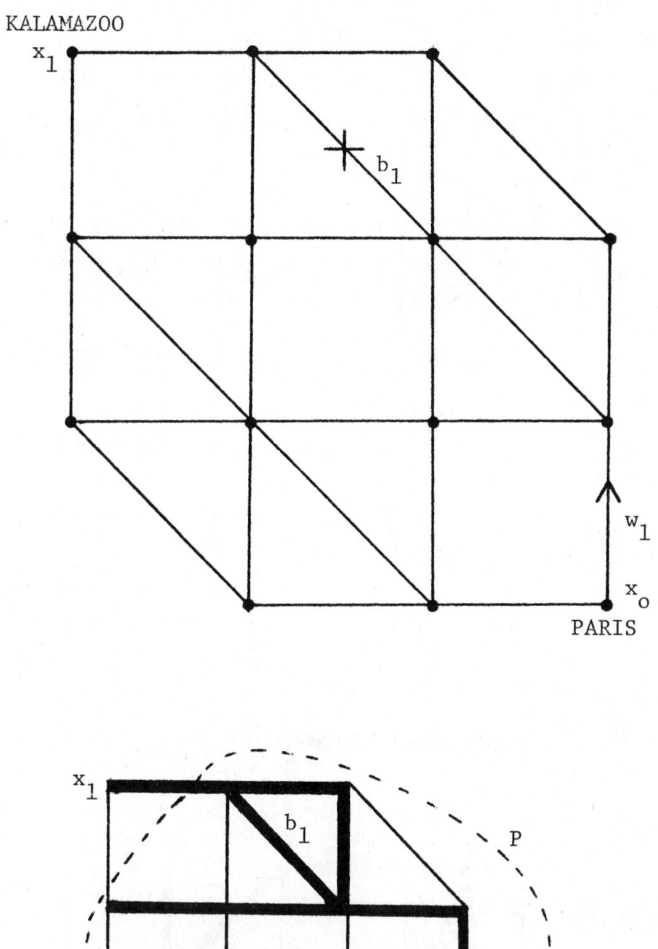

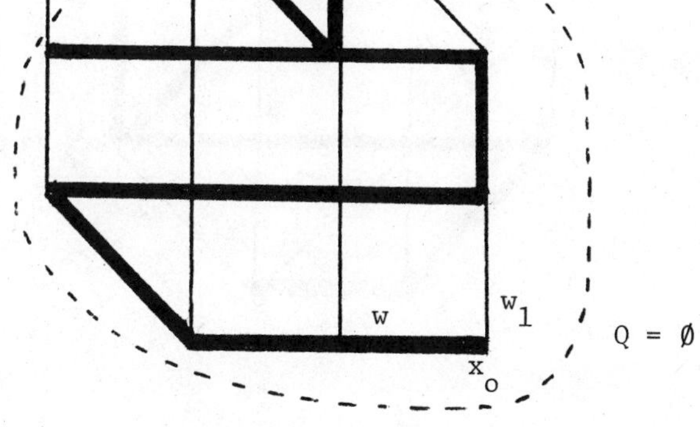

Figure 1.

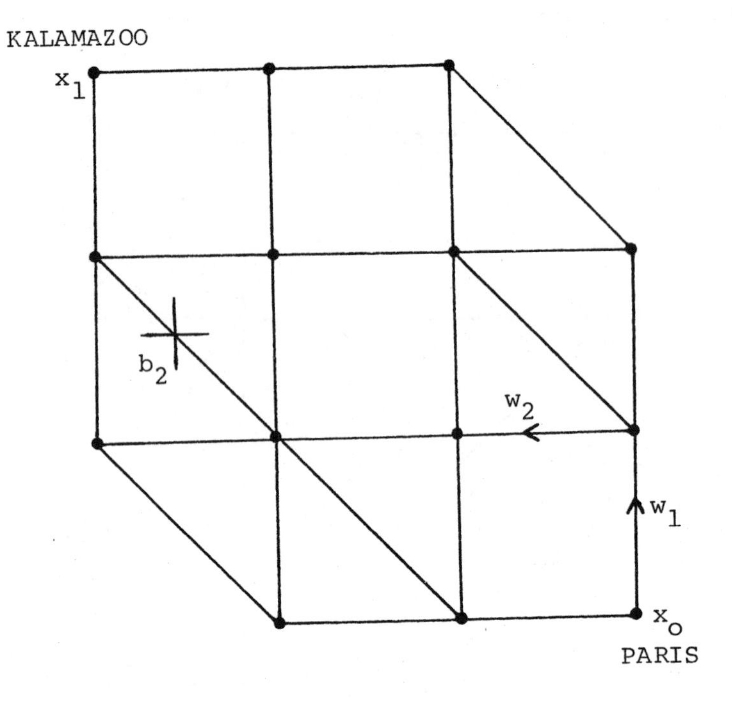

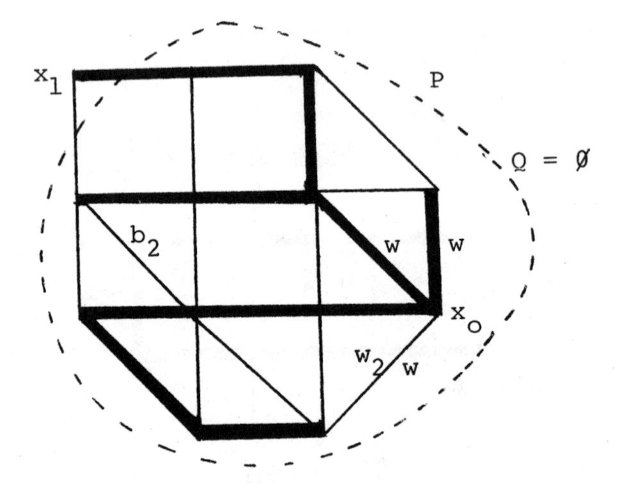

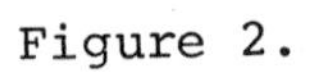

Figure 2.

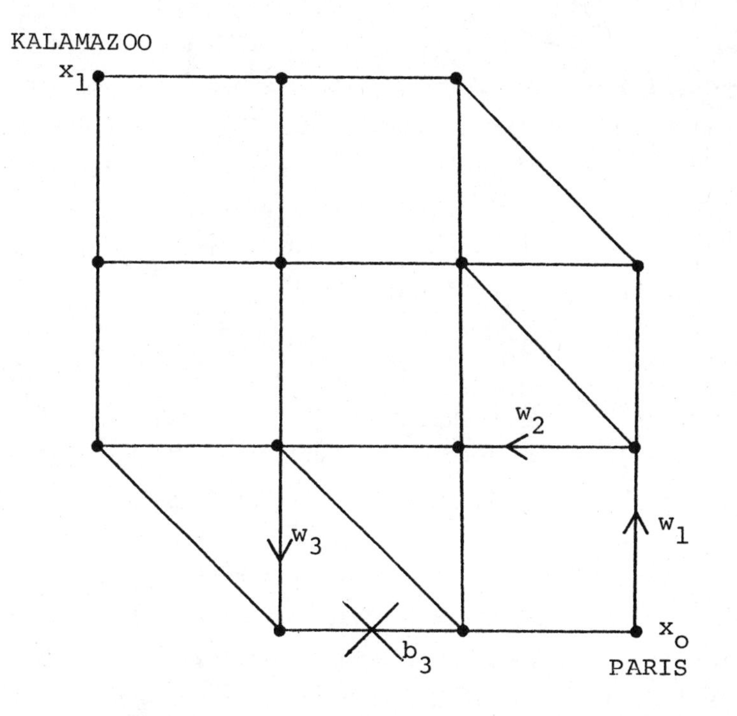

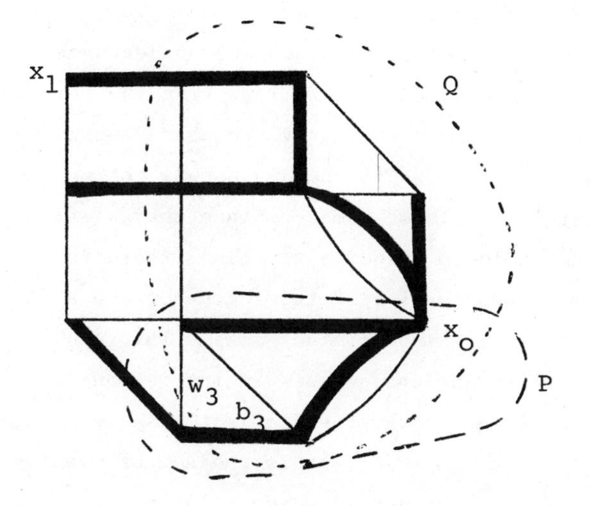

Figure 3.

a directed cocycle separating x_0 from x_1. The Graphic (or Shannon) and Cographic Directed Switching Games are dual in the theory of oriented matroids; the One-Way Game is self-dual. However Strategy A is not dualizable. A specific strategy can be given [3].

A strategy for the Directed Switching Game on a general oriented matroid would be dualizable.

Conjecture

The Directed Switching Game on an oriented matroid M *with respect to a given element* e *is winning for Short playing if (and only if) there exists a block of* M *spanning but not containing* e.

The "only if" part follows from Lehman theory. If this conjecture is true then the classifications of both the Directed Switching Game and the Signing Game (generalizing the One-Way Game) on an oriented matroid with respect to a given element are identical to that of the Lehman Switching Game on the underling matroid with respect to the same element.

We mention some other problems and conjectures:

In the undirected case it follows from the Deletion/Contraction lemma that Lehman theory provides characterizations of winning and losing positions in the Shannon Switching Game (a position is a partially played game). Since the Deletion/Contraction Lemma does not hold in the directed case, there is no evident way to decide from Theorems 1 or 2 the status of a position different from the starting one in directed games. This problem is equivalent to characterize the set of winning responses to a given move of the opponent. Even the simplest case seems difficult: given a block minus an edge what is the set of winning first moves for Short ? To give an actual example let us consider the 4×3 grid of Figure 4.

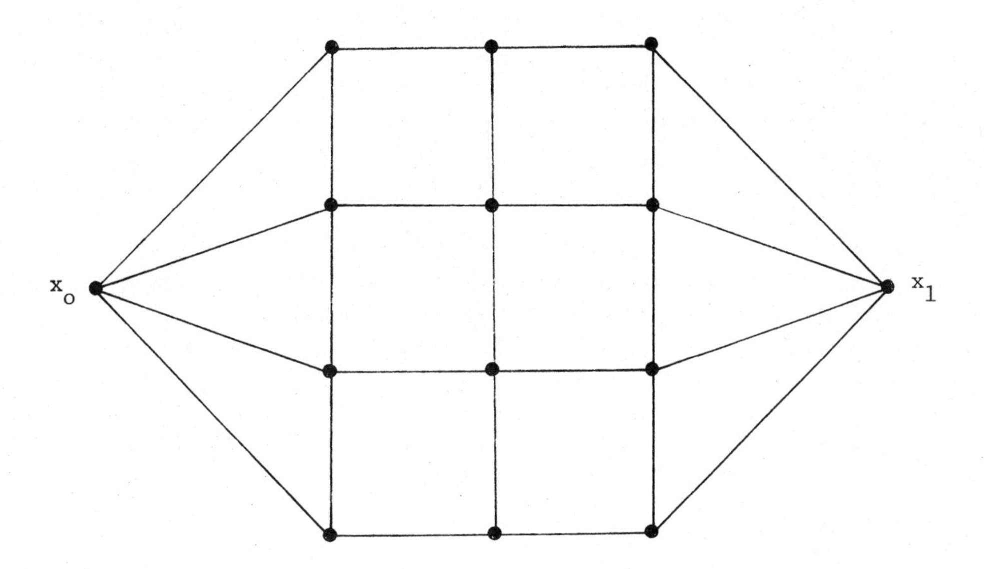

Figure 4.

It can be checked directly that the losing first moves of Short are exactly the vertical edges. All other edges directed from left to right are winning. Conjecture: this property holds for the $n \times (n-1)$ grid for any integer $n \geq 3$ (the case $n = 3$ is also easily checked).

N.B. The winning edges given by Strategy A are all edges incident to x_0 or x_1. In the undirected case all edges are winning.

Since there are polynomial algorithms to construct blocks, the classifications of the Directed Shannon Switching Game and of the One-Way Game, and also the computation of Strategy A are polynomials problems. We suspect that the decision problem for positions in these two games might be NP-complete.

References

1. A. Lehman, A solution of the Shannon Switching Game, J. Soc. Ind. Appl. Math., 12 (1964) 687-725.
2. Y. O. Hamidoune, M. Las Vergnas, Jeux de commutation orientés sur les graphes et les matroides, C. R. Acad. Sci. Paris Ser. I, 298 (1984) 497-499.
3. Y. O. Hamidoune, M. Las Vergnas, Directed Switching Games on Graphs and matroids, 51 pp. submitted to J. Comb. Theory B.

GRAPH THEORETIC APPROACHES TO FINITE MATHEMATICAL STRUCTURES

Frank Harary[1]
University of Michigan

Dedicated to George Polya on the occasion of his 97th birthday

Abstract

The relevance of graph theoretic concepts and methods for other finite mathematical structures was suggested by achievement and avoidance games. These games played on graphs stemmed at first from ramsey theory, which is already known to relate to number theory, group theory, geometry, and other branches of mathematics. We now formulate for any finite mathematical structure X and a given property P which X may enjoy, not only the ramsey number of X with respect to P, but also colorability, connectivity, factorization, packing and mispacking, and covering. These concepts display a remarkable similarity in such an abstract setting.

1. From games on graphs to other finite structures

In general we follow the notation and terminology of [10] for graphs $G = (V,E)$ except for the words, node and edge.

In 2-person achievement and avoidance games, the players are A = Alpha and B = Beta, with A always moving first. In the 2-color achievement game with objective graph F, the number of isolated nodes in the playing board is $p = r(F)$, the ramsey number of F. The players A and B alternately add to

[1] Western Michigan University Visiting Scholar, 4-8 June 1984

the board new green and red edges respectively (these are complementary colors) and the player who first completes graph F in his color is the winner. In the corresponding avoidance game this player is the loser. Neither game can end in a draw by definition of the ramsey number.

This is a special case of playing an instance of a theorem [11] which asserts that its hypothesis implies the existence of some structure. The players alternately make moves which build the hypothesis of the theorem. In the achievement game the first player who reaches the conclusion wins; in the avoidance game he loses. Of course the outcome can never be a draw as the hypothesis implies the conclusion. These do not have to be 2-color games. An example of a 1-color game is provided by the instance that the number p of nodes in the following well-known result is 5. A King in a tournament is a node which can reach every other node in one or two steps.

Redei's Theorem [22] Every tournament has a King.

The Game of Kingmaker.

Here the initial playing board has five isolated nodes u_i. Both players move as follows: draw a new edge $u_i u_j$, then orient it by putting an arrowhead near the center of the edge to facilitate visibility. The player who first makes a King wins.

Observation The winner of this game is A whose only strategy required is to avoid carelessness and move so that B cannot form a King on his next move, and this A can do when $p = 5$.

Unsolved Problems Who wins the corresponding avoidance game, Kingbreaker, when $p = 5$? What is the winning strategy? What are the resolutions of this achievement and avoidance pair of games when $p = 6$? for arbitrary p?

A few other theorems which can be played as games are now listed.

Theorem A (Bruckheimer, Bryan, Muir [4]) For each partition $X = X_1 \cup X_2$ of a nontrivial finite group X, at least one of the subsets X_i generates X.

Theorem B (Erdös, Szekeres [7]) For each positive integer n, any reordering of the sequence $(1, 2, \cdots, n^2 + 1)$ must contain a monotonic subsequence (either increasing or decreasing) of length $n + 1$.

Theorem C (van der Waerden [23]) For any two positive integers n and t, there exists an integer m such that for every partition of $S = \{1, 2, \cdots, m\}$ into t subsets, at least one of the subsets contains an arithmetic progression having n terms.

Of course Theorem C is playable by two players when we take $t = 2$. An analysis of games suggested by Theorem B has been initiated in Harary, Sagan, West [20]. Many other theorems which lend themselves to games can be found in the comprehensive book on ramsey theory by Graham, Rothschild, Spencer [8].

Unsolved Problems. For each game suggested by a theorem, devise an optimal strategy to decide whether A or B is the winner. Also determine how much of the hypothesis can be discarded and still construct the goal under game conditions.

Much research can be done in this area. In a sense, concerning each existence theorem on finite structures, there are two research tasks:

(1) discover and prove the theorem,

(2) devise its strategies and determine the winners.

2. Conditional connectivity and colorability

The connectivity $\kappa(G)$ is the minimum cardinality $|S|$ of a subset $S \subset V = V(G)$ such that $G - S$ is disconnected or trivial. On replacing V in the sentence above by $E = E(G)$, one has the edge-connectivity $\lambda(G)$. We now write $\kappa(G) = \kappa(G,V)$ and $\lambda(G) = \kappa(G,E)$.

When graph G satisfies property P, we write $G \varepsilon P$. The conditional connectivity $\kappa(G,V: P)$ is defined in [13] as the minimum $|S|$ such that $S \subset V$ and $G - S$ is disconnected with each of its components $G_i \varepsilon P$. Similarly $\kappa(G,E: P)$ denotes the conditional edge-connectivity.

These concepts are mentioned in Bollobás [2, Chap. 1].

Some possible examples of properties P in this context include:

(1) bipartite, (2) acyclic, (3) cyclic ≡ not acyclic
(4) unicyclic, (5) cycle, (6) path, (7) star,
(8) complete, (9) trivial, (10) nontrivial.

There is a concise way to describe some of these properties. Let μ be a generic unspecified graphical invariant taking on non-negative integral values n. Then the stipulation that $\mu = n$ or $\mu \leq n$ or $\mu \geq n$ characterizes certain properties P. For example, among the ten properties listed above, one may regard as alternative formulations, with c denoting the number of cycles of G:

(1) $\chi \leq 2$, (2) $c = 0$, (3) $c \geq 1$,
(4) $c = 1$, (9) $p = 1$, (10) $p \geq 2$.

The properties (5) - (8) specify a graphical structure for each component of $G - S$.

The chromatic number $\chi(G)$ is the minimum n such that there exists a partition $V = S_1 \cup \cdots \cup S_n$ for which every induced subgraph $\langle S_i \rangle = G_i$ satisfies the property that $K_2 \subset G_i$, i.e., that G_i is discrete (totally disconnected). When K_2 is replaced by K_3 above, the resulting invariant was studied by Harary and Kainen [18], and when K_2 is replaced by the path

P_n, by Chartrand, Geller, Hedetniemi [6]. On generalizing from these cases, we define [16] the node conditional chromatic number $\chi(G,V\colon P)$ to mean that every $G_i \in P$.

The chromatic index (or edge chromatic number) $\chi'(G)$ is similarly the smallest n such that there exists a partition $E = S_1 \cup \cdots \cup S_n$ for which every $\langle S_i \rangle = G_i$ satisfies $P_3 \not\subset G_i$. When this last property is taken as a general P, we get the edge conditional chromatic number $\chi(G,E\colon P)$.

Conditional chromatic numbers have been independently rediscovered by Burr, Jacobson [5] and Broere, Mynhardt [3]. They are already mentioned in Bollobás [2, Chap. 5]. Another number $\chi(G,E\colon P)$ was studied by Hakimi [9]. Maximum degree conditions are considered in Harary and Jones [17]. Special cases of conditional colorability include arboricity, point arboricity, linear arboricity, thickness, and biparticity.

3. General connectivity, colorability, factorization, covering, packing, and ramsey functions.

A. General connectivity

The two kinds of connectivity of a graph, $\kappa(G,V)$ and $\kappa(G,E)$, can be generalized to any finite mathematical structure X with respect to a property P as follows. Let Y be a specified subset of X. Then define as in [15] the general connectivity $\kappa(X,Y\colon P)$ as the minimum cardinality $|Z|$ of a subset $Z \subset Y$ such that $X - Z \in P$.

Thus $\kappa(G,V) = \kappa(X,Y\colon P)$ when $X = G$, $Y = N$, and P is the property, "disconnected or trivial". Similarly $\kappa(G,E)$ has $Y = E$ with the same P.

Conditional connectivity for graphs can also be incorporated into this formulation by keeping P for the disconnected property and including the desired property P' as well. For example consider the pulverizing connectivity $\kappa(G,V\colon P')$ where P' stands for the trivial graph; this is expressible by

$\kappa(X,Y: P,P')$ where P' means that for each X_i in the partition $X - Z = U_X$, $|X_i| = 1$.

When $Y = X$, we write more briefly $\kappa(X: P)$. Examples of general connectivity are given by:

(a) X is a finite group and P' means "generates X". This connectivity value is of course $|X|$ minus the maximum order of a maximal subgroups of X.

(b) $X = N_k = \{1, 2, \cdots, k\}$ and P means "contains an n-term arithmetic progression".

5c) $X = N_k$ again but P means "there exist three numbers x,y,z such that $x + y = z$. Two different connectivities result from taking x,y,z as distinct or not necessarily so.

(d) X is a connected graph G, Y is either V or E, and property P is "graceful". These graceful connectivities always exist, provided it is a true conjecture that all trees are graceful.

B. General colorability

For any finite mathematical structure X and for any property P which X or its substructures may enjoy, we define the general chromatic number $\chi(X: P)$ as the minimum n for which there is a partition of X into n parts X_i such that for all i, $X_i \in P$. The maximum such value of n corresponds for graphs to the "achromatic number" which was extended to other graphical invariants in [14].

To have $\chi(X_i: P)$ specialize to the chromatic number and the chromatic index of a graph, one must specify a subset Y of X and then define $\chi(X,Y: P)$ as a corresponding partition of Y.

C. General factorization

This is certainly not a new concept as it is simply a partition of a set into subsets having a given property.

Symbolically, a factorization $F(X: P)$ is a partition of X into subsets X_i such that each $X_i \varepsilon P$.

A factorization of a graph $G = (V,E)$ is just a partition of $E = \cup E_i$. Akiyama and Kano [1] studied conditional factorization of graphs in which each spanning subgraph $G_i = (V, E_i)$ has prescribed degree conditions or prescribed structure.

In an isomorphic factorication of G, all the subgraphs G_i are isomorphic to each other. A survey of results and unsolved problems on isomorphic factorizations is presented by Harary and Robinson [19]. Isomorphic factorizations can also be investigated for other finite structures. They are a method of presenting a combinatorial design, as pointed out by Harary and Wallis [21].

D. General covering

For a given structure X and property P, the covering number $c(X: P)$ is the smallest number of substructures of X satisfying P whose union is X. Of course these substructures need not be disjoint (but this is not excluded).

E. General packing

Again given X and P, the packing number $pac(X: P)$ is the maximum number of disjoint substructures satisfying P whose union U is contained in X; thus $X - U$ does not have any subset $Y \varepsilon P$.

Following the max-min viewpoint of [14], we define the mispacking number $pac^-(X: P)$ is the minimum number of disjoint substructures satisfying P whose union $U \subset X$ but $X - U \not\supset Y$ such that $Y \varepsilon P$.

F. General ramsey functions

Recall that the "classic" ramsey numbers for graphs, written $r(m,n)$ in [10, p. 16], are now denoted by $r(K_m, K_n)$

following the development of "generalized ramsey theory for graphs" as surveyed in [12].

Recall also that for three graphs, $G \to F, H$ means that for each partition of $E(G)$ into two parts, one part contains F or the other part contains H. Then the ramsey number $r(F,H)$ is the minimum p such that $K_p \to F,H$. For brevity, $r(F)$ stands for $r(F,F)$. For the purpose of extending this study abstractly to other finite mathematical structures, we now take a new look at $r(F)$.

Define a new function $\text{ram}(G,F)$ as the maximum n such that for all n-partitions of $E(G)$, at least one part contains F. As this is an unfamiliar formulation we pause to illustrate it using for G various K_p and letting $F = K_3$. Clearly $\text{ram}(K_3, K_3) = \text{ram}(K_4, K_3) = \text{ram}(K_5,K_3) = 1$ but $\text{ram}(K_6, K_3) = 2$. Thus for any graph F with no isolated nodes, the ramsey number $r(F)$ is the smallest p such that $\text{ram}(K_p, F) = 2$.

Analogously, given a structure X and a property P we define the ramsey function $\text{ram}(X: P)$ as the maximum n such that for every partition $X = X_1 \cup \cdots \cup X_n$, at least one $X_i \in P$.

4. Symbolic confrontation of general invariants

We now write symbolically all of the topics of the preceding section using $\exists$ and $\forall$ for the quantifiers, $\dot{\cup}$ for a partition (into disjoint nonempty subsets) and a colon for "such that". First we list all the symbolic formulations and then point out similarities and differences.

Factorization

(1) $F(X: P) \equiv (\exists X = \dot{\cup}_n X_i)(\forall i)X_i \in P.$

Connectivity

(2) $\kappa(X: P) = \min n: \ (\exists Z \subset X): |Z| = n \ \wedge$

$$(\exists X - Z = \dot{\cup} X_i)(\forall i)X_i \in P.$$

Packing numbers

(3) $\text{pac}(X: P) = \max n: (\exists Y \notin P)(\exists X - Y = \dot{\bigcup}_n X_i)$

$$(\forall i)\ X_i \in P \ \wedge\ X_i \text{ is P-minimal.}$$

(3⁻) $\text{pac}^-(X: P) = \min n: (\exists Y \notin P)(\exists X - Y = \dot{\bigcup}_n X_i)$

$$(\forall i)\ X_i \in P \ \wedge\ X_i \text{ is P-minimal.}$$

Covering number

(4) $c(X: P) = \min n: (\exists X = \bigcup_n X_i)(\forall i) X_i \in P.$

Chromatic number

(5) $\chi(X: P) = \min n: (\exists X = \dot{\bigcup}_n X_i)(\forall i) X_i \in P.$

(5⁺) $\chi^+(X: P) = \max n: (\exists X = \dot{\bigcup}_n X_i)(\forall i) X_i \in P.$

Ramsey function

(6) $\text{ram}\ (X: P) = \max n: (\forall X = \dot{\bigcup}_n X_i)(\exists i) X_i \in P.$

Actually (1) is not an equation but rather a statement that a factorization is a partition into parts with a given property. Still, (1) is precisely the same as the chromatic number, (5), but without the prefix, "the minimum n such that".

The equation (2) is most similar in appearance to the packing and mispacking number $(3, 3^-)$ with the small but essential difference that in (2), $Z \subset X$ and in (3), $Y \notin P$.

If in (3), set Y is empty and we drop the final P-minimal condition, we get precisely the equation (5^+) for the general achromatic number.

The covering number (4) and the chromatic number (5) are almost identical, the only difference being the dot over the union sign in (5). This stipulates a partition of X with the X_i disjoint while in (4) the X need not be disjoint.

Perhaps the most interesting confrontation is between the chromatic number (5) and the ramsey function (6). These are precisely opposite in that each is obtained from the other by interchanging both <u>max</u> with <u>min</u> and the two quantifiers $\forall$ and $\exists$.

References

[1] J. Akiyama, M. Kano, M-J. Ruiz, Tiling finite figures consisting of regular polygons, These proceedings.

[2] B. Bollobás, Extremal Graph Theory. Academic Press, London (1978).

[3] I. Broere, C. Mynhardt, Generalized colorings of graphs, These proceedings.

[4] M. Bruckheimer, A. C. Bryan, A. Muir, Groups which are the union of three subgroups. Amer. Math. Monthly 77(1970) 52-57.

[5] S. Burr, M. Jacobson, On inequalities involving vertex-partition parameters of a graph. To appear.

[6] G. Chartrand, D. Geller, S. Hedetniemi, A generalization of the chromatic number. Proc. Cambridge Philos. Soc. 64(1968) 265-271.

[7] P. Erdös, G. Szekeres, A combinatorial problem in geometry. Compositio Math. 2(1935) 463-470.

[8] R. L. Graham, B. Rothschild, J. Spencer, Ramsey Theory. Wiley, New York (1981).

[9] S. L. Hakimi, Further results on a generalization of edge coloring. These proceedings.

[10] F. Harary, Graph Theory. Addison-Wesley, Reading (1969).

[11] F. Harary, Achievement and avoidance games designed from theorems. Rend. Sem. Mat. Fis. Milano 51(1981) 163-172.

[12] F. Harary, Generalized ramsey theory I to XIII. The Theory and Applications of Graphs (G. Chartrand et al., eds.) Wiley, New York (1981) 373-390.

[13] F. Harary, Conditional connectivity. Networks 13(1983) 347-357.

[14] F. Harary, Maximum versus minimum invariants for graphs. J. Graph Theory 7(1983) 275-284.

[15] F. Harary, General connectivity. Springer Lecture Notes Math. 1073(1984) 83-92.

[16] F. Harary, Conditional colorability in graphs. Graphs and Applications (F. Harary, J. S. Maybee, eds.) Wiley, New York (1985) 127-136.

[17] F. Harary, K. Jones, Conditional colorability II: Bipartite variations. Congr. Numer. To appear.

[18] F. Harary, P. Kainen, On triangular colorings of a planar graph. Bull. Calcutta Math. Soc. 69(1977) 393-395.

[19] F. Harary, R. Robinson, Isomorphic factorizations X: Unsolved problems. J. Graph Theory 9 (1985). To appear.

[20] F. Harary, B. Sagan, D. West, Computer-aided analysis of monotonic sequence games. Atti Accad. Peloritana dei Pericolanti, Sci. Fis. Mat. Nat., to appear.

[21] F. Harary, W. Wallis, Isomorphic factorizations II: Combinatorial designs. Congr. Numer. 19(1978) 13-28.

[22] L. Rédei, Ein kombinatorischer Satz. Acta Litt. Sci. Szeged 7(1934) 39-43.

[23] B. L. van der Waerden, Beweis einer Baudetschen Vermuting. Nieuw. Arch. Wisk. 15(1927) 212-216.

DRAWINGS OF GRAPHS AND MULTIPLE CROSSINGS

Heiko Harborth

Technische Universität Braunschweig, West Germany

ABSTRACT

Realizations of a graph in the plane are considered where two lines have at most one point in common, either an endpoint or a crossing. For graphs with 2m vertices at most m-fold crossings are possible. It is proved, that the maximum number of m-fold crossings is 2 for m = 3 and m = 4, and at least 2 in general.

1. Introduction

A drawing D(G) of a simple graph G shall denote the following realization of G in the plane: The vertices of G are different points in the plane, and edges between two vertices are simple Jordan curves between the corresponding points in such a way, that two curves, also called lines, have at most one point in common, either an endpoint or a (simple) point of intersection, called crossing. Thus all cases of Figure 1 are forbidden.

Figure 1.

Maximum parts of lines without crossings in its interior are called arcs. Then by points together with crossings, by arcs, and by polygons, which are the nonoverlapping regions of the plane with arcs as sides, there is associated a cell-complex with any D(G). Two drawings of G shall be called isomorphic if and only if there exists an incidence-preserving one-to-one correspondence between their points, crossings, arcs, and polygons.

In general it is hopeless to characterize all drawings of a graph. Only a few numbers C(G) of nonisomorphic (simple) drawings are known: $C(K_n) = 1, 2, 5, 121$ for $n = 3, 4, 5, 6$, respectively, $C(K_{2,2}) = 2$, $C(K_{3,2}) = 6$, $C(K_{3,3}) = 102$ [4,7]. Partial results are known on some properties of drawings, so as maximum and minimum numbers of crossings [1,2], parity of numbers of crossings [4], maximum and minimum numbers of lines without crossings [3,7,8], lines with a maximum number of crossings [6], or triangles in drawings [9].

Here drawings shall be of interest where multiple crossings are allowed. Some first results concerning the maximum number of crossings having the maximum multiplicity m for a graph with 2m vertices are given.

2. Results

Since a threefold crossing determines 6 points, the smallest graph of interest has 6 vertices.

Theorem 1. Any graph G with 6 vertices has at most one drawing with the maximum number 2 of threefold crossings.

This theorem was proved in [5]. The unique $D(K_6)$ is shown in Figure 2. Thus only one of 121 drawings of K_6 exists with two contractible triangles, that is, where all three vertex points are crossings. The unique $D(C_6)$ in Figure 3 with 2 threefold crossings can be completed by unique lines to all possible drawings of graphs with 6 points and with the maximum 2 of threefold crossings.

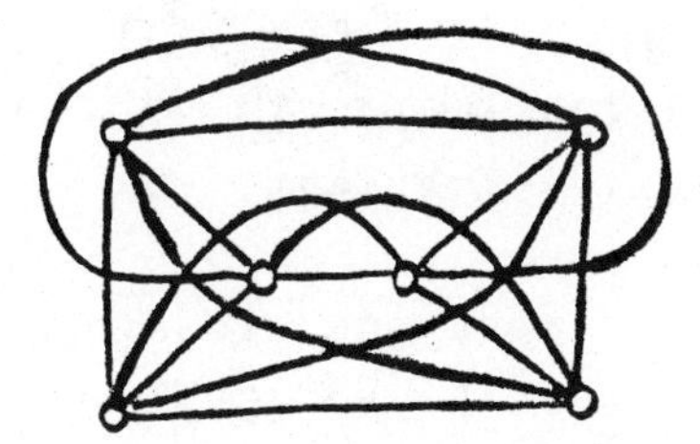

Figure 2.

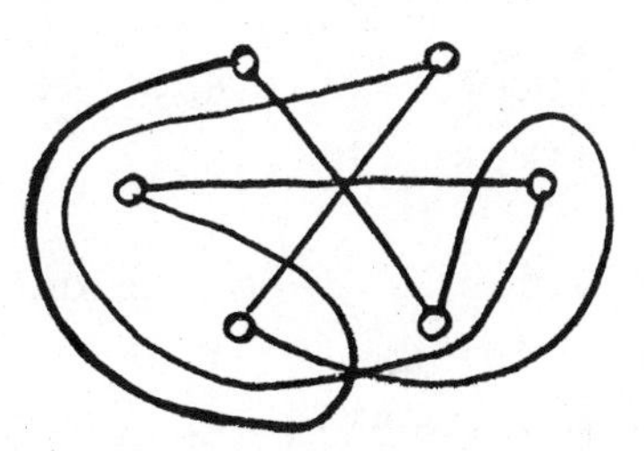

Figure 3.

Theorem 2. Any drawing of a graph with 8 vertices has a maximum number 2 of fourfold crossings. There are exactly 4 nonisomorphic $D(C_8)$ with 2 fourfold crossings.

Theorem 3. For graphs with 2m vertices ($m \geq 3$) there exist drawings with two m-fold crossings.

A trivial upper bound is 2m-1, since the K_{2m} has 2m-1 linear factors. It remains open to find a drawing with 2m points and three or more m-fold crossings.

3. Proofs

To prove Theorem 2 the points determining one fourfold crossing are labelled counter-clockwise by 1,2,...,8 in such a way that (12), (13), or (14) is one line of smallest distance with a second fourfold crossing. Then this crossing is determined by one of the following quadruples of lines:

1.	(12)(34)(56)(78)	12.	(12)(37)(48)(56)
2.	(12)(34)(57)(68)	13.	(12)(38)(45)(67)
3.	(12)(34)(58)(67)	14.	(12)(38)(46)(57)
4.	(12)(35)(46)(78)	15.	(12)(38)(47)(56)
5.	(12)(35)(47)(68)	16.	(13)(24)(57)(68)
6.	(12)(35)(48)(67)	17.	(13)(25)(47)(68)
7.	(12)(36)(45)(78)	18.	(13)(26)(47)(58)
8.	(12)(36)(47)(58)	19.	(13)(26)(48)(57)
9.	(12)(36)(48)(57)	20.	(13)(27)(46)(58)
10.	(12)(37)(45)(68)	21.	(13)(28)(46)(57)
11.	(12)(37)(46)(58)	22.	(14)(27)(36)(58)

Since four points determine at most one crossing, (12) and (56), (13) and (57), so as (14) and (58) cannot intersect oneanother, so that the cases 1., 12., 13., 16., 19., 21., and 22. are impossible.

If both fourfold crossings lie on one line, then the deletion of this line yields the unique drawing $D(C_6)$ of Figure 3. Conversely, there is only one possibility up to symmetry to insert in $D(C_6)$ of Figure 3 two new points such that their connecting line intersects both threefold crossings. Figure 4 shows the unique possibility where two fourfold crossings lie on one line. Thus the quadruples 6., 9., 10., 11., 12., 18., and 19. are discussed.

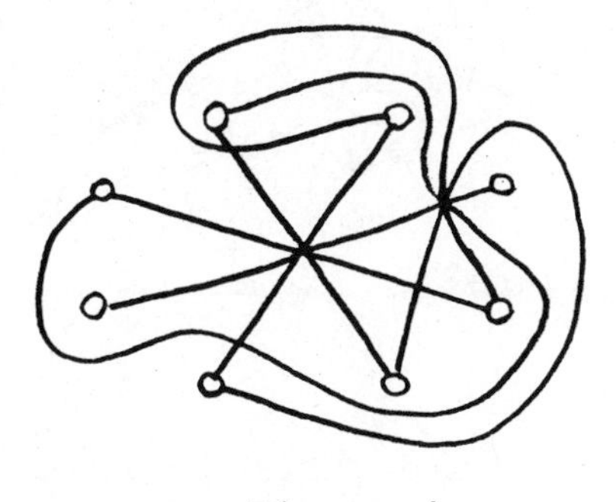

Figure 4.

Addition of 2, 2, 5, 5 (mod 8) to the labels of the points in cases 4., 7., 13., 20. carries these quadruples over to 2., 3., 3., 17., respectively.

Figures 5, 6, and 7 show all different drawings up to symmetry of a line of distance 1, 2, and 3, respectively.

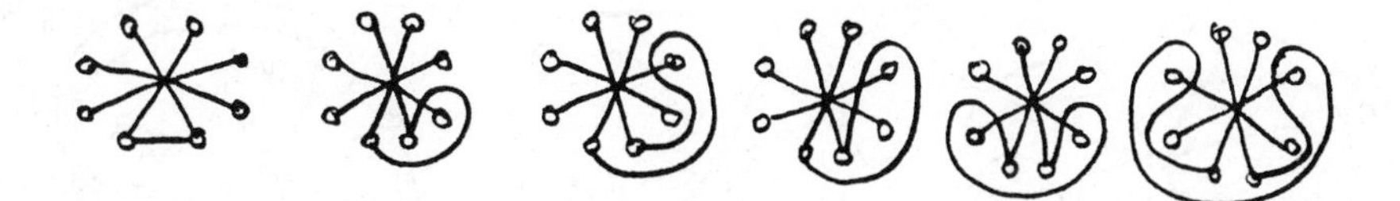

Figure 5.

Figure 6.

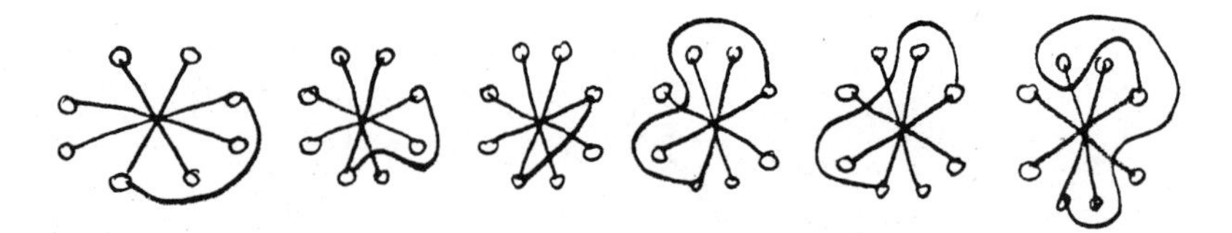

Figure 7.

If all possibilities are checked to combine these types of lines in the remaining cases 2., 3., 5., 8., 14., and 17., then there are only six $D(C_8)$, one in case 2., and five in case 5. (Figures 8 and 9). All $D(C_8)$ of Figure 9 are isomorphic to the corresponding $D(C_8)$ of Figure 8. Thus the four $D(C_8)$ of Figures 4 and 8 are all possible $D(C_8)$ with two fourfold crossings. It is easily checked, that the missing lines of a K_8 can be inserted in an allowed way.

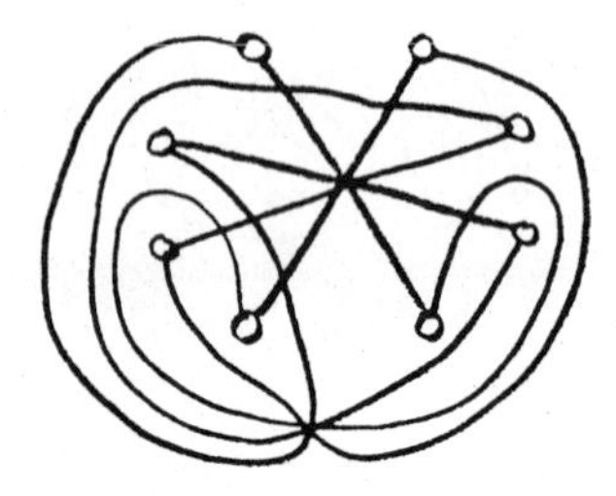

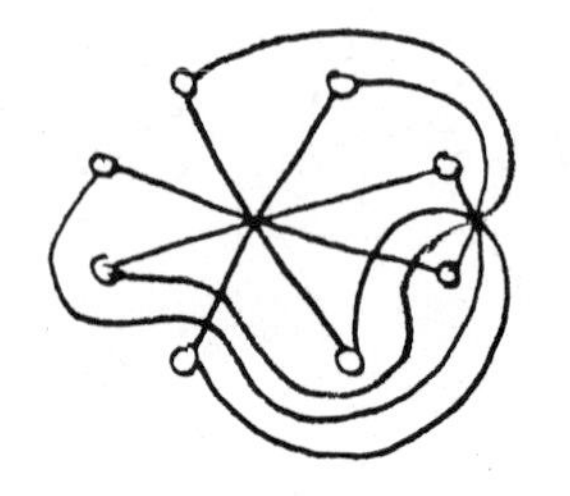

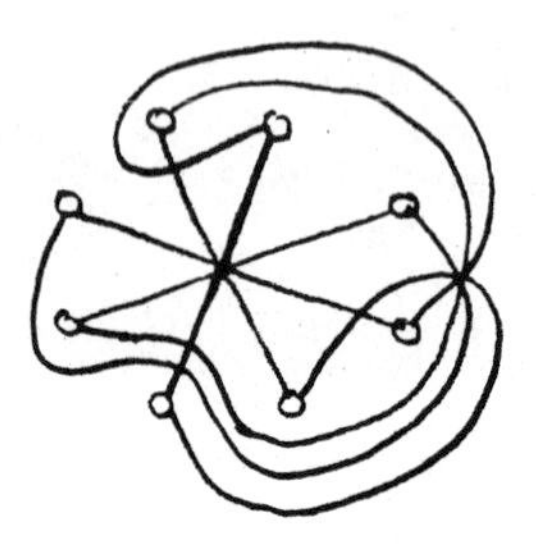

Figure 8.

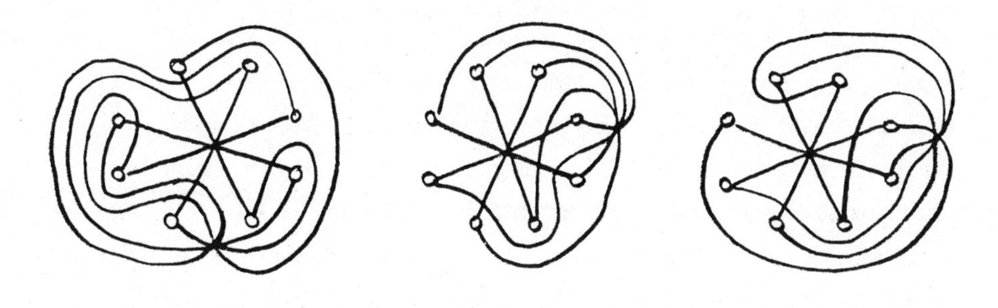

Figure 9.

If a third fourfold crossing exists, the two $D(C_8)$ of Figures 4, 8, and 9 can be combined. This, however, is impossible which in the most cases can be seen already if one additional line, for example the second or fourth line of Figure 5, is tried to insert.

The proof of Theorem 3 is given by Figure 10, which is a generalization of the unique $D(C_6)$, and of the third drawing in Figure 8. All missing lines of $D(K_n)$ can be drawn. Similar generalizations are possible for the other drawings of Figures 8, 9, and 4.

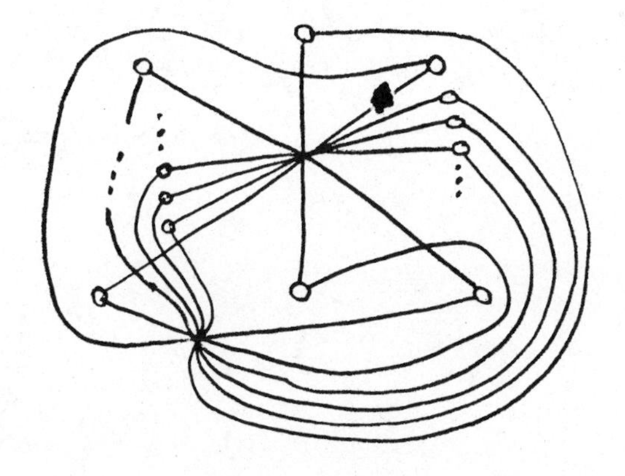

Figure 10.

Finally, Figure 11 presents a nice class of drawings $D(C_n)$ for $n = 4m+2$, $m \geqq 1$: Let one $(2m+1)$-fold crossing be the intersection of all $2m+1$ main diagonals of a regular $(4m+2)$-gon. Any further line beginning at an even labelled point $2r$ intersects the diagonal of the counter-clockwise next point $2r+1$, and runs outside and clockwise back to the opposite point of $2r+2$. Then the outside region is a $(2m+1)$-gon which can be contracted to form the second $(2m+1)$-fold crossing. All missing lines of a K_{4m+2} are possible to draw, however, these drawings dont allow a third $(2m+1)$-fold crossing.

Perhaps it can be conjectured, that never three m-fold crossings occur in a drawing of a graph with 2m vertices.

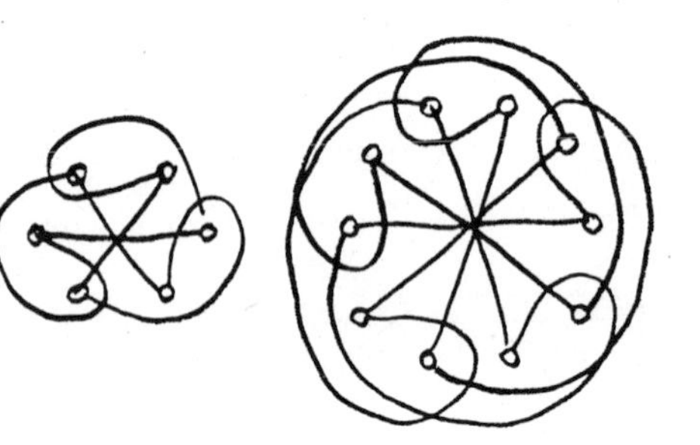
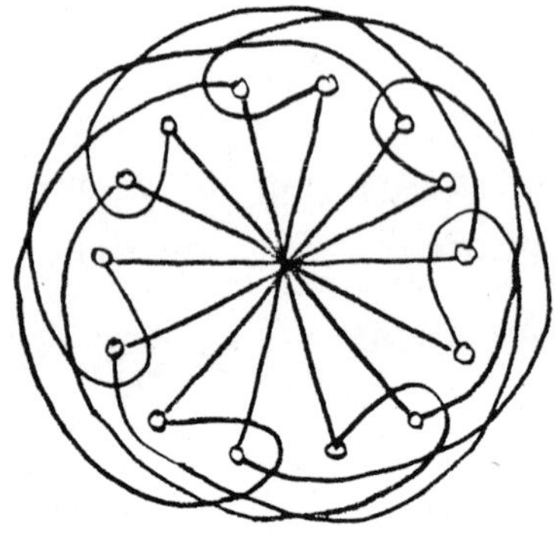

Figure 11.

REFERENCES

1. P. Erdös, and R.K. Guy, Crossing number problems, Amer. Math. Monthly, 80(1973), 52-58.

2. H. Harborth, Über die Kreuzungszahl vollständiger, n-geteilter Graphen, Math. Nachr., 48(1971), 179-188.

3. H. Harborth, and I. Mengersen, Edges without crossings in drawings of complete graphs, J. Combinatorial Theory (B), 17(1974), 299-311.

4. H. Harborth, Parity of numbers of crossings for complete n-partite graphs, Math. Slov., 26(1976), 77-95.

5. H. Harborth, Aufgabe 750, Elem. Math., 31(1976), 121-122.

6. H. Harborth, Crossings on edges in drawings of complete multipartite graphs, Combinatorics (Proc. Fifth Hungarian Colloq., Keszthely 1976), Colloq. Math. Soc. János Bolyai, 18, North-Holland, Amsterdam, 1978, 539-551.

7. I. Mengersen, Darstellungen des vollständigen Graphen ohne kreuzungsfreie Kanten, Diplomarbeit, TU Braunschweig, 1973.

8. I. Mengersen, Die Maximalzahl von kreuzungsfreien Kanten in Darstellungen von vollständigen n-geteilten Graphen, Math. Nachr., 85(1978), 131-139.

9. I.Mengersen, Dreiecke in Darstellungen des vollständigen Graphen, Abh. Math. Sem. Univ. Hamburg, 50(1980), 147-156.

DOMINATION IN TREES: MODELS AND ALGORITHMS

Sandra Hedetniemi *

Clemson University

Stephen Hedetniemi * **

Clemson University

Renu Laskar **

Clemson University

ABSTRACT

A wide variety of domination related parameters of graphs are reviewed, along with the algorithms that have been designed to evaluate these parameters for trees. A new bipartite domination algorithm is also included.

----

* Research supported in part by the National Science Foundation under Grant MCS-8307832

** Research supported in part by the National Science Foundation under Grant #ISP-8011451 (EPSCoR)

I. Introduction

Although a variety of problems involving the placement of queens on a chessboard were considered as early as the 1850's (cf. [1],[2]), it wasn't until many years later that these problems were formulated mathematically by people such as Konig (1936) [30], Berge (1958) [5] and Ore (1962) [36]. A set of vertices D in a graph G=(V,E) is a *dominating set* if every vertex in V-D is adjacent to (or dominated by) at least one vertex in D. The *domination number of G*, denoted $\gamma(G)$, equals the minimum number of vertices in a dominating set of G. During the past 25 years the study of dominating sets in graphs has emerged as a significant area of research not only in graph theory, but in combinatorial optimization and analysis of algorithms as well.

Today it is well known that the problem of determining the domination number of an arbitrary graph is NP-complete [21] and remains NP-complete when restricted to bipartite graphs [16]. However, linear time algorithms have been designed for solving a wide variety of domination problems for trees. In this paper we review most of this algorithmic work and we present as open problems several new domination problems on trees. We also offer a new bipartite domination algorithm for trees. Limitations of space unfortunately prevent us from even mentioning a great deal of work that is closely related to that reported herein. To those authors whose work we are not able to mention here, we offer our apologies; it is only by limiting our discussion to trees that we can keep this paper as short as it is.

II. Domination-related algorithms on trees

1. Domination

As far as we have been able to tell, the first attempt to design a domination algorithm specifically for (directed) trees was made by Daykin and Ng in 1966 [15]. In our reading of their algorithm, however, we discovered an infinite class of trees that their algorithm does not handle correctly. At the same time no quick fix of their algorithm to handle these trees is readily apparent. Unaware of Daykin and Ng's attempt, Cockayne, Goodman and Hedetniemi [10] designed a linear algorithm for computing the domination number of an arbitrary tree in 1975.

2. Optional domination

Let each vertex of a tree be labelled either 'free', 'bound' or 'required'. Each 'bound' vertex must either be dominated or be a member of a dominating set D; each 'required' vertex must be a member of the dominating set; and a 'free' vertex need not be dominated or be a member of the dominating set D. The *mixed domination number* is the minimum number of vertices in a set D which contains all 'required' vertices and dominates or contains all 'bound' vertices. The domination number of a graph G equals the mixed domination number of G, when all vertices of G are labelled 'bound'. The Cockayne, Goodman, Hedetniemi algorithm [10] actually computes the mixed domination number of an arbitrary tree.

3. Weighted domination

Let the vertices v of a tree have associated with them a nonnegative cost c(v), and let the edges (u,v) also have an associated cost w(u,v). A dominating set D defines one or more spanning star subgraphs SSG(D), each of which consists of a set of vertex disjoint stars, at the center of each star is a vertex of D and every vertex in V-D is adjacent to exactly one vertex in D. The cost of an SSG equals the sum of the costs of the vertices in the dominating set together with the costs of the associated edges. Natarajan and White [35] in 1978 designed a linear algorithm for finding a minimum cost SSG of an arbitrary tree. Natarajan and White's algorithm also generalizes to weighted, directed trees.

4. Independent domination

The *independent domination number of a graph G*, denoted i(G), equals the minimum number of vertices in a dominating set D, such that no two vertices in D are adjacent, i.e. D is also an independent set of vertices. It is easy to see that every maximal independent set of vertices is a minimal dominating set. Thus, i(G) equals the minimum number of vertices in a maximal independent set. A linear algorithm for computing i(T) for an arbitrary tree T was designed in 1977 by Beyer, Proskurowski, Hedetniemi and Mitchell [7].

5. Vertex independence

The *independence number of a graph G*, denoted $\beta_0(G)$, equals the maximum number of vertices in an independent set in G. Equivalently, $\beta_0(G)$ equals the maximum number of vertices in an independent dominating set. Linear algorithms for computing $\beta_0(T)$ for an arbitrary tree T were designed, independently, by Daykin and Ng in 1965 [15] and Mitchell in 1977 [32](see also [34]). Daykin and Ng also provided a weighted β_0-algorithm, as did Cockayne and Hedetniemi, independently, in 1976 [11].

6. Vertex cover

The *vertex covering number of a graph G*, denoted $\alpha_0(G)$, equals the minimum number of vertices in a set S which covers (dominates) every edge of G. Because of Gallai's classical theorem [20], that for any nontrivial, connected graph G with p vertices, $\alpha_0(G) + \beta_0(G) = p$, any algorithm for computing $\beta_0(G)$, in effect, computes $\alpha_0(G)$ as well. In fact, the complement of a β_0-set of vertices is an α_0-set. Daykin and Ng's algorithm, therefore, suffices to compute $\alpha_0(G)$ for any tree T. In 1977, Mitchell [32] specifically designed an α_0-algorithm for an arbitrary tree T.

7. Edge domination and independent edge domination

The *edge domination number of a graph G*, denoted $\gamma'(G)$, equals the minimum number of edges in a set F such that every edge in E-F is adjacent to at least one edge in F. A linear algorithm for computing $\gamma'(T)$ for any tree T was designed in

1977 by Mitchell and Hedetniemi [33]. An interesting corollary of this algorithm was the realization that for any nontrivial tree T, $\gamma'(T) = i'(T)$, where $i'(T)$ equals the independent edge domination number of T.

8. Total domination

The *total domination number of a graph G*, denoted $\gamma_t(G)$, equals the minimum number of vertices in a set D such that every vertex in V (rather than in V-D) is adjacent to at least one vertex in D. A linear algorithm for computing $\gamma_t(T)$ for an arbitrary tree T was designed in 1983 by Laskar, Pfaff, S. M. Hedetniemi and S. T. Hedetniemi [31].

9. Matching

The *matching number* (or *edge independence number*) of a graph G = (V,E), denoted $\beta_1(G)$, equals the maximum number of edges in a subset of E, no two of which are adjacent. Equivalently, $\beta_1(G)$ equals the maximum number of edges in a minimal, independent edge dominating set. A linear algorithm for computing $\beta_1(T)$ for an arbitrary tree T was designed in 1975 by Mitchell, Hedetniemi and Goodman [34] and later generalized in 1976 by Goodman, Hedetniemi and Tarjan [22] to both unweighted and weighted b-matchings.

10. Edge cover

The *edge covering number of a graph G* = (V,E), denoted $\alpha_1(G)$, equals the minimum number of edges in a set F such that every vertex in V is incident with at least one edge in F. Because of the second of Gallai's classical theorems [20], that for any nontrivial, connected graph G with p vertices, $\alpha_1(G) + \beta_1(G) = p$, any algorithm for computing $\beta_1(G)$, in effect, computes $\alpha_1(G)$ as well. In fact, Gallai's proof of this theorem provides a procedure for converting a β_1-set of edges S into an α_1-set. Let U be the set of vertices not incident to any edges in S. Form a set of edges W by placing one edge incident to each vertex in U in W. Then the set $S \cup U$ is an α_1-set of T. Therefore, the Goodman, Hedetniemi, Tarjan matching algorithm [22] suffices to compute $\alpha_1(T)$ for any tree T. In 1977, Mitchell [32] specifically designed an α_1-algorithm for an arbitrary tree T.

11. R-domination

Associate with each vertex v_i of a graph G a pair of nonnegative integers (a_i, b_i). A subset of vertices B of V is said to *R-dominate* V if and only if for every vertex v_i in V, either (i) there is a vertex b in B such that $d(v_i, b) \leq a_i$, or (ii) there is a vertex v_j in V such that $d(v_i, v_j) + b_j \leq a_i$. Here, $d(v_i, v_j)$ denotes the distance between vertices v_i and v_j in G. In 1976, Slater [39] designed a linear algorithm for finding a minimum R-dominating set in an arbitrary forest.

12. Greedy domination

Suppose you specify a tree T and positive integers k and d, and you want to find the maximum number of vertices in T which can be dominated, within a distance of d, by a set of vertices of size k. In 1981, Hsu [27] designed an $O(p^3d)$ algorithm for solving this 'greedy' domination problem for an arbitrary tree T with p vertices.

13. Locating-dominating

Given a dominating set D of a graph G = (V,E), for each v in V-D, let S(v) denote the set of vertices in D which are adjacent to v. The set D is called a *locating-dominating set* if for any two vertices v,w in V-D, S(v) ≠ S(w). In [40] Slater gives a linear algorithm for finding a minimum cardinality locating-dominating set in an arbitrary acyclic graph.

14. Kernel

A *kernel* of a directed graph G = (V,A) is a set of vertices K such that no two vertices in K are joined by an arc and for every vertex u not in K, there exists a vertex v in K and a directed arc (u,v) in A. A kernel, in effect, is an independent dominating set in a directed graph. A theorem in Berge [6, p. 311] states that a directed graph without circuits has a kernel and this kernel is unique. Furthermore, the proof of this theorem implicitly gives an algorithm for finding such a kernel.

15. Open irredundance

Let N(v) denote the set of vertices adjacent to a vertex v, and let N[v] = N(v) ∪ {v}. A set of vertices I is *irredundant* if for no vertex v in I is N[v] contained in the union of the sets N[u], for u in I - {v}. A set of vertices O is *open irredundant* if for no vertex v in O is N(v) contained in the union of the sets N[u], for u in O - {v}. Intuitively, a set O is open irredundant if for every vertex v in O there exists at least one vertex not in O which is adjacent to v but to no other vertex in O. In 1983, Farley and Proskurowski [19] designed a linear algorithm for finding the maximum cardinality of an open irredundant set in an arbitrary tree T.

16. P-centers

Associate with each vertex v a nonnegative number w(v) (the weight of v) and with each edge e a positive number l(e) (the length of e). Let $X_p = \{x_1, x_2, \ldots, x_p\}$ be a set of p points on G, where by a point on G we mean a point along any edge, which may or may not be a vertex of G. The distance $d(v, X_p)$ between a vertex v of G and the set X_p is defined by $d(v, X_p) = \min \{d(v, x_i) \mid x_i \text{ in } X_p\}$ where $d(v, x_i)$ is the length of a shortest path in G between vertex v and point x_i. Let $F(X_p) = \max \{w(v)\, d(v, X_p) \mid v \text{ in } V\}$ and let X^*_p be a set of points such that $F(X^*_p)$ is a minimum. Then the set X^*_p is called the (*absolute*) *p-center of G* and the value $F(X^*_p)$ is called the (*absolute*) *radius*. If X_p and X^*_p are restricted to be sets of vertices of G, then X^*_p is called a *vertex p-center* and $F(X^*_p)$ is called the *vertex p-radius*. In 1979, Kariv and Hakimi [28] presented an assortment of algorithms for finding p-centers and absolute p-centers of arbitrary graphs, and specialized algorithms for finding p-centers and absolute p-centers in trees. Among these algorithms were an O(n) algorithm for finding a minimum dominating set of radius r, and an $O(n^2 \lg n)$ algorithm for finding a vertex or absolute p-center in a vertex weighted tree with n vertices.

17. b-domination

Let the vertices of a graph G be $\{v_1, v_2, \ldots, v_p\}$ and let the degrees of these vertices be $d_1, d_2, \ldots, d_p$, respectively. Associate with each vertex v_i a bound b_i, where $0 \le b_i \le d_i$. Let $b = (b_1, b_2, \ldots, b_p)$. A set of vertices D is a *b-dominating set* if for every vertex v_i in V-D, $| D \wedge N(v_i)| \ge b_i$. The *b-domination number for a given vector b* equals the minimum number of vertices in a b-dominating set. In 1984, Hedetniemi, Laskar and Pfaff [26] designed a linear algorithm for computing the b-domination number of an arbitrary tree.[1]

18. ve domination

A vertex v can be said to dominate each vertex adjacent to v and each edge incident to v. An edge e can be said to

----

[1]It has come to our attention that E. J. Cockayne and B. Shepherd (University of Victoria) have recently designed a b-domination algorithm for trees as well.

dominate each edge adjacent to e and the two vertices contained in e. The *ve domination number of a graph G* = (V,E), denoted $\gamma_{ve}(G)$, equals the minimum cardinality of a subset W of V U E which dominates V U E - W. It is easy to see that for any connected graph G, $\gamma_{ve}(G) = \gamma(T(G))$, where T(G) denotes the total graph of G (cf. [4, p. 197]). In 1984, S. M. Hedetniemi, S. T. Hedetniemi and R. Laskar [24] designed a linear algorithm for computing $\gamma_{ve}(T)$ for an arbitrary tree T.

Several other domination-related parameters, that have been defined for arbitrary graphs, have immediate algorithmic solutions when restricted to trees. Limitations of space preclude a full discussion of these here, but they include:

19. neighborhood number $n_0(G)$ defined by Sampathkumar and Neeralagi [37]; for any tree T, $n_0(T) = \gamma(T)$;
20. star partition number $\gamma^*(G)$ defined by Walikar [41]; for any tree T, $\gamma^*(G) = \gamma(T)$;
21. looseness, l(G), defined by Cockayne and Hedetniemi [12]; for any nontrivial tree T, l(T) = 2;
22. clique domination $\gamma_\omega(G)$ (cliques to dominate all vertices); $\gamma_\omega(T)$ can be determined from $\gamma_t(T)$;
23. clique covering $\alpha_\omega(G)$ (cliques to contain all vertices); for any tree T, $\alpha_\omega(T) = \alpha_1(T)$;
24. connected domination $\gamma_c(G)$, defined by Sampathkumar and Walikar [38]; $\gamma_c(T)$ = p - #endvertices for any tree T with p

≥ 3 vertices;

25. maximum minimal domination (Γ) and maximum maximal irredundance (IR); Cockayne, Favaron, Payan and Thomason [9] showed that for any bipartite graph G, $\beta_0(G) = IR(G) = \Gamma(G)$.

III. Open problems

26. Irredundance

The *irredundance number of a graph G*, denoted ir(G), equals the minimum number of vertices in a maximal irredundant set (cf. 15 above). The problem of computing the irredundance number of a tree is perhaps the most widely studied, unsolved domination problem for trees.

27. Uncapacitated b-domination

We remove from the definition of b-domination (cf. 17 above) the restriction that $b_i \leq d_i$. An *uncapacitated b-dominating set* is a multiset of vertices D such that for every vertex v_i in V-D, $|D \wedge N(v_i)| \geq b_i$. In this case, a vertex v, say of degree 3, may have to be dominated say 6 times. This can be accomplished, for example, by assigning two copies of each vertex adjacent to v to the set D. One is interested in finding a minimum cardinality multiset which is an uncapacitated b-dominating set.

28. Efficient domination

A dominating set D of a graph G = (V,E) is *efficient* if for every vertex v in V-D, $|N(v) \wedge D| = 1$, i.e. each vertex not in

D is adjacent to exactly one vertex in D. It is easy to see that not every tree has an efficient dominating set, eg. there is a tree with five vertices that does not have one. However, if a tree T does have an efficient dominating set D, then it can be seen that $\gamma(T) = i(T) = |D|$. The problem is to construct an algorithm for deciding if an arbitrary tree has an efficient dominating set. Notice that efficient total dominating sets may also be defined in the same way. Bange, Barkauskas and Slater who defined this concept [3], actually developed a constructive characterization of the class of trees having two disjoint efficient dominating sets.

29. Path domination

The *path domination number of a graph G*, denoted $\gamma_p(G)$, equals the minimum number of vertex disjoint paths in a set P such that every vertex not contained in a path in P is adjacent to a vertex in at least one path in P. Although several algorithms for finding paths in trees or decomposing trees into paths have been designed, to our knowledge an algorithm for finding the path domination number has not been designed.

IV. A new domination algorithm for trees

30. Bipartite domination

Let G = (V,E) be a connected, bipartite graph and let V1 and V2 be the partite subsets of V. The *bipartite domination number of G*, denoted $\gamma_b(G)$, equals the minimum number of vertices in a subset D of V1 which dominates all of the vertices in V2. It is, first of all, easy to see that the problem of determining $\gamma_b(G)$ for an arbitrary bipartite graph G is NP-complete. We will show that the domination problem (for arbitrary graphs) is transformable to the bipartite domination problem. Let G be an arbitrary graph with vertices $V = \{v_1, v_2, ..., v_p\}$. Let the literals $N = \{N_1, N_2, ..., N_p\}$ represent the closed neighborhoods of the vertices $v_1, v_2, ..., v_p$, respectively. Construct a bipartite graph $H(G) = (V \cup N, F)$, where V1 = N and V2 = V as follows: there is an edge in F between v_i and N_j if and only if vertex v_i is an element of $N[v_j]$. It is easy to see that $\gamma(G) = \gamma_b(H(G))$. Thus, any bipartite domination algorithm can be used to solve the original domination problem, which is known to be NP-complete.

If a bipartite graph G is a tree T, however, a linear algorithm can be designed to compute $\gamma_b(T)$, as follows.

Assume that the tree T is rooted at some vertex, call it vertex 1. We represent T by a parent array, as illustrated in Figure 1; that is if parent(i) = j then $j < i$ (the label of the parent of a vertex is less that the label of the vertex itself). We assume that the root vertex has no parent, i.e. parent(1) = 0. Let the vertices be classified as 1-vertices or 2-vertices, depending on whether they belong to the set V1 or V2, respectively. This classification can be accomplished in O(p) time if necessary.

** label each vertex initially **

label every 2-vertex 'B' (for 'bound', i.e. it must be dominated)

label every 1-vertex 'F' (for 'free', i.e. it may be used)

** process the vertices from vertex p downto vertex 2 **

for u ← p downto 2 do

v ← parent(u)

re-label v according to the following table:

if u is a 1-vertex and v is a 2-vertex; and u and v are labelled as follows:

u = \ v =	'F'	'B'
'F'	∅	∅
'R'	∅	'F'

: then relabel v

if u is a 2-vertex and v is a 1-vertex; and u and v are labelled as follows:

u = \ v =	'F'	'R'
'F'	∅	∅
'B'	'R'	∅

: then relabel v

** In the tables above '∅' means do nothing **

** process vertex 1 **

if vertex 1 is labelled 'B' then label any child of 1 as 'R'

The set of 1-vertices labelled 'R' (required) at the end of the algorithm constitutes a minimum cardinality bipartite dominating set.

Limitations of space preclude a detailed discussion of this algorithm, but details concerning its correctness and O(p) complexity can easily be inferred from the similarities between this algorithm and the Cockayne, Goodman, Hedetniemi tree domination algorithm in [10].

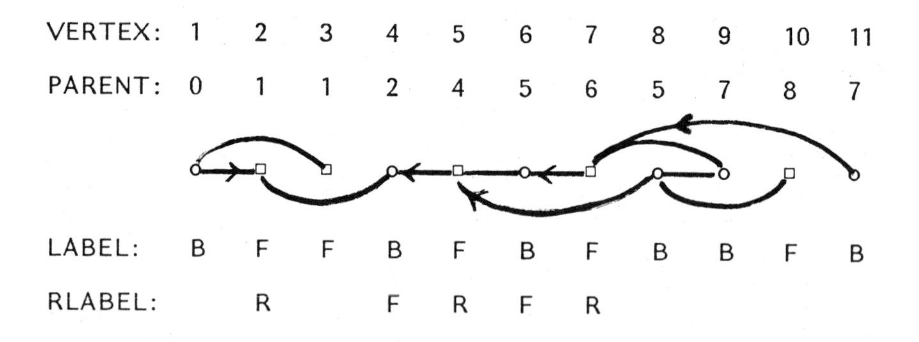

Figure 1

The tree in Figure 1 illustrates the bipartite domination algorithm. Circled vertices belong to V2; boxed vertices belong to V1. All vertices are labelled 'B' or 'F' initially. The algorithm proceeds from the leaves to the root (right-to-left). The arrows indicate which vertices cause their parent to be relabelled. Finally, the root vertex labelled 'B' at the last step of the algorithm causes one of its children (i.e., 2) to be relabelled 'R'.

V. Conclusions

With the possible exception of the problem of determining the irredundance number of a tree, it appears that most domination problems can be solved algorithmically for trees in linear time. It also appears that generalizations of these algorithms to other classes of graphs are not easily obtained. Some limited successes in designing domination algorithms for other classes of graphs have been obtained as follows: Farber for strongly chordal graphs [17]; Kikuno, Yoshida and Kakuda for series-parallel graphs [29]; Corneil and Keil for fixed-k-trees [13]; Farber and Keil for permutation graphs [18]; Booth and Johnson for directed path graphs [8]; Hedetniemi, Laskar and Pfaff for cacti [25]; Corneil and Perl for cographs [14]; and Gurevich, Stockmeyer and Vishkin for graphs that differ from trees by only a fixed number of edges [23]; to mention a few.

VI. Bibliography

1. W. Ahrens, Mathematische Unterhaltungen und Spiele, Leipzig, 1901.

2. Rouse W. Ball, Mathematical Recreations and Problems, London, 1892.

3. D.W. Bange, A. E. Barkauskas and P. J. Slater, Disjoint dominating sets in trees, Tech. Rept. SAND78-1087J, Sandia Laboratories, 1978.

4. M. Behzad, G. Chartrand and L. Lesniak-Foster, Graphs and Digraphs, Wadsworth, Belmont, CA, 1979.

5. C. Berge, Theory of Graphs and its Applications, Methuen, London, 1962.

6. C. Berge, Graphs and Hypergraphs, North Holland, Amsterdam, 1973, 303-324.

7. T. Beyer, A. Proskurowski, S. Hedetniemi and S. Mitchell, Independent domination in trees, *Proc. Eighth S. E. Conf. on Combinatorics, Graph Theory and Computing*, Utilitas Mathematica, Winnipeg, 1977, 321-328.

8. K. S. Booth and J. H. Johnson, Dominating sets in chordal graphs, *SIAM J. Comput.*, *11,1* (1982), 191-199.

9. E. J. Cockayne, O. Favaron, C. Payan and A. Thomason, Contributions to the theory of domination, independence and irredundance in graphs, *Discrete Math.*, *33*(1981), 249-258.

10. E. J. Cockayne, S. E. Goodman and S. T. Hedetniemi, A linear algorithm for the domination number of a tree, *Infor. Processing Lett.*, *4*(1975), 41-44.

11. E. J. Cockayne and S. T. Hedetniemi, A linear algorithm for the maximum weight of an independent set in a tree, *Proc. Seventh S. E. Conf. on Combinatorics, Graph Theory and Computing*, Utilitas Mathematica, Winnipeg, 1976, 217-228.

12. E. J. Cockayne and S. T. Hedetniemi, Towards a theory of domination in graphs, *Networks*, *7*(1977), 247-261.

13. D. Corneil and J. M. Keil, A dynamic programming approach to the dominating set problem on k-trees, manuscript, 1983.

14. D. G. Corneil and Y. Perl, Clustering and domination in perfect graphs, *Discrete Appl. Math.*, to appear.

15. D. E. Daykin and C. P. Ng, algorithms for generalized stability numbers of tree graphs, *J. Austral. Math. Soc.*, *6*(1966), 89-100.

16. A. K. Dewdney, Fast Turing reductions between problems in NP; Chapter 4: Reductions between NP-complete problems, Rept. 71, Dept. Computer Science, Univ. Western Ontario, 1981.

17. M. Farber, Domination, independent dominating and duality in strongly chordal graphs, *Discrete Appl. Math.*, to appear.

18. M. Farber and J. M. Keil, Domination on permutation graphs, manuscript, 1983.

19. A. M. Farley and A. Proskurowski, Computing the maximum order of an open irredundant set in a tree, manuscript, 1983.

20. T. Gallai, Uber extreme Punkt-und-Kantenmengen, *Ann. Univ. Sci. Budapest Eotvos Sect. Math.*, *2*(1959), 133-138.

21. M. R. Garey and D. Johnson, Computers and Intractability: A Guide to the Theory of NP-Completeness, W. H. Freeman, San Francisco, 1979.

22. S. E. Goodman, S. T. Hedetniemi, and R. E. Tarjan, B-matching in trees, *SIAM J. Comput.*, *5*(1976), 104-107.

23. Y. Gurevich, L. Stockmeyer and U. Vishkin, Solving NP-hard problems on graphs that are almost trees and an application to facility location problems, Rept. No. RC9348, IBM Research, Yorktown Heights, N.Y., 1982.

24. S. M. Hedetniemi, S. T. Hedetniemi, and R. Laskar, VE-domination in trees, in preparation.

25. S. T. Hedetniemi, R. Laskar and J. Pfaff, A linear algorithm for the domination number of a cactus, Tech. Rept. 433, Dept. Math. Sciences, Clemson Univ., 1983.

26. S. T. Hedetniemi, R. Laskar and J. Pfaff, b-domination in trees, in preparation.

27. Wen-Lian Hsu, On the maximum coverage problem on trees, Tech. Rept., Dept. Industrial Eng. and Management Sci., Northwestern University, Nov. 1981.

28. O. Kariv and S. L. Hakimi, An algorithmic approach to network location problems I: the p-centers, *SIAM J. Appl. Math.*, *37* (1979), 513-538.

29. T. Kikuno, N. Yoshida and Y. Kakuda, A linear algorithm for the domination number of series-parallel graphs, *Discrete Appl. Math.*, *5*(1983), 299-311.

30. D. Konig, Theorie der endlichen und unendlichen graphen, Chelsea, New York, 1950.

31. R. Laskar, J. Pfaff, S. M. Hedetniemi and S. T. Hedetniemi, On the algorithmic complexity of total domination, *SIAM J. Discrete Appl. Math.*, to appear.

32. S. L. Mitchell, Linear Algorithms on Trees and Maximal Outerplanar Graphs: design, complexity analysis and data structures study, Ph. D. Thesis, Univ. Virginia, 1977.

33. S. L. Mitchell and S. T. Hedetniemi, Edge domination in trees, *Proc. Eighth S. E. Conf. Combinatorics, Graph Theory and Computing*, Utilitas Mathematica, Winnipeg, 1977, 489-509.

34. S. Mitchell, S. Hedetniemi and S. Goodman, Some linear algorithms on trees, *Proc. Sixth S. E. Conf. Combinatorics, Graph Theory and Computing*, Utilitas Mathematica, Winnipeg, 1975, 467-483.

35. K. S. Natarajan and L. J. White, Optimum domination in weighted trees, *Infor. Processing Lett.*, *7*(1978), 261-265.

36. O. Ore, Theory of Graphs, Amer. Math. Soc. Colloq. Publ. 38, Providence, 1962.

37. E. Sampathkumar and P. S. Neeralagi, The neighborhood number of a graph, manuscript, May 1983.

38. E. Sampathkumar and H. Walikar, The connected domination number of a graph, *J. Math. Phy. Sci.*, *13*(1979), 607-613.

39. P. J. Slater, R-domination in graphs, *J. Assoc. Comput. Mach.*, *23*(1976), 446-450.

40. P. J. Slater, Domination and location in acyclic graphs, manuscript, May 1983.

41. H. B. Walikar, On star-partition number of a graph, manuscript, June, 1979.

CONNECTED PLANAR GRAPHS WITH THREE OR MORE ORBITS

Joan P. Hutchinson
Smith College

Lucia B. Krompart
Rochester, MI

ABSTRACT

The cardinalities of the orbits of a graph with V vertices under the action of its automorphism group give a partition of V. In general any partition, except for the partitions of 2,3,4 and 5 into singletons, can arise. Here we present necessary and sufficient conditions for a partition with three parts to arise as the orbit sizes of a connected planar graph; the case of two parts has been established previously. In this paper we demonstrate the sufficiency of these conditions and include some results on partitions with $k>3$ parts. In a subsequent paper the necessity of these conditions will be proved.

1. Introduction

Planar graphs and their automorphism groups have been intensively studied both because of the importance of their geometric and algebraic properties and because of the historical significance and aesthetic appeal of graphs such as those arising from the Platonic solids and tesselations of the sphere. We begin with an abstract automorphism question and quickly find that the Platonic solids and their related graphs play a key role in the answer.

Let G be a simple graph. Then Aut(G) denotes the group of all adjacency-preserving bijections between the vertices. The orbits of G are the transitivity classes of vertices with respect to Aut(G), and the cardinalities of the orbits give a partition of V, the number of vertices of G. The basic question which we address is which partitions of an integer arise as the cardinalities of orbits of a simple connected planar graph.

The study of this question was initiated in [4], where it was shown that every partition, except (1,1), (1,1,1), (1,1,1,1) and (1,1,1,1,1), can arise as the orbit sizes of a planar graph and of a connected graph. However, the situation for connected planar graphs is more complex and interesting.

We call a graph a $(x_1,x_2,...,x_k)$-graph if the cardinalities of its orbits are the numbers $x_1 \leq x_2 \leq \ldots \leq x_k$, and we call a partition representable if there is a simple connected planar $(x_1,x_2,...,x_k)$-graph.

It is clear that for each integer n, (n) is representable by the simple cycle of length n, when $n \geq 3$, and by a single vertex or edge for n=1 or 2. The pairs that are representable have been characterized in [6] as follows.

THEOREM A. Except for (1,1), (x_1,x_2) is representable if and only if

a. x_1 or (if x_1 is even) $x_1/2$ divides x_2, or

b. $(x_1,x_2) \in \{(6,8), (12,20)\}$.

Furthermore, every (6,8)-graph (respectively (12,20)-graph) is the dual of the line graph of the cube (respectively the dodecahedron), possibly with additional edges.

We denote these two special graphs by dL(C) and dL(D) and present a planar drawing of each in Figure 1.

The conditions of Theorem A do not extend simply to a characterization of representable k-tuples or even triples. For example, the triple (2,10,15) can be divided into pairs (2,10) and (10,15), and the triple (3,5,15) into (3,15) and (5,15). All four pairs

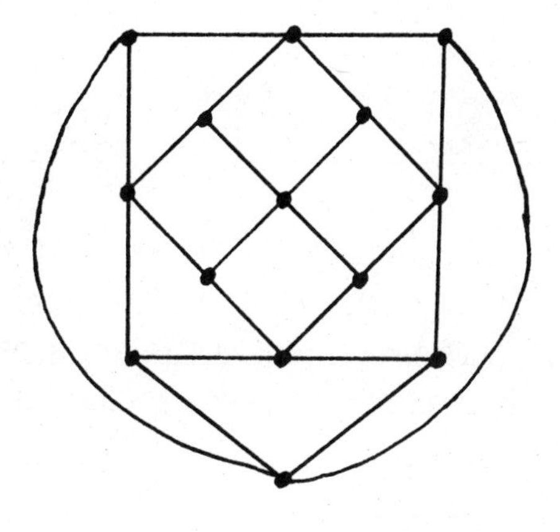

dL(C)

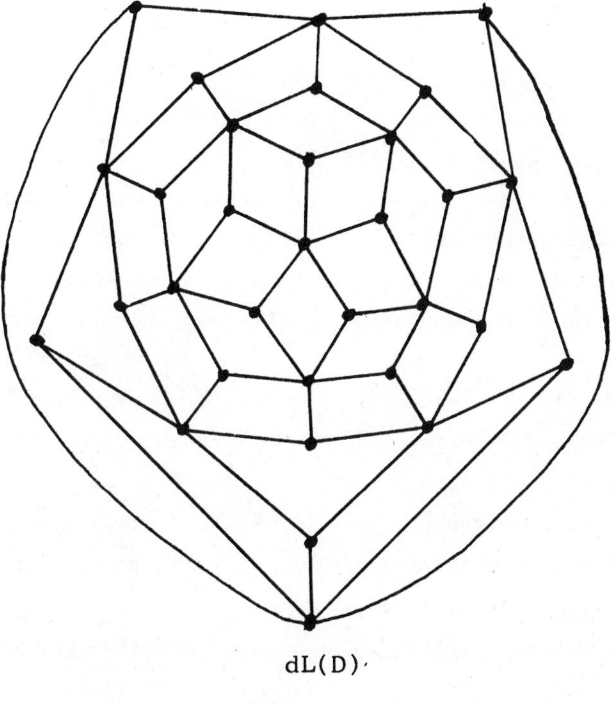

dL(D)

Figure 1.

satisfy Theorem A, but (2,10,15) is representable whereas (3,5,15) is not.

Our principal result is the following.

THEOREM 1. *With two exceptions (x_1,x_2,x_3) is representable if and only if*

a. x_2 is divisible by x_1 or (if x_1 is even) by $x_1/2$, and x_3 is divisible by x_1 or (if x_1 is even) by $x_1/2$, or

b. $(x_1,x_2,x_3) \in \{(6,8,6m), (6,8,8m), (12,20,12m), (12,20,20m), (12,20,30m), (8,12,6m)$ for some positive integer $m\}$.

The two exceptional cases are that (1,1,1) is not representable nor is $(2,x_2,x_3)$ when x_2 and x_3 are odd and x_2 does not divide x_3.

Note that we include the case when x_1 divides x_2 and $x_1/2$ divides x_3, or vice versa. In this paper we will show the sufficiency of the conditions of Theorem 1; in a future paper [5] we will prove the necessity.

Some other extensions of Theorem A are known. For example, in [6] the representable pairs for graphs embedded on the torus are characterized, and more generally for each surface of genus g it is shown that there is a finite set F_g of pairs of numbers such that if (x_1,x_2) is representable by a graph of genus g, then either x_1 or $x_1/2$ divides x_2 or $(x_1,x_2) \in F_g$. The question for edge orbit cardinalities is considered in [3]. The main result there is that there is a simple connected graph with two edge orbits of sizes y_1 and y_2 if and only if $y_1=1$ and y_2 is even, or y_1 and y_2 are not relatively prime. Furthermore, (y_1,y_2) gives the edge orbit sizes of a simple connected planar graph if and only if in addition to the previous conditions either y_1 or y_2 is even or y_1 divides y_2.

In Section 2 we discuss edge-transitivity and show that crucial subgraphs of three-orbit planar graphs are edge-transitive. In Section 3 we demonstrate the sufficiency of Theorem 1 by using edge-transitive graphs to construct the necessary examples. We conclude with information about representable k-tuples.

2. Background

Let G be a graph possibly with multiple edges. Then the edge automorphism group, Aut*(G), is the group of all incidence-preserving bijections between the edges. Every automorphism in Aut(G) induces an edge automorphism in Aut*(G). A graph with one edge orbit under Aut*(G) is called *edge-transitive*.

Before establishing the connection between planar graphs with three vertex orbits and edge-transitive planar graphs, we will mention two useful results.

The first is the graph theoretic backbone of our work, Euler's Formula: $V - E + F = 2$, where V, E and F are respectively the number of vertices, edges and faces of a connected graph embedded in the plane. A corollary of this result is that for simple planar bipartite graphs the average degree, $2E/V$, is bounded by $2E/V \leq 4 - 8/V < 4$. A proof of these facts can be found in [1,8].

The second result, a characterization, can be found in [2;7,p.76].

THEOREM B. *The connected planar edge-transitive graphs are:*

1. *the cycles*
2. *the Platonic graphs: the tetrahedron, cube, octahedron, dodecahedron, and icosahedron*
3. *the line graph of the cube and of the dodecahedron*
4. *the above with each edge replaced by k edges, $k \geqslant 1$, and then subdivided by a single vertex*
5. *the dual of the line graph of the cube and of the dodecahedron*
6. *the complete bipartite graphs $K_{1,n}$ and $K_{2,n}$*
7. *the above with each edge replaced by $k > 1$ edges.*

The following lemma will provide the connection between planar graphs with three vertex orbits and edge-transitive planar graphs.

LEMMA 2. *Let H be a simple planar graph with two vertex orbits O_1 and O_2 such that every edge joins a vertex of O_1 with a vertex of*

O_2. If $|O_1| \neq |O_2|$, then H is edge-transitive.

Proof: Without loss of generality suppose $|O_1| < |O_2|$ and that each vertex of O_i has degree d_i, i=1,2. Since the number of edges in H equals $d_1|O_1| = d_2|O_2|$, $d_2 < d_1$. By Euler's Formula the average degree of H is less than four. Thus $d_2 < 4$.

If d_1 and d_2 are relatively prime, consider O, an edge orbit of H with j_i edges incident with each vertex of O_i, i=1,2. The number of edges in O equals $j_1|O_1| = j_2|O_2|$. Thus $j_1/j_2 = |O_2|/|O_1| = d_1/d_2$. Since $j_i \leq d_i$ and d_1/d_2 is reduced to lowest terms, $j_i=d_i$, i=1,2, and H is edge-transitive.

If d_1 and d_2 are not relatively prime, then because $d_2 < 4$, d_2 divides d_1. If $d_2=1$, then H is $|O_1|$ copies of $K_{1,m}$, $m=|O_2|/|O_1|$. If $d_2=2$, replace every pair of edges incident with a vertex of O_2 by a single edge and obtain a vertex- and edge-transitive planar graph H'. H is a subdivision of H'. We consult Theorem B and search for candidates for H', a vertex- and edge-transitive graph. In all possible cases, when H is formed by subdividing the edges of H', we find H to be edge-transitive. If $d_2=3$, then $d_1=3k$, k>1, and $k|O_1| = |O_2|$. The average degree of H is $(d_1|O_1| + d_2|O_2|)/(|O_1| + |O_2|) = 6k/(1 + k) \geq 4$ for k>1, a contradiction.

Lemma 2 is needed to see that planar graphs with three vertex orbits are composites of edge-transitive graphs. The lemma is also crucial to the proof of the necessity of Theorem 1 [5]; some of the flavor of this connection is shown in the next corollary. Suppose G is a simple connected planar graph with three vertex orbits O_i, i=1,2,3. Note that if one vertex of O_i is adjacent to some vertex of O_j, $i,j \in \{1,2,3\}$, then every vertex of O_i is adjacent to every vertex of O_j and vice-versa. Such a pair of orbits is called <u>adjacent</u>, and when O_i and O_j are adjacent orbits we define G_{ij} to be the subgraph of G with vertex set $O_i \cup O_j$ and with all edges of G that join a vertex of O_i with a vertex of O_j.

COROLLARY 3. Given a simple connected planar graph G with three vertex orbits, suppose O_i and O_j are adjacent orbits of G. If $|O_i| \neq |O_j|$, then G_{ij} is edge-transitive.

Proof: Because every automorphism of G induces an automorphism on G_{ij}, G_{ij} has no more than two vertex orbits. If G_{ij} has one orbit, then it is regular. If d is the degree of a vertex in G_{ij}, the number of edges in G_{ij} is $d|O_i| = d|O_j|$. Thus $|O_i| = |O_j|$, a contradiction. If G_{ij} has two orbits, they are O_i and O_j, and by Lemma 1 G_{ij} is edge-transitive.

3. Examples

We now present examples which demonstrate the sufficiency of the conditions of Theorem 1, calling attention to those graphs which contain interesting edge-transitive subgraphs. We systematically consider all triples (x_1,x_2,x_3) that satisfy Condition a or b of Theorem 1 and in each case produce a planar graph with three orbits of sizes x_1, x_2 and x_3.

Suppose $x_1=1$. If $x_3 \geq 3$ form the edge-transitive graph K_{1,x_2} and join the vertex of degree x_2 to every vertex in a cycle of length x_3. If $x_3=2$, form a triangle and join one vertex to x_2 additional vertices. (See Figure 2.) If $x_1=x_2=x_3=1$, the triple is not representable.

Suppose $x_1=2$. If x_2 or x_3 is even, say x_2, form K_{2,x_3} and join $x_2/2$ additional vertices to each vertex of degree x_3. (See Figure 3a.) If x_2 divides x_3, form K_{2,x_2} and join x_3/x_2 additional vertices to every vertex of degree two. (See Figure 3b.)

Thus we assume $x_1 \geq 3$. Suppose x_1 divides x_2 and x_3. Then (x_1,x_2,x_3) can be realized by one of the following graphs. Form x_1 copies of the edge-transitive graph $K_{1,m}$, $m=x_2/x_1 + x_3/x_1$. If $x_3/x_1 \geq 3$, in each $K_{1,m}$ join x_3/x_1 vertices of degree one in a cycle; if $x_3/x_1=2$, in each $K_{1,m}$ join a pair of vertices of degree one by an edge. To connect the graph join the x_1 vertices of degree m in a cycle. (See Figures 4a and 4b.) If $x_1=x_2=x_3$, form a simple cycle of

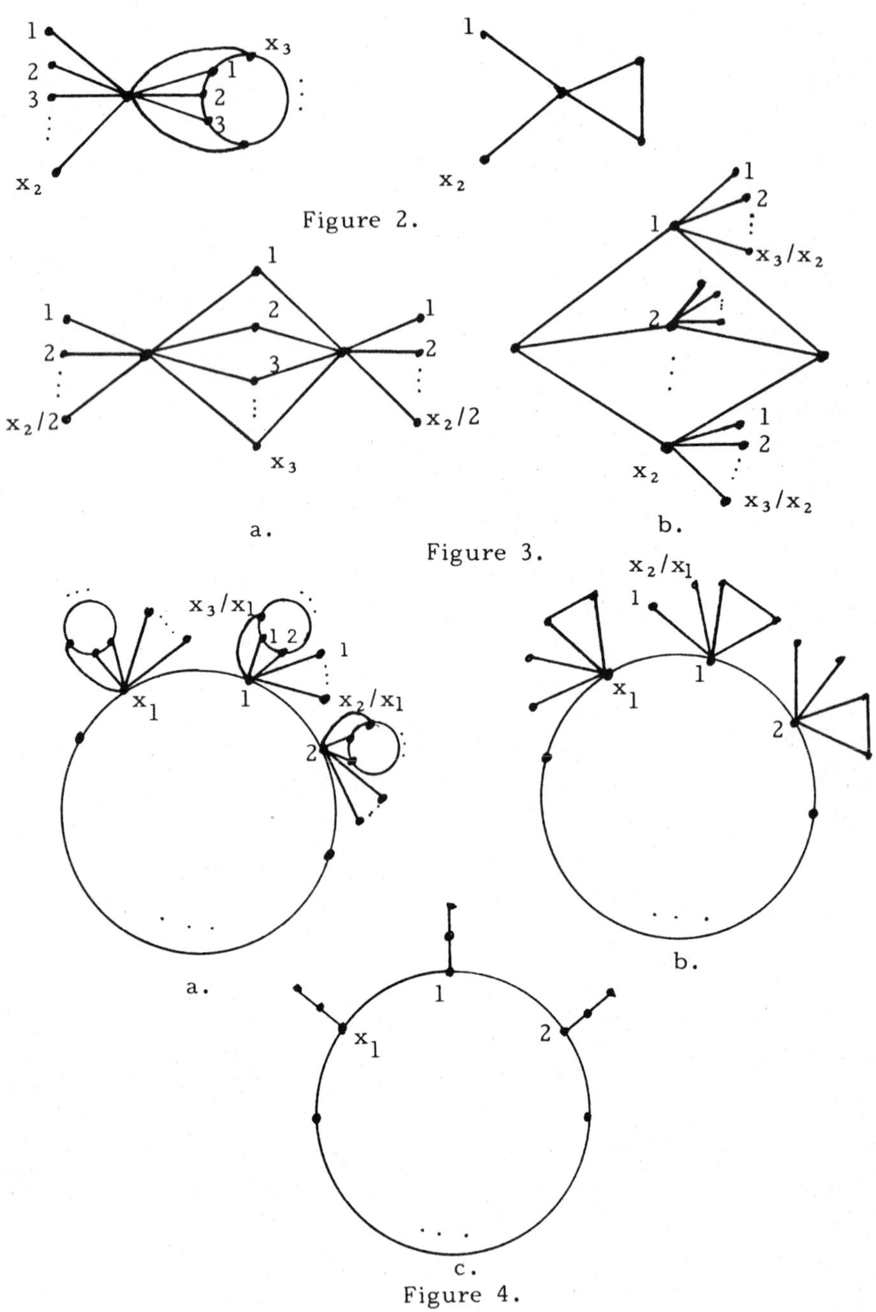

Figure 2.

Figure 3.

Figure 4.

x_1 vertices and to each vertex attach a path of (edge) length two. (See Figure 4c.)

Suppose x_1 divides x_2 and $x_1/2$ divides x_3. Form an x_1 cycle C, replace every other edge by $x_3/(x_1/2) = 2x_3/x_1$ multiple edges, and subdivide each multiple edge by a vertex of degree two. (Call this graph $G(x_1,x_3)$.) Then join x_2/x_1 additional vertices to every vertex of the cycle C. (See Figure 5.) When x_1 divides x_3 and $x_1/2$ divides x_2, reverse the roles of x_2 and x_3.

Suppose $x_1/2$ divides x_2 and x_3. If $x_2 \neq x_3$, begin as in the last paragraph by forming $G(x_1,x_3)$. Then replace every untouched edge of C by $2x_2/x_1$ subdivided multiple edges. (See Figure 6a.) If $x_2=x_3$, form $G(x_1,x_3)$ and join an additional vertex to each vertex of degree two. (See Figure 6b.) Note that all the above graphs are based on the edge-transitive graphs $K_{1,n}$ and $K_{2,n}$.

If $(x_1,x_2,x_3) = (6,8,6m)$ or $(6,8,8m)$ (respectively $(12,20,12m)$ or $(12,20,20m)$), form the dual of the line graph of the cube dL(C) (respectively dodecahedron dL(D)). (See Figure 1.) The graph dL(C) (respectively dL(D)) is edge-transitive and has two vertex orbits, O_1 and O_2, $|O_1|=6$ and $|O_2|=8$ (respectively $|O_1|=12$, $|O_2|=20$). If $x_3=6m$ (resp. 12m), join m additional vertices to each vertex in O_1; if $x_3=8m$ (resp. 20m), join m additional vertices to each vertex in O_2.

If $(x_1,x_2,x_3) = (12,20,30m)$ form dL(D). It is not only edge-transitive but also face-transitive under Aut(G) (i.e. every automorphism maps a closed walk giving the boundary of a face to another face boundary). It has thirty faces, and the boundary of each face contains two vertices from O_1 and two vertices from O_2. Place m vertices in each face and join each to the two vertices from O_1 as in Figure 7a (where the case m=2 is shown).

If $(x_1,x_2,x_3) = (8,12,6m)$, subdivide each edge of the cube. Like dL(D) this graph is edge- and face-transitive. Place m vertices in each of its six faces and join them to the orbit of size twelve

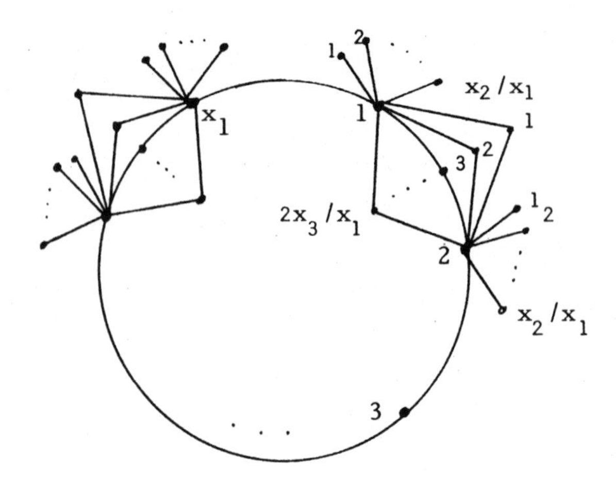

Figure 5.

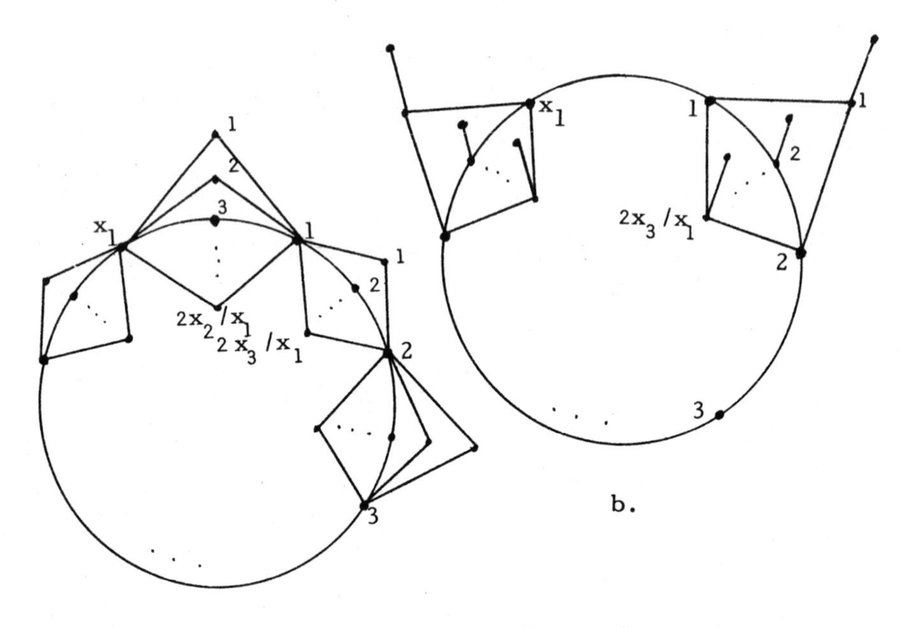

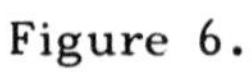
Figure 6.

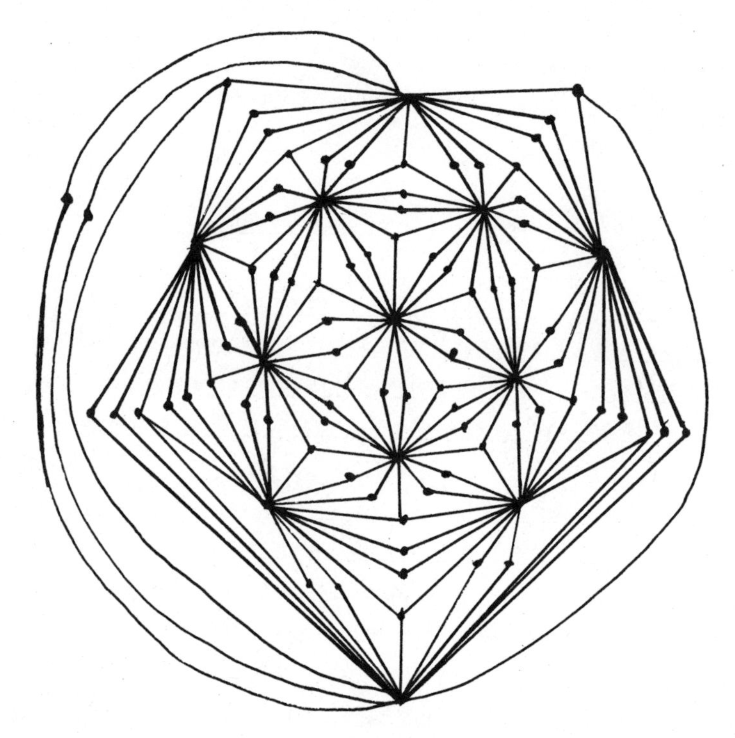

Figure 7a.

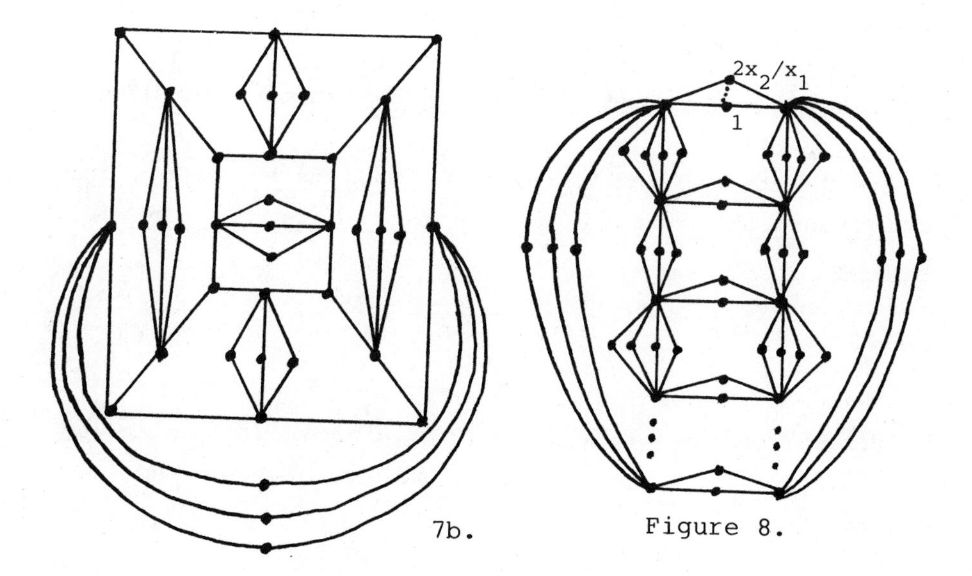

7b.

Figure 8.

as in Figure 7b (with m=3).

These constructions generalize to show that the following k-tuples are representable.

PROPOSITION 4.

1. *Every k-tuple* $(1,x_2,x_3,\ldots,x_k)$ *is representable except for the k-tuples of all ones with* $2\leqslant k\leqslant 5$.

2. *A k-tuple* $(2, x_2,x_3,\ldots,x_k)$ *is representable if the odd entries* $\{x_{i1},x_{i2},\ldots,x_{ij}\}$ *are each divisible by* x_{i1}.

3. *A k-tuple* $(x_1,x_2,\ldots,x_k)$ *with* $x_1\geqslant 3$ *is representable if all but at most two entries are divisible by* x_1 *and these two are divisible by* $x_1/2$.

COROLLARY 5. *A k-tuple of the form* $(1, x_2,\ldots,x_k)$ *is representable if and only if it is not a k-tuple of all ones with* $2\leqslant k\leqslant 5$.

For interest we present a family of graphs with four orbits with numerical patterns not observed in graphs with three or fewer orbits.

PROPOSITION 6. (x_1,x_2,x_3,x_4) *is representable if* x_1 *is divisible by 4 and for each* $i>1$, x_i *is divisible by* $x_1/2$.

Proof: See Figure 8 for $x_1\geq 8$ and $x_3\neq x_4$. In the other cases a modification of this graph or of that in Figure 6 will work.

We conclude with a conjecture.

CONJECTURE *If G is a connected planar graph with orbit partition* $(x_1,x_2,\ldots,x_k)$ *then either*

a. *G contains a subdivision of the cube, the icosahedron, the dodecahedron, or the dual of the line graph of the cube or of the dodecahedron, or*

b. *for each* $i>1$, x_i *is divisible by* x_1 *or (if* x_1 *is even) by* $x_1/2$.

We would like to thank M.O.Albertson, K.L.Collins and E. Gethner for their help with this work and to acknowledge gratefully the support of the National Science Foundation in the Undergraduate Research Participation Program, grant #SPI 79 26984.

REFERENCES

1. M.Behzad, G.Chartrand and L.Lesniak-Foster, Graphs and Digraphs, Prindle, Weber and Schmidt, Boston, 1979.

2. H.Fleischner and W. Imrich, Transitive planar graphs, Math. Slovaca, 29(1979), 97-105.

3. E.Gethner and J.P.Hutchinson, Connected graphs with complementary edge-orbits, Ars Combinatoria, 12(1981), 135-146.

4. F.Harary and G.McNulty, The orbital partitions of graphs, J. Combin. Inform. System Sci., 5(1980), 131-133.

5. J.P.Hutchinson and L.B.Krompart, Partitions that arise from connected planar graphs with three orbits, to appear.

6. J.P.Hutchinson and G.McNulty, Connected graphs of genus g with complementary orbits, Discrete Math., 45(1983), 255-275.

7. L.Lovasz, Combinatorial Problems and Exercises, North-Holland, Amsterdam, 1979.

8. A.T.White, Graphs, Groups and Surfaces, North-Holland, Amsterdam, 1973.

RELATING METRICS, LINES AND VARIABLES DEFINED

ON GRAPHS TO PROBLEMS IN MEDICINAL CHEMISTRY

Mark Johnson

The Upjohn Company

ABSTRACT

A structure-activity variable maps each chemical structure to a real number representing the activity of the associated compound in some biological test system. Medicinal chemistry involves the study of structure-activity variables. Molecular structures are often represented as connection tables when this structure-activity data is stored in the computer. Since a graph is implicit in each connectivity table, graph theory should find many applications in medicinal chemistry.

Metric notions implicit in the study of structure-activity variables are used to motivate the need for a metric defined on graphs. A metric which Bogart (4) defined on the space of asymmetric relations is then extended to the space of graphs.

Notions of neighborhoods, lines, and graph variables mapping graphs to the real line are defined. Each notion is related to problems arising in structure-activity studies. Related graph-theoretic problems are noted as they arise.

1. Introduction

Applications of graph theory to medicinal chemistry are quite diverse. Graph and subgraph recognition is implicit in computerized substructure searching and chemical fragment generation. Various indices defined on graphs are being proposed as predictors of chemical properties (6,7). Notions of distance are used in compound searching and structure-activity studies (2,3,8).

The distance notions in (2) and (3) do not constitute metrics on the space of graphs. This study introduces a metric on the space of graphs. With a metric, one can define an analog of a line in the space of graphs and an interesting family of variables that map graphs to the real line. These notions will be rigorously defined and accompanied by applications to medicinal chemistry and by problems in graph theory arising from these applications.

2. Drug-receptor interactions in medicinal chemistry.

Before and after working its biological effect, a drug or compound usually undergoes a diverse and complex series of interactions, many of which are often poorly understood. It is widely held that the effect itself is elicited by the drug interacting with a macromolecule in a steric specific manner often likened to the fit of a key (drug) in a lock (receptor site of a macromolecule).

Figure 1 illustrates the idea; it and its description is largely taken from the overview (5) of receptor site theory. In Figure 1 we see acetylcholine binding to the enzyme acetylcholinesterase by means of an electrostatic bond between the + and - charges, by means of hydrophobic bonds involving the two CH_3 groups attached to the nitrogen (N), the $COCH_3$ group, and the CH_2CH_2 group, and by means of a hydrogen bond between the hydrogen on the drug and the oxygen on the enzyme.

Obviously, the receptor site (the portion of the macromolecule that interacts with the drug) and the drug must be similar to each other in a complementary way. The pockets of one must correspond to protuberances of the other. By the same token, drugs which are similar to one another (have related groups in corresponding positions) are often expected to elicit similar activities.

This brief description grossly oversimplifies drug-receptor interactions. Some compounds bind to the receptor site, but do not elicit the desired activity. Sometimes slight modifications of a drug remarkably alter its activity. A compound may perfectly fit the receptor site, but may never reach it because of intervening interactions. A metabolite, rather than the compound itself, may actually be interacting at the receptor site.

Oversimplified as it is, this minimal understanding underlies some important questions that arise in medicinal chemisry. For example, when a compound with an interesting activity is found, one often wishes to know what compounds in a set of easily accessible compounds are similar to that active compound. Or given two active, but structurally diverse compounds, one wishes to know what they have in common. Or given activities on a number of related compounds, one wishes to develop structure-activity relationships that predict the activities of such compounds. These prolems will, subsequently, be stated mathematically, but first a metric on the space of graphs must be defined.

Figure 1. A drug-receptor site interaction showing the complimentarity in the pockets and protuberances of the two molecules.

ACETYLCHOLINE

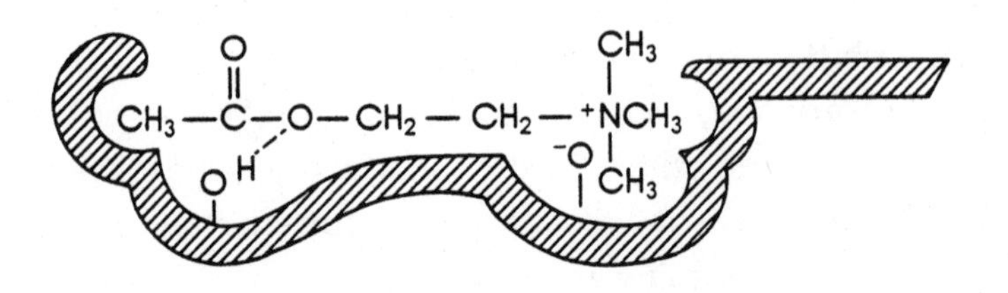

ACETYLCHOLINESTERASE

Figure 2. Three segments with common end points. The + and - signs indicate edge additions and deletions.

A line segment containing a maximal intersection.

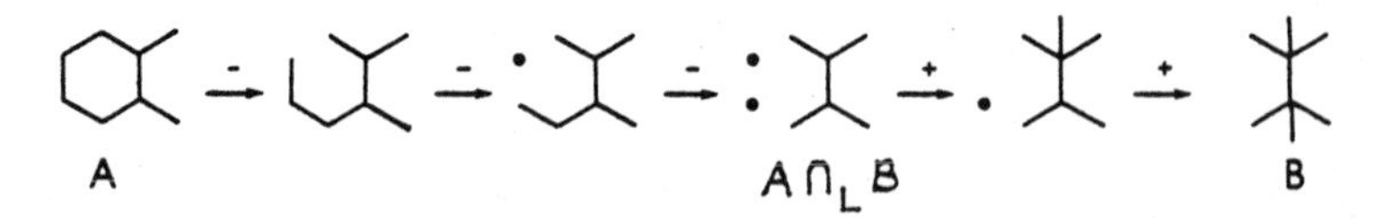

A line segment of connected graphs containing a minimal union.

$A \cup_L B$

A line segment of connected graphs containing no minimal unions or maximal intersections.

3. A metric on the space of graphs.

The metric developed here seems to be a simple extension of the metric Bogart (4) developed for the space of asymmetric relations defined on a fixed set of points.

Let $A = (V,E)$ be any graph. Call $|V| + |E|$ the **cardinality** of A and denote it by $|A|$.

Let $g:V \rightarrow N$ denote any injective function from V to the natural numbers. We will call g a layout. Define $g[V] = \{g(v)|v \varepsilon V\}$ and $g[E] = \{g(v)g(v')|vv' \varepsilon E\}$. Let g[A] denote the graph (g[V], g[E]). Call g[A] the **image graph** of A with respect to g. If $B = (V',E')$ is a subgraph of A, then g[B] will denote the subgraph (g[V'],g[E']) of g[A].

Let $B = (V',E')$ be another graph with image graph h[B] with respect to layout h. Let L = (g,h) and define $A \cup_L B$ to be the graph with vertex set $g[V] \cup h[V']$ and edge set $g[E] \cup h[E']$. Similarly, define $A \cap_L B$ by replacing $\cup$ by $\cap$ in each case. Call L a **bilayout** and call $A \cup_L B$ and $A \cap_L B$ the **union** and **intersection** of A and B **with respect to** L, respectively.

Define d(A,B) by

$$d(A,B) = \min \{|A \cup_L B| - |A \cap_L B|\} \quad (1)$$

where the minimum is taken over all h. By a straightforward proof, it can be shown that equation 1 is invariant to the choice of g.

If a bilayout, L = (g,h), is a solution of 1, we will call it a **minimal bilayout,** and we will refer to $A \cup_L B$ and $A \cap_L B$ as a **minimal union** and a **maximal intersection** of A and B, respectively. If L = (g,h) is a minimal bilayout of A and B, then it can be shown that

$$\begin{aligned} d(A,B) &= |A \cup_L B| - |A \cap_L B| \\ &= 2|A \cup_L B| - |A| - |B| \\ &= |A| + |B| - 2|A \cap_L B| \end{aligned}$$

By considering the following graphs

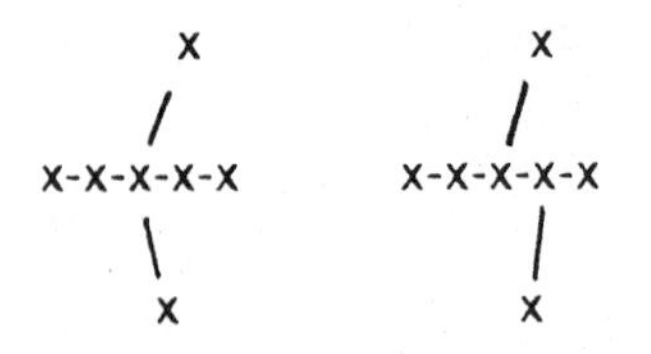

one can readily show that there exist minimal bilayouts $L=(g,h)$ and $K=(g',h')$ for which $A\cap_L B \neq A\cap_K B$ and for which $A\cap_L B \simeq A\cap_K B$ and $A\cup_L B \neq A\cup_K B$. This nonuniqueness should be borne in mind when Figure 2 is discussed.

The following proposition will be used in proving that the function defined by equation 1 is a distance function.

Proposition 1: Let A, B, and C be graphs. If B is not isomorphic to a subgraph of A, then there exists a proper subgraph G of B such that

$$d(A,G) + d(G,C) \leq d(A,B) + d(B,C).$$

Proof: Assume B is not isomorphic to a subgraph of A. Let $L=(g,h)$ be any minimal bilayout and denote $A\cap_L B$ by S. It can be shown that h[B] has either a vertex or edge not possessed by S.

Assume there exists a vertex h(v) of h[B] that is not a vertex of S. Then, $S = S\text{-}h(v)$. It follows that

$$\begin{aligned} d(A,B\text{-}v) &\leq |g[A]| + |h[B\text{-}v]| - 2\,|S\text{-}h(v)| \\ &= |g[A]| + |h[B]| - 1 - n(v) - 2|S| \\ &= d(A,B) - 1 - n(v) \end{aligned}$$

where n(v) is the number of edges in B adjacent to v.

To obtain a similar inequality for d(B-v,C), note that equation 1 is invariant to the choice of g. Thus, there exists a layout f such that $K = (h,f)$ is a minimal bilayout of B and C. Let $T = B \cap_K C$.

If h(v) is a vertex of T, then $|T| = |T-h(v)| + (1 + m(v))$ where m(v) is the number of edges in T adjacent to h(v). It follows that

$$d(B-v,C) \leq |B-v| + |C| - 2|T-h(v)|$$

$$= |B| - 1 - n(v) + |C| - 2|T| + 2(1 + m(v))$$

$$\leq d(B,C) + 1 + n(v)$$

since $m(v) \leq n(v)$. Thus,

$$d(A,B-v) + d(B-v,C) \leq d(A,B) + d(B,C).$$

Now if h(v) is not a vertex of T, then $T-h(v) = T$. It follows that

$$d(B-v,C) \leq |B-v| + |C| - 2|T-h(v)|$$

$$= |B| - 1 - n(v) + |C| - 2|T|.$$

$$= d(B,C) - 1 - n(v).$$

In this case,

$$d(A,B-v) + d(B-v,C) \leq d(A,B) + d(B,C) - 2(1 + n(v)).$$

In either case, the desired inequality holds.

Assume, secondly, that every vertex of h[B] is also a vertex of S. Then there must be an edge $e = vv'$ such that h(v)h(v') is not an edge of S. The above argument can be repeated to show that

$$d(A,B-e) + d(B-e,C) \leq d(A,B) + d(B,C) \blacksquare$$

Corollary 1: Let A, B and C be any graphs, then

$$d(A,C) \leq d(A,B) + d(B,C).$$

Proof: Case 1, assume B is not a subgraph of both A and C. By Proposition 1 and the symmetry of d(A,B) in (1), there exists proper subgraph B_1 of B such that

$$d(A,B_1) + d(B_1,C) \leq d(A,B) + d(B,C).$$

Similarly, if B_1 is not a subgraph of both A and C, there exists a proper subgraph B_2 of B_1 such that

$$d(A,B_2) + d(B_2,C) \leq d(A,B_1) + d(B_1,C)$$

$$\leq d(A,B) + d(B,C).$$

Thus, we can define a sequence $B_0,\ldots,B_n$ of graphs such that (1) $B_0 = B$, (2) B_i is a proper subgraph of B_{i-1} but B_i is not a subgraph of both A and C for $i = 1,\ldots,n-1$, and (3) B_n is a proper subgraph of B_{n-1} and is also a subgraph of both A and C. Moreover,

$$d(A,B_n) + d(B_n,C) \leq d(A,B) + d(B,C).$$

Consequently, we need only establish the triangle inequality for case 2.

Case 2, assume B is a subgraph of both A and C. Clearly, $A \cap_L C \simeq B$ for any minimal bilayout L = (g,h). It follows that

$$d(A,C) = |A| + |C| - 2|B|$$

$$= d(A,B) + d(B,C) \blacksquare$$

Let [A] denote the set of graphs which are isomorphic to A. Let G denote the space of such isomorphism classes. Define the function $d': G \times G \rightarrow R$ by d' ([A], [B]) = d (A,B). It is easy to show that d' is well-defined and satisfies the properties of a metric.

The first question in section A now translates into finding those graphs, B, for which $d(A,B) \langle \varepsilon$ where ε is some positive integer. But other problems also arise. Call X an δ-spanning set of a set Y of graphs if for every graph $G \varepsilon Y$, there exists a graph $G' \varepsilon X$ such that $d(G,G') < \delta$ Translating this to medicinal chemistry, if the compounds within an ε distance of an active compound are of interest and if it is not economical to test all of them, then a minimal δ-spanning set of compounds would be of interest. Obviously, implementing solutions to these problems will require algorithms for determining minimal layouts and minimal spanning sets.

Topologically, this metric space may not be interesting since the metric induces the discrete topology. However, statistical analyses of data involving the activities of compounds require that the problem be isometrically embedded in Euclidean n-space. The problem arises whether or not this metric space can be isometrically embedded in Euclidean n-space for any n.

4. Lines.

Call three graphs, A_i, i = 1,2,3, **collinear** if

$$d(A_{p1},A_{p3}) = d(A_{p1},A_{p2}) + d(A_{p2},A_{p3})$$

for some permutation p_1, p_2 and p_3 of the set {1,2,3}. In this case, A_{p2} will be said to be **between** A_{p1} and A_{p3}. A **line**, L, will be defined to be a maximal set of graphs for which any three graphs in L are collinear. A **segment** with **endpoints** A and B is any maximal set of graphs which is a subset of a line and such that any graph in the segment lies between A and B.

Figure 2 suggests some of the variety of the segments having identical endpoints. These segments were generated by adding to g[A] edges in $A \cup_L B$ that were not in g[A] and deleting from g[A] edges in g[A] that were not in $A \cap_L B$. General propositions that the sequence of graphs so generated always lie on a segment connecting A and B have not yet been developed.

Segments provide the chemist with a means of separating a set of compounds into those compounds relevant to an issue from those which are not. Segments also provide an ordering and metric on the selected compounds.

Figure 3 illustrates two consequences of these two tools. The data for this figure come from Acton and Stone (1). The compounds were selected from a set of 27 related compounds by choosing those compounds that lie on a line containing the simplest compound and the compound with the greatest proportion of sweetness in its taste.

The first graph illustrates how segments can be used to assess the statistical variation in a set of data and to graphically depict how progressive structural changes influence the activity of the associated compounds. The second graph illustrates how the ordering implicit in a segment suggests to the chemist which of these progresive changes are determining the activity by means of a particular physical-chemical property and which are not. (To appreciate the latter point, remove the lines and the chemical structures that suggest the linear ordering and note how the relevance of the physical-chemical property disappears.)

These figures suggest that techniques for determining the graphs lying between two graphs and the line segments connecting two graphs should prove useful. "Parallel" and "orthogonal" line segments should also prove important to data analysis and design in this area. These ideas suggest the possibility of a "geometry" of graphs.

5. Graph variables

Figure 3 introduced an empirically defined variable, proportion sweetness, defined on the space of compounds. Mathematical methods of defining variables on molecular structures are being defined by theoretical chemists based upon mathematical models of the interaction between a drug and a macromolecule. In both cases, concepts and constraints outside the scope of graph theory are involved. This section

Figure 3. Linear orderings imposed on plots of proportion sweetness of some aldoximes versus two predictive variables.

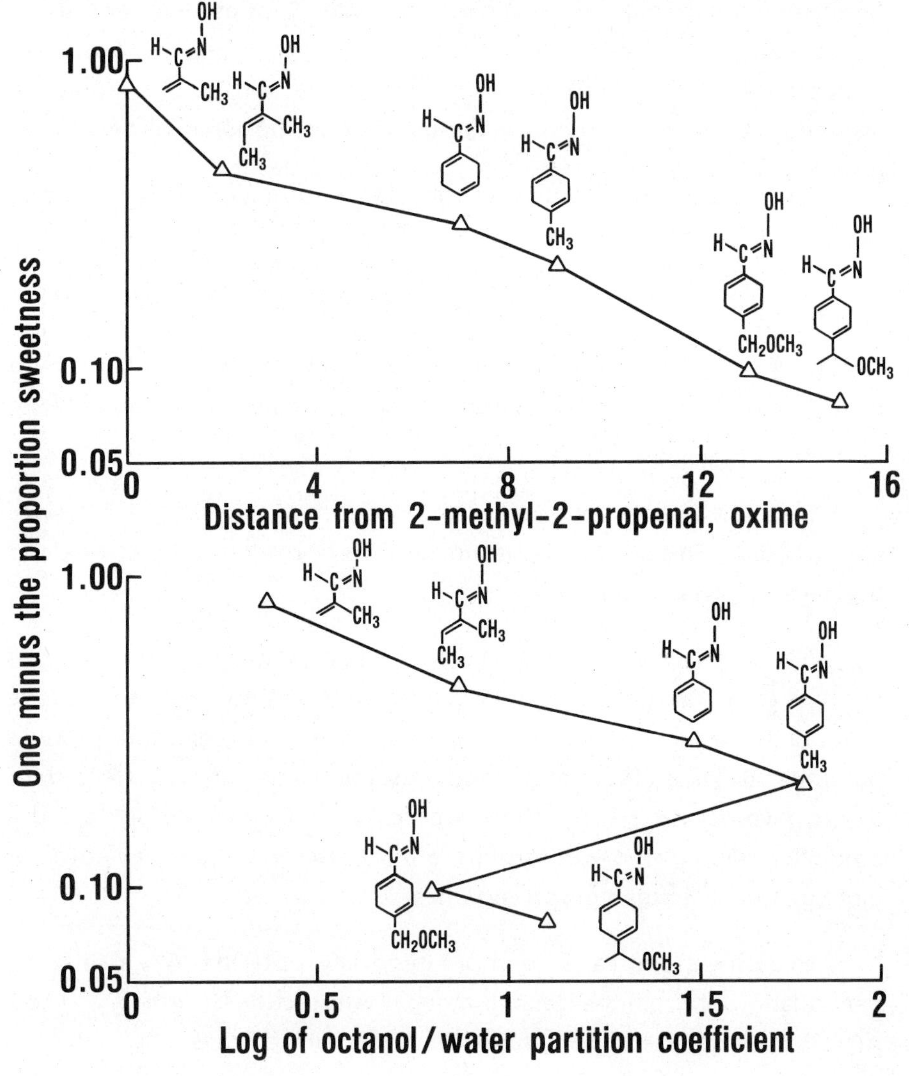

introduces the concept of a variable defined on graphs that lies within the scope of graph theory.

Recall how in Figure 1 the active compound resembled the receptor site in a complementary way. This implies the most active compound is, in a sense, a negative snapshot of the receptor site. Suppose we denote this active compound by ξ. We might assign activities to the other compounds such that the closer the compound is to ξ, the higher its assigned activity. We begin by making this simple notion explicit.

Let ξ be any fixed graph and consider the function $V: G \rightarrow R$ defined by

$$V([A];[\xi]) = |A \cap_L Y|^2/(|A| * |\xi|) \qquad (2)$$

where $L = (g,h)$ is any minimal bilayout. We shall write $V(A;\xi)$ for $V([A];[\xi])$, and simply $V(A)$ if ξ is understood. The function V is an instance of what we shall call a graph variable.

Equation 2 can be viewed as the product of two ratios, $|A \cap_L \xi|/|A|$ and $|A \cap_L \xi|/|\xi|$. This product is maximized when both ratios equal one and the maximum is achieved if $A = \xi$.

Equation 2 was used as a predictive variable for the taste potencies of the sweet compounds in Action and Stone. With the most active compound used as a substitute for ξ, a correlation coefficient of 0.40 was obtained. Based on these predictions, another compund was selected for ξ that gave rise to a correlation coefficient of 0.60.Obviously, a more predictive relationship would require either better "estimates" of ξ, or more generally defined graph variables.

To define graph variables more generally, note that $A \cap_L Y$ selects a particular graph from the class of minimal intersections of A and ξ. Thus, $A \cap_L Y$ can be replaced by any function $\ominus$ mapping G^2 to G.

A reasonable generalization of 2 now follows readily. Let $\xi = (\xi_1,\ldots,\xi_n)$ be any n-dimensional vestor of graphs. Let $\theta = (\theta_1\ldots,\theta_m)$ be any m-dimensional vector of functions mapping G^{n+1} to G. Let $\lambda = (\lambda_1,\ldots,\lambda_{k,})$ be any k-dimensional vector of parameters. A **graph variable** will be any variable $V: G \rightarrow R$ of the form

$$V(A;\xi,\lambda) = f(|\theta_1(A,\xi)|,\ldots,|\theta_m(A,\xi)|;\lambda) \tag{3}$$

where $f(\cdot,\lambda)$ is any function mapping Euclidean m-space to the real line.

Equation 3 presents the data analyst with a broad range of descriptive models of structure-activity variables which associate real numbers with underlying structure. In the structure-activity variable of Figure 2, the real number was proportion sweetness and the underlying structure was the drug-receptor complex. In the model defined by equation 2, the real number was a predictor of taste potency, and the underlying structure was a pair of graphs.

In a graph variable such as the one in equation 2, the medicinal chemist needs procedures for finding the optimal structure ξ, i.e. the structure of the receptor site. This purpose of finding the structure of the receptor site is a special case of determining structures from the values of graph variables defined on those structures. The first chemists must have faced problems of this type. Their success in determining chemical structures from measurements suggests that the structures of graphs are knowable from graph variables. Since graph variables are defined in terms of the cardinalities of particular graphs and since the cardinality of a graph is simply its distance from the null graph, it is natural to ask if this metric space determines the structure of every graph.

ACKNOWLEDGEMENTS

I would like to express my appreciation to John Schultz for introducing me to the problem of structure-activity studies to the referee for many helpful comments. This work was funded by The Upjohn Company.

REFERENCES

1 E. M. Acton and H. Stone, Potential new artificial sweetner from study of structure-taste relationships, **Science**, 193 (1976), 584-586.

2 G. W. Adamson and J. A. Bush, A comparison of the performance of some similarity and dissimilarity measures in the automatic classification of chemical structures, **J. Chem. Inf. Comput. Sci.**,15 (1975), 55-58.

3. A. T. Balaban, A. Chiriac, A. Motoc and Z. Simon, **Steric fit in quantitative structure-activity relations**, Springer-Verlag, Berlin, 1980.

4. K. P. Bogart, Preference structures II: Distances between asymmetric relations, SIAM J. Appl. Math. 29 (1975) 254-262.

5. J. E. Gearien, Receptor site theory, in **Principles of Medicinal Chemistry** , ed by W. E. Foye, Lea & Febiger, Philadelphia, 1981, 129-139.

6. L. B. Kier and L. H. Hall, The nature of structure-activity relationships and the relation to molecular connectivity, **Eur. J. Med. Chem.**, 12 (1977), 307-312.

7. M. Randic, On characterization of molecular branching, **J.Amer. Chem. Soc.** ,97 (1975), 6609-6615.

8. A. J. Stuper, W. E. Brugger and P. C. Jurs, **Computer assisted studies of chemical structure and biological function** , John Wiley & Sons, New York, 1979.

[a,b]-FACTORIZATIONS OF NEARLY BIPARTITE GRAPHS

Mikio KANO
Akashi Technological College

ABSTRACT

We prove that a graph G with the property that any two odd cycles have a vertex in common is [a,b]-factorable if and only if G is an [an,bn]-graph for some positive integer n.

1. Introduction

We deal with finite graphs which may have multiple edges but have no loops. All notation and definitions not given here can be found in [3].

Let G be a graph with vertex set $V(G)$ and edge set $E(G)$, and let H be a subgraph of G. For a vertex x of H, we denote by $deg_H(x)$ the degree of x in H. Let a and b be integers such that $0 \le a \le b$. Then a graph G is called an [a,b]-*graph* if $a \le deg_G(x) \le b$ for all $x \in V(G)$, and an [a,b]-*factor* of a graph is its spanning subgraph F such that $a \le deg_F(y) \le b$ for all $y \in V(F)$. If all the edges of a graph G can be decomposed into [a,b]-factors $F_1, \ldots, F_n$ of G, then the union $F_1 \cup \ldots \cup F_n$ is called an [a,b]-*factorization* of G, and G is said to be [a,b]-*factorable*.

We first present some known results on [a,b]-factorizations of graphs. D. de Werra [9] proved that a bipartite graph G is [a,b]-factorable if and only if G is an [an,bn]-graph for some $n>0$. It is shown in [6] that a graph G is [2a,2b]-factorable if and only if G is a [2an,2bn]-graph for some $n>0$, which is an extension of Petersen's theorem ([8],[3,p.166]). Some sufficient conditions for a graph to be [a,b]-factorable can be found in [1], [4] and [6]. A list of results on factorizations of graphs and related topics can be found in [2].

The purpose of this paper is to prove the following theorem, which is an extension of the result mentioned above due to D. de Werra.

Theorem Let a and b be integers such that $0\le a\le b$, and let G be a graph with the property that any two odd cycles have at least one vertex in common. Then G is [a,b]-factorable if and only if G is an [an,bn]-graph for some positive integer n.

2. Proof of Theorem

We begin with some definitions. Let G be a graph. For a subset S of V(G), we write G-S for the graph obtained from G by deleting the vertices in S together with their incident edges. Similarly, for a subset X of E(G), let G-X denote the graph obtained from G by deleting the edges in X. If S and T be disjoint subsets of V(G), then we denote by $e_G(S,T)$ the number of edges of G joining S and T. Let g and f be integer-valued functions defined on V(G). Then a (g,f)-*factor* of G is a spanning subgraph F of G such that $g(x)\le \deg_F(x)\le f(x)$ for all $x\in V(G)$. It is trivial that if G has a (g,f)-factor, then $g(x)\le f(x)$ for every vertex x of G. A graph G is said to possess the *odd cycle property* if any two odd cycles either have at least one vertex in common, or are joined by an edge.

Lemma 1. (Folkman and Fulkerson [5]) Let G be a graph with the odd cycle property, and g and f be integer-valued functions defined on $V(G)$ such that $g(x) \le f(x)$ for all $x \in V(G)$. Then G has a (g,f)-factor if and only if

$$\delta(S,T) = \sum_{t \in T} \{\deg_G(t) - g(t)\} + \sum_{s \in S} f(s) - e_G(S,T) \ge 0$$

for all disjoint subsets S and T of $V(G)$.

Lemma 2. Let θ be a real number such that $0 \le \theta \le 1$. Let G be a graph with the odd cycle property, and g and f be integer-valued functions defined on $V(G)$. If

$$g(x) \le \theta \cdot \deg_G(x) \le f(x) \quad \text{for all } x \in V(G),$$

then G has a (g,f)-factor.

Proof By Lemma 1, it is sufficient to show that g and f in Lemma 2 satisfy the condition in Lemma 1. Let S, $T \subset V(G)$ such that $S \cap T = \phi$. Then we have

$$\delta(S,T) \ge \sum_{t \in T} (1-\theta)\deg_G(t) + \sum_{s \in S} \theta \deg_G(s) - e_G(S,T)$$
$$\ge (1-\theta)e_G(T,S) + \theta e_G(S,T) - e_G(S,T) = 0.$$

Hence the condition holds (see [8]).

Proof of Theorem Let G denote a graph with the property that any two odd cycles have a vertex in common. It is obvious that if G is $[a,b]$-factorable, then G is an $[an,bn]$-graph for some integer $n>0$. We now prove by induction on n that every $[an,bn]$-graph G can be decomposed into n $[a,b]$-factors. Set $\theta=1/n$, and define two functions g and f on $V(G)$ as follows:

$g(x)=f(x)=a$ if $\deg_G(x)=an$,
$g(x) \le \theta \deg_G(x) \le f(x)$ and $f(x)-g(x)=1$ if $an<\deg_G(x)<bn$; and
$g(x)=f(x)=b$ if $\deg_G(x)=bn$.

Then g, f and θ satisfy the conditions in Lemma 2. Hence G has a (g,f)-factor F. It is immediate that if $an<\deg_G(x)<bn$, then $a(n-1)<\deg_{G-E(F)}(x)<b(n-1)$. Thus $G-E(F)$ is an $[a(n-1), b(n-1)]$-graph, and we conclude by the induction hypothesis that

G can be decomposed into n [a,b]-factors.

Note that a subgraph of a graph with the odd cycle property does not always possess the odd cycle property, and that there exists an [an,bn]-graph which is not [a,b]-factorable and has the odd cycle property for every a and b if at least one of a and b is odd.

REFERENCES

1. J. Akiyama and M. Kano, Almost regular factorization of graphs, *J. of Graph Theory* (to appear).
2. J. Akiyama and M. Kano, Factors and factorizations of graphs —— A survey, *J. of Graph Theory* (to appear).
3. M. Behzad, G. Chartrand and L. Lesniak-Foster, *Graphs & Digraphs*, Prindle, Weber and Schmidt, Boston 1979.
4. H. Era, Semiregular factorizations of regular graphs, *Graphs and Applications* (Proceeding of the first Colorad graph theory symposium) Wiely, 1984.
5. J. Folkman and D.R. Fulkerson, Flows in infinite graphs, *J. of Combinatorial Theory*, 8 (1970), 30-44.
6. M. Kano, [a,b]-factorization of a graph, *J. of Graph Theory* (to appear).
7. M. Kano and A. Saito, [a,b]-factors of graphs, *Discrete Math.*, 47 (1983), 113-116.
8. J. Petersen, Die Theorie der Regularen Graph, *Acta Math.*, 15 (1891), 193-220.
9. D. de Werra, Multigraphs with quasiweak odd cycles, *J. of Combinatorial Theory (B)*, 23 (1977), 75-82.

THE COMPLEXITY OF PEBBLING FOR TWO CLASSES OF GRAPHS

Maria M. Klawe

IBM San Jose Research Laboratory

ABSTRACT

It is known that the problem of determining the number of black pebbles needed to pebble an acyclic directed graph is PSPACE complete, though there is a straightforward polynomial time algorithm for trees. In this paper, we consider the complexity of determining the black pebble number for the class of spreading graphs, and a subclass, nice graphs. In particular we show that the polynomial algorithm for trees can be extended to nice graphs, and that for spreading graphs the problem is in NP.

1. INTRODUCTION

Let G be an acyclic directed graph. A vertex y is said to be an immediate predecessor of a vertex x if there is an edge from y to x. The black pebble game is played on G according to the following set of rules.

(1) A black pebble may be removed at any time.

(2) A black pebble may be placed on a vertex if all of its immediate predecessors have pebbles.

(3) If all the immediate predecessors of a vertex have pebbles, a black pebble may be slid from one of the immediate predecessors onto the vertex.

A configuration is a subset B of vertices of G, representing the subset of vertices of G which hold black pebbles. A black pebble strategy is a finite sequence of configurations $B_0,...,B_n$ on G such that $B_0 = \phi$, and such that for each $i < n$ the configuration B_{i+1} is obtained from applying one of the rules to B_i. The space requirement of a strategy is defined as $\max_{0 \le i \le n} |B_i|$, or in other words, the maximum number of pebbles in any configuration in its sequence. The black pebble number of a vertex is the minimum of the space requirements of strategies which end with the configuration having a single black pebble on that vertex. We will denote the black pebble number of a vertex x by b(x). Similarly the black pebble number of a configuration, B, is the minimum of the space requirements of strategies which end with B. Slightly abusing notation for the sake of simplicity, we also denote the black pebble number of B by b(B). A vertex in G is called an input [output] if it has no incoming [outgoing] edges. It follows immediately from the definitions that if x is an input then $b(x) = 1$.

The black pebble game models register allocation during a deterministic evaluation of a straight line program. More precisely, the inputs to the straight line program and the computations performed in the straight line program are represented by vertices in the acyclic directed graph, and there is an edge from y to x if the input represented by y or the result of the computation represented by y is used in the computation represented by x. Placing a pebble on a vertex which represents a computation models storing the result of that computation in a register. Thus the black pebble number of the configuration of output vertices represents the minimum number of registers needed to evaluate the straightline program deterministically. The black pebble game and its variants have many applications in computer science including comparisons of programming languages, code generation and optimization for compilers, time-space tradeoffs for a wide variety of problems, and as a simulation tool in relating complexity classes. A detailed survey of of pebbling results is given in Pippenger [8].

Gilbert, Lengauer and Tarjan[3] proved that the problem of determining whether $b(x) \leq k$ is PSPACE complete for x a vertex of an arbitrary acyclic directed graph. In this paper we consider the restriction of this problem to two special classes of graphs.

If a graph has the property that the black pebble number of each vertex is a polynomial time computable function of the black pebble numbers of its immediate predecessors, then since all inputs have black pebble number 1, the obvious recursive algorithm computes all the black pebble numbers in polynomial time. We now explore a possible candidate for such a polynomial time computable function.

For any set B let B[j] be the vertices in B with black pebble number at least j. For each vertex set B we define its measure M(B) as $\max\{j - 1 + |B[j]| : B[j] \neq \phi\}$. It is not hard to see that $b(B) \leq M(B)$ since M(B) is the space requirement of the following strategy which achieves B. First use an optimal black pebble strategy to place a black pebble on the vertex in B with the highest black pebble number. Leaving that pebble on its vertex, now use an optimal black pebble strategy to black pebble the vertex in B with the second highest black pebble number. Continue in this way to black pebble the vertices in B in decreasing order of black pebble number until all have black pebbles. We will refer to this as the independent strategy for pebbling B. It is easy to see that the independent strategy is not always optimal since it may be more efficient to pebble some of the vertices in B jointly rather than independently. Let P(x) denote the set of immediate predecessors of x. From the rules of the black pebble game it is obvious that b(x) = b(P(x)) for every x, and hence we always have $b(x) \leq M(P(x))$. However, if G is a tree, then b(x) = M(P(x)) for all vertices x, yielding a polynomial time algorithm as described above. In this paper we will show that another class of graphs, nice graphs, also has this property, and hence the same polynomial time algorithm.

We say that y is a predecessor of x if there is a directed path from y to x in G, and a proper predecessor if in addition $y \neq x$. An acyclic directed graph G is nice if it has the following properties.

(a) If y and z are immediate predecessors of a vertex x then b(y) = b(z).

(b) If y and z are distinct immediate predecessors of a vertex x then y is not a predecessor of z.

(c) If $x_1,\ldots,x_k$ are vertices such that x_i is not a predecessor of x_j whenever $i \neq j$, then there exist vertex disjoint paths $L_2,\ldots,L_k$ containing no predecessors of x_1 such that L_i is a path from an input to x_i for each $i = 2,\ldots,k$.

It is natural to question the interest in determining black pebble numbers of vertices of nice graphs, since the definition seems to involve already knowing the black pebble numbers. However, if for example the predecessor subgraphs of two immediate predecessors y and z of a vertex x are isomorphic, then we know that b(y) = b(z) without necessarily knowing the value of b(y).

Pyramid graphs and their generalizations are interesting examples of nice graphs which are not trees. A pyramid graph of height m has m levels containing m,m-1,...,2,1 vertices respectively, arranged so that each non-input vertex has incoming edges from the two vertices immediately on its left and right in the level below. A pyramid of height 7 is shown in Figure 1. Another way of visualizing pyramid graphs is as triangular fragments of directed two-dimensional rectilinear lattices. In [1] Cook proved that the black pebble number of the apex of a pyramid of height m is m. Roughly speaking, the purpose of this result was to provide "evidence" that there is a problem which can be solved in polynomial time but not polylog space. Because of their symmetric structure as part of a rectilinear grid, pyramid graphs appear with some frequency in computer science applications, both as computation graphs and as connection graphs for networks of processors. They are perhaps the simplest example of an acyclic directed graph which is highly "non-tree like".

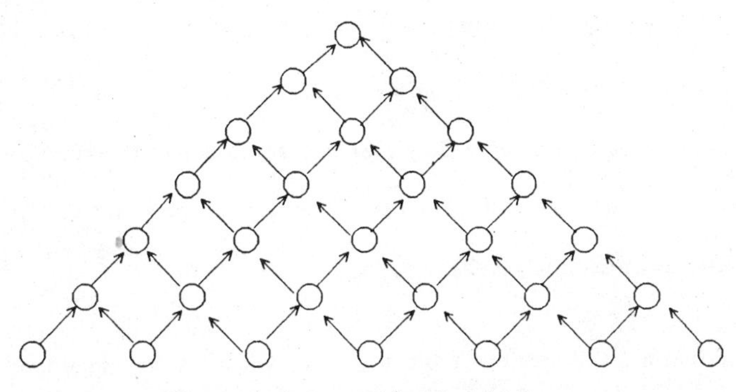

Figure 1. A pyramid of height 7.

Nice graphs were introduced in Klawe[4] in order to prove a lower bound of $b(x)/2 + 1$ for the black and white pebble number of x (a nondeterministic variant of the black pebble number), for x a vertex of a pyramid. In fact this lower bound was proved for a substantially larger class, spreading graphs, lending more evidence to the conjecture that this bound applies to vertices of all graphs.

For spreading graphs, we will prove that the problem of determining the black pebble number of each vertex given the black pebble numbers of all its proper predecessors is in NP, and hence there is a nondeterministic polynomial time algorithm to compute black pebble numbers of vertices of spreading graphs. Comparing this result with the PSPACE completeness result for arbitrary graphs suggests that it is unlikely that the conjecture that the black and white pebble number is bounded by 1/2 the black pebble number for vertices of arbitrary graphs can be settled by reducing it to spreading graphs. We leave the definition of spreading graph to the next section as it is rather complicated.

2. A LOWER BOUND ON THE BLACK PEBBLE NUMBER OF A CONFIGURA

The results in this paper are obtained with the use of a cost function C(B) which bounds b(B) from below whenever B is a configuration on a spreading graph. This cost function is the black pebble analogy of the black and white pebble cost function c(B,W) defined in Klawe [4]. Since the proof that $C(B) \leq b(B)$ for all configurations of spreading graphs is so similar to the analogous proof in [4] we will omit virtually all the details. We begin this section with the definitions of C(B) and spreading graphs.

We say that a set of vertices Y blocks a vertex x if for every path L from an input to x we have $Y \cap L \neq \phi$. Moreover we say that Y blocks a set of vertices X if Y blocks every vertex in X. It is easily verified that blocking is a transitive property, i.e. if Z blocks Y and Y blocks X then Z blocks X. Thus for example if Y blocks P(x) then Y blocks x itself.

The cost function is defined as $C(B) = \min\{M(Y)\text{: Y blocks B}\}$. The intuitive explanation of why C(B) might be a lower bound for b(B) is as follows. As mentioned above, M(B) is the space requirement of the independent pebble strategy for B, but this strategy will not be optimal whenever it is easier to pebble the vertices of B jointly than independently. By considering sets which block B rather than B itself, this problem is avoided, at least for spreading graphs.

It is obviously from the definition that for any graph G, if B is a single vertex x then $C(B) \leq b(B) = b(x)$ since x blocks itself. However, even for configurations B with two vertices this no longer necessarily holds when G is not a spreading graph, as will be shown in an example at the end of this section.

We now give the definition of spreading graph. Since the definition is quite complicated, the reader may wish to to refer to [4] where these concepts are dealt with in more detail.

We call a set V of vertices tight if for every vertex v in V, the set $V \setminus \{v\}$ does not block v. We use $\langle V \rangle$ to denote the set of vertices which are blocked by V, and if V is tight, for each x in $\langle V \rangle$ we define $V_x = \{v \in V\text{: there is a path L from an input to x such that } L \cap V = \{v\}\}$. Let $P^*(x)$ denote the predecessors of x. A set V is said to be connected if it is tight, and if the graph with vertices $\langle V \rangle$ and edges $\{(x,y)\text{: } x,y \in \langle V \rangle \text{ and } P^*(x) \cap \langle V_x \rangle \cap P^*(y) \cap \langle V_y \rangle \neq \phi\}$ is a connected graph. For any set X of vertices we define $b_j(X) = \min\{|Y|\text{: Y blocks X[j] and } Y = Y[j]\}$. Finally, a graph G is called a spreading graph if for every connected set V of G and $j \geq 0$ such that $\langle V \rangle[j] \neq \phi$ we have $|V| \geq b_j(\langle V \rangle) + j - \min\{b(v)\text{: } v \in V\}$.

A (rather long) proof that nice graphs are spreading graphs is given in Klawe[4, Theorem 3.7]. We will find the following example of a connected set useful in the next section.

REMARK 2.1. If V is tight and $V = V_x$ for some x in $<V>$, then V must be connected.

Proof. Since $V \subset P^*(x) \cap <V_x>$ and $V \cap P^*(y) \cap <V_y> \neq \phi$ for all y in $<V>$, every vertex in $<V>$ is adjacent to x so clearly the graph on $<V>$ is connected.

□

The following lemma gives the property of spreading graphs which is crucial in proving that the cost function is a lower bound on black pebble numbers of configurations on spreading graphs. This lemma will also be used in obtaining the black pebble number algorithms for nice graphs and spreading graphs.

LEMMA 2.2. If B is a configuration on a spreading graph, then there is a set Y such that Y blocks B, $C(B) = M(Y)$, and $|Y| \leq |B|$.

This lemma is proved in a completely analogous manner to Theorem 3.7 in [4]. Rather than repeat its long and involved proof here, we merely point out the small ways in which the proofs differ. First of all, the relation $<<$ is replaced by $\leq\leq$ defined as follows. We say that $X \leq\leq Y$ if for each j there is some $i \leq j$ such that $j + |X[j]| \leq i + |Y[i]|$. Lemma 3.4 is replaced by

"if $X \cap Z = Y \cap Z = \phi$ and $X \leq\leq Y$ then $M(X \cup Z) \leq M(Y \cup Z)$",

which has a very similar proof. Lemma 3.5 remains the same, and Lemma 3.6 is replaced by the following statement.

"Suppose B and Y are vertex subsets of a spreading graph such that Y blocks B and Y is connected. Then there is a set Y′ such that Y′ blocks B with Y′ $\leq\leq$ Y and $|Y'| \leq |B|$."

The proof this is essentially the same as that of Lemma 3.6 but slightly simpler. With these substitutions, it is easy to see how to modify the proof of Theorem 3.7 [4] to obtain Lemma 2.2.

We now prove the theorem which will yield our lower bound.

THEOREM 2.3. If $B_0, B_1, \ldots, B_n$ is a black pebble strategy on a spreading graph, then for each i we have $C(B_i) \leq \max_{0 \leq j \leq i} |B_j|$.

Proof. The proof is by induction on i. This is clearly true for i = 0, so assume that $i \geq 1$ and that $C(B_{i-1}) \leq \max_{0 \leq j \leq i-1} |B_j|$. Let Y block B_{i-1} such that $C(B_{i-1}) = M(Y)$ and $|Y| \leq |B|$. If the move from B_{i-1} to B_i consists of anything other than placing a pebble on an input vertex, it is easy to see that Y also blocks B_i, and hence $C(B_i) \leq C(B_{i-1})$ and we are done. Thus we may assume that $B_i = B_{i-1} \cup \{y\}$ where y is an input. Let $Y' = Y \cup \{y\}$. Clearly Y′ blocks B_i, so by the induction hypothesis it suffices to show that $M(Y') \leq \max\{M(Y), |B_i|\}$ since $C(B_i) \leq M(Y')$. We have $Y'[j] = Y[j]$ for each $j > 1$ and $Y'[1] = Y \cup \{y\}$ so $M(Y') = \max\{M(Y), 1 + |Y[1]|\} \leq \max\{M(Y), 1 + |B_{i-1}|\} \leq \max\{M(Y), |B_i|\}$ as desired. □

As an immediate corollary we have:

COROLLARY 2.4. If B is a configuration on a spreading graph then $c(B) \leq b(B)$

We close this section with an example of a graph with a configuration B such that $c(B) > b(B)$. Let G be the graph shown in Figure 2, and let $B = \{y,z\}$. All the input vertices have black pebble number 1, w and x have black pebble number 4, and y and z have black pebble number 5. Moreover it is easy to see that $b(B) = 5$. However $C(B) = 6$ since $M(B) = 6 = M(\{w,x,y,z(1),z(2)\}) = M(w,x,y(1),y(2),z(1),z(2))$.

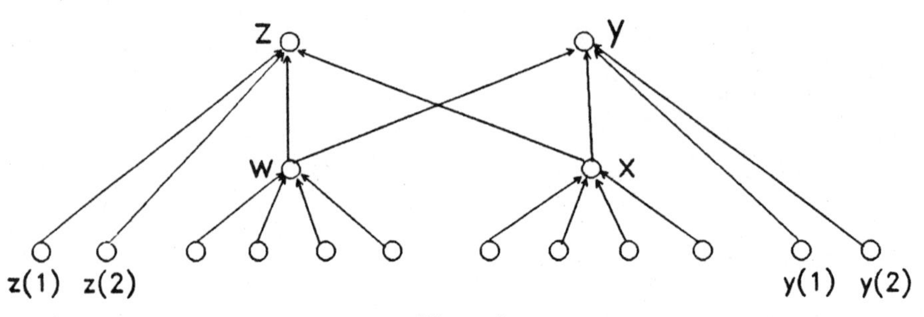

Figure 2.

3. COMPLEXITY OF THE BLACK PEBBLE NUMBER

We are now ready to prove our claims on the complexity of determining the black pebble number for vertices of nice graphs and spreading graphs.

PROPOSITION 3.1. If x is a vertex of a spreading graph, then $b(x) = C(P(x))$.

Proof. We first claim that $b(x) = C(\{x\})$. By Lemma 2.2 there is a vertex y such that y blocks x and $C(\{x\}) = M(\{y\}) = b(y)$. Moreover by 2.1 the set $\{y\}$ is connected and hence we have $1 = |\{y\}| \geq b_{b(x)}(<\{y\}>) + b(x) - b(y) \geq 1 + b(x) - b(y)$. Thus $C(\{x\}) = b(y) \geq b(x)$. Also $C(\{x\}) \leq b(x)$ by Corollary 2.4 so $C(\{x\}) = b(x)$ as claimed. Next, since every set which blocks $P(x)$ must also block x, we have $C(\{x\}) \leq C(P(x))$. Finally, since $b(x) = b(P(x))$ we have $C(\{x\}) \leq C(P(x)) \leq b(P(x)) = b(x) = C(\{x\})$ and hence $b(x) = C(P(x))$. □

COROLLARY 3.2. If x is the vertex of a spreading graph, then the problem of determining whether $b(x) \leq k$ is in NP.

Proof. We sketch a nondeterministic polynomial time algorithm. Let $x(1),\ldots,x(n) = x$ be the predecessors of x ordered in increasing order of maximum length of directed path from an input, and let i be maximal such that x(i) is an input. Note that the order of the vertices guarantees that if x(m) is a proper predecessor of x(j), then $m < j$. For $1 \leq j \leq i$, set $b(x(i)) = 1$. For $j = i+1$ to n, guess the set V(j) of proper predecessors of x(j) such that $M(V(j)) = C(P(x(j))) = b(x(j))$ and set $b(x(j)) = M(V(j))$. □

PROPOSITION 3.3. If x is a vertex of a spreading graph which satisfies conditions (b) and (c) of the definition of nice graph, then $b(x) = M(P(x))$.

Proof. By Proposition 3.1 it suffices to show that $C(P(x)) = M(P(x))$. By Lemma 2.2 there is a set W which blocks P(x) with $C(P(x)) = M(W)$ and $|W| \leq |P(x)|$. Without loss of generality we may assume that W is tight. Clearly it is enough to show that $M(W) \geq M(P(x))$. By properties (b) and (c) in the definition of a nice graph, it is easy to see that if P is any subset of P(x), and V blocks P then V must have cardinality at least $|P|$. This implies that for each j with $P(x)[j] \neq \phi$ we have $b_j(P(x)) = |P(x)[j]|$ since obviously P(x)[j] blocks P(x)[j] and $(P(x)[j])[j] = P(x)[j]$. By Remark 2.1 we see that W is connected since $W = W_x$. Thus we have $|W| \geq |P(x)[j]| + j - \min\{b(v): v \in W\}$ for each j. Rearranging and taking the maximum over j yields $M(P(x)) = \max\{j - 1 + |P(x)[j]| : P(x)[j] \neq \phi\} \leq \min\{b(v): v \in W\} - 1 + |W| \leq M(W)$, and we are done. □

COROLLARY 3.4. There is an algorithm to compute the black pebble numbers of vertices of spreading graphs satisfying conditions (b) and (c) of nice graphs, whose running time is O(m log m), where m is the number of edges in the graph. For nice graphs the algorithm runs in O(m) time.

Proof. Given the black pebble numbers of the vertices in P(x), we can compute $b(x) = M(P(x))$ in time $O(|P(x)| \log |P(x)|)$. (For nice graphs M(P(x)) can be computed in time $O(|P(x)|)$ because of condition (a) in the definition.) Since m is the sum of $|P(x)|$ over all the vertices of the graph, the black pebble numbers of vertices can be computed in increasing order of maximum length of path from an input, in time O(m log m) [O(m) for nice graphs]. □

REFERENCES

[1] S. Cook, An observation on time-storage trade off, J. Comput. System Sci. 9(1974), 308-316.

[2] S. Cook and R. Sethi, Storage requirements for deterministic polynomial time recognizable languages, J. Comput. System Sci. 13(1976), 25-37.

[3] J. Gilbert, T. Lengauer, and R. Tarjan, The pebbling problem is complete for polynomial space, SIAM J. Comput. 9(1980), 513-524.

[4] M. Klawe, A tight bound for black and white pebbles on the pyramid, Proceedings 24th Ann. Symp. Found. Comp. Sci., 1983, 410-419, to appear in JACM.

[5] T. Lengauer and R. Tarjan, The space complexity of pebble games on trees, Inform. Process. Lett. 10(1980), 184-188.

[6] M. Loui, The space complexity of two pebble games on trees, Lab. for Comp. Sci. Tech. Report 133, MIT, Cambridge, MA (1979).

[7] F. Meyer auf der Heide, A comparison of two variations of a pebble game on

graphs, Theoret. Comput. Sci. 13(1981), 315-322.

[8] N. Pippenger, Pebbling, IBM Research Report RC 8258 (1980).

COMPATIBLE MATCHINGS IN BIPARTITE GRAPHS

Clyde P. Kruskal* and Douglas B. West
University of Illinois

ABSTRACT

In a bipartite graph with vertices X,Y indexed as $x_1, \ldots, x_m$ and $y_1, \ldots, y_n$, a *compatible matching* is a matching with no pairs of edges $x_i y_j$ and $x_k y_l$ having $i<k$ and $j>l$. We show that any graph with N edges must contain a compatible matching with at least $\lceil \sigma - \sqrt{\sigma^2 - N} \rceil$ edges, where $\sigma = (m+n)/2$ is the average of the part-sizes. This is best possible; we exhibit graphs where equality holds. When $\sigma - \sqrt{\sigma^2 - N}$ is an integer, we characterize and enumerate the extremal graphs. We also obtain extremal results for the number of compatible matchings needed to cover the edges, and integral min-max relations for both parameters. We describe correspondences with problems in perfect graphs, partially ordered sets, and Young tableaux.

*Research supported in part by a grant from IBM Corporation.

1. Introduction

Matching in bipartite graphs is a well-studied subject in graph theory; we consider a variant, called *compatible matching,* first studied by Borodin and Hopcroft [BH]. Given a fixed ordering $x_1, \ldots, x_m$ and $y_1, \ldots, y_n$ of the vertices in the two parts, a *compatible matching* or *c-matching* is a matching in which crossing edges do not appear. Having a matching but no crossings forbids pairs of edges $x_i y_j$ and $x_k y_l$ with $i \leq k$ and $j \geq l$. Let α_c be the size of the maximum c-matching.

Compatible matchings can be motivated by layout problems. The vertices on the two sides of the graph can represent fixed terminals, and non-crossing edges represent wires that can appear on the same layer. Let χ_c be the minimum number of c-matchings needed to cover the edges, called the *compatible chromatic index.* If wires must keep to a single layer, then χ_c is the number of layers needed to solve the wiring problem.

The extremal problem for c-matchings arose from the computational problem of merging. Borodin and Hopcroft [BH] proved a special case of the lower bound on α_c using the pigeonhole principle, and used it to prove a lower bound on the complexity of parallel merging. They showed that n processors require (at least) $\frac{1}{2}\log_2\log_2 n + \Omega(1)$ comparison steps to merge two sorted lists of n elements each into one sorted list. Using the general result here, Kruskal [Kr] improves this by showing that $\log_2\log_2 n + \Omega(1)$ comparison steps are required. This result is tight, since [Kr] also presents an algorithm that uses only $\log_2\log_2 n + O(1)$ comparison steps.

We consider extremal values of α_c and χ_c in terms of the number of edges in the graph. Let **G** be the class of bipartite graphs with N edges and parts of size m and n, and let $\sigma = (m+n)/2$. We show that any graph in **G** must contain a c-matching with at least $\lceil \sigma - \sqrt{\sigma^2 - N} \rceil$ edges, and we exhibit graphs where equality holds. The extremal graphs are far from unique; we describe and enumerate them when $\sigma - \sqrt{\sigma^2 - N}$ is an integer (§6). We obtain similar extremal results for χ_c (§5).

To understand the bound, consider the case $m = n$. Then $\sigma - \sqrt{\sigma^2 - N} = n - n\sqrt{\beta}$, where $\beta = 1 - N/n^2$ is the fraction of edges omitted.

The proportion of vertices unmatchable can be as high as the square root of the proportion of edges missing. The extremum can be attained by using all the edges $x_i y_j$ with $\min\{i,j\}=1$, then all those with $\min\{i,j\}=2$, and so on, until N edges have been included (see §2).

We give two proofs that graphs in **G** have $\alpha_c \geq \lceil \sigma - \sqrt{\sigma^2 - N} \rceil$. In §3 we partition the edges of $K_{m,n}$ into classes of compatible edges, where the edges $x_i y_j$ with a fixed value of $i-j$ belong to the same class. A careful application of the pigeonhole principle yields the result. This generalizes the argument given by Borodin and Hopcroft [BH] for the case $m=n$, $N=\frac{3}{4}n^2$.

In §4 we take a different approach by considering a dual covering problem. This is suggested by the corresponding extremal problem for arbitrary matchings, which is a standard exercise in elementary graph theory. By the Konig-Egervary Theorem, the size of the largest matching in a bipartite graph equals the smallest number of vertices needed to cover all the edges. To forbid a large matching, concentrate the edges at few vertices. N edges can be concentrated at a set of $\lceil N/\max\{m,n\} \rceil$ vertices and no fewer, so any graph with N edges has a matching of size at least $\lceil N/\max\{m,n\} \rceil$.

The vertex cover is, equivalently, a set of "stars" covering the edges; no two edges in a matching can belong to a single star. To forbid large c-matchings, we need to describe the appropriate dual covering object, and put as many edges as possible in the union of few of these. Let a *twist* be a set of edges such that each edge intersects every other (possibly at an endpoint). Twists and c-matchings intersect in at most one edge, a fact that yields two natural dual pairs of packing and covering problems using the additional parameters θ_c and ω_c. In particular, the largest c-matching size (α_c) is bounded by the minimum number of twists needed to cover the edges (θ_c), and the largest twist size (ω_c) is bounded by the minimum number of c-matchings required to cover the edges (χ_c).

In fact, equality holds for both pairs of problems in every bipartite graph; this follows from translating these problems into vertex packing and covering problems in a class of perfect graphs called permutation graphs. These can be defined in the context of the channel routing problem defined earlier. If each terminal gets exactly one wire, then the terminal labels on one side form a permutation of those on the other according to the connections. Vertices are adjacent in the corresponding graph if and only if the labels appear in opposite order on the two sides, i.e. if the line segments between the two pairs of terminals intersect. A graph is a *permutation graph* if it has such a representation. Permutation graphs are the cocomparability graphs of two-dimensional partial orders, and are well-known to be perfect. (Permutation graphs and perfect graphs are discussed in great detail in [Go].)

A bipartite graph $G \in \mathbf{G}$ can be transformed into a permutation graph $P(G)$ on N vertices so that the compatible matchings of G correspond to independent sets in $P(G)$ and the twists of G correspond to cliques in $P(G)$ (see details in §4). The fact that $\alpha=\theta$ and $\chi=\omega$ in perfect graphs implies $\alpha_c(G)=\theta_c(G)$ and $\chi_c(G)=\omega_c(G)$.

§5 contains extremal results about the compatible chromatic index χ_c, again using the dual problem, which is the size of the maximum twist. For both parameters, the lower bounds are relevant to a problem in partially ordered sets, as noted in §5. The extremal results can be summarized as follows, where G is an arbitrary graph in $\mathbf{G}$.

THEOREM 1. Best possible bounds on $\alpha_c(G)$ and $\theta_c(G)$ are

$$\lceil \sigma-\sqrt{\sigma^2-N}\,\rceil \le \alpha_c(G)=\theta_c(G) \le \min\{N,m,n\}.$$

THEOREM 2. Best possible bounds on $\chi_c(G)$ and $\omega_c(G)$ are

$$\max\{\lceil N/n\,\rceil, \lceil m+n-2\sqrt{mn-N}\,\rceil\} \le \chi_c(G)=\omega_c(G) \le \min\{N,m+n-1\}.$$

In both cases, the upper bounds are trivial. As soon as there are $\min\{m,n\}$ edges available, there are graphs with c-matchings of size $\min\{m,n\}$. With fewer edges, all edges can be placed compatibly. No c-matching exceeds size $\min\{m,n\}$. Similarly, for the upper bound in Theorem 2, there are twists with up to $m+n-1$ edges, but no more, since there are only $m+n-1$ possible $j-i$ differences for $x_i y_j$, and no pair of edges with the same difference appear in a single twist.

Finally, we mention a related parameter with more flexibility, which has been considered from a purely graph-theoretic viewpoint. Given a graph, place the vertices on a circle in some order, and fill in the edges of the graph as chords in noncrossing classes. The minimum number of classes required, over all vertex orderings, is the *book thickness* [BK] or *pagenumber* [BS] of the graph. Book thickness has been determined for complete graphs and some complete bipartite graphs [BK], and is at most 7 for planar graphs [BS,He], which may not be best possible. The present context is obtained by considering a single fixed ordering of the vertices.

2. The smallest maximum c-matching

We construct a graph G_N on N edges whose largest c-matching has size $\lceil \sigma-\sqrt{\sigma^2-N}\,\rceil$. Order the edges $\{x_i y_j\}$ of $K_{m,n}$ lexicographically on $(\min\{i,j\},\max\{i,j\})$, and in addition put $x_i y_j \le x_j y_i$. This yields a total order L on the edges. Let the edges of G_N be the first N edges in this ordering. The fact that $\alpha_c(G_N)=\lceil \sigma-\sqrt{\sigma^2-N}\,\rceil$ follows most easily from the dual twist-covering problem.

LEMMA 1. $\alpha_c(G_N)=\lceil \sigma-\sqrt{\sigma^2-N}\,\rceil$.

Proof. It suffices to show G_N a twist-covering of size $r=\lceil \sigma-\sqrt{\sigma^2-N}\,\rceil$ and a c-matching of size r. If $\min\{i,j\}\le r$ for every edge $x_i y_j$, the former is easy: let twist T_i consist of the edges $x_i y_n, \ldots, x_i y_{i+1}, x_i y_i, x_{i+1} y_i, \ldots, x_m y_i$. The edges $x_i y_i$ form the desired c-matching if r is the smallest value such that $\min\{i,j\}\le r$ for every edge.

To show this, it suffices that $f(r-1)<N\le f(r)$, where $f(r)=rm+rn-r^2=2r\sigma-r^2$ is the number of pairs i,j with $\min\{i,j\}\le r$. By the quadratic formula, $-r^2+2r\sigma-N\ge 0$ for positive r if and only if $r\ge \sigma-\sqrt{\sigma^2-N}$, and similarly $f(r-1)<N$ if and only if $r-1<\sigma-\sqrt{\sigma^2-N}$. □

3. Pigeonhole proof of lower bound

By the computation above, $f(r-1)<N\le f(r)$ if and only if $r=\lceil \sigma-\sqrt{\sigma^2-N}\,\rceil$. Hence the lower bound is equivalent to:

LEMMA

2. Any subgraph of $K_{m,n}$ with more than $(r-1)(2\sigma-r+1)=f(r-1)$

edges has a c-matching of size r.

Proof. Let G be an arbitrary graph in **G**. We partition the edges of G into $m+n-1$ compatible classes; all the edges in a given class can appear together in a single c-matching. We show that a graph with N edges must have at least $\lceil\sigma-\sqrt{\sigma^2-N}\,\rceil$ edges in some single class.

Partition the edges into classes C_j, where C_j consists of all edges of the form $x_i y_{i+j}$; any subset of a single class constitutes a c-matching. The extreme classes C_{-m+1} and C_{n-1} consist of single edges $x_m y_1$ and $x_1 y_n$. Each successive class with smaller displacement has one more edge, until size $\min\{m,n\}$ is reached. Discard all classes with less than r edges; these contain a total of $2\sum_{i=1}^{r-1} i = r(r-1)$ edges. This make sense because the formula $\sigma-\sqrt{\sigma^2-N}$ always yields $r\le\min\{m,n\}$.

Discarding these $2r-2$ classes leaves $2\sigma-2r+1$ classes of size at least r. Since G has at most $r(r-1)$ edges in the discarded classes, it has more than $f(r-1)-r(r-1)=(r-1)(2\sigma-2r+1)$ edges in the remaining classes. By the pigeonhole principle, some class must contain r edges of G. $\square$

4. Lower bounds using duality

As discussed in the introduction, we can bound the size of a c-matching by covering the edges of G with fewest twists. To show this gives $\alpha_c(G)$ exactly, we must interpret c-matchings and twist coverings as independent sets and clique coverings in perfect graphs.

To transform G into a permutation graph, expand each vertex v into $d(v)$ vertices of degree 1, each getting one of the edges, and index these vertices in opposite order from their neighbors in the other part. This ensures that all the edges incident to a single vertex in G cross in the new graph. A set of edges is compatible if and only if it corresponds to an independent set in the intersection graph of the edges in the new drawing. This intersection graph $P(G)$ is a permutation graph, as described in the introduction. Permutation graphs are perfect, which means that $\alpha(H)=\theta(H)$ for all induced subgraphs H (see [Go] for an extensive exposition on permutation graphs). Hence we have $\alpha_c=\theta_c$. This leads to another proof of the main result:

Alternate proof of lower bound. The minimum α_c is a nondecreasing function of N. Hence it suffices to determine the largest number of edges in a graph with $\alpha_c = r$. As noted above, this is the same as finding the maximum size union of r twists in $K_{m,n}$. We need only consider maximal twists. Every maximal twist has $m+n-1$ edges, beginning with $x_1 y_n$ and successively increasing the subscript on x by 1 or decreasing the subscript on y by 1 until reaching the edge $x_m y_1$. This yields a bijection between maximal twists and lattice paths from $(1,n)$ to $(m,1)$ that always move one step down or to the right; call this the *lattice correspondence.* We must maximize the number of lattice points in the union of r such paths.

Assume $r \le \min\{m,n\}$. Starting from $1,n$, there are $i+1$ lattice points on the diagonal reached after i steps, for $i<r$. Similarly for the last r diagonals near $m,1$. The other $m+n+1-2r$ diagonals have at least r points each. Since each path hits each diagonal exactly once, the maximum number of points in the union is $2\sum_{i=0}^{r-1} i + r(m+n+1-2r) = 2r\sigma - r^2$. This is achieved, for example, by the twists in §2. □

5. Compatible edge-colorings

Having determined the largest and smallest α_c for graphs with N edges, we can also ask for the extremal values of χ_c.

Proof of Theorem 2. Since $P(G)$ is perfect, we have $\chi_c = \omega_c$. The upper bound on ω_c was given in §1. Achieving the lower bound is almost as easy; as before, a canonical edge-ordering works. The minimum value of χ_c is nondecreasing in N, so maximizing the union of r compatible classes is an equivalent problem. Let $d = j - i - (|m-n|/2)$. Intuitively, we want to choose edges $x_i y_j$ in increasing order of $|d|$, since this fills up the largest compatible classes first. These classes correspond to the diagonals in the lattice correspondence of §4.

To show that this greedy ordering gives the largest union of r compatible classes, partition $K_{m,n}$ into the twists T_i of §2. In the lattice correspondence, each consists of a one vertical and one horizontal segment. Since twists and compatible classes intersect in at most one edge, T_i contributes at most $\min\{r,|T_i|\}$ to the union of r compatible classes. Each T_i intersects the $|T_i|$ displacement classes with minimal $|d|$, so the union of the r displacement classes with minimal $|d|$ achieves the bound.

Let $g(r)$ be the maximum size of the union of r compatible classes. To compute $g(r)$, we must determine the size of these displacement classes. Suppose $m \geq n$, and let $\rho=(m-n)/2$. If $r \leq m-n+1$, then $g(r)=rn$. When r is larger, we successively add classes of size $n-1,n-1,n-2,n-2,\cdots$. If $r-(m-n+1)=t$ is even, this yields $g(r)=rn-2\sum_{i=1}^{t/2} i=rn-(t/2+1)(t/2)$. Since $t/2=(r-1)/2-\rho$, this can be written symmetrically in m and n as $r\sigma-(r^2-1)/4-\rho^2$. To get $g(r)$ when t is odd, evaluate this for $r-1$ and add $n-[(r-1)-(m-n+1)]/2-1 = n-r/2+\rho$. This yields $g(r)=r\sigma-r^2/4-\rho^2$. To summarize, let $\epsilon(r)=0$ for r odd, $\epsilon(r)=1$ for r even. Then

$$g(r)=\begin{cases} rn & r \leq m-n+1 \\ r\dfrac{m+n}{2}-\dfrac{r^2-\epsilon(r)}{4}-(\dfrac{m-n}{2})^2 & r>m-n+1 \end{cases}$$

To get the lower bound on χ_c, we invert $g(r)$ to express r in terms of N. In particular, $\min_{G \in \mathbf{G}} \chi_c(G)=r$ if and only if $g(r-1)<N \leq g(r)$. Using the quadratic formula again yields the pleasant formula $\chi_c \geq \lceil m+n-2\sqrt{mn-N}\,\rceil$ for $N>n(m-n+1)$, $\chi_c \geq \lceil N/n \rceil$ for $N<n(m-n+1)$. $\square$

The functions f and g are relevant to the theory of chains and antichains in posets. A compatible class translates via the lattice correspondence into a collection of incomparable elements (an *antichain)* in the ordering on lattice points that puts $(i,j) \leq (k,l)$ if $i \geq k$ and $j \leq l$. Twists translate into totally ordered sets *(chains).* The lattice points in $[1,m] \times [1,n]$, under this ordering, form the direct product of two chains (totally ordered sets) $C_m \times C_n$. Maximum unions of r twists or c-matchings become maximum unions of r chains or antichains. The twist decomposition $\{T_i\}$ of §2 becomes the standard symmetric chain decomposition of $C_m \times C_n$ found in [dBTK], [GK], and elsewhere. We do not know whether the simple formulas f and g given here for the maximum size union of r chains or antichains in

$C_m \times C_n$ have been recorded before.

6. The number of extremal graphs

In Section 4 we showed that the maximum number of edges in a graph with $\alpha_c(G)=r$ is $f(r)$. This is achieved precisely when G is a maximum union of r twists. In the lattice correspondence, such graphs correspond to using all points on the r diagonals nearest $(1,n)$, all points on the r diagonals nearest $(m,1)$, and taking r distinct points from each intermediate diagonal. Geometrically, this means these graphs are in one-to-one correspondence with choices of r disjoint monotone lattice paths from $\{(1,n-r+1),\ldots,(r,n)\}$ to $\{(m-r+1,1),\ldots,(m,r)\}$. Let the bottom-most and top-most paths be the "first" and "last" paths.

Each path in such a configuration takes $m-r$ steps to the right and $n-r$ steps down, and can be encoded by recording the step numbers on which down steps occur. Record these numbers column by column in a rectangular array; note that the columns are strictly increasing. Each of the paths reaches the same diagonal after j steps. The paths are disjoint if and only if after j steps the ith path has taken at least as many down steps as the $i+1$st path, for all i and j. In terms of the corresponding array, this means that the step on which path $i+1$ goes down for the kth time must be at least as late as the step on which path i goes down for the kth time. Hence row k of the array must be nondecreasing, for all k. Conversely, any array of numbers satisfying these conditions corresponds to an acceptable set of paths. Hence the maximum unions of r twists (or maximum graphs with $\alpha_c=r$) are in one-to-one correspondence with column-strict tableaux with r columns, $n-r$ rows, and entries at most $m+n-2r$.

These tableaux have been enumerated. More generally, column-strict tableaux are positive integer arrays with nondecreasing rows and strictly increasing columns, having λ_i numbers in the ith row, with $\lambda_1 \geq \cdots \geq \lambda_m$. For the number of such tableaux having a bound on the entry size, Littlewood [Li] originally obtained a generating function via algebraic methods (see also [St]); Remmel and Whitney [RW] give a bijective proof of this. Bender and Knuth [BK] obtained an explicit formula. The computation now appears as a nontrivial homework exercise in [Kn].

We suspend our notational conventions temporarily to describe the enumerative result in standard tableau notation. Suppose the number of parts (rows) is m, and the upper limit on entry values is N. The trick is to add zeros to the shape λ so the shape has $m=N$ parts. Then the number of column-strict tableaux is the determinant of a matrix whose ijth entry is $\binom{\lambda_j+m-j}{m-i}$. This can be converted to $V(\lambda_1+m-1,\lambda_2+m-2,\ldots,\lambda_m)/(m-1)!\cdots 0!$ by row operations, where $V(x_1,\ldots,x_n)$ denotes the Vandermonde determinant, which is the determinant of the matrix whose ijth entry is x_j^i. It is well-known that $V(x_1,\ldots,x_n)=\prod_{i<j}(x_j-x_i)$. The formulas for the number of tableaux are proved by induction on $\Sigma\lambda_i$ and the number of rows.

Consider the special case of a rectangular shape with m rows, r columns (i.e. $\lambda_i=r$), and upper limit N on tableau entries, where $N\geq m$. The formula becomes

$$V(r+N-1,\ldots,r+N-m,N-m-1,\ldots,0)/(N-1)!\cdots 0!.$$

Using the expression for V as a product of differences, the differences between the first m terms yield $(m-1)!\cdots 0!$, the differences between the last $N-m$ terms yield $(N-m-1)!\cdots 0!$, and the differences between the first m and last $N-m$ terms yield $\frac{(r+N-1)!}{(r+m-1)!}\cdots\frac{(r+N-m)!}{r!}$. Collecting all this together, canceling $(N-m-1)!\cdots 0!$ from the numerator and denominator, and introducing m factors of $(N-m)!$ in the numerator and denominator, we can rewrite the answer in terms of binomial coefficients as

$$\prod_{i=1}^{m}\frac{\binom{r+N-i}{N-m}}{\binom{N-i}{N-m}}$$

In the context of counting the extremal graphs, which is the same as counting the maximum-sized unions of r chains in the product of an m-chain with an n-chain, we want $n-r$ rows, r columns, and entries at most $m+n-2r$. Substituting these values for m, r and N in the expression above yields

$$\prod_{i=1}^{n-r} \frac{\binom{m+n-r-i}{m-r}}{\binom{m+n-2r-i}{m-r}}$$

Acknowledgement

We thank C.L. Liu for lively discussions on enumerating the extremal graphs.

References

[BK] E.A. Bender and D.A. Knuth, Enumeration of plane partitions, *J. Combinatorial Theory (A)* 13(1972), 40-54.

[BH] A. Borodin and J.E. Hopcroft, Routing, merging, and sorting on parallel models of computation, *Proc. 14th ACM Symp. Th. Comp.* (1982), 338-344.

[BK] F. Bernhart and P.C. Kainen, The book thickness of a graph, *J. Combinatorial Theory (B)* 27(1979), 320-331.

[BS] J.F. Buss and P.W. Shor, On the pagenumber of a planar graph, preprint.

[dBTK] N. deBruijn, C. Tengbergen, and D. Kruyswijk, On the set of divisors of a number, *Nieuw Arch. Wiskunde* 23(1951), 191-193.

[Go] M.C. Golumbic, *Algorithmic graph theory and perfect graphs,* Academic Press (1980)

[GK] C. Greene and D.J. Kleitman, Strong versions of Sperner's theorem, *J. Combinatorial Theory (A)* 20(1976), 80-88.

[Kn] D.A. Knuth, *The Art of Computer Programming, Vol III.* Addison-Wesley (1973), Chapter 5.

[Kr] C.P. Kruskal, Optimal parallel merging, in preparation.

[Li] D.E. Littlewood, *The Theory of Group Characters, 2nd ed.* Oxford Univ. Press, London (1950).

[RW] J.B. Remmel and R. Whitney, A bijective proof of the hook formula for the number of column strict tableaux with bounded entries. *Europ. J. Comb.* 4(1983), 45-63.

[St] R.P. Stanley, Theory and applications of plane partitions, part 2. *Studies in Appl. Math.* 50(1971), 259-279.

A LINEAR TIME ALGORITHM FOR FINDING AN OPTIMAL DOMINATING SUBFOREST OF A TREE

E. L. Lawler*
University of California at Berkeley

P. J. Slater
University of Alabama at Huntsville

ABSTRACT

A linear-time algorithm is presented for the following problem. Given a tree T and two nonnegative integers $b(v), c(v)$, for each vertex v, find a subforest $H \subseteq T$ with a minimum number of components, subject to the constraints that (i) each vertex v not in H is adjacent to (dominated by) at least $b(v)$ vertices in H, and (ii) each vertex v of H has degree at most $c(v)$. Special cases of this problem are ordinary domination ($b(v) = 1$, $c(v) = 0$, for all vertices), b-domination ($b(v)$ arbitrary, $c(v) = 0$), "path domination" ($b(v) = 1$, $c(v) = 2$), and "clique domination" ($b(v) = 1$, $c(v) = 1$). By choosing $b(v) = 0$ or $b(v) > d(v)$, where $d(v)$ is the degree of v in T, one can also make v either a "free" or a "required" vertex.

*Research supported in part by NSF Grant MCS-8311422.

1. Introduction

Let $G = (V,E)$ be a graph. A subset $D \subseteq V$ is a *dominating set* if each vertex in $V - D$ is adjacent to (or dominated by) at least one vertex in D. It is NP-complete to determine a minimum size dominating set even for the restricted class of bipartite graphs. However, a minimum dominating set can be found in linear time for the still more restricted class of trees [1].

One generalization of the domination problem is the *b-domination* problem. Each vertex $v \in V$ is assigned a nonnegative integer $b(v)$. The object is to find a minimum size subset $D \subseteq V$ such that each vertex in $V - D$ is adjacent to at least $b(v)$ vertices in D. It is known that this problem can also be solved in linear time for the special case of trees [2].

A still further generalization is as follows. In addition to the nonnegative integers $b(v)$, let each vertex $v \in V$ be assigned a nonnegative integer $c(v)$. The object is to find a subgraph $H \subseteq G$ with a minimum number of components, subject to the constraints that (i) each vertex v not in H is adjacent to at least $b(v)$ vertices in H and (ii) each vertex v of H has degree at most $c(v)$.

In this paper we describe a linear time algorithm for solving this generalized domination problem in the special case that G is a tree. (H is then necessarily a subforest.) Note that as special cases we have the ordinary domination problem ($b(v) = 1$, $c(v) = 0$, for all vertices v), the b-domination problem ($b(v)$ arbitrary, $c(v) = 0$), the *path domination* problem ($b(v) = 1$, $c(v) = 2$), and the clique domination problem ($b(v) = 1$, $c(v) = 1$). Note also that if $b(v) = 0$ then v is a "free" vertex and if $b(v) > d(v)$, where $d(v)$ is the degree of v, then v is a "required" vertex. It follows that the *optional* domination problem is also a special case [1].

2. The Algorithm

With reference to an arbitrarily chosen root, we process the vertices of T in "postorder", using depth-first search. When it comes time to process a given vertex v we have already processed its children (if any) and, by assumption, determined the vertices and edges of H that are contained in the subtrees rooted to each of these children. Accordingly, we must decide whether or not to add v to H and, if v is added to H, which of the edges between v and its children are to be added as well.

We choose to add v to H if and only if one or more of the following conditions hold:

(1) v is the root of T and fewer than $b(v)$ of its children are in H, or v is not the root and fewer than $b(v) - 1$ of its children are in H.

(2) v has a child $u \notin H$ with fewer than $b(u)$ of its children in H. (Note that such a child must have at least $b(u) - 1$ of its children in H, by (1).)

(3) $c(v) > 0$ and v has a child $u \in H$ that is incident to fewer than $c(u)$ edges in H.

If v is added to H, then we add as many edges (v,u) to H as possible, being constrained by only the minimum of $c(v)$ and the number of children $u \in H$ that are incident to fewer than $c(u)$ edges already chosen to be in H. (If the number of children u eligible to be incident to edges (v,u) in H is greater than $c(v)$, we make an arbitrary selection of $c(v)$ edges to such children.)

Depth-first search is well-known to require linear time, and the operations specific to this algorithm are clearly possible to perform within linear time.

3. An Example

Consider the tree shown in Figure 1, in which the vertices are numbered in postorder with respect to the root. The ordered pair $(b(v), c(v))$ is indicated next to each vertex v. The algorithm obtains a dominating subforest as shown in Figure 2.

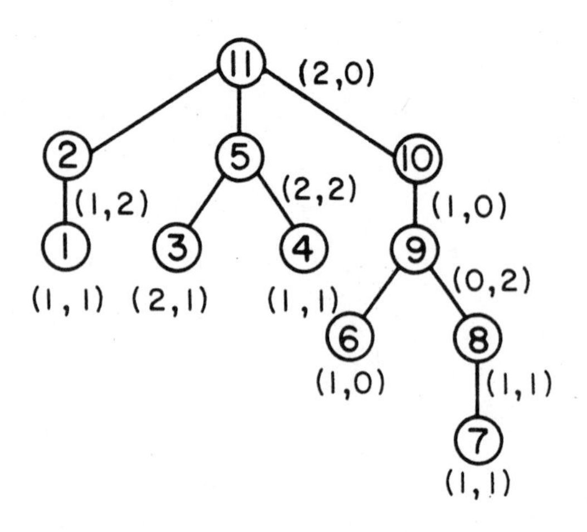

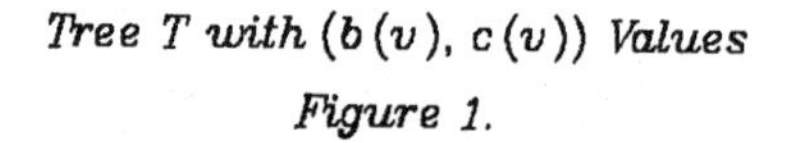

Tree T with $(b(v), c(v))$ Values
Figure 1.

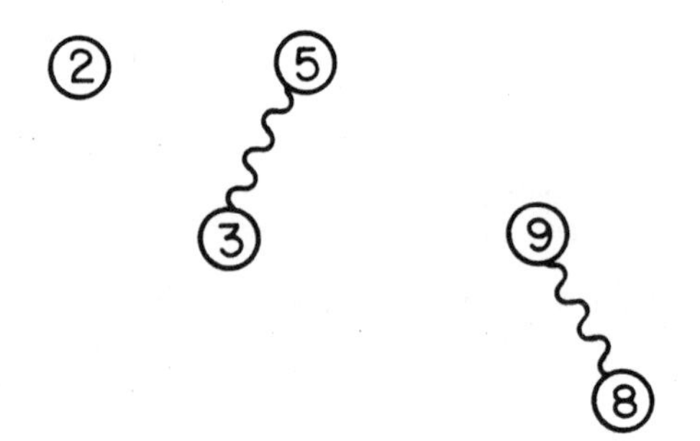

Minimum-Component Dominating Subforest
Figure 2.

4. Proof of Validity

We believe it is evident that the algorithm finds a feasible dominating subforest, so we concentrate only on the question of optimality. Let H_A be a subforest determined by the algorithm and let H_O be any other feasible dominating subforest. We shall show by a sequence of rearrangements, none of which increases the number of components, that H_O can be transformed into H_A, thereby proving that H_A has a minimum number of components.

Consider the vertices of T in the same (post)order they are processed by the algorithm. Suppose for vertex v, that H_O has exactly the same vertices and edges as H_A in each of the subtrees rooted to its children. There are several possible cases to consider. The reader should verify each rearrangement indicated below does not increase the number of components of H_O and leaves H_O a feasible subforest.

Case 1. $v \not\in H_A$ and $v \not\in H_O$. Then no rearrangement of H_O is required at v. If v is the root of T, then $H_O = H_A$ and we are done.

Case 2. $v \not\in H_A$ and $v \in H_O$.

(a) If v is the root of T then remove v from H_O (reducing the number of components of H_O). Now $H_O = H_A$.

(b) v is not the root of T and has parent w. If $w \not\in H_A$ or if $w \in H_O$ then remove v from H_O, together with the edge (w, v), if it is in H_O. If $w \in H_A$ and $w \not\in H_O$, then remove v from H_O and add w.

Case 3. $v \in H_A$ and $v \not\in H_O$. This can occur only if v has a child $u \in H_O$ that is incident to fewer than $c(u)$ edges in H_O. Add v to H_O, and also add to H_O those edges (v, u) that are in H_A.

Case 4. $v \in H_A$ and $v \in H_O$. Remove from H_O all edges incident to v that are in H_O but not in H_A (including possibly edge (w, v), where w is the parent of v) and add to H_O all edges that are in H_A but not in H_O. (Vertex w may have to be added to H_O, if (w, v) is in H_A but not in H_O.) Note that the number of edges incident to v in H_A is necessarily at least as large as the number of edges incident to v in H_O, so this rearrangement cannot increase the number of components of H_O.

REFERENCES

[1] E. J. Cockayne, S. E. Goodman and S. T. Hedetniemi, "A Linear Algorithm for the Domination Number of a Tree", *Infor. Processing Lett.*, 4 (1975) 41-44.

[2] S. T. Hedetniemi, private communication, 1984.

[3] S. M. Hedetniemi, S. T. Hedetniemi and R. Laskar, "Domination in Trees: Models and Algorithms", these proceedings.

TOWARD A MEASURE OF VULNERABILITY

II. THE RATIO OF DISRUPTION

Marc J. Lipman
Indiana University-Purdue University
at Fort Wayne

Raymond E. Pippert
Indiana University-Purdue University
at Fort Wayne

ABSTRACT

We consider a parameter which optimizes the "effort needed to achieve a certain amount of disconnection" in a graph. The parameter is compared with other measures of the vulnerability of a graph, including some traditional measures - vertex and edge connectivity.

I. Introduction

The word "vulnerability," representing a parameter for graphs, has been interpreted in several different ways, such as connectivity [1,3], edge-connectivity [3], and the minimum number of edges whose removal increases the diameter [2]. The intention of these interpretations seems plain: "Vulnerability" should be a measure of how easily some aspect of a graph can be destroyed.

Consider Figure 1. Both graphs have connectivity and edge-connectivity equal to 2. But the removal of a 2-vertex (or 2-edge) cutset in a) leaves almost all of the graph intact, whereas the removal of a 2-vertex (or 2-edge) cutset in b) cuts the graph in half, a far more serious matter in (for example) a communication network.

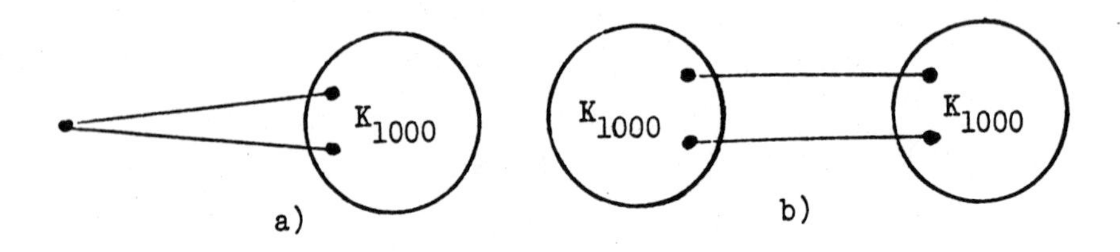

FIGURE 1

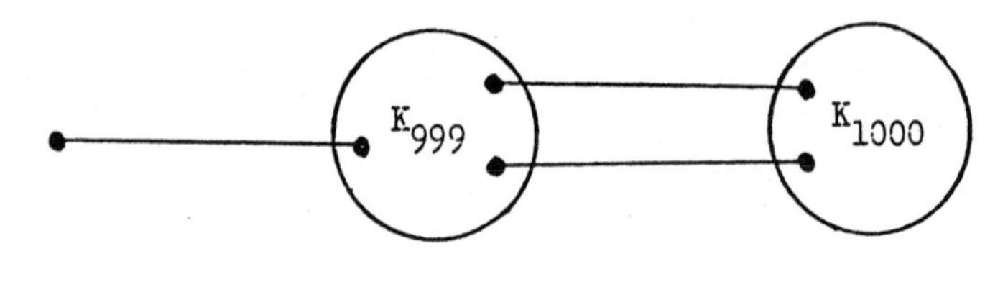

FIGURE 2

Consider Figure 2. The graph has connectivity and edge-connectivity equal to 1, but removing either 2 vertices or 2 edges can achieve considerably greater pairwise disconnection in this graph than can be achieved by the removal of a single vertex or edge.

For both of these reasons we would like to attempt to quantify "vulnerability" as a relative, as well as an absolute parameter.

Let $V = V(G)$ be the vertex set of G, and let $E = E(G)$ be the edge set of G. To simplify notation we let $|V| = p$, $h = \lfloor p/2 \rfloor$, and $h' = \lceil p/2 \rceil$ throughout.

If λ is the edge connectivity of a simple graph G, then λ is the cardinality of a minimum set of edges whose removal separates at least one vertex from the rest of G. For $1 \leq r \leq h$, let λ_r denote the cardinality of a minimum set of edges whose removal separates a set of at least r but at most h vertices from the rest of G. Thus $\lambda = \lambda_1$. Let $R = r/(p - r)$, the ratio of the separated pieces of G.

Definition: The ratio of disruption of G, denoted $\rho(G)$ or ρ, is given by

$$\rho = \max_{1 \leq r \leq h} \left\{ \frac{R}{\lambda_r} \right\}.$$

An r which achieves ρ will be denoted r^*. Similarly, a λ_{r^*} will be denoted λ^*.

II. Elementary observations

Proposition 2.1: Let G and H have the same vertex set, and $E(G) \subset E(H)$. Then $\rho(G) \geq \rho(H)$.

proof: It is enough to observe that λ_r in G is always at most λ_r in H. □

Thus adding edges cannot increase ρ, which corresponds to our intuitive feeling of "strengthening" a graph by adding adjacencies. That is, supergraphs are (and should be) less vulnerable.

We now calculate ρ for several families of graphs.

Proposition 2.2: a) $\rho(K_p) = 1/h'^2$;

b) $\rho(C_p) = h/2h'$;

c) $\rho(P_p) = h/h'$;

d) $\rho(K_{1,n}) = 1/h'$.

proof: It is straightforward to calculate λ_r and hence R/λ_r for all values of r.

a) $\lambda_r = r(p - r)$. Therefore $R/\lambda_r = 1/(p - r)^2$, so $r^* = h$.

b) $\lambda_r = 2$ for all r. Therefore $r^* = h$.

c) $\lambda_r = 1$ for all r. Therefore $r^* = h$.

d) $\lambda_r = r$. Therefore $R/\lambda_r = 1/(p - r)$, so $r^* = h$. □

<u>Proposition 2.3</u>: $\rho(K_{m,n})$ is determined by the values and parity of m and n as follows:

$$\rho(K_{m,n}) = \begin{cases} 2/mn & m,n \text{ even} \\ 2/(mn - 1) & m,n \text{ odd} \\ 2h/mnh' & \text{one of } m,n \text{ odd} \end{cases}$$

If m = n, the formulas become

$$\rho(K_{n,n}) = \begin{cases} 2/n^2 & n \text{ even} \\ 2/(n^2 - 1) & n \text{ odd} \end{cases}$$

proof: Suppose $m \le n$, and $r = \alpha + \beta$ where α vertices lie in the m-set and β vertices lie in the n-set. Separating r vertices via α & β requires $\alpha(n - \beta) + \beta(m - \alpha)$ edges.

Case 1: $\alpha > m/2$. Separating r vertices via $(\alpha - 1)$ & $(\beta + 1)$ requires $(\alpha - 1)(n - \beta - 1) + (\beta + 1)(m - \alpha + 1)$ $= \alpha(n - \beta) + \beta(m - \alpha) + (2\beta - n) + (m - 2\alpha) + 2$ edges. Since $\alpha > m/2$, it follows that $\beta < n/2$, so that $2\beta - n < 0$, $m - 2\alpha < 0$ and the number of edges required to separate r vertices via $(\alpha - 1)$ & $(\beta + 1)$ is no larger than the number of edges needed to separate r vertices via α & β.

Case 2: $\alpha \leq m/2$ and $r = \alpha + \beta < h$. We compute ratios, ρ_t for r via α & β and ρ_t' for r + 1 via α & $(\beta + 1)$.

$$\rho_t = \frac{\alpha + \beta}{[(m - \alpha) + (n - \beta)]\,[\alpha(n - \beta) + \beta(m - \alpha)]}$$

$$\rho_t' = \frac{\alpha + \beta + 1}{[(m - \alpha) + (n - \beta - 1)]\,[\alpha(n - \beta) + \beta(m - \alpha) + (m - 2\alpha)]}$$

Then

$$\frac{\rho_t'}{\rho_t} = \frac{[\alpha + \beta + 1][(m - \alpha)(n - \beta)][\alpha(n - \beta) + \beta(m - \alpha)]}{[\alpha + \beta][(m - \alpha)(n - \beta - 1)][\alpha(n - \beta) + \beta(m - \alpha) + (m - 2\alpha)]}$$

$$= \frac{(m - \alpha)(n - \beta)}{(m - \alpha)(n - \beta - 1)} \cdot \left[\frac{[\alpha + \beta + 1][\alpha(n - \beta) + \beta(m - \alpha)]}{[\alpha + \beta][\alpha(n - \beta) + \beta(m - \alpha) + (m - 2\alpha)]}\right]$$

$$> \frac{[\alpha + \beta + 1][\alpha(n - \beta) + \beta(m - \alpha)]}{[\alpha + \beta][\alpha(n - \beta) + \beta(m - \alpha) + (m - 2\alpha)]}$$

$$= \frac{(\alpha + \beta)(\alpha n + \beta m - 2\alpha\beta) + [\alpha n + \beta m - 2\alpha\beta]}{(\alpha + \beta)(\alpha n + \beta m - 2\alpha\beta) + [(\alpha + \beta)(m - 2\alpha)]} .$$

Since $\alpha n \geq \alpha m$, and so $\alpha n \geq \alpha m - 2\alpha^2$, it follows that $\alpha n + \beta m - 2\alpha\beta \geq \alpha m - 2\alpha^2 + \beta m - 2\alpha\beta$ so the numerator of $\frac{\rho_t'}{\rho_t}$ is greater than the denominator, ie: $\frac{\rho_t'}{\rho_t} > 1$.

Therefore, by Cases 1 and 2, ρ is achieved when $\alpha \leq m/2$ and $r = \alpha + \beta = h$.

Case 3: $\beta > m/2$. A computation similar to the one used in Case 1 implies that the ratio is no worse (and possibly better) if β is reduced, that is if $r = h$ is achieved via $(\alpha + 1)$ & $(\beta - 1)$.

Thus one value of r^* is h, and for $r^* = h = \alpha + \beta$, $\alpha = \lfloor m/2 \rfloor$. If m + n is even, $r^* = (m + n)/2$ and $R = 1$. If m + n is odd, $r^* = (m + n - 1)/2$ and $R = (m + n - 1)/(m + n + 1)$. If m, n are both odd, $\lambda^* = (mn - 1)/2$. Otherwise $\lambda^* = mn/2$. □

All the preceding graphs achieve ρ at the maximum $R = h/h'$. The graph in Figure 3 has $p = n + 3$. Via e, $\lambda^* = 1$, $r^* = 2$, $R = 2/(n + 1)$, and $\rho = 2/(n + 1)$.

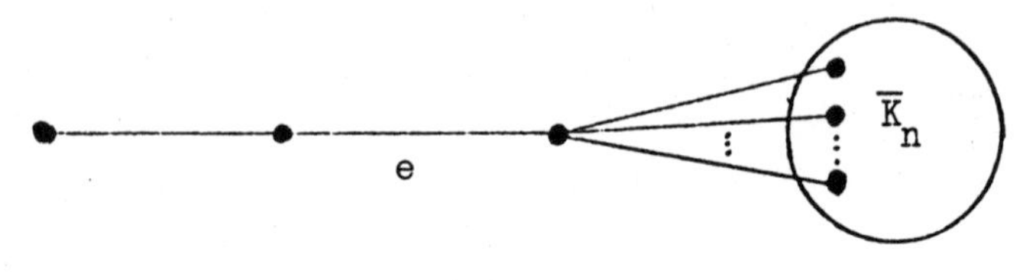

FIGURE 3

Proposition 2.4: The minimum value for ρ is $\rho(K_p) = 1/h'$. The maximum value of ρ is 1. For example, an even path has $\rho = 1$, and a graph composed of 2 large complete graphs connected by a single edge has $\rho = 1$.

Given the edge connectivity vector, $\overline{\lambda} = (\lambda_1, \lambda_2, \ldots, \lambda_h)$, it is straightforward to calculate all the ratios R/λ_r, and hence ρ. If $r^* \neq h$, then λ_{r^*+1} is defined and since $r^*/(p - r^*) < (r^* + 1)/(p - r^* - 1)$, it follows that $\lambda_{r^*+1} > \lambda^*$. That is, λ^* occurs just before a "jump" in the vector $\overline{\lambda}$. Further, suppose $\rho \leq 1/\ell$. Since $R \leq 1$ always, it follows that $\lambda^* \leq \ell$. In particular, $\lambda \leq \lambda^* \leq \ell$. That is, $\lambda \leq 1/\rho$. The graph in Figure 4 shows that there are no other restrictions between λ and ρ : For every $\lambda \leq h$, $\rho = 1/h$.

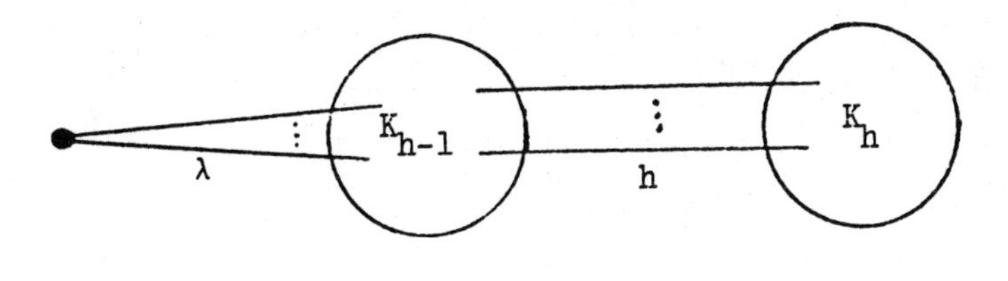

FIGURE 4

III. The Ratio of disruption

As Figure 5 shows, in general ρ can be achieved in more than one way for a given graph.

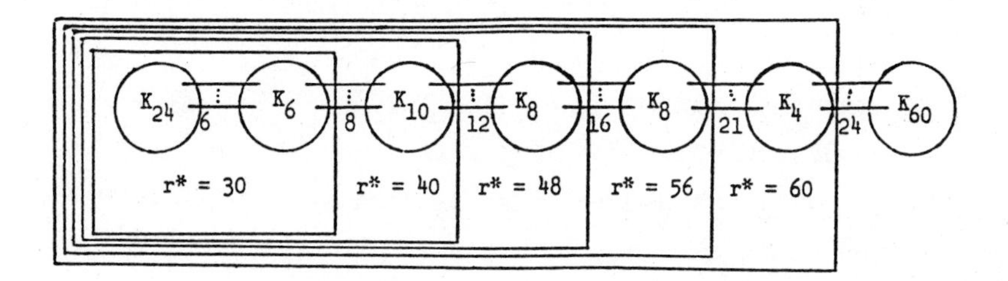

$$p = 120, \ \rho = \frac{1}{24} = \frac{24}{96 \cdot 6} = \frac{30}{90 \cdot 8} = \frac{40}{80 \cdot 12} = \frac{48}{72 \cdot 16} = \frac{56}{64 \cdot 21} = \frac{60}{60 \cdot 24}$$

FIGURE 5

For any partition achieving ρ, $V = A \cup B$, $|A| = r^*$, G is a spanning subgraph of a graph of the form in Figure 6.

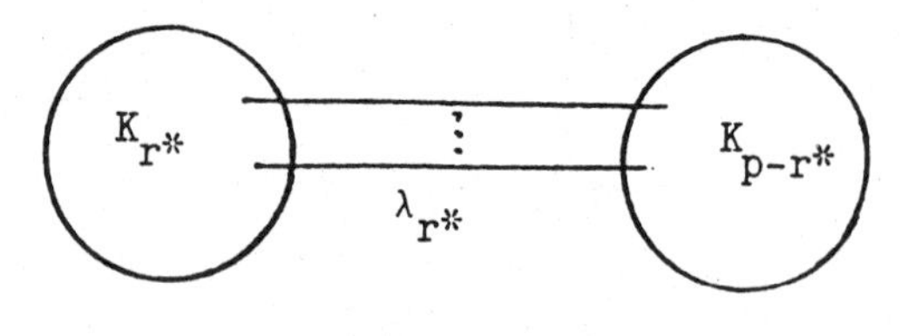

FIGURE 6

Suppose $\rho = R/\lambda^*$ where $R = m/n$. Then, from the above remark, $|E|$ is maximum for G as in Figure 6. If $p = k(m + n)$, then $|E| \leq \binom{km}{2} + \binom{kn}{2} + \lambda^*$ which simplifies to

$$|E| \leq \frac{(m^2 + n^2)p^2}{2(m + n)^2} - \frac{p}{2} + \lambda^*.$$

<u>Definition</u>: A graph G is <u>ρ-minimal</u> if for every edge $e \in E$, $\rho(G - e) > \rho(G)$.

<u>Proposition 3.1</u>: If every edge is in a set of λ^* edges, for some λ^*, then G is ρ-minimal.

proof: Obvious from the definition. □

<u>Corollary 3.1</u>: If G is edge-transitive, then G is ρ-minimal. Thus K_p, C_p, $K_{m,n}$ are all ρ-minimal. Similarly the Petersen graph is ρ-minimal. It is straightforward to compute ρ(Petersen) = 1/5 with $\lambda^* = 5$. The graph of Figure 7 is ρ-minimal even though not every edge in G is in a set of edges achieving ρ. Thus the converse of Proposition 3.1 is false.

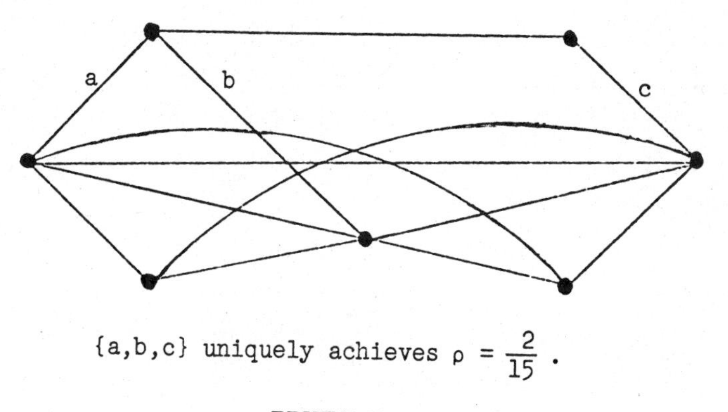

$\{a,b,c\}$ uniquely achieves $\rho = \frac{2}{15}$.

FIGURE 7

IV. Forbidden Values

There are constraints among p, ρ, and λ^*. If $p = k(m + n)$ and $\rho = m/n\lambda^*$ for particular values of λ^* and $R = m/n$, then $p = (\rho n\lambda^* + n) = kn(\rho\lambda^* + 1)$. Thus $n(\rho\lambda^* + 1)$ divides p.

Example: Suppose $\rho = 2/11$ via $\lambda^* = 3$. Then $R/\lambda^* = \rho$, so $R = \lambda^*\rho = 6/11$. Thus $p = k(m + n)$ where $m/n = 6/11$, so that 17 divides p.

Since $\lambda^* \leq 1/\rho$, there is a natural upper bound on possible values for λ^*. Further, for each possible value of λ^* there exists a necessary divisibility condition on p if λ^* is achieved.

Example continued: Suppose $\rho = 2/11$. Then $\lambda^* \leq 11/2$ so $\lambda^* \leq 5$. The following table displays the divisibility conditions imposed on p by ρ and λ^*:

λ^*	R	p is divisible by
1	2/11	13
2	4/11	15
3	6/11	17
4	8/11	19
5	10/11	21

Therefore, unless p is divisible by at least one from the set $\{13,15,17,19,21\}$, then $\rho = 2/11$ is impossible for that p.

We will denote the set of enabling divisors for ρ by $S(\rho)$. Thus the result of the preceding example is that $S(2/11) = \{13,15,17,19,21\}$.

Example: The table for $\rho = 1/6$ is as follows:

λ^*	R	p is divisible by
1	1/6	7
2	2/6 = 1/3	4
3	3/6 = 1/2	3
4	4/6 = 2/3	5
5	5/6	11
6	6/6 = 1 = 1/1	2

Since 2 divides p if 4 does, the divisibility condition imposed by 4 is redundant. Therefore $S(1/6) = \{2,3,5,7,11\}$.

Suppose that t is prime and $\rho = 1/t$. Then all of the fractions $R = s/t$ are in lowest terms for $s = 1,2,\ldots,t - 1$. Therefore $S(1/t) = \{t + 1, t + 2, \ldots, t + (t - 1), 2\}$.

A corollary to the above considerations is the following:

Proposition 4.1: If p is odd and $t \geq p$ for t prime, then $\rho(G) \neq 1/t$ for any graph G with $|V| = p$.

Suppose that $\rho = s/t$ is in lowest terms. If $s = 1$, then via $\lambda^* = t$, $R = 1$ is possible and $2 \in S(\rho)$. If $s \neq 1$, then $R = 1$ is not possible and $2 \notin S(\rho)$. That is, $2 \in S(\rho)$ if and only if $1/\rho$ is an integer.

Similarly, $3 \in S(\rho)$ if and only if $R = 1/2$ is possible. This in turn reduces to the condition $1/\rho$ is an even integer.

On the other hand, $5 \in S(\rho)$ if and only if either $R = 1/4$ or $R = 2/3$ is possible. These potential ratios reduce to the respective conditions "3 divides $1/\rho$" and "3 divides $2/\rho$."

Thus, the precise conditions that determine membership in $S(\rho)$ can in principle be computed, but are complex.

V. Conclusion

There are variations and extentions of the definition of the ratio of disruption. For example, one might consider weighted graphs and weighted ratios. For the graph of Figure 8, $\rho = 1/2$ with $R = 2/2$ and $\lambda^* = 2$. If x has weight 2 and y, z, w have weight 1, the total weight is 5, and $\rho = 1/3$ with $R = 2/3$ and $\lambda^* = 1$.

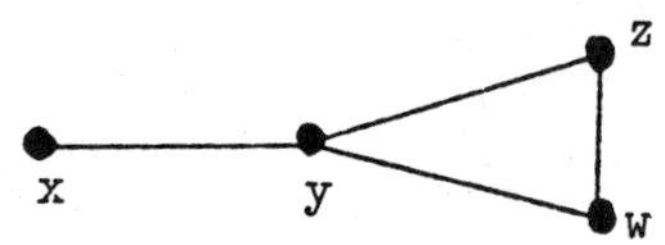

FIGURE 8

The authors would like to make a case for keeping the term "vulnerability" generic. Different applications require different specific parameters, all of which measure something approximating "vulnerability."

REFERENCES

1. Amin, Ashok T. and Hakimi, Louis "Graphs with Given Connectivity and Independence Number or Networks with Given Measures of Vulnerability and Survivability," IEEE Transactions on Circuit Theory, Vol. CT-20, No. 1 January 1973 pp. 2-10.

2. Bernard, J. C., Bond, J., Paoli, M., Peyrat, C. "Graphs and Interconnection Networks: Diameter and Vulnerability," Surveys in Combinatorics, 1983, E. Keith Lloyd, ed. London Math Society Lecture Notes, #82, Cambridge University Press (1983) pp. 1-30.

3. Boesch, F. T. "Lower Bounds on the Vulnerability of a Graph," Networks, Vol. 2(1972) pp. 329-340.

CUBIC GRAPHS AND THE FOUR-COLOR THEOREM

Feodor Loupekine
18 Bardwell Court, Oxford, OX2 6SX, England

John J. Watkins
The Colorado College, Colorado Springs, CO 80903

ABSTRACT

There have been many reformulations of the four-color theorem. In 1880 Tait showed this theorem to be equivalent to the statement that every bridgeless cubic planar graph is 3-edge-colorable. We show it to be equivalent to the stronger statement that every bridgeless cubic graph with crossing number less than 2 is 3-edge-colorable. Thus, for example, the crossing number of the Petersen graph is at least 2, since it is not 3-edge-colorable.

In 1898 Heawood recast the four-color theorem in terms of linear congruences by labeling the vertices of a cubic graph +1 or -1. We similarly cast it in terms of linear congruences by labeling the 'angles' of a cubic graph 0 or 1. In addition we study a hierarchy of coloring properties that a cubic graph may possess concerning vertex colorings and coverings by disjoint sets of even circuits.

1. Introduction

Although the four-color map problem dates from 1852, it did not become widely known until 1878 due to the efforts of Arthur Cayley. In 1879 Kempe published his 'proof' and Tait followed quickly in 1880 with a 'proof' of his own. In [8] Tait provided a necessary condition which he believed to be true. His idea was to turn a map coloring problem into an edge coloring problem for cubic graphs. The following theorem is proved in [2,pp.26-27].

<u>Theorem 1</u> (Tait). The four-color theorem is equivalent to the statement that every bridgeless cubic planar graph is 3-edge-colorable.

What Tait did not realize is that without 'bridgeless' and 'planar' this theorem is false. That 'bridgeless' is required is shown by the graph in Figure 1, and that 'planar' is required is shown by the Petersen graph in Figure 2.

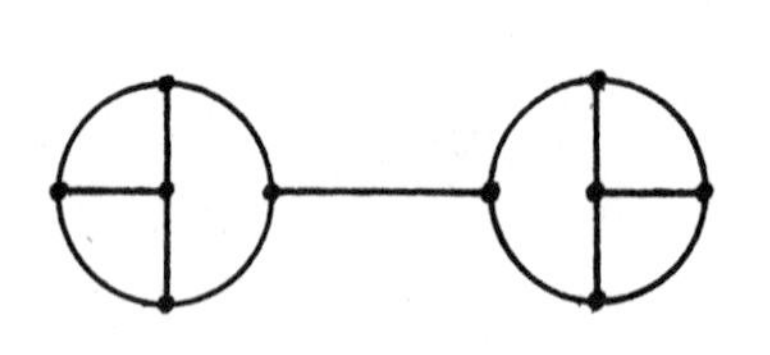

Figure 1

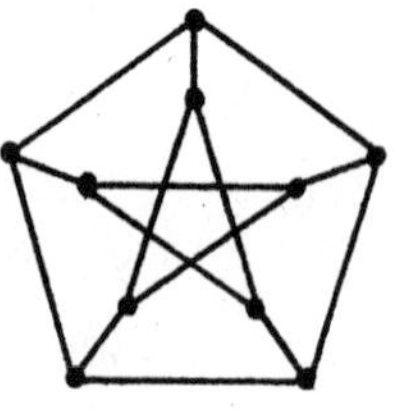

Figure 2

Heawood found the flaw in Kempe's proof of the four-color theorem in 1890 and provided his own algebraic approach in 1898 [3] by relating map coloring to solving a system of congruences modulo 3.

<u>Theorem 2</u> (Heawood). A bridgeless cubic planar graph is 3-edge-colorable if and only if its vertices can be labeled either +1 or -1 in such a way that the sum of the labels around any region is divisible by 3.

In the following theorem (see [7,pg.111]) planarity plays no role.

Theorem 3 (Tait, Petersen). A cubic graph G is 3-edge-colorable if and only if there is a set of disjoint cycles of even length which cover the vertices of G.

2. A Generalization of Tait's Theorem

An easy consequence of Theorem 3 is the parity lemma (see [1], [5]) which involves the notion of cutting edges into two pieces. This leaves two *semi-edges*, each with a vertex at one end and no vertex at the other end. A *cut* consists of a series of such 'edge-cuts' which disconnects the graph into two pieces.

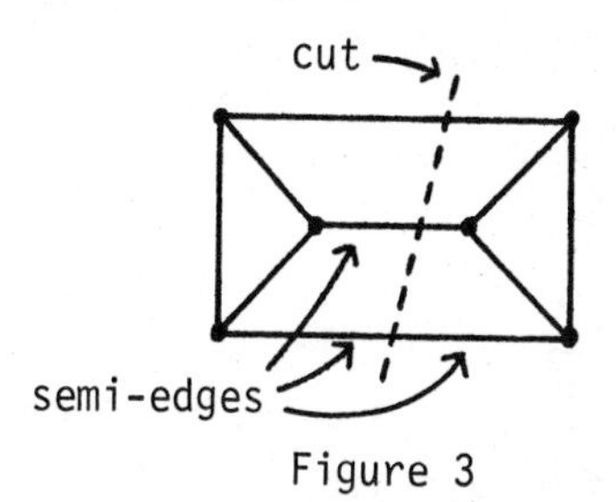

Figure 3

The Parity Lemma. Let G be a cubic graph that has been 3-edge-colored with colors 1,2 and 3. If a cut splits n_i edges of color i for i = 1,2,3, then

$$n_1 \equiv n_2 \equiv n_3 \equiv n_1+n_2+n_3 \pmod 2.$$

This follows immediately by considering the edges colored 1 and 2, similarly 1 and 3, and 2 and 3. They form a covering set of disjoint even cycles.

The case in which we are interested is when four edges have been split. Then $n_i \equiv 0 \pmod 2$, and $n_i = 0,2$ or 4 for each i. So either all four edges were previously the same color, or there were two of one color and two of another. We can now state and prove our generalization of Tait's theorem.

Theorem 4. The four-color theorem is equivalent to the statement that every bridgeless cubic graph with crossing number less than 2 is 3-edge-colorable.

Proof. Since the crossing number of a planar graph is 0, this condition is sufficient by Tait's theorem. Now assume the four-color theorem, and let G be a bridgeless cubic graph with crossing number less than 2. If G is planar, then it is 3-edge-colorable by Tait's theorem. Hence, assume that its crossing number is 1 and that G has been drawn in the plane with a single crossing; this is illustrated in Figure 4. Now insert a square at the crossing to create a planar cubic graph G*, as shown in Figure 5.

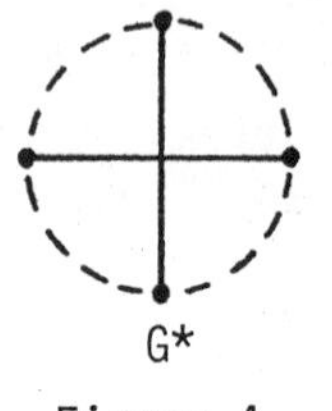

Figure 4

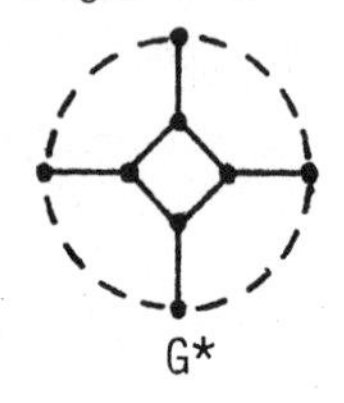

Figure 5

G* is a map, and we can color the regions of this map with four colors A,B,C and D. We can then color the edges of G* with the colors a,b and c as follows: if an edge separates colors A and D, or B and C, color it a; if an edge separates colors B and D, or A and C, color it b; if an edge separates colors C and D, or A and B, color it c. It is easy to see that this standard procedure based on the Klein four-group yields a 3-edge-coloring of G*.

By the parity lemma there are four possible arrangements of the colors of the edges joining the inserted subgraph to the rest of G*. These are shown in Figure 6.

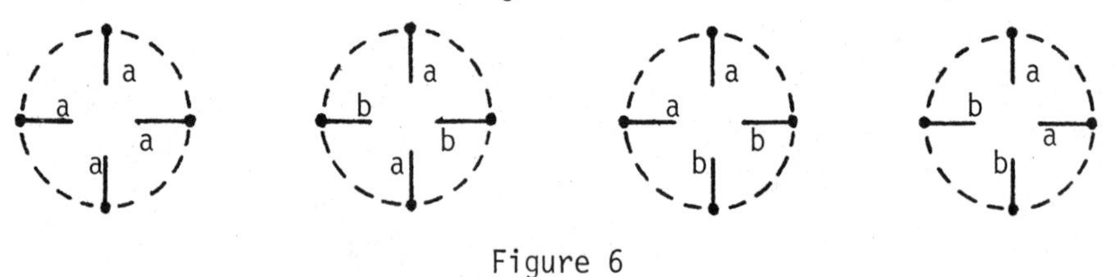

Figure 6

Of these, the first two give rise to a 3-edge-coloring of G if one simply rejoins the semi-edges to their original state. The third and fourth arrangements are equivalent. It suffices, therefore, to show that the third arrangement can be used to 3-edge-color G. By possibly renaming the colors of the regions, we may assume that the regions are colored as in Figure 7.

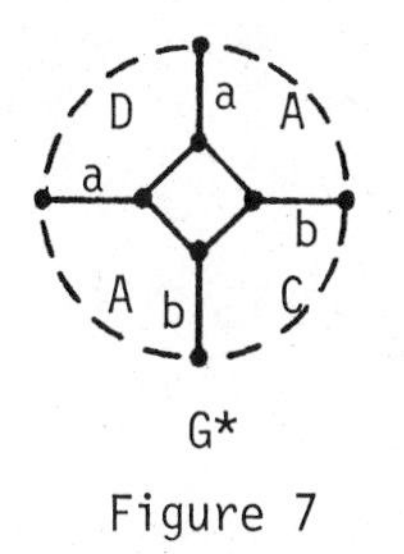

Figure 7

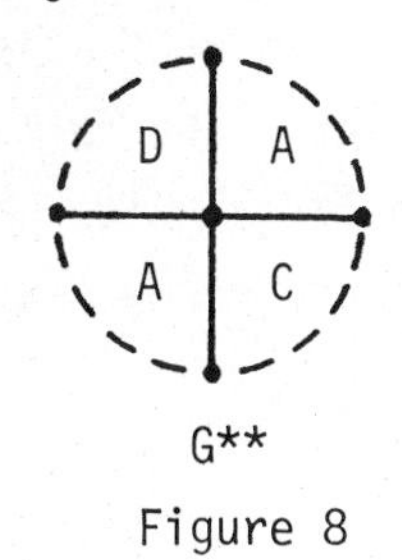

Figure 8

Next, we contract the square down to a single vertex. This gives us a graph G** together with a 4-coloring of its map, as shown in Figure 8. We now use a Kempe-chain argument. There cannot be both a D-C chain from the region colored D to the region colored C and an A-B chain between the regions colored A in G**.

If there is no D-C chain, begin at the region colored C and do a D-C color switch. The result is a 4-coloring of the map of G**, as in Figure 9. If there is no A-B chain, begin with the left-hand region colored A and do an A-B color switch. The result is shown in Figure 10.

Figure 9

Figure 10

In each of these two cases we can now use the map coloring to color the edges of G**. This is also shown, respectively, in Figures 9 and 10. With the exception of the center vertex, this gives a successful 3-edge-coloring of G**. But if we delete the center vertex and rejoin the semi-edges as they were originally, we have a 3-edge-coloring of G. This completes the proof.

3. Angle Coloring and the Y-Property

An *angle coloring* of a cubic graph is a labeling of the

'angles' of the graph with 0 or 1. Since an angle can be formally defined as a pair of adjacent edges, this makes sense even for non-planar graphs. The following theorem is in the spirit of Heawood vertex labeling. With each vertex we associate six angles, three 'interior' and three 'exterior', as in Figure 11.

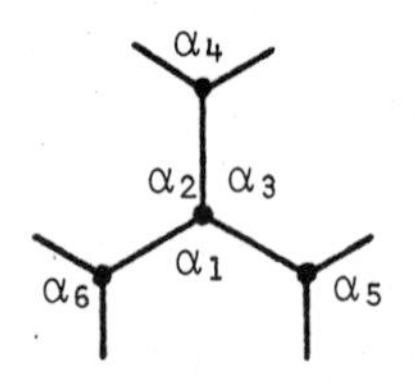

Figure 11

Theorem 5. A cubic graph is 3-edge-colorable if and only if there is an angle coloring of the graph such that at each vertex the six associated colors satisfy the congruence

$$\alpha_1 + \alpha_2 + \alpha_3 - \alpha_4 - \alpha_5 - \alpha_6 \equiv 2 \pmod 4 \quad .$$

Proof. Since the labels are restricted to 0 and 1, this means that the sum is either +2 or -2. Let G be a cubic graph that is 3-edge-colorable. By Theorem 3 there is a covering set of disjoint even cycles. For each cycle, label (in pairs) the angles with only one edge along the cycle 0 and 1 alternately, as in Figure 12.

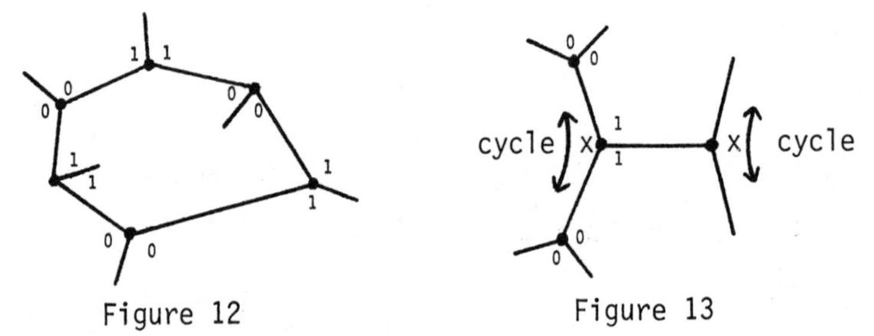

Figure 12

Figure 13

The remaining unlabeled angles come in pairs at either end of an edge joining two of the cycles. For each pair, color the two angles x, where x is either 0 and 1. This is illustrated in Figure 13 at a vertex where the first two angles were labeled 1. Here the sum is x+1+1-x-0-0 = 2. If the first two angles were labeled 0, then the sum would be x+0+0-x-1-1 = -2. Thus, we have the desired angle coloring.

To prove the converse, assume that G is a cubic graph with an angle coloring that satisfies the system of congruences. We inflate each vertex to a triangle to get a new graph G*, and use the angle coloring of G to color the vertices of G* as shown in Figure 14. Since triangles do not affect 3-edge-colorability, G is 3-edge-colorable if and only if G* is. The original congruences that hold at each vertex of G for the angles now hold at each triangle of G* for the vertices.

Figure 14

Figure 15

For a given triangle in G* there are essentially four possible labelings of the three vertices using 0 and 1. However, since we can exchange 0 and 1, we discuss only the two triangles as labeled in Figure 15.

For the first triangle in Figure 15 to satisfy the congruence, the three vertices adjacent to it must all be labeled 1 (so the sum is -2). This means that each vertex in the triangle has one neighbor that shares its label and two neighbors with the opposite label.

For the second triangle in Figure 15 to satisfy the congruence, exactly two of the vertices adjacent to it must be labeled 1. In this case we contract the triangle back to a single vertex with label 0. This vertex also has one neighbor that shares its label and two neighbors with the opposite label.

In this way we now have a graph G** which is 3-edge-colorable if and only if G is. A cubic graph is said to have the *Y-property* if its vertices can be labeled 0 or 1 so that for each vertex there is exactly one adjacent vertex with the same label. So G** has the Y-property. We are finished with our proof as soon as we show that a graph with the Y-property is 3-edge-colorable. This is the next theorem.

Theorem 6. A cubic graph with the Y-property is 3-edge-colorable.

Proof. We shall construct a covering set of disjoint even cycles. Assume that the vertices of the graph have been labeled with 0 and 1 in such a way that each vertex has exactly one neighbor with the same label. Begin at a vertex v_0 labeled, say, 0. Then proceed to an adjacent vertex labeled 1. Continue in this way to 'grow' a cycle labeled 0 and 1 alternately. The Y-property guarantees that we can always continue and that we never cross the growing path. Hence, we must finally close the cycle at v_0. This cycle is even since the 0s and 1s alternate.

We now repeat this process for any vertices not already covered by an even cycle. Again the Y-property prevents any of the cycles from intersecting. In this way, we get a covering set of disjoint even cycles and so, by Theorem 3, the graph is 3-edge-colorable.

That the Y-property is a stronger condition than 3-edge-colorability is shown by the graph in Figure 16. This graph is 3-edge-colorable, but does not have the Y-property. If it were to have the Y-property, the outer three vertices could not all have the same label. This quickly leads to a contradiction.

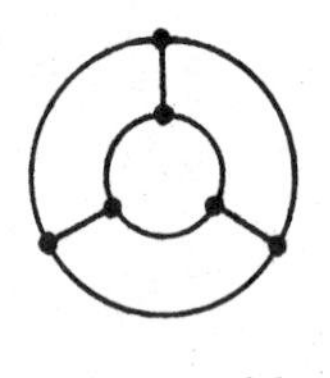

Figure 16

4. Other Related Properties

A covering set of disjoint even cycles is called *uniform* if any cycle in the graph intersects the covering set in an even number of edges.

Theorem 7. A cubic graph has the Y-property if and only if there is a uniform covering set of disjoint even cycles.

Proof. Assume that the graph has the Y-property. As in the proof of Theorem 6, construct a covering set of even cycles whose vertices are alternately labeled 0 and 1. Edges that connect the cycles have the same label at both ends. The number of changes along an arbitrary cycle between 0 and 1 is even, and these changes can occur only along a cycle from the covering set. Hence, a cycle meets the covering set in an even number of edges.

Conversely, assume that there is a uniform covering. Take one of the cycles and label its vertices alternately 0 and 1. Then take a vertex adjacent to this cycle. Label it with the same label as the vertex in the cycle to which it is adjacent. Then label its cycle with 0 and 1 alternately. Continue in this fashion until all of the vertices have been labeled. It is easy to see that the graph has the Y-property if we can show that all of the edges that are not in one of the covering cycles have the same label at each end. Suppose that we have just, for the first time, labeled a vertex that is adjacent to a vertex in another covering cycle with a different label. This gives us a cycle with exactly one change from 0 to 1 that does not occur along a covering cycle. This cycle, therefore, meets the covering cycles in an odd number of edges contrary to hypothesis. This completes the proof.

We have seen that Y-colorability is a stronger property than 3-edge-colorability. There is a still stronger property. A cubic graph is *evenly 4-vertex-colorable* if one can 4-color its vertices so that the neighbors of any vertex have three different colors. It has been pointed out by Douglas Woodall that the number of vertices in such a graph must be divisible by 4.

Theorem 8. A cubic graph that is evenly 4-vertex-colorable has the Y-property.

Proof. Suppose that the graph has been evenly 4-vertex-colored with colors A,B,C and D. Label the vertices colored A or B with 0 and those colored C or D with 1.

That the converse to this theorem is false is illustrated by the graph with 12 vertices in Figure 17. That it has the Y-property is demonstrated by the labeling. On the other hand, if we begin at any vertex with an arbitrary color and try to 4-color the vertices evenly, we soon reach a contradiction.

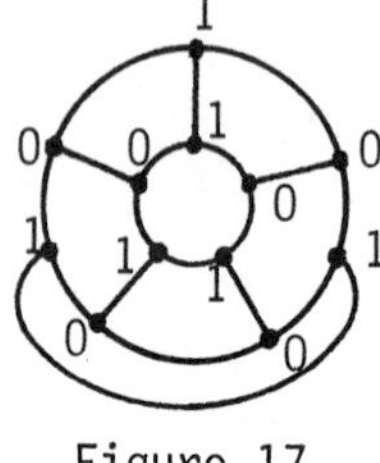

Figure 17

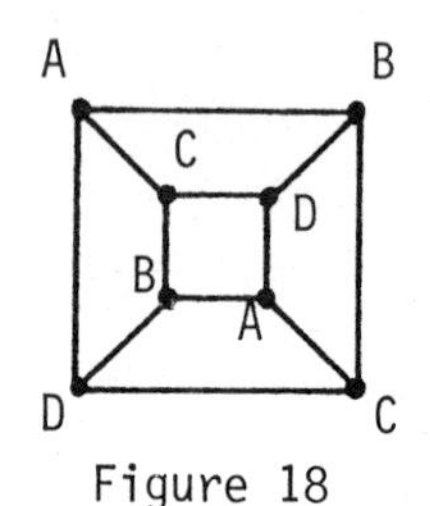

Figure 18

We now return to Heawood's vertex labeling. An easy corollary of Theorem 2 is that, if the number of edges around each region of a bridgeless planar cubic graph is divisible by 3, then it is 3-edge-colorable (label each vertex +1). This is the basis of an algorithm given by Ringel in 1959 [6] for 3-edge-coloring such a cubic graph. Begin with any edge and label the edges with colors a,b and c, always going in cyclic order clockwise around each vertex.

Theorem 9. A bridgeless cubic planar graph in which the number of edges around each region is divisible by 3 is evenly 4-vertex-colorable.

Proof. Use Ringel's procedure to 3-edge-color the graph with colors a,b and c. Color any vertex D and then use the Klein four-group to color the remaining vertices. Since the edges attached to any vertex are colored a,b and c, the three adjacent vertices must have three different colors.

The converse is false, as the graph in Figure 18 shows. Heawood continued to try to use his vertex labeling to solve the four-color problem. In 1932 he devoted a lengthy article to proving the following proposition [4]. We conclude with a short proof.

Theorem 10 (Heawood). If the vertices of a bridgeless cubic planar graph are labeled +1 or -1, it is impossible to have two adjacent regions whose boundary sum is non-zero (modulo 3) while all other regions have zero sum (modulo 3).

Proof. Suppose that this situation is possible. This is depicted in Figure 19 where the two non-zero sums are x and y. As we take the sum over all regions the total must be 0 (modulo 3), since each vertex label is counted 3 times. Hence, $x+y \equiv 0 \pmod 3$. Incidentally, this shows that there cannot be a single region with a non-zero sum.

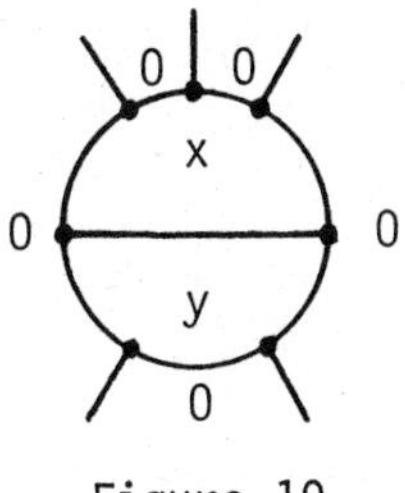

Figure 19

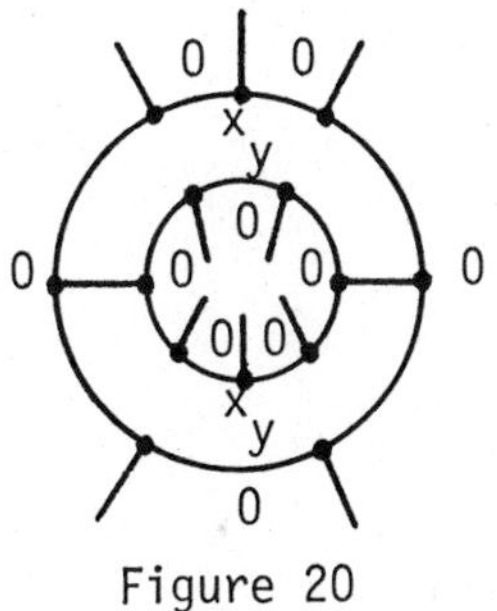

Figure 20

Now cut the center edge, turn the graph 'inside out', and splice it into the original graph upside down. This is shown in Figure 20. The boundary sum for all regions is now 0, since $x+y \equiv 0 \pmod 3$. Hence, the regions of this map can be colored with four colors, but then so can the regions of the original map. This in turn gives a 3-edge-coloring that is consistent with the labeling of the vertices with +1 or -1. We deduce that $x \equiv y \equiv 0 \pmod 3$.

ACKNOWLEDGEMENT

This paper was made possible by the kind hospitality of the Mathematical Institute, Oxford University.

REFERENCES

1. B. Descartes, Network colourings, *Math. Gazette*, 32(1948), 67-69.

2. S. Fiorini and R.J. Wilson, *Edge-Colourings of Graphs*, Research Notes in Math. 16, Pitman, London, 1977.

3. P.J. Heawood, On the four-colour map theorem, *Quart. J. Pure Appl. Math.*, 29(1898), 270-285.

4. P.J. Heawood, On extended congruences connected with the four-colour map theorem, *Proc. London Math. Soc. (2)*, 33(1932), 252-286.

5. R. Isaacs, Infinite families of nontrivial graphs which are not Tait colorable, *Amer. Math. Monthly*, 82(1975), 221-239.

6. G. Ringel, *Farbungsprobleme auf Flächen und Graphen*, Berlin, 1959.

7. T.L. Saaty and P.C. Kainen, *The Four-Color Problem*, McGraw-Hill, New York, 1977.

8. P.G. Tait, Remarks on the colouring of maps, *Proc. Roy. Soc. Edinburgh,* 10(1880), 501-503, 729.

AN EFFECTIVE APPROACH TO SOME PRACTICAL CAPACITATED TREE PROBLEMS

V. V. MALYSHKO
Byelorussian State University
Minsk, USSR

ABSTRACT

The practical task of minimizing cable connections between a central computer and a field of heliostats in the design of a solar power system can be modeled as an NP-hard case of the Geometric Capacitated Tree Problem. A polynomial approximation algorithm with a proven absolute performance ratio for the Capacitated Tree Problem has been constructed. We develop a computer package of programs based on this approximation algorithm and a heuristic procedure for local optimization. Computational results confirm the relative efficiency of this package for solving real tasks of large dimensions.

1. Introduction.

A hierarchical multilevel control system of a solar power station includes a central computer, a set of minicomputers of unknown cardinality, a set of n heliostats and a cable network connecting the above components [2, 3, 14]. In the two-level control system, for example, every heliostat is connected with a microprocessor; every microprocessor controls at most k (> 0) heliostats and must be connected with the central computer. The problem is to synthesize a network of minimum sum cost (including the cost of microprocessors and cable connections).

Let $G(V,E)$ be a complete weighted graph, in which the vertex set V models a point set of heliostats locations or computing facilities ($|V| = n + 1$), and where the edge set E consists of possible communication lines, the edge weight being equal to the distance between the corresponding points. Let the central computer be located at the vertex $v_0 \in V$. Suppose that every microprocessor of cost α controls at most k heliostats, the cost of a unit length of cable is β, and the edge $e \in E$ has the length $c(e)$. For every triple $v, u, w \in V$, the triangle inequality in Euclidean space assures us that $c(v,u) + c(u,w) \geq c(v,w)$.

Then the synthesis of the two-level system with minimum sum cost may be reduced to the construction of a spanning tree $T(V,\bar{E})$ of a graph $G(V,E)$ such that:

$$\text{if the edge } (v, v_0) \in \bar{E}, \text{ then } 2 \leq \deg v \leq k; \tag{1}$$

$$\text{if the edge } (v, v_0) \notin \bar{E}, \text{ then } \deg v = 1; \tag{2}$$

$$\beta \cdot \sum_{e \in \bar{E}} c(e) + \alpha \cdot \deg v_0 \to \min. \tag{3}$$

The structures of required spanning trees T^2 and T^3 for two- and three -level systems are shown in Figures 1 and 2.

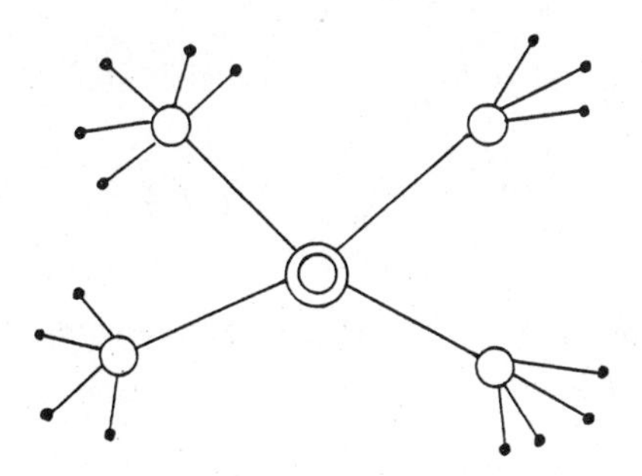

FIGURE 1.

A capacitated spanning tree T^2. The location of central computer marks by ◎ (first level), the location of microprocessors - by ○ (second level), the location of heliostats - by ● .

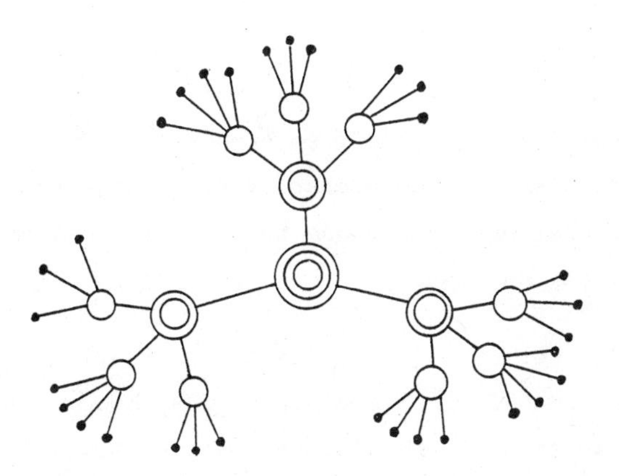

FIGURE 2

A capacitated spanning tree T^3. The location of the central computer marks by ◎ (first level), the location of microprocesors of the second level marks by ◎ , the microprocessors of the third level-by ○ , the heliostats - by ●

The second term in the sum (3) is a linear function, and

$$\left\lceil \frac{n}{k} \right\rceil \le \deg v_0 \le \left\lceil \frac{n}{2} \right\rceil .$$

So the optimization problem (1-3) can be reduced to the sequential construction of spanning trees under conditions (1,2), and objective function

$$\sum_{e \in \bar{E}} c(e) \to \min \qquad (4)$$

and $\deg v_0 = \left\lceil \frac{n}{k} \right\rceil, \left\lceil \frac{n}{k} \right\rceil + 1, \cdots, \left\lceil \frac{n}{2} \right\rceil$. Then we must choose the tree $T^*(V,\bar{E}^*)$ which satisfies condition (3).

2. Problem Complexity.

The problem (1, 2, 4) is a partial case of the NP-hard Geometric Capacitated Spanning Tree Problem [7].

Define the dissociation number dis(T) of a tree T as the smallest number of nodes whose removal produces only isolated vertices and edges.

THEOREM 1 [12]. Let $D(n) = \max\{dis(T)\}$ for all trees with at most n vertices. If $D(n) \ge c \cdot n^{\varepsilon}$ for some $c, \varepsilon > 0$, then the construction of the minimum spanning tree T is an NP-hard problem.

A tree T^2 may be divided into isolated vertices by removing vertices, adjacent to v_0. The number of such vertices is at least $p = \left\lceil \frac{n}{k} \right\rceil$. A tree T^3 may be decomposed by removing v_0 and vertices adjacent with the leaves of T^3. In both cases the number of removable vertices is equal to c n. Thus the problem (1, 2, 4) is NP-hard. It is clear that this problem for a multilevel system is NP-hard as well. The complexity of analogical problems for different objective functions was studied in [5].

3. Algorithms for Finding Minimum Spanning Trees with Degrees Restrictions.

In this section we briefly discuss three alternative approaches to the problem (1, 2, 4):

1) an exact method which reduces the problem to a $p = \left\lceil \frac{n}{k} \right\rceil$-median problem for a weighted graph [6] and solves it by using the simplex method and brand-and-bound;
2) an approximate polynomial algorithm on the base of reduction of the considered problem to the p-median problem on the minimum spanning tree of the initial graph [13];
3) an approximate algorithm of local optimization [15].

ALGORITHM 1

The problem (1, 2, 4) can be formulated as a binary integer linear program [4]. Let

$$x_{ij} = \begin{cases} 1, & \text{if } (v_i, v_j) \in \bar{E}, \\ 0, & \text{otherwise} \end{cases}$$

and $c_{ij} = c(v_i, v_j)$. Then the problem is:

$$\sum_{j=1}^{n} \sum_{i=0}^{j-1} c_{ij} \cdot x_{ij} \to \min \tag{5}$$

$$\sum_{i=j+1}^{n} x_{ji} + \sum_{i=1}^{j-1} x_{ij} \geq 1, \quad j = 1, 2, \cdots, n; \tag{6}$$

$$\sum_{i=j+1}^{n} x_{ji} + \sum_{i=0}^{j-1} x_{ij} \leq k, \quad j = 1, 2, \cdots, n; \tag{7}$$

$$\sum_{i=1}^{n} x_{io} = p; \tag{8}$$

$$x_{ij} \leq x_{io}, \quad i = 1, 2, \cdots, n; \quad j = 1, 2, \cdots, n; \tag{9}$$

$$x_{ij} \in \{0,1\}, \; i = 1, 2, \cdots, n; \quad j = 1, 2, \cdots, n. \qquad (10)$$

Problem (5-10) is a binary integer linear program with n^2 variables. It can be solved exactly by a directed branch and bound procedure, or by the simplex method with relaxation of the constraint (10) to $0 \le x_{ij} \le 1$. When n is large, then both methods may prove inefficient in finding a solution. We used the simplex-program [10] only for solving the problems of small dimensions with the purpose of comparison of the optimal solution with the approximated one, which can be found by any of the next heuristic algorithms.

ALGORITHM 2

The main idea of this Algorithm consists of constructing the minimum spanning tree $T'(V,E')$ [13]; approximate solving of the p-median problem on a tree $T'(V,E')$; connecting every vertex-median with at most k leaves on the base of postorder traversal of $T'(V,E')$ [1].

Step 1. Construct a minimum spanning tree T' of the graph G using Prim's algorithm. Let vertex v_0 be the root of T'.

Step 2. Employing the postorder traversal of T', assign to each vertex $v_i \in T'$ the weight $W(v_i)$ which equals the number of its descendants. Allocate the median v_0^* in the vertex v_0. Let $\ell = 1$ and $S = \phi$.

Step 3. Remove the edges adjacent to v_ℓ^* and let $v_{j1}, v_{j2}, \cdots, v_{jt}$ be the other end-vertices of these edges. Include the set of subtrees $T_1, T_2, \cdots, T_t$ with roots $v_{j1}, v_{j2}, \cdots, v_{jt}$ in S.

Step 4. Choose the subtree T_0^* from S such that:

$$W(v_{j0}) = \max_{T_i \in S} \{W(v_{ji})\}.$$

Let $S = S - T_0^*$.

Step 5. In a subtree T_0^* find the vertex v_γ such that

$$\Delta(v_\gamma) = \min_{v_i \epsilon T_0^*} \{\Delta_{(v_i)}\}, \text{ where}$$

$$\Delta(v_i) = |2\, W(v_i) - W(v_{j0})|.$$

Then v_γ is the vertex with approximately equal numbers of ancestors and descendants.

Step 6. Allocate the median $v_{\ell+1}^*$ in the vertex v_γ. If $p > 1$, then let $\ell = \ell + 1$ and return to Step 3.

Step 7. Let $\{v_1^*, v_2^*, \cdots, v_p^*\}$ be the set of allocated medians. Again, using postorder traversal of T', connect any noncentral median with any vertex v_j which has not yet been considered and is a descendant of that median.

Step 8. Connect any vertex not yet considered with the nearest non-central median which has less than k adjacent vertices. Connect all noncentral medians $v_1^*, v_2^*, \cdots, v_p^*$ with the central median v_0^* and stop.

THEOREM 2 [9]. The time complexity of Algorithm 2 is $O(n^2)$.

It should be noted that the algorithms for finding a perfect matching in bipartite graphs can be used for the realization of Steps 7 and 8 with time complexity $O(n^{5/2})$.

Let T be the set of all possible minimum spanning trees for the given minimization problem (1, 2, 4) whose instance is denoted by I and

$$\rho_{min} = \min_{(v_i,v_j)\varepsilon T'\varepsilon \mathcal{T}}, \qquad \rho_{max} = \max_{(v_i v_j)\varepsilon T'\varepsilon \mathcal{T}} \{c_{ij}\},$$

$$c_{max} = \max_{v_i,v_j\varepsilon V} \{c_{ij}\}, \qquad c_{min} = \min_{v_i,v_j\varepsilon V} \{c_{ij}\} .$$

Let OPT(I) and A(I) be the values of optimal and approximate solutions for instance I respectively. Then the absolute performance ratio [7] for the approximate algorithm A is determined by the following formulas:

$$R_A = \inf\{r \geq 1: \ R_A(I) = \frac{A(I)}{OPT(I)} \leq r \ \text{ for all instances I}\}.$$

THEOREM 3 [9]. When the points of the set V are uniformly distributed on the plane, the absolute performance ratio of Algorithm 2 is

$$R_2 = \frac{\rho_{max}}{\rho_{min}} \cdot \frac{(k+1)^2}{2k} . \tag{11}$$

ALGORITHM 3

This algorithm is a modified procedure of local optimization 15. The algorithm proceeds by starting with a given set of p-meidans and vertex assignments to those medians. The main phase of the algorithm is to determine whether switching of two vertex assignments can lead to an overall reduction in the total distance. Algorithm 3 executes all possible switchings and can involve as many steps as the number of possible median sets, namely $\binom{n}{p}$. Thus Algorithm 3 is not polynomially bounded.

THEOREM 4. The absolute performance ratio of Algorithm 3 is

$$R_3 = 1 + \frac{c_{max} - c_{min}}{\rho_{min}} . \tag{12}$$

Proved estimations of the time complexity and accuracy of Algorithms 2 and 3 and the analysis of their average case behaviors (see Section 4) give the possibility of determining the structure of the developed computer program package for the problem solution (1, 2, 4). The initial solution is found on the base of Algorithm 2. This solution is defined more precisely by Algorithm 3.

4. Computational Results.

FORTRAN codes were written to perform Algorithms 2 and 3. Processing was done on the University of Michigan's Amdahl 470/v8 computer (see Table 1) and the Institute of Mathematics Byelorussian USSR Academy of Science EC-1060 computer (see Table 2).

PROBLEM DIMENSIONS	ALGORITHM 1			ALGORITHM 2			ALGORITHM 3		
	CPUs	ITERATIONS	VALUE A_1	CPUs	VALUE A_2	$\frac{A_2}{A_1}$	CPUs	VALUE A_3	$\frac{A_3}{A_1}$
n = 28, p = 3	35,9	618	5783,0	0,15	7469	1,29	0,32	6788	1,18

PROBLEM DIMENSIONS	c(T')	ALGORITHM 2			ALGORITHM 3		
		CPUs	VALUE A_2	$\frac{A_2}{c(T')}$	CPUs	VALUE A_3	$\frac{A_3}{c(T')}$
n = 100, p = 5	6844,6	4,1	24437	3,57	10,3	21612	3,16

TABLE 1.

PROBLEM NUMBER	$p=\lceil n/k \rceil$	ALGORITHM 2		ALGORITHM 3		$\frac{A_2}{A_3} \cdot 100\%$
		VALUE A_2	$\frac{A_2}{c(T')}$	VALUE A_3	$\frac{A_3}{c(T')}$	
1.	5	26289,4	2,98	25920,6	2,94	1,4
2.	6	25794, 9	2,93	24974,9	2,83	3,3
3.	7	26218,4	2,97	24319,3	2,75	7,2
4.	8	25776,1	2,91	22875,4	2,58	11,3
5.	9	24672,2	2,79	21518,3	2,43	12,8
6.	10	23206,5	2,63	19685,7	2,22	15,3

TABLE 2.

The vertices of the graph $G(V,E)$ have randomly been located uniformly in the field 1000 meters by 1000 meters. For the case $n = 28$, $k = 10$ (i.e., $p = 3$), ten problems were solved by Algorithms 1, 2, 3 consecutively. For the case $n = 100$, $k = 20$ (i.e. $p = 5$), the realization cost of Algorithm 1 was expensive. So the values of the approximate solutions A_2 and A_3 due to Algorithms 2 and 3 were compared with the length of the minimum spanning tree $c(T')$ instead of the linear program solution. Table 1 contains average characteristics for Algorithms 1, 2, 3 on ten problems. For both cases the average accuracy of Algorithm 3 was better than for Algorithm 2, but its time requirements was greater.

Table 2 contains average accuracy (on three problems) of Algorithms complex (Algorithm 2 plus Algorithm 3) for the case $n = 100$, $p = 5, 6, \cdots, 10$ and $c(T') = 8880.9$.

Computational results corroborate the theoretical formula (11) of accuracy dependence on the value p.

5. Conclusions.

The problem of synthesis of the capacitated minimum spanning trees in the design of a solar power system has been presented. The developed approach to this NP-hard problem provides a good approximate solution and acceptable computer time expenditure. Theoretical analysis and computer experiments corroborated the possibility of using our methods for solving problems of large dimensions.

Acknowledgement. The author wishes to thank Professor J. R. Birge (Department of Industrial and Operations Engineering of the University of Michigan) for his participation in the initial stage of this work.

REFERENCES

1. A.V. Aho, J. E. Hopcroft and J. D. Ullman, The Design and Analysis of Computer Algorithms. Addison-Wesley, Reading, Mass. (1974).
2. G. E. Ahromenko, et al. The control of a field of heliostates of solar power plants. Heliotechnics (USSR) 4(1980) 16-22.
3. J. R. Birge, V. V. Malyshko and V. A. Pediko, Structure optimization in control system for the heliostats field of power station. Heliotechnics (USSR) (1984) to appear.
4. J. R. Birge and V. V. Malyshko, Methods for a network design problem in solar power systems. J. Comput. Operat. Res. (1984) to appear.
5. P. M. Camerini, G. Galbiati and F. Maffioli, Complexity of spanning tree problems: Part 1. European J. Operat. Res. 5(1980) 346-352.
6. N. Christofides, Graph Theory, An Algorithmic Approach. Academic Press, New York (1977).
7. M. R. Garey and D. S. Johnson, Computer and Intractability: A Guide to the Theory of NP-Completeness. Freeman, San Francisco (1979).

8. V. V. Malyshko and N. G. Shebeko, Computer programs package for optimal planning of a communication network in a solar power station. VIII USSR Symposium on Software Systems for Solving Optimal Planning Problems, Narva (1984) 151-152.
9. V. V. Malyshko, An approximate polynomial algorithm for construction capacitated spanning trees in large graphs. Izv. Akad. Nauk SSSR Tehn. Kibernet. (Moskow) (1985) to appear.
10. B. A. Murtagh and M. A. Saunders, MINOS User's Guide, Systems Optimization Laboratory. Technical Report 77-9, Stanford (1977).
11. C. H. Papadimitriou, The complexity of the capacitated tree problem. Report TR-21-76, Center for Research in Computing Technology, Harvard University, Cambridge (1976).
12. C. H. Papadimitriou and M. Yannakakic, The complexity of restricted spanning tree problems J. Assoc. Comput. Machinery 2 (1980) 285-309.
13. R. C. Prim, Shortest connection networks and some generalizations. Bell. Syst. Tech. J. 36 (1957) 1389-1401.
14. O. T. M. Smith and P. S. Smith, A helioelectric farm, solar energy: international progress (T. N. Veziroglu, ed.) Pergamon Press, New York (1980) 1368-1407.
15. M. B. Teitz and P. Bart, Heuristic methods for estimating the generalized vertex median of a weighted graph. Operations Research 16 (1968) 955-961.

CONCURRENT FLOW AND CONCURRENT CONNECTIVITY IN GRAPHS

David W. Matula
Southern Methodist University

ABSTRACT

A concurrent flow function $f:P \to [0,1]$ of throughput z on the paths of a graph $G = (V,E)$ has $\sum f(\{p|p$ has end vertices $i,j \in V\}) = z$ for every vertex pair $i,j \in V$ and $\sum f(\{p|p$ is incident to $e \in E\}) \leq 1$ for every edge $e \in E$. The concurrent connectivity $\zeta(G)$ is the maximum throughput of any concurrent flow in G. We derive bounds on $\zeta(G)$ sufficient to compute this parameter for several classes of graphs including K_n, C_n, P_n, $K_{i,j}$, regular k-partite graphs, and trees. We describe a Mengerian type path/cut duality relating concurrent connectivity to a critical k-partite cut dually characterized by a maximum elongation of distance across the cut.

I. Introduction and Summary.

Concurrent flow in networks is readily illustrated by problems such as traffic flow in road networks and message transfer in packet-switched telecommunications networks. For a graph model of these problems the vertices represent entities (cities, computers) in which there exists a flow of traffic between all pairs of vertices that must be hosted concurrently. The edges correspond to traffic channels (roads, communication lines) which are capacitated, and paths represent potential traffic routings for flow between the corresponding end vertices.

For the graph model described herein we assume all edges have the same capacity (not unrealistic for many communication problems) and that the concurrent flows provide the same level of traffic between each pair of vertices concurrently. The general network problem allowing varying capacities over edges and differing traffic flow between vertex pairs is developed in our more comprehensive paper [Ma85]. The restriction to graphs succinctly illustrates the combinatorial structure of problems related to concurrent flow and allows us both to define the concept of the "concurrent connectivity" of a graph and formulate a Mengerian type concurrent-flow/separating-edge-set dual characterization of concurrent connectivity.

To describe a concurrent flow we assign a fractional flow to each path of the graph G so that the sum of the flows on all paths between each and every pair of end vertices is the same value (termed the "throughput") where the sum of flows on all paths containing any given edge is at most unity. Figure 1 illustrates a concurrent flow for each of four different graphs. Paths with nonzero flows are tabulated.

The parameter $\zeta(G)$ is introduced as the maximum throughput of concurrent flow in a graph G and termed the concurrent connectivity of G. In section II we provide an alternative combinatorial multigraph interpretation of $\zeta(G)$; specifically, $\zeta(G)$ is equal to the largest ratio k/m, for which the multigraph formed by taking m distinct copies of each edge of the n vertex graph G possesses a set of $k\binom{n}{2}$ edge-disjoint paths where exactly k paths have the same end vertices for each of the $\binom{n}{2}$ vertex pairs of G. We prove two upper bounds on $\zeta(G)$ sufficient to compute this parameter for many classes of graphs including complete graphs, paths, cycles, cubes, trees,

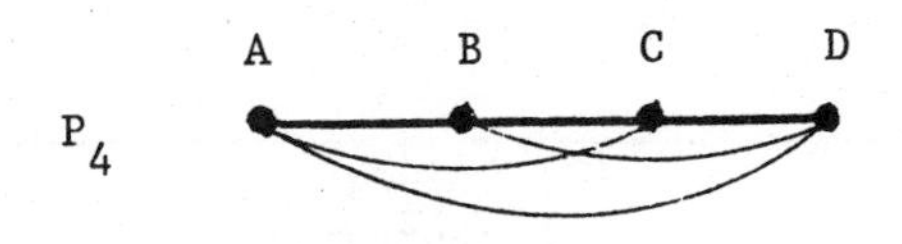

AB,BC,CD,ABC,BCD,ABCD @1/4

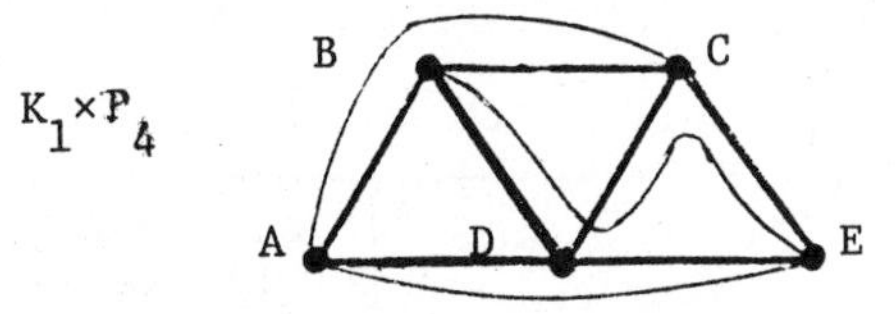

AB,AD,BC,BD,CD,CE
DE,ABC,ADE,BDCE @1/2

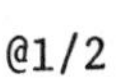

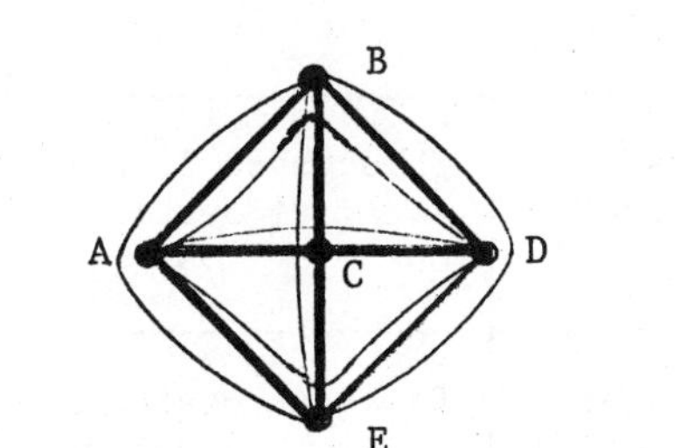

AB,AC,AE,BC,BD,CD,CE,DE @2/3
ABD,AED,BAE,BDE @1/6
ACD,BCE @1/3

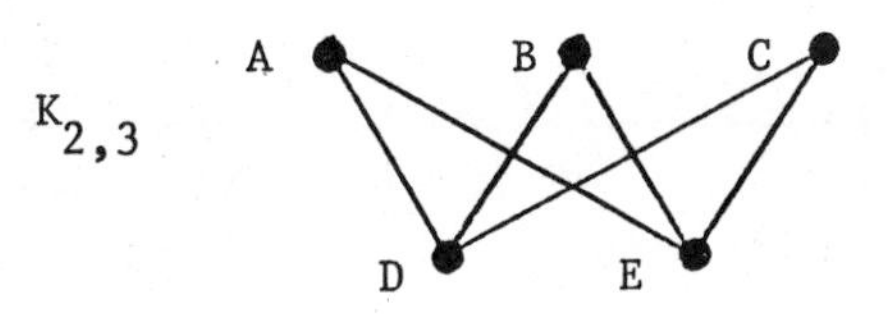

AD,AE,BD,BE,CD,CE @3/7
ADB,ADC,AEB,AEC,BDC,BEC @3/14
DAE,DBE,DCE @1/7

Figure 1. Concurrent flows yielding throughputs of 1/4 for P_4, 1/2 for $K_1 \times P_4$, 2/3 for W_5, and 3/7 for $K_{2,3}$.

complete bipartite and regular complete k-partite graphs.

In section III we introduce the notion of the "elongation" of a distance function. We describe the "maximum concurrent flow/ maximum elongating distance" duality theorem and finally the constraining concurrent-flow-critical k-partite cut of a graph characterized by this duality theory. Our results on duality for concurrent flow theory are summarized in Table 1 by analogue to important aspects of the elegant single commodity network flow theory originated by Ford and Fulkerson [FF62]. This paper contains a few proofs of certain aspects of the theory sufficient to illustrate the development. The reader is referred to our more comprehensive paper [Ma85] for full proofs in the generalized context of concurrent flow in networks.

Flow Type / Properties	Single Commodity (Source-Sink) Flow	Concurrent (All Pairs) Flow
Objective	Maximize value of source-to-sink flow	Maximize throughput of concurrent flow
Cut Inequality	Flow value less than or equal to capacity of any cut	Throughput less than or equal to density of any cut
Critical Saturated Edges	Edges saturated by every maximum flow contain a minimum cut	Edges saturated by every maximum concurrent flow constitute a k-partite cut
Duality Theorem	The maximum value of any flow equals the minimum capacity of any cut	The maximum throughput of any concurrent flow equals the reciprocal of the maximum elongation of any distance
Critical Cut Dual Characterization	A cut exists with i) all edges saturated by every maximum flow ii) minimum capacity	A k-partite cut exists with i) all edges saturated by every maximum concurrent flow ii) some maximum elongating distance ascribes non-zero distance to precisely the edges of this cut
Proof/Algorithm Paradigm	Flow augmentation proves duality and provides an efficient algorithm	Flow swapping proves duality and provides an efficient algorithm

Table 1: Analogous results in the duality theories of single commodity network flow [FF62] and concurrent flow.

II. Concurrent Flow and Sparsest Cuts in Graphs

Let P denote the set of all non-trivial paths of the graph $G=(V,E)$. For every $e \in E$, P_e shall denote the set of all paths containing the edge e; for every distinct pair $i,j \in V$, P_{ij} shall denote the set of all paths between the end vertices i,j. Each path $p \in P$ is then a member of the unique P_{ij} determined by its end vertices, and p is also contained in k of the sets P_e when p has length k in G. Our goal is to assign a fractional flow to each path of the graph G so that the minimum sum of the flows on all paths between any pair of end vertices is as large as possible subject to the following condition. The concurrent hosting of all flow should result in no more than a unit sum for the flows on all paths passing through any particular edge.

A concurrent flow function $f:P \to [0,1]$ of throughput z in the graph $G=(V,E)$ associates a real valued flow $0 \le f(p) \le 1$ with every path p of the graph G such that $\sum_{p \epsilon P_{ij}} f(p) = z$ for all distinct end-vertex pairs $i,j \in V$, where $\sum_{p \epsilon P_e} f(p) \le 1$ for every edge $e \in E$. The problem of determining the largest throughput achievable by any concurrent flow function in a graph G is then expressed by the following linear program.

Maximum Concurrent Flow Problem (MCFP)

Given the graph $G=(V,E)$ with variables $f(p) \ge 0$ for each path $p \in P$ and the unrestricted variable z, determine the maximum value of z subject to the constraints:

$$
\begin{aligned}
z - \sum_{p \epsilon P_{ij}} f(p) &= 0 \qquad \text{for all distinct } i,j \in V, \\
\sum_{p \epsilon P_e} f(p) &\le 1 \qquad \text{for all } e \in E .
\end{aligned}
\tag{1}
$$

The Maximum Concurrent Flow Problem (MCFP) for a graph G shall mean the linear program (1), which has a solution for any graph G since $z = 0$, $f(p) = 0$ for all p, is always feasible and $z \leq 1$. For any optimal solution $\hat{f}:P \to [0,1]$, $\hat{z}$ of the MCFP for G, $\hat{f}$ is termed a <u>maximum concurrent flow</u> function for G, and the maximum value $\hat{z}$ of the throughput is termed the <u>concurrent connectivity</u>, $\zeta(G)$, of G.

<u>Lemma 1</u>: (Rationality Lemma) There exists a rational-valued maximum flow function for any graph G, and furthermore:

(i) $\zeta(G) = 0$ iff G is disconnected or trivial,

(ii) $\zeta(G) = 1$ iff G is a non-trivial complete graph,

(iii) $\zeta(G)$ is a rational value with $0 < \zeta(G) < 1$ whenever G is connected but not complete.

<u>Proof</u>: The linear program MCFP has all coefficients either zero or unity and has the feasible solution $z = 0$, $f(p) = 0$ for all $p \in P$. Then by linear programming theory [DA63] there exists an optimal solution where all variables, $\hat{f}(p)$ for all p and $\hat{z}$ are rational. Conditions (i), (ii), (iii) follow immediately from consideration of the path set P of the respective graphs. □

In the single commodity network flow theory of Ford and Fulkerson the Integrity Theorem (see p. 19 of [FF62]) states that if the capacities of the edges of a network are integral-valued, then there exists an integral-valued maximum (source-to-sink) flow function with an integral-valued maximum flow value v. If the integral-capacitated network is replaced by a multigraph having c(e) distinct edges in place of each edge of capacity c(e), then the Integrity Theorem allows us to identify the maximum of v edge-disjoint paths between the source and sink vertices of the corresponding multigraph. Furthermore, when the network capacities are all unity so that the corresponding multigraph is simply a graph, the maximum flow value v then corresponds by Menger's Theorem to the edge connectivity of the graph. There is an analogous multigraph interpretation of a maximum concurrent flow function that provides an alternative combinatorial graph-theoretic definition of the concurrent connectivity $\zeta(G)$.

The graph $G = (V,E)$ is termed _(k,m)-concurrently connected_ if the multigraph $G^{[m]}$ formed by taking m distinct copies of each edge of E possesses a set $P^{[m]}$, termed a _(k,m)-connecting set_, of $k\binom{|V|}{2}$ edge-disjoint paths with exactly k of the paths having end vertices i,j for each distinct vertex pair $i,j \in V$.

Lemma 2: For any graph G,

$$\zeta(G) = \max\left\{\frac{k}{m} \,\middle|\, G \text{ is } (k,m)\text{-concurrently connected}\right\}. \tag{2}$$

Proof: Let $P^{[m]}$ be a (k,m)-connecting set for the graph $G = (V,E)$. If j paths of $P^{[m]}$ traverse the same vertex sequence (using distinct copies of the multiple edges), then assign the flow $f(p) = j/m$ to the corresponding path $p \in P$ of G. The resulting flow function $f:P \to [0,1]$ is feasible for the MCFP yielding throughput $z = k/m$, so $k/m \le \zeta(G)$.

Let $\hat{f}:P \to [0,1]$, $\zeta(G)$ be an optimal solution to the MCFP for $G = (V,E)$. By Lemma 2 we may assume $\hat{f}$ is rational-valued, so let $\hat{f}(p) = i_p/j_p$ with $\gcd(i_p,j_p) = 1$ for every $p \in P$. Let m be the least common multiple of the set $\{j_p\}_{p\in P}$, so then $m\hat{f}(p)$ is integral for all $p \in P$, and from (1) $m\zeta(G)$ is also integral. Form the $(m\zeta(G),m)$-connecting set $P^{[m]}$ for the multigraph $G^{[m]}$ by including $m\hat{f}(p)$ edge disjoint paths (using distinct copies of the edges along the path) in $P^{[m]}$ for every path $p \in P$ of the underlying graph G. This demonstrates that G is $(m\zeta(G),m)$-concurrently connected, verifying that the maximum in (2) is attained. □

Analogous to the well-known cut capacity bound on maximum flow in single commodity network flow theory [FF62], we shall derive a cut inequality bound on maximum concurrent flow that will enable us to determine $\zeta(G)$ for many classes of graphs without requiring any direct solutions of the rather large linear program MCFP.

For the graph $G=(V,E)$, let $(A,\bar{A})$ denote the set of all edges of E having one end vertex in the nonvoid proper subset $A\subset V$, and the other end vertex in $\bar{A}=V-A$.

Lemma 3: For any graph G, the concurrent connectivity $\zeta(G)$ satisfies

$$\zeta(G) \le \frac{|(A,\bar{A})|}{|A||\bar{A}|} \quad \text{for any cut } (A,\bar{A}) \text{ of } G. \tag{3}$$

Proof: Let $\hat{f}$ be a maximum concurrent flow in $G=(V,E)$. For any cut $(A,\bar{A})$ of G, let $P_A \subset P$ be the set of paths of G having one end vertex in A and the other end vertex in $\bar{A}$. Then since

$\sum_{p\varepsilon P_{ij}} \hat{f}(p) = \zeta(G)$ for any $i \neq j \varepsilon V$,

$$\sum_{p\varepsilon P_A} \hat{f}(p) = \sum_{i\varepsilon A, j\varepsilon \bar{A}} \left(\sum_{p\varepsilon P_{ij}} \hat{f}(p) \right) = \sum_{i\varepsilon A, j\varepsilon \bar{A}} \zeta(G) = |A||\bar{A}|\zeta(G) .$$

Now every path $p\,\varepsilon\,P_A$ contains at least one edge of the cut $(A,\bar{A})$, so also

$$\sum_{p\varepsilon P_A} \hat{f}(p) \le \sum_{e\varepsilon(A,\bar{A})} \left(\sum_{p\varepsilon P_e} \hat{f}(p) \right) \le \sum_{e\varepsilon(A,\bar{A})} 1 = |(A,\bar{A})| ,$$

and we obtain (3). □

The right-hand side of inequality (3) will be termed the density of the cut $(A,\bar{A})$ in G, denoted $\text{den}(A,\bar{A})$, so

$$\text{den}(A,\bar{A}) = \frac{|(A,\bar{A})|}{|A||\bar{A}|} . \tag{4}$$

A cut $(A,\bar{A})$ of minimum density in G will be termed a sparsest cut of G. If we display the adjacency matrix for G in block diagonal form as in Figure 2, $\text{den}(A,\bar{A})$ then corresponds to the relative number of nonzero entries in (each of) the off diagonal shaded blocks of Figure 2.

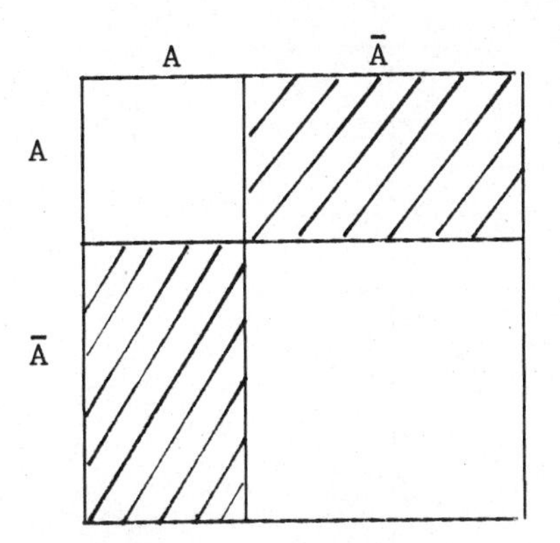

Figure 2: Parition of adjacency matrix by the cut $(A,\bar{A})$

Verification that a specific concurrent flow is maximum for a graph G is often derived from the following observation.

Observation: Suppose some concurrent flow $f:P \to [0,1]$ is demonstrated and a cut $(A,\bar{A})$ identified where every path $p \in P$ with $f(p) > 0$ crosses $(A,\bar{A})$ at most once, and where $\sum_{p\in P_e} f(p) = 1$ for each edge in $(A,\bar{A})$. Then for the corresponding throughput z we obtain $z|A||\bar{A}| = |(A,\bar{A})|$, so by Lemma 3 we conclude that f is a maximum concurrent flow, $(A,\bar{A})$ is a sparsest cut, and $\zeta(G) = z = \text{den}(A,\bar{A})$.

Employing this observation in the examples of Figure 1, note that $\text{den}(\{A,B\},\{C,D\}) = 1/4$ for the graph P_4, $\text{den}(\{A\},\{B,C,D,E\}) = 1/2$ for the graph $K_1 \times P_4$, and $\text{den}(\{A,B\},\{C,D,E\}) = 2/3$ for the wheel W_5, verifying that the corresponding throughputs 1/4, 1/2, and 2/3 are each maximal. The observation is not sufficient for $K_{2,3}$, where by enumeration the sparsest cut has density 1/2, strictly greater than the throughput 3/7 illustrated in Figure 1. In fact $K_{2,3}$ is the smallest graph (and only graph on five or fewer vertices) for which the inequality (3) is strict for all cuts. The results in Table 2 all follow readily from Lemma 3 using the preceding observation.

Graph: G	Concurrent Connectivity: $\zeta(G)$
Complete Graph: K_n, $n\geq 2$	1
Cycle: C_n	$8/n^2$ for n even $8/(n^2-1)$ for n odd
Path: P_n	$4/n^2$ for n even $4/(n^2-1)$ for n odd
m-Cube: Q_m ($n=2^m$ vertices)	$1/2^{m-1}$
Star: $K_{1,n}$	$1/n$
Tree (n vertices)	$1/\max_{(i,j)\varepsilon E} \{\lvert A_i\rvert \lvert A_j\rvert\}$ where A_i, A_j is the vertex partition determined by removing edge (i,j)

Table 2: Values of the concurrent connectivity, $\zeta(G)$, for several classes of graphs where the concurrent connectivity is equal to the density of the sparsest cut.

Lemma 4: The density of the sparsest cut is equal to the concurrent connectivity for all graphs indicated in Table 2 with values for $\zeta(G)$ as shown.

Proof: For every graph K_n, P_n, $K_{1,n}$, odd cycle C_n, and any tree there is a unique shortest path between every pair of vertices. Sending the largest possible identical flow on all such shortest paths in each specified graph with normalization so that $\sum_{e\varepsilon P_e} f(p) \leq 1$ achieves the throughput values shown in Table 2, which are readily seen to be maximal in each case by inspection using the preceding observation.

For the even cycle and m-dimensional cube Q_m, the vertices may be partitioned into two equal parts by an appropriate cut of only two edges for C_n, and 2^{m-1} edges for Q_m, yielding the bounds $\zeta(C_n) \leq 8/n^2$ for n even and $\zeta(Q_m) \leq 1/2^{m-1}$. Employing the symmetry of these graphs, concurrent flows are readily illustrated achieving these bounds. □

Supplementing the "cut bound" on concurrent connectivity of Lemma 3 is the following "distance bound".

<u>Lemma 5</u>: For any graph $G = (V,E)$, with $V = \{1,2, \ldots, n\}$,

$$\zeta(G) \leq |E| / \sum_{i<j} d_{ij} , \tag{5}$$

where d_{ij} is the distance (path length) between vertices i,j in V.

<u>Proof</u>: Any path $p \in P$ of length k with flow $f(p) > 0$ consumes $kf(p)$ units of capacity of the edges, where in total the edges provide $|E|$ units of capacity. So for any concurrent flow f of throughput z, we obtain $z \sum_{i<j} d_{ij} \leq |E|$, hence (5). □

The graphs for which the distance bound (5) yields an equality are the "edge-path-regular" graphs which have been extensively investigated in [MD80]. An <u>edge-path-regular-graph</u> is characterized by the property that some list of paths, allowing multiple copies of paths, exists where (i) every path is a shortest path between its end vertices, (ii) every pair of vertices are the end vertices of the same number, say k, of paths, and (iii) each edge occurs in the same number, say m, of paths of the list. Note that by associating a flow of 1/m with each path of such a list we then obtain $z = k/m = |E| / \sum_{i<j} d_{ij}$. In [MD80] it is shown that any connected edge-symmetric graph is edge-path-regular, which implies the complete graphs K_n, cycles C_n, stars $K_{1,n}$, and m-dimensional cubes Q_m are graphs for which the concurrent connectivity satisfies both inequalities (3) and (5)

with equality. The complete bipartite graph $K_{i,j}$ and regular complete k-partite graph $K_{i,i,\ldots,i}$ are both edge-symmetric so from (5)

$$\zeta(K_{i,j}) = ij/(i^2+j^2+ij-i-j) \qquad \text{for any } i,j \geq 1, \tag{6}$$

$$\zeta(K_{i,i,\ldots,i}) = (|V|-i)/(|V|+i-2) \text{ for any } |V|/i \geq 2,\ i \geq 1. \tag{7}$$

From (6) we obtain $\zeta(K_{2,3}) = 3/7$ confirming the optimality of the concurrent flow illustrated in Figure 1, and showing that inequality (3) can be strict.

The "cut bound" of Lemma 3 and "distance bound" of Lemma 5 are both simple special cases of a general duality result for concurrent connectivity involving the notion of a "k-partite cut" characterized by a maximum elongating distance function on the graph G.

III. Elongating Distance and k-partite Cuts in Graphs

The elegant max-flow min-cut theorem [FF62] of single commodity network flow theory establishes the dual concept of the existence of a minimum cut set of edges whose capacity is the limiting value of the maximum flow. We now describe an analogous dual to maximum concurrent flow also embodied in the form of a critical separating set of edges. The metric identifying this separating edge set is, however, not simply its capacity. This separating edge set is characterized by an "elongating" distance function maximizing the ratio of the total distance between all vertex pairs to the total distance on just the edges of the separating set.

A _distance function_ $d: V \times V \to R$ on the graph $G = (V,E)$ has

$$\begin{array}{lll} \text{i)} & d(i,i) = 0, & \text{for all } i \in V, \\ \text{ii)} & d(i,j) = d(j,i) \geq 0 & \text{for all } i,j \in V, \\ \text{iii)} & d(i,j) + d(j,k) \geq d(i,k) & \text{for all } i,j,k \in V, \end{array} \tag{8}$$

where notationally $d(e) = d(i,j)$ for $e \in E$, $e = (i,j)$. The elongation of d, denoted by $\eta(d)$, with $V = \{1,2,\ldots,n\}$, is given by

$$\eta(d) = \frac{\sum_{i<j} d(i,j)}{\sum_{e \in E} d(e)} \tag{9}$$

whenever $\sum_{e \in E} d(e) > 0$. For any connected non-trivial graph G, the elongation of G is defined by

$$\eta(G) = \max\{\eta(d) \mid d \text{ is a distance function on } G\}, \tag{10}$$

and any $\hat{d}: V \times V \to R$ with $\eta(\hat{d}) = \eta(G)$ is a maximum elongating distance function on G.

For any k-part partition $A_1, A_2, \ldots, A_k$ of the vertices of the graph $G = (V,E)$, we term the set of all edges having end vertices in different parts of the partition a k-partite cut of G and denote this by $(A_1, A_2, \ldots, A_k)$. For any distance function $d: V \times V \to R$, the equivalence classes of vertices $A_1, A_2, \ldots, A_k$ under the relation $u \equiv v$ if and only if $d(u,v) = 0$ determine the d-separating k-partite cut $(A_1, A_2, \ldots, A_k)$. Note that $d(e) > 0$ if and only if $e \in (A_1, A_2, \ldots, A_k)$, and we may write $d(A_i, A_j) = d(u,v)$ for $u \in A_i$, $v \in A_j$. Then the elongation of d may be interpreted as the elongation (of flow) across the d-separating k-partite cut $(A_1, A_2, \ldots, A_k)$ in the sense that

$$\eta(d) = \frac{\sum_{1 \le i < j \le k} |A_i||A_j| d(A_i, A_j)}{\sum_{e \in (A_1, A_2, \ldots, A_k)} d(e)}. \tag{11}$$

The problem of determining a maximum elongating distance (MEDP) can be expressed as a linear program utilizing (8) and the added normalization constraint $\sum_{e \in E} d(e) = 1$, with the objective to maximize $\sum_{i<j} d(i,j)$. The rationality of the solution of this linear program is sufficient to prove the following integrity lemma.

Lemma 6: Every connected graph G has an integral-valued maximum elongating distance function which assumes nonzero values on exactly the edges of some k-partite cut.

We now state our duality theorem and prove the simpler inequality portion.

Theorem 7 [Max-Concurrent Flow/Max-Elongating Distance]:

For any graph G, the maximum throughput of any concurrent flow is equal to the reciprocal of the maximum elongation of any distance on G, i.e.

$$\zeta(G) = 1/\eta(G) \,. \tag{12}$$

Proof: Given any $d: V \times V \to R$ and any $p \in P$, let $d(p)$ denote the sum of the distances on the edges of p, hence $d(p) \geq d(i,j)$ for $p \in P_{ij}$. For any concurrent flow $f: P \to [0,1]$ of throughput z, the expression $\sum_{p \in P} d(p)f(p)$, which may be regarded as the "volume of flow" in G, satisfies using (1),

$$\sum_{p \in P} d(p)f(p) \geq \sum_{i<j} d(i,j) \sum_{p \in P_{ij}} f(p) = z \sum_{i<j} d(i,j) \,. \tag{13}$$

Furthermore,

$$\sum_{p \in P} d(p)f(p) = \sum_{e \in E} d(e) \sum_{p \in P_e} f(p) \leq \sum_{e \in E} d(e) \,. \tag{14}$$

It follows that

$$z \leq \sum_{e \in E} d(e) \Big/ \sum_{i<j} d(i,j) \tag{15}$$

for any nonzero distance d and concurrent flow f on the connected graph G, so $\zeta(G) \leq 1/\eta(G)$. For the proof that equality can be attained see [Ma85]. □

Note that the inequality (15) with $d(i,j) = 1$ for $i \in A$, $j \in \bar{A}$ and $d(i,j) = 0$ otherwise, yields Lemma 3 as a corollary; and with $d(i,j)$ the standard shortest path distance on G we obtain Lemma 5 as a corollary.

We now describe a unique k-partite cut of G that illustrates the duality inherent in concurrent flow theory in that this cut is characterized both by its relation to a maximum concurrent flow and to a maximum elongating distance on G.

For any graph $G = (V,E)$, the edge $e \in E$ is <u>saturated</u> by the concurrent flow $f:P \to [0,1]$ whenever $\sum_{p \in P_e} f(p) = 1$. The edge $e \in E$ is termed <u>critical</u> for concurrent flow if every maximum concurrent flow function $\hat{f}:P \to [0,1]$ saturates e.

<u>Theorem 8</u>: For any connected graph $G = (V,E)$ the set of critical edges for maximum concurrent flow constitute a unique k-partite cut $(A_1, A_2, \ldots, A_k)$, where $(A_1, A_2, \ldots, A_k)$ is also the unique $\hat{d}$-separating k-partite cut having a maximum number k of parts for a maximum elongating distance function $\hat{d}$ on G.

<u>Proof</u>: We show that the set C of critical edges forms a k-partite cut. For $e \in E - C$, there is a corresponding maximum concurrent flow $\hat{f}_e$ with $\sum_{p \in P_e} \hat{f}_e(p) < 1$ since e is not critical. Then $\hat{f}:P \to [0,1]$ defined by $\hat{f}(p) = \sum_{e \in E-C} \hat{f}_e(p) / (|E| - |C|)$ is a maximum concurrent flow saturating only the critical edges of G. The maximality of $\hat{f}$ means there can be no spanning tree of unsaturated edges, so C is a separating set of edges of G. Let $A_1, A_2, \ldots, A_k$ be the vertex sets of the components of $G - C$, where then $k \geq 2$ is uniquely determined. Now $(A_1, A_2, \ldots, A_k) \subset C$. For any $e \in C - (A_1, A_2, \ldots, A_k)$, the vertices incident to e would be in a common part A_i and hence be joined by a path of edges unsaturated by $\hat{f}$. Some flow on a path $p \in P_e$ with $\hat{f}(p) > 0$ could then be diverted off e to follow the subpath of unsaturated edges while maintaining the optimality of throughput, implying e is not critical. Therefore $C = (A_1, A_2, \ldots, A_k)$, proving the flow saturation part of the theorem. For the balance of the proof see [Ma85]. □

Ford and Fulkerson demonstrated [FF62] practical advantages of their flow augmentation paradigm vis-a-vis general linear programming techniques in providing both a simple proof of the max-flow min-cut theorem and an efficient algorithm for determining a maximum flow and dual minimum cut for single commodity network flows. We can prove Theorems 7 and 8 from general linear programming duality theory, however, here also a flow manipulation paradigm seems propitious. In [Ma85] we develop a flow swapping procedure for concurrent flow and utilize this procedure to prove duality results such as Theorems 7 and 8. The flow swapping paradigm provides an algorithm dually determining both sequences of concurrent flows and elongating distances each approaching optimality. Flow swapping thus provides promise for efficiently computable approximate solutions to relatively large MCFP's [BM84]. The MCFP is also amenable to efficient solution by linear programming techniques that employ special network structure [He83,Pa84] when the formulation (1) of the MCFP is not too large, e.g. up to about 30-vertex graphs should be manageable.

We further note that the MCFP can be reformulated in "node-arc form" as a linear program with number of rows and columns polynomially bounded in $|V|$ and $|E|$. Thus the corresponding decision problem for the MCFP is in the class P of polynomially-bounded-time decision problems. It is an open question whether or not determining the sparsest cut can be resolved in polynomially-bounded time.

There are diverse areas of immediate application of concurrent flow duality theory of which we mention two. Studies of routing in packet-switched telecommunications networks [FC71,FGK73] employ a model with constraints similar to that of the MCFP, where a non-linear objective serves to minimize total delay given stochastically determined queueing delay on each edge. The duality theory of concurrent flow presents much useful information on the nature of delay and congestion as the packet-switched network asymptotically approaches full capacity (and corresponding infinite delay).

Cluster analysis seeks to determine a partition or hierarchy of partitions of objects into clusters given similarity data on object pairs [SS73]. By associating similarity with edge capacity in the MCFP, the uniquely determined partition defined by the critical k-partite cut gives a partition into clusters with many properties potentially useful to cluster interpretation derived from concurrent flow duality theory [Ma83].

References

[BM84] Biswas, J. and Matula, D. W., Heuristic Approaches for the Maximum Concurrent Flow Problem, submitted for publication.

[DA63] Dantzig, G. B., Linear Programming and Extensions, Princeton Univ. Press, Princeton, 1963.

[FF62] Ford, L. R. and Fulkerson, D. R., Flows in Networks, Princeton Univ. Press, Princeton, 1962.

[FGK73] Fratta, L., Gerla, M. and Kleinrock, L., The Flow Deviation Method: An Approach to Store-and Forward-Communication Network Design, Networks 3, 1973, pp. 97-133.

[FC71] Frank, H. and Chou, W., Routing in Computer Networks, Networks 1, 1971, pp. 99-112.

[He83] Helgason, R. V., Solving the Maximum Concurrent Flow Problem Using Linear Programming and Network Codes, presented at ORSA/TIMS National Meeting, Orlando, Florida, 1983.

[Ma83] Matula, D. W., Cluster Validity by Concurrent Chaining, in Numerical Taxonomy, Felsenstein, J., ed., NATO ASI Series G No. 1, Springer-Verlag, New York, 1983.

[Ma85] Matula, D. W., Maximum Concurrent Flow in Networks, in preparation.

[MD82] Matula, D. W. and Dolev, D., Path-Regular Graphs, Tech. Report 82-CSE-3, Computer Science Dept., Southern Methodist University, June 1980.

[Pa84] Patty, B. W., The Basis Suppression Method for Linear Programs with Special Structure Excluded by an Objective Side Column, Ph.D. Thesis, Dept. of Operations Research, Southern Methodist University, 1984.

[SS73] Sneath, P. H. A. and Sokal, R. R., Numerical Taxonomy, W. H. Freeman, San Francisco, 1973.

A LINEAR ALGORITHM FOR TOPOLOGICAL BANDWIDTH IN DEGREE THREE TREES

Zevi Miller
Department of Mathematics and Statistics
Miami University
Oxford, Ohio 45056

Abstract

Let G be a graph, and f a one-one map of G into the positive integers. The bandwidth of G is $B(G) = \min_f \max\{|f(x) - f(y)|: xy \in E(G)\}$, where the max is taken over all edges xy in G and the min over all maps f. $B(G)$ is related to the matrix bandwidth $B(M)$ for a symmetric matrix M, and knowledge of the latter parameter is important for the efficient execution of certain matrix operations. The problem of determining $B(G)$ for arbitrary G was shown by Papadimitriou to the NP-complete, and subsequently it was proved NP-complete even when $G \in \Omega$, where Ω is the set of trees with maximum degree 3. Let $B^*(G)$ be the minimum possible bandwidth of any subdivision of G, i.e., any graph obtained from G by inserting degree 2 points along edges of G. We present a $O(n)$ algorithm for computing $B^*(G)$ when $G \in \Omega$.

1. Introduction

Let $G = (V,E)$ be a finite graph with no loops or multiple edges. A one-one map $f:V(G) \to Z$ of the vertex set of G into the integers will be called a layout of G. We let $|f| = \max\{|f(v) - f(w)|: vw \in E(G)\}$, and we define the bandwidth of

G, <u>B(G)</u>, to be $B(G) = \min_f |f|$. For convenience, we adopt the convention $B(K_1) = 1$, where K_1 is the singleton graph.

The problem of determining B(G) for any G has importance in the theory of sparse matrices. Given an $n \times n$ matrix $A = (a_{ij})$, define a graph G(A) by letting $V(G(A)) = \{1, 2, \cdots, n\}$ and declaring that ij is an edge iff $a_{ij} \neq 0$ or $a_{ji} \neq 0$. Suppose we can find a symmetric permutation of the rows and columns of A so that all nonzero entries lie within the first k super-diagonals or the first k subdiagonals about the main diagonal, i.e., there exists a permutation matrix P such that PAP^t has all its nonzero entries within a symmetric $(2k + 1)$-band about the main diagonal. Then clearly $B(G(A)) \leq k$. Conversely if $B(G(A)) \leq k$, then the stated P exists. The problem of finding the P that yields the smallest k, equivalently, finding B(G(A)), is therefore important in efficiently operating on and storing matrices, e.g., for Gaussian elimination, systolic processes, etc. Relevant work can be found in [1,3, 9].

Recently the importance of bandwidth for complexity theory has been demonstrated in some work of Sudborough and Monien [11, 15]. The main kind of result has been that when certain well known graph problems, which are complete for some complexity class C, are restricted to graphs with bounded bandwidth, the resulting problem is complete for space bounded or simultaneous space and time bounded subclasses of C. Of course the bandwidth problem for arbitrary graphs [13], and even for trees of maximum degree 3 [6], is known to be NP-complete.

The problem of given a graph G to determine whether $B(G) \leq k$ when k is fixed was proved polynomial time solvable in [14], with an improvement in [7].

The subject of this paper is the following topological version of bandwidth, attributed in [2] to R. Graham. For any graph G, let S(G) be the set of all subdivisions of G, i.e., the set of all graphs obtainable from G by finitely many compositions of the operation, called elementary subdivision, of replacing some edge xz by two edges xy and yz, where y is a vertex not previously in the graph. Obviously any $H \in S(G)$ is homeomorphic with G and hence has the same underlying topology as G. The topological bandwidth of G, $B^*(G)$, is defined by $B^*(G) = \min\{B(H): H \in S(G)\}$.

Topological bandwidth is related to sparse matrices as follows. Given the system $Ax = b$, construct an equivalent system by replacing some term $a_{ij}x_j$ by the new variable y and add a new equation $a_{ij}x_j = y$. This operation is equivalent to inserting a point of degree 2 in the edge ij of G(A). Clearly $B^*(G(A))$ is the smallest k such that for some system $A'x = b'$ equivalent to $Ax = b$ by a sequence of the above operations there is a permutation matrix P such that $PA'P^t$ has the $2k + 1$ band form. The new system may be more efficient to work with if it is not too large, i.e., if the number of additional degree 2 vertices is not too large.

Another application is in regarding a graph G as a circuit with gates (corresponding to points of G) laid out linearly for automation purposes [12, 16]. Inserting degree 2 points along edges of G corresponds to inserting drivers along wires of the circuit. Hence $B^*(G)$ is the minimum length of the longest connection over all linear layouts allowing drivers. This minimum gives a lower bound on the time delay in transmission between elements of the circuit.

The cutwidth, cw(G), of a graph G is defined as the minimum, over all layouts f, of $\max_i |\{uv \in E(G): f(u) \leq i < f(v)\}|$. The relation between $B^*(T)$ and cw(T) for trees T was explored by F.R.K. Chung in [5]. It was shown that cw(T) and $B^*(T)$ can be arbitrarily far apart, but that $cw(T) \leq B^*(T) + \log_2 B^*(T) + 2$. Recently Yannakakis has found an $O(n \log n)$ algorithm for cw(T) for arbitrary trees T [17].

The main result of this paper is a $O(n)$ algorithm for computing $B^*(T)$ where T is any tree of maximum degree 3. First we develop a $O(n \log n)$ algorithm which contains all the essential graph theoretic ideas. Some additional data structure then leads to the $O(n)$ solution. Our result is a contrast to the NP-completeness of bandwidth for trees of maximum degree 3, and hence shows that topological structure "alone" is not the reason for this NP-completeness.

We remark that an $O(n \log n)$ algorithm for $B^*(T)$ in degree three trees T is obtained independently by Makedon, Sudborough, and Papadimitriou in [10]. They show that the topological bandwidth problem for arbitrary graphs is NP-complete, and they explore relations with pebbling and searching.

We follow standard graph theoretic notation, as may be found in [2] or [8] for example. In particular, if $G = (V,E)$ is a graph and H is a subgraph of G, then $G - H$ is the graph obtained from G by deleting all points of H and all edge of G incident on those points.

2. The Main Algorithm

Let P be a path joining two end vertices of T. A subtree α of T is called a _hanging from P_ if it is one of the connected components of $T - P$. Thus each such α contains a unique vertex x_α such that $x_\alpha y \in E(T)$ for some $y \in P$. We denote by $Z(P)$ the set of all hangings from P. Let $\|P\|$, the _norm of P in T_, be defined by $\|P\| = 1 + \max\{B^*(\alpha): \alpha \in Z(P)\}$. To emphasize the role of T, we sometimes write this as $\|P\|_T$. We let $K_{1,2}$ (resp. $K_{1,3}$) be the tree with a central point joined to two (resp. three) endpoints.

We begin with some lemmas.

Lemma 1. For any tree T there exists a layout of T satisfying $|f| = B(T)$ such that $f^{-1}(\min f(x))$ and $f^{-1}(\max f(x))$ are both end vertices of T.

Proof: Let g be a layout of T such that $|g| = B(T)$. Let $p = g^{-1}(\min g(x))$ and $q = g^{-1}(\max g(x))$. We may suppose that at least one of p and q is not an end vertex. Let P be the path in T joining p and q, and let Y_p (resp. Y_q) be the set of branches at p (resp. q) having empty intersection with P. Now define a new layout g_1 of T as follows.

$$g_1(x) = \begin{cases} g(x), & x \in V(T)\setminus[Y_p \cup Y_q], \\ 2g(p) - g(x), & x \in Y_p, \\ 2g(q) - g(x), & x \in Y_q. \end{cases}$$

Notice that g_1 just reflects Y_p and Y_q about p and q respectively, and $|g_1| = |g|$. If $g_1^{-1}(\min g_1(x))$ and $g_1^{-1}(\max g_1(x))$ are both end vertices, then we are done. If not, then repeat the above process with g_1 in place of g. Eventually we obtain a layout for T of the required kind. □

A path P joining two end vertices of the tree T is called a backbone of T. If f is a layout of T such that $z_1 = f^{-1}(\min f(x))$ and $z_2 = f^{-1}(\max f(x))$ are both end vertices of T, then the backbone P joining z_1 and z_2 is called a backbone of f. We also say that f has backbone P.

Let $T^* \in S(T)$, and let P be a backbone of T. A layout f of T^* will be called a t-layout of T (t for topological). We call P a t-backbone of T if there exists such an f and T^* satisfying

(i) $|f| = B^*(T)$, and

(ii) the path joining $f^{-1}(\min f(x))$ and $f^{-1}(\max f(x))$ corresponds to P under the subdivision operation (so that $f^{-1}(\min f(x))$ and $f^{-1}(\max f(x))$ are endpoints of T^* corresponding to the endpoints of P).

Corollary 1.1. Every tree T has a t-backbone.

Proof: let f be a t-layout of T with $|f| = B^*(T)$. By Lemma 1 we can find a t-layout g such that $g^{-1}(\min g(x))$ and $g^{-1}(\max g(x))$ are end vertices v_1 and v_2 (respectively) while $|g| = B^*(T)$. The path in T joining the vertices v_1 and v_2 is the required t-backbone. □

Given a layout f of G and $H \subseteq G$, the layout of H induced by f is the layout f_H: $V(H) \rightarrow \{1, 2, \cdots, |V(H)|\}$ which maps the points of H in the same order as they are mapped by f.

Lemma 2. Let T be a non-trivial tree of maximum degree 3, let P at t-backbone of T, and let $Z(P) = \{B_1, B_2, \cdots, B_k\}$. Then $B^*(T) = \|P\| \equiv 1 + \max B^*(B_j)$.

Proof: We start with the lower bound $B^*(T) \geq \|P\|$.

Consider a t-layout of T, say f: $T^* \rightarrow Z$ for some $T^* \varepsilon S(T)$, such that $|f| = B^*(T)$. Let $B_i^* \varepsilon S(B_i)$, $1 \leq i \leq k$, be the maximal hanging of P^* corresponding to B_i. Let f_i be the layout of B_i^* induced by f. Clearly $|f_i| \geq B(B_i^*) \geq B^*(B_i)$, so there exist $c, d \; \varepsilon \; V(B_i^*)$ (say with $f(c) < f(d)$) such that $f_i(d) - f_i(c) \geq B^*(B_i)$. After translating f_i to correspond to the the restriction of f to B_i^*, we find that there is an edge $ab \; \varepsilon \; E(P^*)$ such that wlog either

(i) $a < f_i(c)$ and $b > f_i(d)$, or

(ii) $f_i(c) < a < f_i(d)$.

In either case we get $|f| \geq 1 + B^*(B_i)$, so by the arbitrariness of i we get $|f| \geq \|P\|$.

The bound $B^*(T) \leq \|P\|$ follows from the layout indicated in Figure 1. For each i, $1 \leq i \leq k$, we take a layout f_i of some $B^* \varepsilon S(B_i)$ satisfying $|f_i| = B^*(B_i)$. Assuming wlog that the B_i occur along P in indicial order, we place the layouts f_i along Z in the same order so that f_i and f_{i+1} have no overlap for all i. We then lay out some $P^* \varepsilon S(P)$ so that consecutive points on P^* are $\|P\|$ units apart, $\min\{P^*\} < \min f_1(x)$, and $\max\{P^*\} > \max f_k(x)$ (where we identify P^* with points of Z). Of course this requires us to "squeeze in" points of P^* into the range of the layouts f_i. We make room for this by iteratively translating values $f_i(x)$ greater than the squeeze in point one unit to the right. Since consecutive points on P^* are $\|P\|$ units apart, and since $B^*(B_i) < \|P\|$, we will never have to stretch any pair $f_i(x)f_i(y)$, $xy \in E(B_i^*)$, more than $\|P\|$ units apart. This defines a layout $f\colon T^* \to Z$ for some $T^* \in S(T)$. By construction we have $|f(x) - f(y)| \leq \|P\|$ for any $xy \in E(P^*) \cup (\cup E(B_i^*))$. Now for each i there is a unique edge $w_i p_i \in E(T)$, $w_i \in B_i$, $p_i \in P$. We choose a $w_i \in B_i^*$ (to correspond to w_i and B_i), and then find a $p_i \in P^*$ (to correspond $p_i \in P$) that is at most distance $\|P\|$ from w_i. Letting these $w_i p_i$ be the remaining edges of T^*, we get $|f| \leq \|P\|$ as required. □

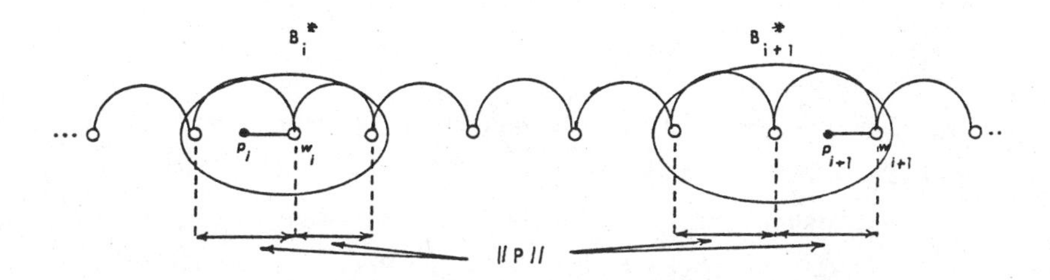

Figure 1. The optimal t-layout of T from Lemma 2.

In order to make use of the lemma in the algorithm we need some further notation.

Let T be a tree of maximum degree 3. If the vertices of T are numbered in the order of a depth-first search (DFS) starting at r, then the resulting numbers can be used to define the usual relations on rooted trees: If x and y belong to the same branch at r and the path in T from r to y passes through x, then we write $x \le y$. If $x \le y$ then we call x an <u>ancestor</u> of y and we call y a <u>descendant</u> of x. If y is a descendant of x and $xy \in E(T)$, then y is a <u>son</u> of x and x is a <u>father</u> of y. If $Q \subset V(T)$, then we define <u>head (Q)</u> to be the vertex $x \in Q$ such that $x \le y$ for all $y \in Q\backslash\{x\}$. For $x \in V(T)$, let $B_x(T)$ denote the subtree of T consisting of x and all descendants of x in T. We write B_x instead of $B_x(T)$ when T is understood by context. For $x \in V(T)$ we denote by T_x the tree with vertex set $V(B_x) \cup \{\hat{x}\}$, for some symbol $\hat{x}$, and edge set $E(B_x) \cup \{\hat{x}x\}$. We say that B_x is <u>k-primitive</u> (or just primitive when k is understood) if $B^*(B_x) = k$ and $B^*(B_y) < k$ for all $y \in B_x$. Note that if B_x and B_z are distinct k-primitive branches, then $B_x \cap B_z = \phi$. If P is a path in T, we say that P <u>threads</u> the branch B_x if P passes through x and some end vertex of B_x.

<u>Corollary 2.1</u>.

(a) Let B_x and B_y be branches of T such that $B_x \cap B_y = \phi$ and $B^*(B_x) = B^*(B_y) = B^*(T)$. Then any t-backbone of T threads both B_x and B_y.

(b) If T has three branches B_{x1}, B_{x2}, B_{x3} having pairwise empty intersection, then $B^*(T) \ge 1 + \min_i(B_{xi})$.

<u>Proof</u>:

(a) Let P be a t-backbone of T violating the conclusion. Then at least one of B_x or B_y is contained in some hanging C C of P. Thus, by the lemma, $B^*(T) \ge 1 + B^*(C) \ge 1 + B^*(B_x) > B^*(T)$, a contradiction.

(b) By Lemma 1 we know that T must have a t-backbone P. Now observe that for any t-backbone P of T one of the B_{xi}, say B_{x1}, must be contained in a hanging from P. Hence

$B^*(T) \geq 1 + B^*(B_{x1}) \geq 1 + \min_i B^*(B_{xi})$. □

Corollary 2.2. For any backbone β of T we have $B^*(T) \leq \|\beta\|$. Hence by Corollary 1.1 and Lemma 2, $B^*(T) = \min\{\|\beta\|: \beta$ a backbone of $T\}$.

Proof: Let $Z(\beta) = \{B_j: 1 \leq j \leq k\}$. Using the proof of the upper bound in the lemma, one can construct a t-layout g of T such that $|g| = \|\beta\|$. □

We call a backbone β of T optimal if $\|\beta\|$ is minimal over all backbones of T, and thus by Corollary 2.2 $\|\beta\| = B^*(T)$. For our algorithm it therefore suffices to find an optimal backbone of T and to compute its norm.

Before proceeding to the algorithm, we introduce some notation. Let A and C be trees on which a DFS has been performed with roots x and y respectively. Denote by $(A \vee C)_r$ the tree with $V(T) = V(A) \cup V(C) \cup \{r\}$, $E(T) = E(A) \cup E(C) \cup \{rx,ry\}$. The basic idea in our "main" procedure LABEL(A,C,r) is to use an inductive knowledge of $B^*(A)$, $B^*(C)$, and some additional information (described below) to find an optimal backbone β of $(A \vee C)_r$ and compute $B^*((A \vee C)_r) = \|\beta\|$.

Now let τ be any tree of maximum degree 3 rooted at a point μ. Define h(τ), the highpoint of τ, as follows. If τ is a path with u one of its two endpoints then let $h(\tau) = u$. Otherwise, let u' be the point having two sons which is closest to u. Also, let B_τ be the set of all optimal-backbones β of τ satisfying head(β) $\geq$ u' (a simple argument shows $B_\tau \neq \phi$). We let $h(\tau) = \text{head}(\hat{\beta})$, where $\hat{\beta}$ is an element of B_t such that head(β) $\leq$ head($\hat{\beta}$) for all $\beta \in B_\tau$. The existence of such a $\hat{\beta}$ is seen as follows. If there exists only one

$B^*(\tau)$-primitive branch $B_z(\tau) \subset \tau$, then $h(\tau) = u'$ since we can use for our β any path which threads $B(\tau)$ and passes through u'. If there exist two $B^*(\tau)$-primitive branches B_{z_1}, B_{z_2}, then any $\beta \in B_\tau$, head(β) = q is the point closest to u' on the path from z_1 to z_2. Thus $h(\tau) = q$. For any $s \in \tau$, let $\underline{\tau/s}$ be the tree $\tau - B_s(\tau)$. We will be interested in $\tau/h(\tau)$. We define the <u>signature</u> of τ, <u>sign(τ)</u>, to be the 3-tuple $(B^*(\tau), \nu, h(\tau))$, when ν is the number of $B^*(\tau)$-primitive branches in τ (so, $\nu = 1$ or 2).

We define a nested sequence <u>Tree(τ)</u> of subtrees $\{\tau_i\}$ of τ, where $u \in \tau_i$ for all i, as follows. Let $\tau_1 = \tau$. If τ is path, then Tree(τ) = $\{\tau_1\}$. Otherwise assume inductively that $\tau_1, \tau_2, \cdots, \tau_i$ have been defined. If $h(\tau_i) = u$, then Tree(τ) $= \{\tau_1, \tau_2, \cdots, \tau_i\}$. Otherwise let $\tau_{i+1} = \tau_i/\overline{h(\tau_i)}$. Note that if u has only one son, then $\tau_{|\mathrm{Tree}(\tau)|}$ is a path with endpoint u. We define <u>deck(τ)</u> as the ordered list deck(τ) $= \{\mathrm{sign}(\tau_1), \mathrm{sign}(\tau_2), \cdots, \mathrm{sign}(\tau_{|\mathrm{Tree}(\tau)|})\}$, where sign($\tau$) is top and $\mathrm{sign}(\tau_{|\mathrm{Tree}(\tau)|})$ is the bottom. We remark that when u has two sons (so u = u'), then Tree(τ) $\subset \{(\gamma_i \vee \mu_j)_u\colon \gamma_i \in \mathrm{Tree}(B_{\mathrm{leftson}(u)}), \mu_j \in \mathrm{Tree}(B_{\mathrm{rightson}(u)})\}$. The opposite containment is not true in general.

The procedure LABEL(A,C,r) has deck(A) and deck(C) as input, and deck ($A \vee C$) as output (the subscript r is dropped since it is fixed throughout). In particular, we have $B^*(A)$ and $B^*(C)$ from the top entries of these decks.

We consider several cases, all but one of which are resolved trivially. The exceptional case leads to a recursive call of LABEL on itself. We let Tree(A) $= \{\tau_1, \tau_2, \cdots, \tau_k\}$, and Tree(C) $= \{\mu_1, \mu_2, \cdots, \mu_\ell\}$. Initially we scan sign(τ_1) and sign(μ_1) in deck(A) and deck(C), respectively.

Case 1: $B^*(A) = B^*(C) = b$.

Since $\nu(A) + \nu(C) \geq 2$, we have $\nu(A) + \nu(C) \geq 3$ or $\nu(A) = \nu(C) = 1$. We consider these subcases in turn.

Subcase 1a: $\nu(A) + \nu(C) \geq 3$.

By Corollary 2.1(b) it follows that $B^*(A \vee C) \geq 1 + b$. But any backbone β of $A \vee C$ containing r satisfies $\|\beta\| \leq 1 + b$ since any hanging α of β is a subtree of A or C so by monotonicity of B^* satisfies $B^*(\alpha) \leq b$. Hence by Corollary 2.2 we have $B^*(A \vee C) = 1 + b$. Clearly $\nu(A \vee C) = 1$ since $A \vee C$ is its own $(1 + b)$-primitive branch, and $h(A \vee C) = r$ since (as remarked above) any backbone containing r is optimal. The procedure ends with output $\text{deck}(A \vee C) = (1 + b, 1, r)$.

Subcase 1b: $\nu(A) = \nu(C) = 1$.

Let $B_x \subset A$ and $B_y \subset C$ be the unique b-primitive branches of A and C, respectively. Consider any backbone β of $A \vee C$ which threads both B_x and B_y. Each of its hangings α is either properly contained in B_x or B_y, or contains no b-primitive branches. Hence $B^*(\alpha) \leq b - 1$, so $\|\beta\| \leq b$ and thus $B^*(A \vee C) \leq b$ by Corollary 2.2. The opposite inequality is immediate by monotonicity of B^*. Thus $B^*(A \vee C) = b$. Obviously $\nu(A \vee C) = 2$ since B_x and B_y are are the b-primitive branches, and $h(A \vee C) = r$ since any optimal backbone threading B_x and B_y contains r. The procedure ends with output $\text{deck}(A \vee C) = (b,2,r)$.

Case 2: $b = B^*(A) > B^*(C)$

Subcase 2a: $\nu(A) = 1$

Here we note that any backbone β of $A \vee C$ which threads the unique b-primitive branch of A and contains r satisfies $\|\beta\| \leq b$. As in Subcase 1b, this implies $B^*(A \vee C) = b$ and we end with output $\text{deck}(A \vee C) = (b,1,r)$.

Subcase 2b: $\nu(A) = 2$

Let B_x and B_y be the two b-primitive branches of A. Clearly any backbone β of $A \vee C$ satisfying $\|\beta\| = b$ (if such exists) must thread both B_x and B_y, so that $head(\beta) = h(A)$. In addition, its hanging $(A \vee C)/h(A)$ must have B^* value $\leq b - 1$. In fact, we have $B^*(A \vee C) = b \Leftrightarrow B^*((A \vee C)/h(A)) \leq b - 1$. Since $(A \vee C)/h(A) = (A/h(A)) \vee C$, we can find $B^*((A \vee C)/h(A))$ by recursively calling on $LABEL(A/h(A),C,r) = LABEL(\tau_2,\mu_1,r)$. This call is made after popping $sign(\tau_1) = sign(A)$ from the top of deck(A) so that $sign(\tau_2) = sign(A/h(A))$ is now available for scanning.

If in the output of $LABEL(\tau_2,\mu_1,r)$ we have $B^*(\tau_2 \vee \gamma_1) = b$, then $LABEL(\tau_1,\mu_1,r)$ returns $deck(A \vee C) = (1 + b,1,r)$. If the output gives $B^*(\tau_2 \vee \mu_1) \leq b - 1$, then we return $deck(A \vee C) = deck(\tau_2 \vee \mu_1) \vee (b,2,h(A))$, where the right side is the ordered list whose first member is $(b,2,h(A))$ followed by the ordered list $deck(\tau_2 \vee \gamma_1)$. Note that the latter list is obtained as the output of $LABEL(\tau_2,\mu_1,r)$. The procedure is thereby completed.

We illustrate the Subcase 2b requiring recursion in Figures 2 and 3. The pointers in Figure 3 show the entries to be scanned in the indicated calls to LABEL.

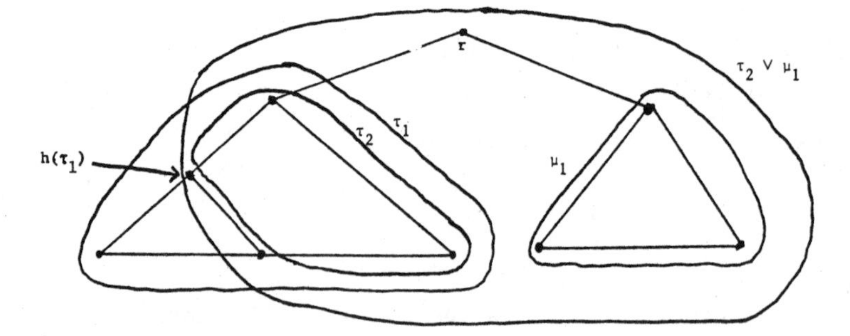

Figure 2. Determining $B^*(\tau_1 \vee \gamma_1)$ from $B^*(\tau_2 \vee \gamma_1)$.

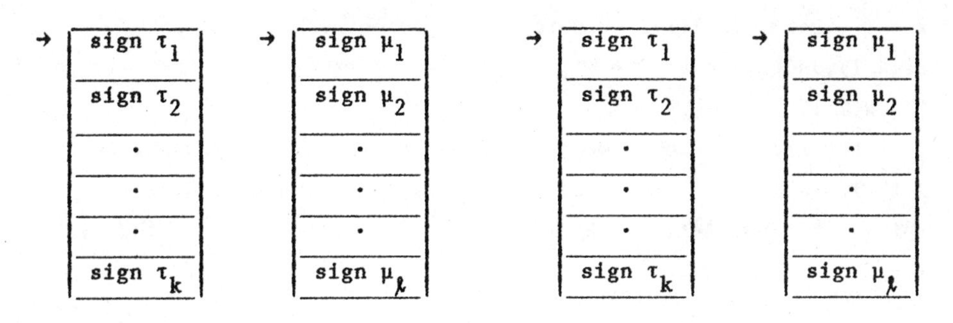

Figure 3. The deck structure in recursion.

The procedure may continue calling itself through possibly $k + \ell$ levels in recursion (after popping a sign(τ_i) or sign(μ_j) in preparation each time) until one of the cases 1 or 2a obtains, where, of course, A and C are replaced by some τ_i or μ_j. At this point the answer cascades back up as in the discussion of Subcase 2b to give the final output of LABEL(A,C,r). We note that at some level one of two decks may be depleted. The handling of this case requires additional inductive information (carried in the signatures) and additions to the algorithm which we omit here for brevity. Details may be found in the full paper.

The validity of the procedure (i.e. the correctness of $B^*(A \vee C)$ and deck($A \vee C$)) is clear in cases 1 and 2a. In Subcase 2b it follows essentially by induction and the relation between $B^*(\tau_1 \vee \mu_1)$ and $B^*(\tau_2 \vee \mu_1)$.

3. Complexity

We now discuss the complexity of procedure LABEL(A,C,r) and the complexity of the resulting algorithm for $B^*(T)$ in trees T of maximum degree 3.

Let T be any tree of maximum degree 3 rooted at a point r. We assume a DFS on T with associated branches B_x for all

$x \in T$. Let $|deck(T)|$ denote the number of signatures in deck(T), so $|deck(T)| = |Tree(T)|$. Our first goal is to estimate $|deck(T)|$.

For the next lemma we remark that a tree T is k-primitive iff it is its own (and hence unique) k-primitive branch, i.e, $B^*(T) = k$ and $B^*(T') < k$ for any proper subtree T' with $r \notin T'$.

Lemma 3. Let T be a k-primitive tree. Then $|T| \geq 2^{k-1}$.

Proof: Let $f_k = \min\{p: |T| = p,\ T$ is k-primitive$\}$. We show that $f_k \geq 2^{k-1}$ by induction on k.

For $k = 1$, the one point tree is k-primitive by convention and has minimum order. Thus $f_1 \geq 1 = 2^0$. Now suppose $f_j \geq 2^{j-1}$ for $j < s$.

Let T be s-primitive and have minimum possible order. We claim T must have at least two $(s - 1)$-primitive branches. Suppose first that T has no $(s - 1)$-primitive branches. Then any backbone β of T through r satisfies $\|\beta\| \leq s - 1$, so $B^*(A) \leq s - 1$, a contradiction. Similarly if T has only one $(s -1)$-primitive branch B_z, then any backbone β of T threading B_z and passing through r satisfies $\|\beta\| \leq s - 1$, again a contradiction. Hence T has at least two $(s - 1)$-primitive branches, so by induction $|T| = f_s \geq 2f_{s-1} \geq 2^{s-1}$. □

A sequence of trees $T_1, T_2, \cdots, T_k$ is called monotone if $B^*(T_i) > B^*(T_{i+1})$, $i \geq 1$.

Lemma 4. For any tree T, Tree(T) is monotone.

Proof: Let $Tree(T) = \{\tau_1, \tau_2, \cdots, \tau_\ell\}$. Now by definition $\tau_i/h(\tau_i) = \tau_{i+1}$, $1 \leq i \leq \ell - 1$, and $B^*(\tau_i) = \|\beta_i\|$ where β_i is a t-backbone of τ_i with $head(\beta_i) = h(\tau_i)$. It follows that

τ_{i+1} is a hanging of β_i in τ_i. Hence $B^*(\tau_{i+1}) \le \|\beta_i\| - 1 = B^*(\tau) - 1$. □

For a tree T, let $|\text{deck}(T)|$ be the number of frames in deck(T). Thus $|\text{deck}(T)| = |\text{Tree}(T)|$.

Corollary 4.1. Let T be a tree with n points. Then $|\text{deck}(T)| = O(\log n)$.

Proof: Since $|\text{deck}(T)| = |\text{Tree}(T)|$, it suffices to show that $|\text{Tree}(T)| = O(\log n)$.

Let $\text{Tree}(T) = \{\tau_1, \tau_2, \cdots, \tau_\ell\}$, and let $B^*(\tau_i) = k_i$. By Lemmas 3 and 4, Tree(T) is monotone and $|\tau_i| \ge 2^{k_i - 1}$ since τ_i must contain a k_i-primitive branch. Since $\tau_i \subset T$ for all i, it follows that $|\text{Tree}(T)| = \ell \le O(\log n)$. □

Lemma 5: Let A and C be trees with $|A| + |C| = n$. Then LABEL(A,C,r) has complexity $O(\log n)$.

Proof: We bound the complexity of LABEL by bounding the maximum number of nested calls on itself and the maximum number of computations per call.

Consider first the number of computations per call. Our operations essentially are checking the number of primitive branches, and defining $\text{deck}(A \vee C)$ as a single entry or as the result of pushing a single entry onto the already known $\text{deck}(A/h(A) \vee C)$. Each of these takes $O(1)$ time, so the call takes $O(1)$ time.

Since the number of calls is bounded by $|\text{deck}(A)| + |\text{deck}(C)|$, we have at most $\log|A| + \log|C| = \log(|A||C|) \le \log(n^2) = O(\log n)$ calls. Hence LABEL(A,C,r) has time $O(\log n)$. □

We can now give an $O(n \log n)$ algorithm for $B^*(T)$. Later we sketch how it can be refined to linear time.

Theorem 1: Let T be a tree of maximum degree 3. Then $B^*(T)$ can be computed in $O(n \log n)$ time.

Proof: First one can reduce to the case where T has a point r of degree 2. We omit this here.

Now view r as the root of T. Now order $V(T)$ in reverse depth first search from endpoints "up" to r (i.e. in postorder). If e is an endpoint, then let $deck(\{e\}) = (1,1,e)$ (where we recall the convention $B^*(\{e\}) = 1$). Now for some $x \in T$, assume inductively that $deck(B_y)$ is known for each of (the at most 2) sons of x. If x has only one son y, then $deck(B_x)$ can be obtained from $deck(B_y)$ using the information on depleted decks alluded to after the algorithm but omitted here for brevity. Suppose then that x has sons y_1 and y_2. We then apply $LABEL(B_{y_1}, B_{y_2}, x)$ and get $deck(B_x)$ as output.

Continuing to work our way up we eventually apply $LABEL(B_{s_1}, B_{s_2}, r)$, where s_i are the sons of r. The output, $deck(B_r) = deck(T)$, has in its topmost entry the required $B^*(T)$. The algorithm then halts.

Clearly there are at most n applications of $LABEL(B_{y_1}, B_{y_2}, x)$, each requiring at most $\log |B_x| \leq \log n$ time. Hence the total time is bounded by $O(n \log n)$. □

4. An $O(n)$ time bound

In this section we describe informally a way of lowering the complexity of our algorithm to $O(n)$.

Let $Tree(A) = \{\tau_i: 1 \leq i \leq \ell\}$ and $Tree(C) = \{\gamma_j: 1 \leq j \leq k\}$. The main idea is to modify procedure $LABEL(A,C,r)$. In the worst case it processes all $|deck(A)| + |deck(C)| \leq \log |A| + \log |C|$ signatures in $deck(A)$ and $deck(C)$, giving a time bound $O(\log n)$, $n = |(A \vee C)_r|$. We indicate how the addition of new decks that point to gaps in the sequences $\{B^*(\tau_i)\}$ and $\{B^*(\gamma_j)\}$ permits a $O(\min\{\log|A|, \log|C|\})$ implementation of the procedure.

As motivation, suppose that LABEL(A,C,r) performs in recursion LABEL(τ_c, γ_d, r), say with $B^*(\tau_c) > B^*(\gamma_d)$. Suppose also that for some $e > c$ we have $B^*(\tau_e) = B^*((\gamma)_d)$ and $B^*(\tau_{i+1}) = B^*(\tau_i) - 1$ for $c \le i \le e - 1$. Then we have the equivalences $B^*(\tau_c \vee \gamma_d) \le B^*(\tau_c) \Leftrightarrow B^*(\tau_{c+1} \vee \gamma_d) \le B^*(\tau_c) - 1 \Leftrightarrow \cdots \Leftrightarrow B^*(\tau_e \vee \gamma_d) \le B^*(\tau_c) - (e - c)$, where we abbreviate $(\tau_i \vee \gamma_j)r$ by $\tau_i \vee \gamma_j$. Of course the falsity of the above inequalities yields $B^*(\tau_c \vee \gamma_d) = B^*(\tau_c) + 1$ in which case deck($\tau_c \vee \gamma_d$) has only one frame. Thus the result of LABEL(τ_e,γ_d,r) telescopes up to decide the result of LABEL(τ_c, γ_d, r). As now constituted, LABEL(A,C,r) performs the $e - c$ operations for achieving the telescoping. Our goal is to do this in O(1) time by using a pair of successive pointers from a separate deck, one to $B^*(\tau_c)$ and the other to $B^*(\tau_e)$. When the inequalities are false, we advance the pointer to $B^*(\tau_c)$ one unit and delete the one to $B^*(\tau_e)$, and when they are true we leave both pointers stationary.

To be precise, we define two additional decks, GAP(A) and GAP(C) as follows. Suppose $R_{jm} = \{\tau_j, \tau_{j+1}, \cdots, \tau_m\}$ is a subsequence of Tree(A) satisfying $B^*(\tau_{i+1}) = B^*(\tau_i) - 1$, $j \le i \le m - 1$, $B^*(\tau_{j-1}) > 1 + B(\tau_j)$, and $B^*(\tau_{m+1}) < B^*(\tau_m) - 1$. We call R_{jm} a <u>run</u> of Tree(A) (and also of deck(A) under the correspondence between deck(A) and Tree(A)). For any $\tau \in$ Tree(A), let <u>run(τ)</u> denote the run of Tree(A) to which τ belongs. For each run R_{jm} there will be two consecutive elements $b(R_{jm})$, $t(R_{jm})$ in the list GAP(A). We let t(jm) point to $B^*(\tau_j)$ in sign(τ_j) and b(jm) point to $B^*(\tau_m)$ in sign(τ_m). If R_{jm} and R_{qr} are runs with $m > q$, then the pointers to R_{jm} come before (i.e. above) those to R_{qr}. The deck GAP(C) is constructed similarly.

Our new procedure $\overline{\text{LABEL}}$(A,C,r) can be described informally as follows. It has input deck(A), deck(C), GAP(A), GAP(C), and output deck($(A \vee C)_r$) and GAP($(A \vee C)_r$). Wlog, we take

$B^*(A) \geq B^*(C)$. For brevity we consider only the first case where $\log|C| \leq \log|A|$, the case $\log|C| > \log|A|$ being done with straightforward but tedious changes. We let bottomdeck(T) be the bottommost entry of deck(T) for a tree T.

I. Work your way up deck(A) from bottomdeck(A) to $\mathrm{sign}(\tau_u)$, where $u = \min\{j: B^*(\tau_j) \leq B^*(\gamma_1)\}$ if $B^*(\tau_\ell) \leq B^*(\gamma_1)$ and $u = \ell$ otherwise. Also work up GAP(A) until you reach $b(\mathrm{run}(\tau_u))$.

II. Perform $\mathrm{LABEL}(\tau_u,\gamma_2,r)$.

III. Construct $\mathrm{deck}((A \vee C)_r)$ and $\mathrm{GAP}((A \vee C)_r)$ as follows.

A. Suppose $B^*(\tau_u \vee \gamma_1) = (1 + B^*(\gamma_1)$, so that $\mathrm{deck}(\tau_u \vee \gamma_1)$ is the single frame $\mathrm{sign}(\tau_u \vee \gamma_1) = (1 + B^*(\gamma_1),\ 1 + B^*(\gamma_1),r)$.

1. Check if there is a run R_{ab} in Tree(A) one of whose members τ satisfies $B^*(\tau) = 1 + B^*(\gamma_1)$. (Note: Such an R_{ab} must be $\mathrm{run}(\tau_u)$ or the run preceeding τ_u in Tree(A).)

2. If there is, then move up to $t(R_{ab})$ in GAP(A) (up ≤ 2 pointers) and follow $t(R_{ab})$ to the top frame $\mathrm{sign}(\tau_a)$ of R_{ab}. Now do $\mathrm{sign}(\tau_a \vee \gamma_1) \leftarrow (B^*(\tau_a) + 1,\ B^*(\tau_a) + 1,r)$. Now $\mathrm{deck}((A \vee C)_r)$ is obtained from deck(A) by replacing $\mathrm{sign}(\tau_a)$ by $\mathrm{sign}(\tau_a \vee \gamma_1)$, letting $\mathrm{sign}(\tau_a \vee \gamma_1)$ be the bottom of the bottom of $\mathrm{deck}((A \vee C)_r)$, and leaving all entries $\mathrm{sign}(\tau_j)$, $1 \leq j \leq a - 1$, in place above $\mathrm{sign}(\tau_a \vee \gamma_1)$. These entries are new regarded as $\{\mathrm{sign}(\tau_j \vee \gamma_1)\}$.

3. If there is no R_{ab}, then append $\mathrm{deck}(\tau_u \vee \gamma_1)$ ($= \mathrm{sign}(\tau_u \vee \gamma_1)$) to deck(A) just below $\mathrm{sign}(\tau_{u-1})$, and delete $\mathrm{sign}(\tau_j)$, $j \geq u$, from deck(A). The new list is $\mathrm{deck}((A \vee C)_r)$. Again,

the entries $\{sign(\tau_j)\}$, $1 \le j \le u - 1$, are left unchanged and are regarded as $\{sign(\tau_j \vee \gamma_1)\}$.

4. $GAP((A \vee C)_r)$ is obtained by making corresponding changes in $GAP(A)$. Suppose for example that (2) holds, and let $run(\tau_a \vee \gamma_1) = R_{\alpha\beta}$ in $deck((A \vee C)_r)$. If $R_{\alpha\beta}$ is not a singleton, then $R_{\alpha\beta}$ corresponds to a run $R_{\alpha(\beta-1)}$ in deck(A) with $\tau_a \vee \gamma_1$ appended at the end. Hence we let $t(R_{\alpha\beta}) = t(R_{\alpha(\beta-1)})$ while $b(R_{\alpha\beta})$ points to $sign(\tau_a \vee \gamma_1)$. Naturally $b(R_{\alpha(\beta-1)})$ (in GAP(A)) is deleted. All elements of GAP(A) corresponding to runs above $R_{\alpha\beta}$ are retained in $GAP((A \vee C)_r)$. We omit here the changes appropriate to (3) since similar changes are described below.

B. Suppose $B^*(\tau_u \vee \gamma_1) = B^*(\gamma_1)$.

Then $deck(A \vee C)_r$ is obtained from deck(A) by attaching $deck(\tau_u \vee \gamma_1)$ to deck(A) just below $sign(\tau_{u-1})$. The elements $sign(\tau_j)$, $j \ge u$, are dropped, so $bottomdeck(\tau_u \vee \gamma_1)$ becomes $bottomdeck((A \vee C)_r)$.

Now $GAP((A \vee C)_r)$ is constructed as follows.

1. Create $GAP(\tau_u \vee \gamma_1)$ by processing $deck(\tau_u \vee \gamma_1)$ top to bottom and forming the pointers appropriate to each run.

2. Attach $GAP(\tau_u \vee \gamma_1)$ to GAP(A) as follows. Let GAP'(A) be the subdeck of GAP(A) of pointers corresponding to $run(\tau_{u-1})$ and all runs above $run(\tau_{u-1})$.

   a. If $B^*(\tau_{u-1}) > 1 + B^*(\gamma_1)$ then just append $GAP(\tau_u \vee \gamma_1)$ to the bottom of GAP'(A) and retain the pointer structure in each.

   b. If $B^*(\tau_{u-1}) = 1 + B^*(\gamma_1)$, then append as above but remove $b(run(\tau_{u-1}))$ from GAP'(A) and

$t(\mathrm{run}(\tau_u \vee \gamma_1))$ from $\mathrm{GAP}(\tau_u \vee \gamma_1)$. (This has the effect of merging the runs $\mathrm{run}(\tau_{u-1})$ and $\mathrm{run}(\tau_u \vee \gamma_1)$ into one in $\mathrm{GAP}(\tau_u \vee \gamma_1)$.

We now examine the complexity of our procedure.

Theorem 2: $\overline{\mathrm{LABEL}}(A,C,r)$ has complexity $O(\min\{\log|C|, \log|A|\})$.

Proof: For brevity we again restrict ourselves to the case $B^*(A) \geq B(^*C)$ and $\log|C| \leq \log|A|$ treated above.

First, I and II each require $O(\log|C|)$ time since each process at most $O(B^*(\gamma_1)) = O(\log|C|)$ frames. The construction of $\mathrm{deck}((A \vee C)_r)$ takes $O(1)$ time since it consists in just appending $\mathrm{deck}(\tau_u \vee \gamma_1)$ below a certain position in $\mathrm{deck}(A)$. This position is accessed in $O(1)$ time since after performing I and II we are just below the pointer in $\mathrm{GAP}(A)$ that leads to this position.

$\mathrm{GAP}(A \vee C)_r)$ is constructed in $O(1)$ time in case A, and in $O(\log|C|)$ time in case B. In A we simply adjust one or two pointers at the position of $\mathrm{GAP}(A)$ reached after I and II. In B we first construct $\mathrm{GAP}(\tau_u \vee \gamma_1)$. This takes $O(\log|C|)$ time since we are just scanning frames $\mathrm{sign}(\tau_i \vee \gamma_j)$ of $\mathrm{deck}(\tau_u \vee \gamma_1)$ all with B^* values $\leq B^*(\gamma_1)$. The number of such frames is at most $B^*(\gamma_1)$ by Lemma 4, and $B^*(\gamma_1) \leq O(\log|C|)$. We then splice together $\mathrm{GAP}(\tau_u \vee \gamma_1)$ and $\mathrm{GAP}'(A)$ by at worst deleting two pointers (the position below which we have already accessed), and this is $O(1)$ time. Hence the total time of $\overline{\mathrm{LABEL}}(A,C,r)$ is $O(\log|C|)$, as required. □

The complexity of our algorithm becomes the following. The last application of LABEL, namely $\mathrm{LABEL}(B_{s_1},B_{s_2},r)$, takes time at most which is $O(\log(\frac{n}{2}))$ by Theorem 2. The level of recursion below required at most two applications of LABEL and this took time $\le O(2\log(\frac{n}{2}))$. The k'th level of recursion takes $\le O(2^{k-1}\log(n/2^k))$. Hence the total time is

$\le O(\sum_{k=1}^{\lfloor \log b \rfloor} 2^{k-1}\log(n/2^k))$. Reversing the sum and letting C be the constant $C = \log(n/2^{\lfloor \log n \rfloor})$, so $C \le \log 2$, we get time $O(nC + \frac{n}{2}(C + \log 2) + \frac{n}{4}(C + \log 4) + \cdots) \le O(n)(\sum(1/2^k))$ $+ O(n)(\sum(\log(2^k)/2^k)) = O(n)$.

Remark: The author thanks I.H. Sudborough and J. Ellis for discussions which inspired the improvement in time complexity to $O(n)$.

References

1. I. Arany, L. Szoda, and W. Smith, "An improved method for reducing the bandwidth of sparse symmetric matrices", Proc. 1971 IFIP Congress, 1246-1250.
2. M. Behzad, G. Chartrand, and L. Lesniak-Foster, Graphs and Digraphs, Prindle, Weber, Schmidt (1979).
3. J. Chvatalova, "On the bandwidth problem for graphs", Thesis, Dept. of Combinatorics and Optimization, Univ. of Waterloo, 1981.
4. P. Chinn, J. Chvatalova, A. K. Dewdney, and N. E. Gibbs, "The bandwidth problem for graphs and matrices", J. of Graph Theory 6 (1982) 223-254.
5. F. R. K. Chung, "On the cutwidth and topological bandwidth of a tree", Technical Report, Bell Laboratories, Murray Hill, New Jersey (1982).
6. M. R. Garey, R. L. Graham, D. S. Johnson, and D. Knuth, "Complexity results for bandwidth minimization", SIAM J. Applied Mathematics 34 (1978) 447-495.

7. E. Gurari and I. H. Sudborough, "Improved dynamic programming algorithms for the bandwidth minimization problem and the min cut linear arrangement problem", Technical report, Dept. of Electrical Engineering and Computer Science, Northwestern University, Evanston, Illinois (1982).
8. F. Harary, Graph Theory, Addison-Wesley, Reading (1969).
9. W. Loui and A. B. Sherman, "Comparative analysis of the Cuthill-McKee ordering algorithms for sparse matrices", SIAM J. Numerical Analysis (1976).
10. F. S. Makedon, C. H. Papadimitriou, and I. H. Sudborough, "Topological bandwidth", to appear.
11. B. Monien and I. H. Sudborough, "Bandwidth constrained NP-complete problems", Proc. 11th Annual ACM Symposium on Theory of Computing (1981) 207-217.
12. T. Ohtsuki, H. Mori, E. A. Kuh, T. Kashiwabara, and T. Fujisawa, "One-dimensional logic gate assignments and interval graphs", IEEE Trans. on Circuits and Systems, CAS-26, 9 (1979) 675-684.
13. C. H. Papadimitriou, "The NP-completeness of the bandwidth minimization problem", Computing 16 (1976) 237-267.
14. J. B. Saxe, "Dynamic programming algorithms for recognizing small bandwidth graphs in polynomial time", SIAM J. on Algebraic and Discrete Methods, 1 (1980) 363-369.
15. I. H. Sudborough, "Bandwidth constraints on problems complete for polynomial time", to appear.
16. S. Trimberg, "Automated chip layout", IEEE Spectrum (1982) 38-45.
17. M. Yannakakis, "A polynomial algorithm for the min cut linear arrangement of trees", 24th Annual IEEE Symposium on Foundations of Computer Science, Tucson, Arizona, (1983).

GENERALIZED COLORINGS OF GRAPHS

Christina M. Mynhardt[1]
University of South Africa

Izak Broere[2]
Rand Afrikaans University

ABSTRACT

Let $F \neq K_1$ be any graph. A graph G is called F-free if F is not an induced subgraph of G. If F and G are graphs, then an F-coloring of G is a partition of V(G) into subsets that induce F-free subgraphs of G. The F-chromatic number $F\langle G\rangle$ of G is the smallest number of color classes needed for an F-coloring of G. In this paper we study F-colorings and the F-chromatic number of graphs. The relationship between F-colorings and cocolorings is indicated and graphs for which $K_{1,2}$-colorings and cocolorings coincide are characterized. The problem of finding a graph G subject to certain restrictions for which $F\langle G\rangle$ is arbitrarily large, is investigated. In this connection, we list several classes of graphs F and H for

---

[1]Research supported in part by the South African Council for Scientific and Industrial Research.

[2]Research supported by the Rand Afrikaans University and the South African Council for Scientific and Industrial Research.

which there exists, for each integer $n \geq 1$, a graph G with $F\langle G\rangle \geq n$ while G does not contain H as induced subgraph.

1. INTRODUCTION

We generally follow the notation and terminology of [1]. In particular, we write $F < G$ $(F \not< G)$ to denote that the graph F is (is not) an induced subgraph of G.

A graph G is called F-*free* if $F \not< G$. If F and G are graphs, then an F-*coloring* of G is a partition of V(G) into subsets (also called *color classes*) that induce F-free subgraphs of G. Such a partition exists if and only if $F \neq K_1$ and consequently we henceforth assume that $F \neq K_1$. The F-*chromatic number* $F\langle G\rangle$ of G is the smallest number of color classes needed for an F-coloring of G. Clearly, $F\langle G\rangle$ exists and is finite for all G.

Other generalizations of colorings of graphs have been studied in, for example, [3], [6], [7], [8], [11] and [12]. We study F-colorings and the F-chromatic number of graphs with particular emphasis on the question of raising $F\langle G\rangle$ with restrictions on G, while F-colorings of outerplanar and planar graphs are discussed in [2].

We begin with the following remarks :

Remarks.

1. For any graphs F and G, $F\langle G\rangle = 1$ if and only if $F \not< G$.
2. If $F_1 < F_2$, then every F_1-coloring of a graph G is also an F_2-coloring of G. Hence $F_2\langle G\rangle \leq F_1\langle G\rangle$ for all G. In particular, if F is nonempty, then $K_2 < F$ and consequently, $F\langle G\rangle \leq K_2\langle G\rangle = \chi(G)$ for all G.
3. If $G_1 < G_2$, then every F-coloring of G_2 induces an F-coloring of G_1. Hence $F\langle G_1\rangle \leq F\langle G_2\rangle$ for all F.
4. Note that $F < G$ if and only if $\bar{F} < \bar{G}$. Therefore, $F\langle G\rangle = \bar{F}\langle \bar{G}\rangle$ for all F and G.

2. Cocolorings and F-colorings

A *cocoloring* of a graph G, introduced by Lesniak-Foster and Straight in [8], is defined as a partition of V(G) into subsets that induce complete or empty subgraphs of G. The smallest number of colors needed for a cocoloring of G is called the *cochromatic number* z(G) of G. There is a no graph F satisfying $F\langle G\rangle = z(G)$ for all G, since for such a graph F we would have $F\langle K_{1,2}\rangle = z(K_{1,2}) = 2$ and hence $F < K_{1,2}$ — no such F satisfies the requirements, however. Taking $F = K_{1,2}$ we do have the following : If G is any graph, then a partition $V_1,\ldots,V_n$ of V(G) is an F-coloring of G if and only if for each i, $i = 1,\ldots,n$, the graph $\langle V_i\rangle$ is the disjoint union of complete graphs. Consequently, every cocoloring of G is an F-coloring of G. We characterize the graphs for which the converse is also true.

Theorem 1.

For any graph G the following statements are equivalent :

(i) Every $K_{1,2}$-coloring of G is a cocoloring of G.

(ii) G is $\overline{K_{1,2}}$-free.

(iii) G is a complete p-partite graph for some p.

Proof.

We show that (i) implies (ii) by contradiction. Suppose that $\overline{K_{1,2}} < G$ and let $\overline{K_{1,2}} \cong \langle \{u,v,w\}\rangle$ for $\{u,v,w\} \subseteq V(G)$. Color G in such a way that {u,v,w} forms one color class, while G-{u,v,w} receives any $K_{1,2}$-coloring in other colors. This coloring is a $K_{1,2}$-coloring of G but not a cocoloring.

To prove that (ii) implies (iii), suppose $\overline{K_{1,2}} \not< G$ and form a partition $V_1,\ldots,V_p$ of V(G) in such a way that V_1 is a maximal independent set of G and for each i, $i = 1,\ldots,p-1$, V_{i+1} is a maximal independent set of $G-(V_1 \cup \ldots \cup V_i)$. If $u \in V_i$ and $v \in V_j$ where $i < j$, then v must be adjacent to some $w \in V_i$. If u and v are nonadjacent, then $\langle \{u,v,w\}\rangle \cong \overline{K_{1,2}}$, which is a contradiction. Hence u and v are adjacent and thus $G \cong K_{n_1,\ldots,n_p}$ with $|V_i| = n_i$, $i = 1,\ldots,p$.

Finally, to show that (iii) implies (i), we consider any $K_{1,2}$-coloring of $G = K_{n_1,\ldots,n_p}$ with p-partition $V_1,\ldots,V_p$. If V is any color class with respect to such a $K_{1,2}$-coloring, then V is the union of complete graphs. Hence $V \subseteq V_i$ for some i or $|V \cap V_i| \leq 1$ for each i, i = 1,...,p. But then $\langle V \rangle$ is empty or complete and therefore the $K_{1,2}$-coloring of G is also a cocoloring of G. □

Corollary 1.

(i) $K_{1,2}\langle K_{n_1,\ldots,n_p}\rangle = \min\limits_{1 \leq j \leq p} \{p, p-j+n_j\}$

(ii) $\overline{K_{1,2}}\langle K_{n_1} \cup \ldots \cup K_{n_p}\rangle = \min\limits_{1 \leq j \leq p} \{p, p-j+n_j\}$.

Proof.

By Theorem 1, $K_{1,2}\langle K_{n_1},\ldots,{n_p}\rangle = z(K_{n_1},\ldots,{n_p})$ and hence (i) follows from Theorem 2 of [8], while (ii) is a restatement of (i) using Remark 4. □

3. Raising $F\langle G\rangle$ with restrictions on G

For all $F(\neq K_1)$, the F-chromatic number $F\langle G\rangle$ can be made arbitrarily high with suitable G. This can be seen inductively by noting that $F\langle F\rangle = 2$ and that if $F\langle G\rangle = k$ and $H = F[G]$, the composition of F and G, then $F\langle H\rangle > k$, as is shown in Lemma 1. (The *composition* $H = F[G]$ of F and G has $V(H) = V(F) \times V(G)$, and two vertices (u_1,u_2) and (v_1,v_2) of H are adjacent if and only if $u_1v_1 \in E(F)$, or $u_1 = v_1$ and $u_2v_2 \in E(G)$.)

Lemma 1.

If F,G and H are graphs such that $F\langle G\rangle = k$ and $H = F[G]$, then $F\langle H\rangle \geq k+1$.

Proof.

Let F,G and H satisfy the hypothesis of the lemma. Since $G < H$, we see that $F\langle H\rangle \geq F\langle G\rangle$. Suppose $F\langle H\rangle = k$, let $M_1, \ldots, M_p$ with $p = |V(F)|$ be the copies of G in H and consider any F-coloring $V_1, \ldots, V_k$ of H. Since $F\langle G\rangle = H$, it is clear that $|V(M_i) \cap V_j| \geq 1$ for each $i = 1, \ldots, p$ and each $j = 1, \ldots, k$. Consequently, $F < \langle V_j \rangle$ for each $j = 1, \ldots, k$ (by the construction of H). This contradiction establishes the lemma. □

It is also known that $\chi(G)$ can be made arbitrarily high with K_3-free graphs G (see [4] and [10]) and even with graphs with arbitrary girth (see [5] and [9] for a probabalistic and constructive proof respectively). The same can be done with z(G) since $z(G) = \chi(G)$ if G is K_3-free (see Theorem 1 of [8]). We now consider the corresponding problem in the present situation :

For which graphs F and H can $F\langle G\rangle$ be made arbitrarily high with graphs G for which $H \not< G$?

The following remarks are relevant :

Remarks.

5. Let $H_1 < H_2$. If G is H_1-free, then G is also H_2-free. Therefore, if $F\langle G\rangle$ can be made arbitrarily high with H_1-free graphs G, then so with H_2-free graphs G.
6. If $H < F$, then $F\langle G\rangle \leq H\langle G\rangle$ (by Remark 2) and $H\langle G\rangle = 1$ if $H \not< G$ (by Remark 1). Therefore, we henceforth assume that F and H are graphs with $H \not< F$.

We explicitly state the aforementioned result of Erdös [5] and Lovász [9] in the following theorem :

Theorem A.

For every two integers $m,n \geq 2$ there exists an n-chromatic graph whose girth exceeds m.

Corollary 2.

Let H be a graph that contains a cycle. For any integer $n \geq 1$ there exists an n-chromatic H-free graph G.

Proof.

Let the girth of a graph G be denoted by g(G) and let n be any positive integer. If n = 1, then K_1 is the desired graph. If $n \geq 2$ then, by Theorem A, there exists an n-chromatic graph G with $g(G) > g(H)$. Clearly, G is H-free. □

We are now ready to present one of our main results, which provides a generalization of the result in [4] and [10].

Theorem 2.

For every pair of positive integers n and k there exists a K_3-free graph $G_{n,k}$ with $K_{1,n}\langle G_{n,k}\rangle = k$.

Proof.

We prove this result by induction on k — note that for all n, $K_{1,n}\langle K_1\rangle = 1$ and $K_{1,n}\langle K_{1,n}\rangle = 2$.

Let $G = G_{n,k}$ be a K_3-free graph with $K_{1,n}\langle G\rangle = k$. We first construct a graph H as follows : Take k-1 copies $G_1, \dots, G_{k-1}$ of G and let $V(G_i) = \{v_{i1}, \dots, v_{ip}\}$ for i = 1, ... ,k-1, where p is the order of G. For each of the p^{k-1} possible choices of one vertex v_{ij} from each G_i, i = 1, ... ,k-1, add (k-1)(n-1)+1 new vertices and join each of these to each particular choice of v_{ij} in G_i, i = 1, ... ,k-1, to form H. Let

$$S = V(H) - \bigcup_{i=1}^{k-1} V(G_i).$$

Clearly, $K_{1,n}\langle H\rangle \geq k$. Moreover, any $K_{1,n}$-coloring of $G_1 \cup \dots \cup G_{k-1}$ in k colors can be extended to a $K_{1,n}$-coloring of H in k colors by choosing, for each $v \in S$, a color not appearing at a neighbor of v. Hence $K_{1,n}\langle H\rangle = k$. Obviously, H is K_3-free. Now consider any $K_{1,n}$-coloring of H in the colors 1, ... ,k and let j be any one of these colors; we may assume that $j = k$. Each color appears at least once in each G_i, $i = 1, \ldots, k-1$ and hence we may choose a vertex w_i of G_i which is colored i, $i = 1, \ldots, k-1$. But then each of the colors $i, \ldots, k-1$ can appear at most $n-1$ times among the $(k-1)(n-1)+1$ new vertices adjacent to $w_1, \ldots, w_{k-1}$ and thus color k appears at least once among these vertices. Hence, with respect to any $K_{1,n}$-coloring of H in k colors, each color is used for the vertices in S.

The required graph $G_{n,k+1}$ can now be constructed by taking n copies $H_1, \ldots, H_n$ of H and by adding one new vertex which, for each $i = 1, \ldots, n$, is joined to each vertex in S_i where $S_i = V(H_i) - \bigcup_{j=1}^{k-1} V(G_j)$. Clearly, $G_{n,k+1}$ is K_3-free. Also, the above property of $K_{1,n}$-colorings of H implies that $K_{1,n}\langle G_{n,k+1}\rangle \geq k+1$. Finally, it is easily seen that $K_{1,n}\langle G_{n,k+1}\rangle \leq k+1$ and hence $K_{1,n}\langle G_{n,k+1}\rangle = k+1$. □

Corollary 3.

For every pair F, H of graphs with $F < K_{1,n}$ and $K_3 < H$ and for every integer $k \geq 1$ there exists an H-free graph G with $F\langle G\rangle = k$. □

In [11], a construction is given showing that $K_p\langle G\rangle$ can be made arbitrarily high with K_{p+1}-free graphs G for $p \geq 2$. Of course, this immediately generalizes to

Corollary 4.

For any pair of integers p and q with $2 \leq p < q$ and with F and H satisfying $F = K_p$ and $K_q < H$, $F\langle G\rangle$ can be made arbitrarily high with H-free graphs G. □

Our next theorem shows that if H contains the path P_n for some $n \geq 4$ as induced subgraph while F is P_n-free, then $F\langle G\rangle$ can be made arbitrarily high using H-free graphs G. We first prove a lemma.

Lemma 2.

If $H \cong P_n$ for some $n \geq 4$ while $P_n \not< F$, then, for any positive integer k there exists an H-free graph G_k with $F\langle G_k \rangle \geq k$.

Proof.

We prove this result by induction on k. Clearly, we may choose $G_1 = K_1$ and $G_2 = F$.

Suppose G_k is H-free with $F\langle G_k \rangle \geq k \geq 2$ and let $G_{k+1} = F[G_k]$. By Lemma 1, $F\langle G_{k+1} \rangle \geq k+1$. We show that $H \not< G_{k+1}$ by contradiction. Let $M_1, \ldots, M_p$ with $p = |V(F)|$ be the copies of G_k in G_{k+1} and suppose $H < G_{k+1}$. By the induction hypothesis, $H \not< G_k$ and therefore in G_{k+1}, at least two copies of G_k must contain vertices of H. On the other hand, if $|V(M_i) \cap V(H)| \leq 1$ for each $i = 1, \ldots, p$, this would imply that $H < F$ which it is not. Hence $|V(M_i) \cap V(H)| \geq 2$ for at least one i; say M_1 has this property. Since H is connected and not entirely contained in M_1, there is a vertex $u \in V(M_i) \cap V(H)$ for some $i = 2, \ldots, p$, which, by the construction of G_{k+1}, is adjacent to each vertex in $V(M_1) \cap V(H)$. Thus, since H is a path, $|V(M_1) \cap V(H)| = 2$; say $V(M_1) \cap V(H) = \{w_1, w_2\}$. If $\langle \{w_1, w_2\} \rangle$ is non-empty, H would contain a triangle, contradicting $H \cong P_n$, $n \geq 4$. Consequently, there is also a vertex $v \in V(M_j) \cap V(H)$ for some $j = 2, \ldots, p$ adjacent to w_1 and w_2 while $u \neq v$. But then $\langle \{w_1, w_2, u, v\} \rangle$ contains a cycle of length four — a contradiction which proves the lemma. □

Theorem 3.

Let F and H be graphs such that for some $n \geq 4$, $P_n < H$ but $P_n \not< F$. For any positive integer k there exists an H-free graph G_k with $F\langle G_k \rangle \geq k$.

Proof.

By Lemma 2, the theorem holds if $H \cong P_n$ and therefore by Remark 5 it also holds for arbitrary graphs H with $P_n < H$. □

Two corollaries to Theorem 3 are interesting. The first follows directly from Theorem 3 since $P_n < C_{n+1}$.

Corollary 5.

Let F and H be graphs such that $C_n < H$ for some $n \geq 5$ while F is C_n-free. For any positive integer k there exists an H-free graph G_k with $F\langle G_k \rangle \geq k$. □

Corollary 6.

For any pair of integers m and n with $n > m \geq 2$ and any integer $k \geq 1$, there exists a P_n-free graph G_k with $P_m\langle G_k \rangle \geq k$.

Proof.

If m = 2, then $P_2\langle G \rangle = \chi(G)$ for all graphs G and we may choose $G_k = K_k$. For $m \geq 3$ and $n > m$, note that $P_n \not< P_m$ and apply Theorem 3.

If we use the construction discussed in the proof of Lemma 2 to find a P_n-free graph G_k with $P_m\langle G_k \rangle \geq k$ for k large, we see that G_k contains complete graphs of large order. Interestingly, if m = 4 we can also raise $P_4\langle G \rangle$ arbitrarily using graphs G which do not contain complete graphs of large order. The following lemma will be useful in proving this.

Lemma 3.

Let F and H be graphs sucht that H contains a cycle and F is H-free. If there exists a constant f with $0 < f \leq 1$ such that $F\langle G \rangle \geq f\chi(G)$ for all H-free graphs G, then $F\langle G \rangle$ can be made arbitrarily large with H-free graphs G.

Proof.

By Corollary 2, $\chi(G)$ can be made arbitrarily large using H-free graphs G and consequently so can $F\langle G \rangle$, by choosing G in such a way that $\chi(G)$ is sufficiently large. □

Theorem 4.

For any integers $k \geq 1$ and $n \geq 3$, there exists a K_n-free graph G with $P_4\langle G\rangle \geq k$.

Proof.

Consider any K_n-free graph G and suppose $P_4\langle G\rangle = m$. Let $V_1,\ldots,V_m$ be a P_4-coloring of G. Then for each $i = 1,\ldots,m$, the graph $\langle V_i\rangle$ contains neither K_n nor P_4 as induced subgraph and consequently, $\chi(\langle V_i\rangle) \leq n-1$ (see, for example, Theorem 11.7, p.238 of [1]). But $\chi(G) \leq \sum_{i=1}^{m} \chi(\langle V_i\rangle) \leq m(n-1)$ and hence $F\langle G\rangle \geq \chi(G)/(n-1)$. The desired result now follows by using Lemma 3. □

Lemma 3 could possibly be used to obtain more results of this nature. However, note that a suitable constant f does not exist for all graphs F :

Theorem 5.

Let F and H be graphs such that H contains a cycle while $H \not\leq F$. If there exists a constant f with $0 < f \leq 1$ such that $F\langle G\rangle \geq f\chi(G)$ for all H-free graphs G, then F is acyclic.

Proof.

Suppose F has a cycle. By Theorem A, if $n \geq 2$ is any integer, there exists an n-chromatic graph G with $g(G) \geq \max\{g(F), g(H)\}$. Clearly, $F \not\leq G$ and hence $F\langle G\rangle = 1$ (by Remark 1). However, G is also H-free and therefore, by the hypothesis of the theorem, $F\langle G\rangle \geq f\chi(G)$ for some constant f. Thus $f \leq 1/\chi(G)$ which can be made arbitrarily small by choosing n large enough. Hence $f = 0$, a contradiction. □

We have now discussed several classes of graphs F and H for which $F\langle G\rangle$ can be made arbitrarily high with H-free graphs G. We conclude this paper with a conjecture which, if true, would give a characterization of those graphs F and H for which this is possible.

Conjecture 1.

For any pair of graphs $F(\neq K_1)$ and H and any integer $k \geq 1$ there exists an H-free graph G with $F(G) \geq k$ if and only if $H \not< F$.

REFERENCES

1. M. Behzad, G. Chartrand and L. Lesniak-Foster, *Graphs and Digraphs*, Prindle, Weber & Schmidt, Boston, 1979.
2. I.Broere and C.M. Mynhardt, Generalized colorings of outer-planar and planar graphs, *Proceedings of the Fifth International Conference on the Theory and Applications of Graphs*, John Wiley & Sons, New York, 1984.
3. G. Chartrand, D.P. Geller and S. Hedetniemi, A generalization of the chromatic number, *Proc. Camb. Phil. Soc.* 64 (1968), 265-271.
4. B. Descartes, Solution to problem 4526, *Amer. Math. Monthly* 61 (1954), 352-353.
5. P. Erdös, Graph Theory and Probability II, *Canad. J Math.* 13 (1961), 346-352.
6. F. Harary, Conditional colorability in graphs, *Graphs and Applications*, Wiley, New York, 1984 (to appear).
7. F. Harary and M. Plantholt, Graphs with the line-distinguishing chromatic number equal to the usual one, *Utilitas Mathematica* 23 (1983), 201-207.
8. L. Lesniak-Foster and H.J. Straight, The cochromatic number of a graph, *Ars. Comb.* 3 (1977), 39-45.
9. L. Lovász, On the chromatic number of finite set-systems, *Acta Math. Acad. Sci. Hungar* 19 (1968), 59-67.
10. J. Mycielski, Sur le coloriage des graphes, *Coll Math.* 3 (1965), 161-162.

11. H. Sachs and M. Schäuble, Konstruktion von Graphen mit gewissen Farbungseigenschaften, *Beiträge zur Graphentheorie* (Kolloquim Manebach, 1967), 131-136, Teubner, Leibzig, 1968.
12. S. Stahl, n-tuple colorings and associated graphs, *J. Combinatorial Theory Ser. B* 20 (1976), 185-203.

AN UPPER BOUND ON THE CHROMATIC INDEX OF MULTIGRAPHS

Takao Nishizeki

Tohoku University

Kenichi Kashiwagi

Mitsubishi Electric Corp.

ABSTRACT

A new upper bound is obtained for the chromatic index $q^*(G)$ of multigraphs G. Let $p(G)$ be a trivial lower bound of $q^*(G)$ defined as follows:

$$p(G) = \mathrm{MAX}_{H \subset G} \left\lceil |E(H)| / \lfloor |V(H)|/2 \rfloor \right\rceil$$

where H runs over all subgraphs of G, E(H) is the set of edges of H, and V(H) the set of vertices of H. Let d(G) be the maximum degree of G. Then the upper bound is expressed as follows:

$$q^*(G) \leq \mathrm{MAX}\{p(G), \lfloor 1.1d(G)+0.8 \rfloor\}.$$

A polynomial algorithm can be extracted from the proof which edge-colors any given multigraph G=(V,E) with at most $\lfloor 1.1q^*(G)+0.8 \rfloor$ colors. The running time is $O(|E|(d(G)+|V|))$ and the storage space is $O(|E|)$.

1. Introduction

The edge-coloring problem arises in many applications, including various scheduling and partitioning problems [FW]. The problem is simply stated: color the edges of a given multigraph G using as few colors as possible, so that no two adjacent edges receive the same color. This minimum number of colors is called the chromatic index of G and denoted by $q^*(G)$.

Let d(G) denote the maximum degree of a vertex. By a well-known result due to Vizing, the chromatic index of a simple graph (that is, one without multiple edges) is at most d(G)+1 [Vi]. The situation for graphs that are not simple, often called multigraphs, is more complicated, and it is this case that will be considered here.

In this paper we obtain a new upper bound for the chromatic index $q^*(G)$ of multigraphs G:

$$q^*(G) \le \mathrm{MAX}\{p(G), \lfloor 1.1d(G)+0.8 \rfloor\}.$$

Here p(G) is a trivial lower bound on $q^*(G)$ defined as follows:

$$p(G)=\mathrm{MAX}_{H\subset G} \lceil |E(H)|/\lfloor |V(H)|/2 \rfloor \rceil$$

where H runs over all subgraphs of G, E(H) is the set of edges of H, and V(H) the set of vertices of H. The upper bound follows a long history of research dating back to Shannon [Sh], who proved the following upper bound: $q^* \le \lfloor 3d/2 \rfloor$. His bound has been successively improved by Goldberg [Go1,Go2] and Andersen [An]. The best known one, given by Goldberg [Go2], is the following:

$$q^*(G) \le \mathrm{MAX}\{p(G), \lfloor (9d+6)/8 \rfloor\}.$$

Our bound improves his bound slightly and all the previous bounds [Sh,Go1,An].

In view of the potential applications, it would be useful to have an efficient algorithm capable of coloring any multigraph G with the minimum number $q^*(G)$ of colors. Unfortunately no such efficient algorithm is currently known. Moreover, Holyer has shown that the edge-coloring problem is NP-hard [Ho], and therefore it seems unlikely that any such polynomial-time algorithm exists. He actually proved that deciding whether a simple graph has chromatic index of 3 or less is NP-complete. This implies that unless P=NP, there does not exist a polynomial-time approximation algorithm that finds an edge-coloring for multigraphs that uses at most $(4/3-\epsilon)q^*(G)$ for any ϵ [NS].

From the proof of our bound a polynomial-time algorithm can be obtained which finds an edge coloring of G that uses at most $(4/3)q^*(G)$ colors; moreover it uses at most $\lfloor 1.1q^*(G)+0.8 \rfloor$ colors. This result improves on the following sequence of algorithms: Nishizeki and Sato's [NS], which achieves the $(4/3)q^*$ bound as well, but uses as many as $\lfloor (5q^*+2)/4 \rfloor$ colors; Hochbaum and Shmoys' [HS] which uses at most $\lfloor (7q^*+4)/6 \rfloor$ colors; and Hochbaum, Nishizeki and Shmoys' [HNS] which uses at most $\lfloor (9q^*+6)/8 \rfloor$ colors. In the terminology of [GJ], the "absolute performance ratio" of our algorithm is 4/3 while the "asymptotic performance ratio" is 1.1. The running time of the algorithm is $O(|E|(d(G)+|V|))$ and the storage space is $O(|E|)$.

2. Preliminaries

This section introduces terminology. Throughout the paper, $G=(V,E)$ denotes a given *multigraph* of vertex set V and edge set E, which may contain multiple edges but no self-loops. Since we deal with only a multigraph, we simply call G a *graph*. An edge-coloring of G with at most q colors is called a *q-coloring*. Denote by uv an edge joining vertices u and v. Assume that a graph G is edge-colored with a set of q colors. Color c in the set is a *missing color* of vertex $v \in V$ (or v *misses* c) if color c is assigned to none of the edges incident to v. Denote by M(v) the set of all the missing colors of v. Denote by C(u,v) the set of colors assigned to the multiple edges joining vertices u and v. An edge colored c is called a *c-edge*. For two colors a and b, the spanning subgraph of G induced by all the edges colored a or b is called an *ab-subgraph*, and denoted by G[a,b]. Each connected component of G[a,b] is either a path or a cycle, in which edges are colored alternately a or b. Such a path (or cycle) is called an *ab-alternating path* (*cycle*). We denote by G[a,b;x] the connected component of G[a,b] containing vertex x. A vertex $x \in V$ is an end of an ab-alternating path if and only if a or $b \in M(x)$. Interchanging the colors a and b of the edges in an ab-alternating path or cycle yields another q-coloring of G. Interchanging the colors of an alternating path or cycle is often called simply *recoloring* a path or cycle. If $a \in M(x)$ for a vertex $x \in V$, then G[a,b;x] is an ab-alternating path starting from x. If all the edges of G except an edge e=xy are colored with q colors, $a \in M(x)$ and $b \in M(y)$, then the ab-alternating path between x and y, if any, is called an *ab-critical path* in particular. Note that every critical path contains an odd number of vertices. The ab-critical path is denoted by P(a,b), and the number of its vertices is denoted by |P(a,b)|. If there is no ab-critical path, |P(a,b)| is defined to be infinite.

3. Upper bound.

Let $q(G)=MAX\{p(G), \lfloor 1.1d(G)+0.8 \rfloor\}$. The following theorem is the main result of this paper.

Theorem 1. Every multigraph $G=(V,E)$ satisfies $q^*(G) \leq q(G)$.

Suppose to the contrary that $q^*(G) > q(G)$ for a graph G, and let G be such a graph with a minimum number of edges. Let $e=xy$ be an arbitrary edge of G. By the assumption $G'=G-e$ can be edge colored with q(G) colors. Then we will show that a q-coloring of G' can be extended to a q-coloring of G, that is, a contradiction occurs.

The following facts can be easily derived.

Fact 1. $d(G) \geq 3$.

Fact 2. $q(G) \geq \lfloor 1.1d(G)+0.8 \rfloor \geq d(G)+1$.

Fact 3. In a q-coloring of G' every vertex has at least one missing color; in particular x and y have at least two.

Fact 4. The ends x and y of the uncolored edge e have no common missing color.

Fact 5. G' has an ab-critical path for any $a \in M(x)$ and $b \in M(y)$.

The following lemma has been proved in [Go2, Lemma 4].

Lemma 1. No two vertices on any critical path have a common missing color.

We obtain the following lemma from the definition of $q(G)$.

Lemma 2. Let S be a set of vertices no two of which have a common missing color. Then we have

(a) If $x, y \in S$ then $|S| \leq 10$; and

(b) If $x \in S$ then $|S| \leq 20$.

Proof. (a) Clearly $|M(x)|, |M(y)| \geq q-d+1$, and $|M(v)| \geq q-d$ for every vertex v other than x and y. Since $\Sigma_{v \in S} |M(v)| \leq q$, we have

$$2(q-d+1) + (|S|-2)(q-d) \leq q,$$

which immediately yields

$$|S| \leq (q-2)/(q-d).$$

Noting that $d \geq 3$ and $q > (1.1d+0.8) - 1$, we easily have $|S| \leq 10$.

(b) similar to (a). Q.E.D.

Lemma 3. Let P(a,b) be the longest critical path among all the q-coloring of G'. Then P(a,b) has at most nine vertices.

Proof. By Lemma 1 no two vertices on P(a,b) have a common missing color. Therefore $|P(a,b)| \leq 10$. Since clearly $|P(a,b)|$ is odd, we have $|P(a,b)| \leq 9$. Q.E.D.

Thus we shall consider the case in which the longest critical path has 9, 7, 5 or 3 vertices. One of the following two lemmas, Lemma 4, has been proved in [Go2, Lemma 5].

Lemma 4. Any two critical paths P(a,b) and P(f,g) contain no two vertices u and v of a common missing color.

Lemma 5. At most 10 vertices are contained in the critical paths in a q-coloring of G'.

Proof. immediate from Lemmas 1, 2 and 4. Q.E.D.

Lemma 6. Let P(a,b) be an ab-critical path in a q-coloring of G', and let $c \in M(x)$. Recolor an ac- or bc-alternating path or cycle containing neither x nor y. Let P(a,b)' be the new ab-critical path in the resulting q-coloring of G. Then

$$|V(P(a,b) \cup P(a,b)')| \leq 10.$$

Let H be the subgraph of G induced by the vertices of the longest critical path P(a,b). An edge of G is said to leave H at vertex z if it joins a vertex z in H and a vertex not in H. The definition of q(G) implies $q \geq \lceil |E(H)|/\lfloor |V(H)|/2 \rfloor \rceil$. Clearly each of the q colors is assigned to at most $\lfloor |V(H)|/2 \rfloor$ edges of H. However H has an uncolored edge xy. Therefore there must be a color c assigned to at most $\lfloor |V(H)|/2 \rfloor - 1$ edges of H. Thus H contains three or more vertices to which no c-edge is incident. Since at most one of them misses color c, G' contains two or more c-edges leaving H. (In particular there exist three or more c-edges leaving H if no vertex in H misses c.)

The following Lemma 7 is implicitely mentioned in [Go2].

Lemma 7. One may assume that an edge colored with a missing color of x or y leaves H.

Lemma 8. Let uu' be a c-edge leaving H, where $u \in V(H)$ and $u' \notin V(H)$. Then G[a,c;u] (or G[b,c;u]) have two or more a-(or b-)edges on P(a,b).

Lemma 9. The longest critical path P(a,b) contains at least five vertices.

Proof. Suppose that P(a,b) contains exactly three vertices. Then by Lemma 7 one may assume that $c \in M(x)$ and two c-edges leave H at y and the intermediate vertex of P(a,b). Then G[b,c;y] contains exactly one b-edge on P(a,b), contradicting Lemma 8. Q.E.D.

By Lemmas 3 and 9 we shall consider three cases, in which the longest critical path has 9, 7, or 5 vertices, respectively. In either case, recoloring some components of two-colored subgraphs, one can derive a q-coloring of G' contradicting one of the lemmas above. Since the complete proof cannot be included for reasons of space, we will give a proof only for the following specific case.

Case 1: the longest critical path P(a,b) contains nine vertices.

Let $P(a,b)=xz_1z_2\ldots z_7y$. By Lemma 7 one may assume that $c \in M(x)$ and a c-edge leaves H. No vertex in P(a,b) except x misses color c, and at least two c-edges leave H. By Fact 5 there must be a cb-critical path P(c,b). Path P(c,b) contains no c-edges leaving H, for otherwise there would exist at least 11 vertices in $P(a,b) \cup P(c,b)$, contradicting Lemma 5. Therefore $V(P(c,b)) \subset V(P(a,b))$. By Lemma 3 P(c,b) contains at most 9 vertices. If $|P(c,b)|=9$, then there could not exist a c-edge leaving H. If $|P(c,b)|=7$, then the component of G[b,c] containing the c-edges leaving H would have exactly one b-edge on P(a,b), contradicting Lemma 8. Thus $|P(c,b)|=$ 5 or 3. We will consider only the following case.

Case 1.1: $|P(c,b)|=3$.

In this case path P(c,b) is xz_1y. Further consider only the following case.

Case 1.1.1: $c \in C(\overline{z_2}, \overline{z_7})$.

In this case all the c-edges leaving H are

incident to z_3, z_4, z_5 or z_6. Since $z_1z_2z_7yz_1$ is an ac-alternating cycle, $G[a,c;z_3]$ contains neither a-edge z_1z_2 nor z_7y. However $G[a,c;z_3]$ contains two or more a-edges on $P(a,b)$ by Lemma 8. Therefore $G[a,c;z_3]$ must contain z_3z_4 and z_5z_6 together with all the c-edges leaving H, consequently $G[a,c;z_3]= G[a,c;z_6]$.

We first show that a c-edge z_3z_3' leaves H. Otherwise, a c-edge joins z_3 with either z_4, z_5 or z_6 since $c\notin M(z_3)$. Suppose first $c\in C(z_3, z_4)$, then $G[a,c;z_6]$ containing two c-edges leaving H would contain exactly one a-edge z_5z_6 on $P(a,b)$, contradicting Lemma 8. Next suppose that $c\in C(z_3,z_5)$, then recoloring $G[a,c;z_3]$ would produce a new ab-critical path having 11 or more vertices, contradicting Lemma 3. Finally suppose $c\in C(z_3,z_6)$, then $G[b,c;z_3]$ would contain exactly one b-edge on $P(a,b)$, contrary to Lemma 8.

Similarly we can show that a c-edge z_6z_6' leaves H at z_6.

Let $P(a,b)'$ be the new ab-critical path produced by recoloring $G[a,c;z_3]$. Then $P(a,b)'$ contains $x,z_1,z_2,z_3,z_3',\ldots,z_6',z_6,z_7$ and y, and consequently $|V(P(a,b)\cup P(a,b)')| \geq 11$, contradicting Lemma 6.

REFERENCES

[AHU] A. V. AHO, J. E. HOPCROFT, and J. D. ULLMAN, The Design and Analysis of Computer Algorithm, Addison-Wesley, Reading, Mass., 1974.

[An] L.D. ANDERSEN, On edge-colouring of graphs, Math. Scand., 40 (1974), pp.161-175.

[FW] S. FIORINI and R. J. WILSON, Edge-Colourings of Graphs, Pitman, London, 1977.

[GJ] M. R. GAREY AND D. S. JOHNSON, Computers and Intractability, W. H. Freeman, San Francisco, 1979.

[G1] M.K. GOLDBERG, On multigraphs with almost maximal chromatic class (in Russian), Discret Analiz, 23 (1973), pp.3-7.

[G2] M. K. GOLDBERG, Edge-coloring of multigraphs: Recoloring technique, Journal of Graph Theory, 8 (1984), pp.123-137.

[HNS] D.S. HOCHBAUM, T. NISHIZEKI and D.S. SHMOYS, Edge coloring multighraphs is easier than it seems, submitted to J. of Algorithms, 1984.

[HS] D.S. HOCHBAUM and D.B. SHMOYS, An asymptotic 7/6-approximation algorithm for edge coloring multigraphs, preprint, 1983.

[Ho] I. J. HOLYER, The NP-completeness of edge-colourings, SIAM J. on Comput., 10 (1980), pp. 718-720.

[NS] T. NISHIZEKI and M. SATO, An algorithm for edge-coloring multigraphs (in Japanese), Trans., Inst. of Elecronics and Communication Eng., J67-D, 4 (1984), pp.466-471.

[Sh] C.E. SHANNON, A theorem on colouring lines of a network, J. Mathematical Phys. 28 (1949), pp.148-151.

[Vi] V.G. VIZING, On an estimate of the chromatic class of a p-graph (in Russian), Discret Analiz 3 (1964), pp.25-30.

Mailing address: Takao Nishizeki, Department of Electrical Communications, Faculty of Engineering, Tohoku University, Sendai 980, Japan.

BANDWIDTHS AND PROFILES OF TREES

Andrew M. Odlyzko
Bell Laboratories
Murray Hill, NJ 07974

and

Herbert S. Wilf
University of Pennsylvania
Philadelphia, PA 19104

1. Introduction

Let f be some assignment of distinct integers to the vertices of a graph G. The *width* $\beta(f)$ of f is

$$\beta(f) = \max_{(u,v)\varepsilon E(G)} f|(u) - f(v)| \qquad (1.1)$$

We will call such a function a *layout*, and if its width is $\leq K$, then it is a K-*layout* of the vertices of G.

The *bandwidth* of the graph G is then

$$\beta(G) = \min_{f} \beta(f). \qquad (1.2)$$

The following problem is known (Papadimitriou [1]) to be NP-complete: *Given a graph G and a positive integer K. Is the bandwidth of $G \leq K$* ? In [2], Garey, Graham, Johnson and Knuth showed that the question remains NP-complete even if G is restricted to the class of trees, and even more, if the trees have no vertex of valence greater than 3. In [4] Saxe gave an elegant algorithm that, for K fixed, decides in polynomial time if a graph has a bandwidth $\leq K$; hence the problem 'really becomes difficult' when K and V become large together.

The authors of [2] raised a number of questions for further investigation, one of which asked for information regarding the distribution of bandwidths of graphs. In [12] Turner has given a number of such results. In this paper we study the distribution of the bandwidths of trees.

We observe first (Theorem 2.1) that almost all (in the sense of labelled trees) trees of n vertices have bandwidths larger than $\sqrt{n}/g(n)$ ($g(n) \to \infty$), and that the *average* bandwidth of trees of n vertices is at least $C\sqrt{n}$ (Theorem 2.2).

Next we will describe a particular layout of the vertices of a tree that will be shown to have a relatively small width. That layout is based on the *layers* of the tree, i.e., the sets of vertices at constant distances from the root. We then study the distributions of the properties of labelled trees that are important in the analysis of this special layout. In particular, we introduce the idea of the *profile* of a tree T. If T is rooted at vertex #1, then let i_r be the number of vertices of T that are at a distance r from the root ($r = 0, 1, 2, \cdots$). The vector i is the profile of T.

We give an exact count (Theorem 2.4) of the labelled trees that have a given profile vector i.

If the *width* of T is the largest entry of the profile vector, then we show that the vertices of a tree T can always be assigned a layout whose width is at most twice the width of T. If $f(n,D)$ is the number of trees of n vertices that have width $\leq D$, we derive a quite explicit generating function (Theorem 3.1) for f, one that is similar to a kind of determinant that was first studied by Major MacMahon [5] in connection with his 'Master Theorem'.

We then consider the asymptotic behavior of $f(n,D)$ for fixed D and large n. The exact determination of this behavior involves a close study of the roots of a determinantal equation that is similar to, but by no means identical to, the characteristic equation of a matrix (it is, for example, of degree $\binom{n+1}{2}$ for an $n \times n$ matrix!). Fortunately, the similarity is sufficiently great that we are able to construct (Section 4, below), virtually intact, a theory of Perron-Frobenius type (see Gantmacher [6]) concerning the location of the roots of this kind of determinantal equation. We call this the theory of the *e-values* and *e-vectors* of a square matrix of positive entries.

With the aid of these root location theorems we will find, rather explicitly, the leading term of the asymptotic behavior of the number of n-trees whose width does not exceed D, for large n and fixed D (Theorem 5.1), and this in turn will lead to estimates of the number of trees of bandwidth $\leq D$ (Theorem 5.2).

In Section 6, we prove that the average width of trees of n vertices is $O(\sqrt{n \log n})$, and therefore that the average bandwidth of trees of n vertices is also $O(\sqrt{n \log n})$.

2. The bandwidths of trees

In our study of the bandwidths of trees we will consider labelled trees only. We must remark that the labels on the vertices of a labelled tree will never be used for any bandwidth determination. Instead, we will always assign a layout to its vertices in some other way.

First we quote the following well known

Lemma 2.1. *In a graph of* n *vertices and diameter* d, *the bandwidth is* $\geq (n-1)/d$. *In a tree* T *of* n *vertices and height* h, *the bandwidth is* $\geq (n-1)/(2h)$.

Now in Rényi and Szekeres [10] (see [8], [11] for further results on the distribution of heights of trees) it is shown that if $g(n) \to \infty$ as $n \to \infty$ then almost all labelled trees T have

$$\text{height}\ (T) < \sqrt{n}g(n)\ . \tag{2.1}$$

This, together with the lemma, yields

Theorem 2.1. *Fix a function* $g(n)$ *such that* $g(n) \to \infty$ *as* $n \to \infty$. *Then almost all labelled trees* T *of* n *vertices have* *bandwidth*$(T) > \sqrt{n}/g(n)$.

The next result gives a slightly sharper lower estimate for the average bandwidth. In Theorem 6.1 we will give the upper bound $O(\sqrt{n \log n})$ for this average.

Theorem 2.2. *The average bandwidth of trees of* n *vertices is larger than* $C\sqrt{n}$.

What can be said in the other direction? To handle this problem we need a good, if not optimal, way to lay out the vertices of a tree. The following scheme comes to mind at once.

Consider vertex #1 to be the root of the tree T. We will assign labels $f(v)$ to the vertices v of T, in increasing order of their distances from the root, with vertices at constant distance from the root receiving a consecutive block of labels.

Hence define the ith layer of the tree to be

$$\mathcal{L}(i) = \{v \in V(T) \mid \text{dist}(v,1) = i\} \quad (i = 0, 1, \cdots). \qquad (2.2)$$

We write $h = h(T)$ for the *height* of T (relative to the root 1), namely

$$h(T) = \max\{i \mid \mathcal{L}(i) \neq \emptyset\}. \qquad (2.3)$$

Further, we write $i_m = |\mathcal{L}(m)|$ $(m = 0, 1, \cdots, h)$.

Now we assign a layout f to $V(T)$. First $f(1) = 1$. Inductively, if layout labels $f(v)$ have been assigned to all vertices $v \in \mathcal{L}(j)$, then the interval

$$[i_0 + \cdots + i_j,\ i_0 + \cdots + i_{j+1} - 1]$$

will be the set of labels that will be used in the layout of layer $\mathcal{L}(j+1)$. Further, these labels will be assigned in an arbitrary manner, say, to fix ideas, in the same order of size as the original labels of these vertices.

In general what estimate for the bandwidth do we get from the above layout? Consider $|f(v') - f(v'')|$ for some edge (v',v'') of T. If $v' \in \mathcal{L}(j)$ and $v'' \in \mathcal{L}(j+1)$, then by construction, the difference $|f(v') - f(v'')|$ is at most $i_j + i_{j+1} - 1$.

Definition. *By the width* $W(T)$ *of a labelled tree* T *we mean*

$$W(T) = \max_m i_m. \tag{2.4}$$

Then we have shown

Theorem 2.3. *The bandwidth* $\beta(T)$ *and the width* $W(T)$ *of a labelled tree* T *satisfy*

$$\beta(T) \le 2W(T) - 1\ . \tag{2.5}$$

This theorem establishes an upper bound for the bandwidth of a tree in terms of the width of the tree. We remark here that in the other direction no such simple bound can exist. To be more precise, we give, for each fixed positive integer K, *an example of a tree whose width is* K *but whose bandwidth is just* 3.

We turn our attention to the study of the widths of trees (for a probabilistic study of the layers of trees see Stepanov [8]). It is convenient to introduce the entire *profile* of a tree, by which we mean the vector $i(T) = (i_0, i_1, \cdots, i_h)$ where $i_m = |\mathcal{L}(m)| (m = 0, h)$. We will use the symbol $|i|$ to mean the sum of the entries of the vector i.

Since $W(T)$ is the largest entry in the profile vector i, we will study $W(T)$ by the first asking for $N(i)$, the number of labelled trees that have profile vector i (note that every such tree has $n = |i|$ vertices).

By induction on the height we prove

Theorem 2.4. *The number* $N(i)$ *of labelled trees with profile vector* i *is*

$$N(i) = (|i| - 1)! \frac{i_0^{i_1}}{i_1!} \frac{i_1^{i_2}}{i_2!} \cdots \frac{i_{h-1}^{i_h}}{i_h!} . \tag{2.6}$$

3. Generating functions

In this section we will develop some generating functions that are relevant to the distribution of trees by widths. One of these will be the multivariate generator of trees by their profile vectors, and it will turn out to satisfy a functional equation (eq. (3.4) below) that generalizes the familiar equation that is the basis for several proofs of Cayley's theorem. The second one (eq. (3.9) below) will be a rather unusual kind that will generate the numbers of trees by their widths. It will be in matrix form, and it will be the starting point for the asymptotic analysis of the growth of the numbers of trees with given width.

First we consider the generating function $T(x)$, in infinitely many variables $x = (x_0, x_1, \cdots)$, that is defined by

$$\begin{aligned} T(x) &= \sum_T \frac{x^{i(T)}}{(|V(T)| - 1)!} \\ &= x_0 + x_0 x_1 + \frac{x_0 x_1^2}{2!} + 2 \frac{x_0 x_1 x_2}{2!} + \cdots , \end{aligned} \tag{3.1}$$

where the sum is over all trees of all sizes, and $i(T)$ is the profile vectors of T.

To write down the functional equation that T satisfies we will use the right shift operator $\Re$ that is defined by

$$\Re\,(x_0, x_1, \cdots) = (1, x_0, x_1, \cdots)$$

We show that the equation that $T(x)$ satisfies is

$$T(\Re\, x) = \exp(T(x)). \tag{3.4}$$

New we look at quite a different kind of generating function, this time for $f(n,D)$, the number of labelled trees of n vertices whose width is $\leq D$, i.e., whose profile vector has all of its entries $\leq D$. By Theorem 2.4 we have the formula

$$f(n,D) = (n-1)! \sum_{h \geq 0} \sum_{\substack{i_1, \cdots, i_{h-1} = 1, D \\ |i| = n-1}} \frac{1^{i_1} i_1^{i_2} \cdots i_{h-1}^{i_h}}{i_1! i_2! \cdots i_h!} \tag{3.5}$$

and our next task will be to find a civilized generating function for this quantity.

Consider the $D \times D$ matrix Y that is defined by

$$Y_{ij} = \frac{i^j t^j}{j!} \quad (i,\ j = 1,\ D) \tag{3.6}$$

where t is an indeterminate. Then we prove

Theorem 3.1. *If D is a fixed positive integer, and if the $D \times D$ matrix Y is defined by (3.6), and if $!(n,D)$ is the number of trees of n vertices and width $\leq D$, then we have the generating function*

$$\sum_{\ell=1}^{D} (I - Y)^{-1}_{1,\ell} = \sum_{n\geq 0} \frac{f(n+1, D)}{n!} t^n \tag{3.7}$$

in which the series converges for small t.

The asymptotic behavior of $f(n,2)$, for example, depends on the zero of $1 - t - 2t^2 + t^3$ that is closest to the origin. It is the positive real number $.5549581\cdots$, and so we have

$$\lim_{n\to\infty} f(n,2)^{1/n} = (.5549581\cdots)^{-1}$$
$$= 1.8019377\cdots \quad .$$

In the general case we have to deal with the generating function (3.7), which is somewhat similar to a type that was first studied by MacMahon in his 'Master Theorem'. For D fixed, the asymptotic behavior of $f(n,D)$ will be governed by the smallest root of the equation

$$\det(I - Y) = 0. \tag{3.9}$$

If we let $t = 1/\lambda$, then (3.9) can be written as

$$\det[\lambda^i \delta_{i,j} - \frac{i^j}{j!}]_{i,j=1}^{D} = 0. \tag{3.10}$$

This equation is reminiscent of the characteristic equation of a matrix, but the appearance of powers of λ on the diagonal destroys the illusion. Equation (3.10) is a polynomial equation of degree $\binom{D+1}{2}$ in λ, and so it has $\binom{D+1}{2}$ roots in the complex plane. We are interested in the one(s) of largest absolute value. We will prove, in the next section, that there is just one root of largest absolute value and it is real and positive. Next we give a minimax characterization of that root, à la Perron-Frobenius.

The $D(D + 1)/2$ roots of the equation (3.10) will be called the e-values of the matrix shown there. The next section is devoted to the study of the e-values of general square matrices with positive entries. In Section 5 the results will be applied to the asymptotics of the number of trees with given width.

4. The e-values of a matrix

By the e-values of a $D \times D$ matrix A we mean the roots $\lambda_1, \cdots, \lambda_N \left(N = \binom{D+1}{2}\right)$ of the polynomial equation

$$\det\left[\lambda^i \delta_{i,j} - a_{i,j}\right]_{i,j=1}^{D} = 0. \tag{4.1}$$

Corresponding to each e-value there is a ≥ 1-dimensional space of e-vectors x that satisfy

$$\sum_{j=1}^{D} a_{i,j} x_j = \lambda^i x_i \quad (i = 1,D).$$

The e-values of the $D \times D$ identity matrix, for example, are all of the jth roots of unity, for each $j = 1,D$.

We will assume henceforth that the entries $a_{i,j}$ of A are strictly positive, and we write

$$\mu = \min_{i,j}\{a_{i,j}\} > 0; \quad M = \max_{i,j}\{a_{i,j}\}. \tag{4.2}$$

Our aim now is to obtain a variational description of the largest e-value of A. If $x = (x_1, \cdots, x_D)$ is some vector, we write

$$I(x) = \{i | 1 \leq i \leq D, \ x_i \neq 0\}. \tag{4.3}$$

Further, we let Q_D denote the 'first quadrant of the unit sphere', defined by

$$Q_D = \{x \mid \|x\| = 1;\ x_i \geq 0\ (\forall i = 1, D)\}. \qquad (4.4)$$

Clearly Q_D is a compact subset of R^D.

For $x \in Q_D$ let

$$r(x) = \min_{i \in I(x)} \left\{ \frac{(Ax)_i}{(x)_i} \right\}^{1/i}. \qquad (4.5)$$

Lemma 4.1. *r is a continuous function of x in Q_D.*

Lemma 4.2. *Let x^* be a point of Q_D at which r attains its maximum value. Then $\forall i = 1, D$: $(x^*)_i > 0$.*

Lemma 4.3. *For every $x \in Q_D$, the number $r(x)$, defined by (4.5), has the property that*

$$r(x) = \max\{r \mid (Ax)_i - r^i x_i \geq 0\ (\forall i = 1, D)\}. \qquad (4.6)$$

Lemma 4.4. *Let x^* give a maximum value to $r(x)$ in Q_D. Then*

$$(Ax^*)_i - r^i(x^*)_i = 0 \quad (i = 1, \cdots, D). \qquad (4.8)$$

Definition. *The principal e-root of a matrix A of positive entries is the number*

$$r = r(A) = \max_{Q_D} r(x) = \max_{Q_D} \min_{i \in I(x)} \left\{ \frac{A(x)_i}{(x)_i} \right\}^{1/i} \quad (4.9)$$

Definition. *An e-root-vertor* x *of* A *is a vector* x *for which* $r(x) = r(A)$.

Theorem 4.1. *Let* A *be a* $D \times D$ *matrix of positive entries. Let* $r = r(A)$ *be its principal e-root. Define*

$$E(\lambda) = E(\lambda;A) = \det \left[\lambda^i \delta_{i,j} - a_{i,j} \right]_{i,j=1}^{D} \quad (4.10)$$

Then

(i) *the number* r *is a root of largest absolute value of the equation* $E(r) = 0$, i.e., r *is an e-value of* A *of largest modulus. An e-root vector corresponding to* r *is an e-vector of* A *belonging to the e-value* r.

(ii) $r = r(A)$ *satisfies the inequalities*

$$\min_{i=1,D} \left\{ \sum_{j=1}^{D} a_{i,j} \right\}^{1/i} \le r \le \max_{i=1,D} \left\{ \sum_{j=1}^{D} a_{i,j} \right\}^{1/i} . \quad (4.11)$$

(iii) *the e-vector* x^* *that corresponds to* r *is unique, apart from a scale factor, and has strictly positive entries only, if the scale factor is suitably chosen.*

(iv) *if* $B_{i,j}(\lambda)$ *denotes the cofactor of the* (i,j) *entry of the matrix* $\{\lambda^i \delta_{i,j} - a_{i,j}\}$ *then all* $B_{i,j}(r)$ *have the same sign, for all* $i,j = 1,D$, *and none are zero.*

(v) $r = r(A)$ *is a simple root of the equation* $E(r) = 0$.

(vi) *the e-value* r *is unrepeated in modulus, that is, if* $\lambda \ne r$ *is an e-value of* A, *then* $|\lambda| < r$.

5. Asymptotics of numbers of trees by widths

The purpose of the development of the theory of e-values, in the previous section, was to permit us to make reasonably precise asymptotic estimates of the growth of $f(n,D)$, the number of trees of n vertices with width not exceeding D.

First recall the generating function (3.7)

$$\sum_{\ell=1}^{D} (I - Y)^{-1}_{1,\ell} = \sum_{n\geq 0} \frac{f(n + 1,D)}{n!} t^n \tag{5.1}$$

in which

$$Y_{i,j} = i^j t^j/j! \quad (i,\ j = 1,\ D). \tag{5.2}$$

Let $\lambda_1, \cdots, \lambda_N$ $(i,\ j = 1,\ D)$ be the e-values of the matrix

$$A = (a_{i,j});\ a_{i,j} = i^j/j! \quad (i,j = 1,D). \tag{5.3}$$

Then the singularities of the rational function of t on the left hand side of (5.1) occur among the reciprocals of the e-values of A. If $\lambda_1 = r(A)$ is the principal e-value of A then λ_1 is simple, and the practical fraction expansions of the left side of (5.1) is of the form

$$\sum_{\ell=1}^{D} (I - Y)^{-1}_{1,\ell} = P(t) + \frac{c_1(D)}{1 - rt} + \sum_{j=2}^{N'} \frac{q_j(t)}{(1 - \lambda_j t)^{m_j}} . \tag{5.4}$$

In the above, $P(t)$ is a polynomial, $N' \leq N$ is the number of *distinct* e-values of A, m_j is the multiplicity of λ_j, and $q_j(t)$ is a polynomial in t $(j = 2,\ N')$.

It is important to observe that $c_1(D) \neq 0$. Indeed, if we multiply (5.4) by $1 - rt$ and let $t \to r^{-1}$ we find that

$$c_1(D) = (rE'(r))^{-1} \sum_{s=1}^{D} B_{s,1}(r) \tag{5.5}$$

where $E'(r)$, $B_{i,j}(r)$ were defined in part (iv) of Theorem 4.1 above. It was shown there that they all have the same sign, and therefore $c_1(D) > 0$. Now we can combine (5.4) and (5.1) to deduce

Theorem 5.1. *For* D *fixed, the number* $f(n,D)$ *of trees with* n *vertices and width* $\leq D$ *satisfies*

$$f(n,D) \sim A(D)r(D)^n(n-1)! \qquad (n \to \infty) \tag{5.6}$$

in which

(a) $A(D) = (rE'(r))^{-1} \sum_{s=1}^{D} B_{s,1}(r)$

(b) $E(x) = \det\{x^i\delta_{i,j} - i^j/j!\}_1^D$

(c) $r = r(D)$ *is the largest root of* $E(x) = 0$

(d) $B_{i,j}(x)$ *is the cofactor of the* (i,j) *entry of the matrix* $\{x^i\delta_{i,j} - i^j/j!\}_1^D$.

We illustrate with $D = 2$. Then $E(x) = x^3 - x^2 - 2x + 1$, $r(2) = 1.8019377...$, $E'(r) = 4.13709...$, $B_{11}(r) = 1.24698...$, $B_{21}(r) = .5$, $A(2) = .234345...$ Hence

$$f(n,2) \sim .234345..(1.8019377..)^n(n-1)! \qquad (n \to \infty) . \tag{5.7}$$

Theorem 5.2. *For a fixed* $K > 0$, *the probability that a tree of* n *vertices has bandwidth* $\leq K$ *is at least*

$$An^{\frac{3}{2}}\left\{\frac{r(\lfloor\frac{K+1}{2}\rfloor)}{e}\right\}^{n} \tag{5.10}$$

where $A = A(K)$.

6. The average width and bandwidth of trees

The main results of this section are the following.

Theorem 6.1. *If* $\bar{W}_n$ *denotes the average width of trees of* n *vertices then*

$$c_1\sqrt{n} < \bar{W}_n < c_2\sqrt{n \log n} .$$

Theorem 6.2. *If* $\bar{b}_n$ *denotes the average bandwidth of trees of* n *vertices then*

$$c_3\sqrt{n} < b_n < c_4\sqrt{n \log n} .$$

Theorem 6.3. *Almost all trees of* n *vertices have width* $< 10\sqrt{n \log n}$, *and therefore have bandwidth* $< 20\sqrt{n \log n}$. *More precisely, we have*

$$\text{Prob}\{\beta(T) \geq 10\sqrt{n \log n}\} = O(\frac{1}{n^{12}}).$$

In Theorems 6.1 and 6.2 the left members of the inequalities have already been proved. In Theorem 2.2 we showed that $\bar{b}_n > c_3\sqrt{n}$, and $\bar{W}_n < c_1\sqrt{n}$ follow at once from Theorem 2.3. That $\bar{b}_n < c_4\sqrt{n \log n}$ will follow, again by Theorem 2.3, from the assertion that $\bar{W}_n < c_2\sqrt{n \log n}$. Finally, Theorem 6.3 is even stronger than that assertion, hence the remainder of this section is devoted to a proof of Theorem 6.3.

Let us write $p(n, m, k)$ for the probability that a tree T of n vertices has exactly m vertices in layer k (i.e., that $(i(T))_k = m$). Further let

$$P_k(x,z) = \sum_{n,m} \frac{n^{n-1}}{n!} p(n, m, k) x^n z^m \quad (k \ge 0). \quad (6.1)$$

Then $P_0(x,z) = zY(x)$, where $Y(x) = \sum_{n\ge1} \frac{n^{n-1} x^n}{n!}$.

In the paper [7], Meir and Moon showed that

$$P_{k+1}(x,z) = x \exp\{P_k(x,z)\} \quad (k \ge 0). \quad (6.2)$$

We propose to estimate the factorial moments of the $p(n, m, k)$. For this purpose define

$$\Gamma(r,k) = \frac{\partial^r}{\partial z^r} P_k(x,z)\Big]_{z=1}. \quad (6.3)$$

Then we have the starting values

$$\Gamma(0,k) = Y \quad (k \ge 0; \quad Y = Y(x))$$

and $\Gamma(0,0) = \Gamma(1,0) = Y$, $\Gamma(r,0) = 0$ $(r \ge 2)$, $\Gamma(r,1) = Y^{r+1} (r \ge 0)$, $\Gamma(1,k) = Y^{k+1} (k \ge 0)$, $\Gamma(2,k) = Y^{k+2}(1 - Y^k)/(1 - Y)$ $(k \ge 0)$.

First we claim that the $\Gamma(r,k)$ satisfy the following recurrence relation.

$$\Gamma(r + 1,k + 1) = \sum_{j=0}^{r} \binom{r}{j} \Gamma(r - j,k + 1)\Gamma(j + 1,k) \quad (6.4)$$

for $r, k \ge 0$.

Now we will use (6.4) to estimate the growth of the $\Gamma(r,k)$. If m is an odd positive integer, we will write $m!! = 1 \cdot 3 \cdot 5 \cdot 7 \cdots m$, and we will put $(-1)!! = 1$. The numbers $m!!$ satisfy the recurrence

$$(2r - 1)!! = \sum_{j=0}^{r-1} \binom{r}{j} (2r - 2j - 3)!!(2j - 1)!! \quad (r \geq 1). \quad (6.6)$$

Lemma 6.1. *For* $0 \leq Y < 1$, $k \geq 0$ *and* $r \geq 1$ *we have*

$$\Gamma(r,k) \leq \frac{(2r - 3)!!Y^{k+r}}{(1 - Y)^{r-1}} . \quad (6.7)$$

7. Remarks and open questions

1^0: We have seen that the average bandwidth of trees of n vertices is $> C\sqrt{n}$ and is $< C\sqrt{n \log n}$. What is the precise form of the principle term of the asymptotic behavior of the average bandwidth? of the average width? We have carried out numerical calculations of the average width of labelled trees of n vertices for $n \leq 30$. The calculations were based on the recurrence formula that results from (3.7) if one multiples by $(I - Y)^{-1}$ and compares the coefficients of like powers of t, namely

$$F(i, r, D) = \sum_{k=1}^{\min(D,r-1)} i^k F(k, r - k, D)/k! \quad (7.1)$$

together with the initial conditions $F(i,1,D) = 1(i = 1,D)$. In terms of these quantities $F(i,r,D)$, the number of trees of n vertices whose width is $\leq D$ is $f(n,D) = (n - 1)!F(1,n,D)$.

These calculations suggest that the average width is $\sim\sqrt{n}$ (e.g., the average width of trees of 25 vertices is 4.9383..).

2^0: In a paper soon to be published, N. Dershowits and S. Zaks prove results that imply that for binary trees the average width and bandwidth are each $\leq \sim\sqrt{\pi n}$. Our thanks go to P. Flajolet for calling this work to our attention.

3^0: The conclusion of our Theorem 6.1 actually holds for more general classes of trees than just labelled trees. In fact the proof goes through without significant change for any of the 'simple families' of trees that were studied by Meir and Moon in [7].

4^0: What is the correct rate of approach of r(D) to e?

References

1. C. H. Papadimitriou, The NP-completeness of the bandwidth minimization problem, Computing 16 (1976), 263-370.
2. M.R. Garey, R. L. Graham,. D.S. Johnson and D. E. Knuth, Complexity results for bandwidth minimization, SIAM J. Appl. Math. 34 (1978), 477-495.
3. Herbert S. Wilf, Mathematics for the Physical Sciences, John Wiley and Sons, New York 1962 (reprinted by Dover, 1978)
4. James B. Saxe, Dynamic-programming algorithms for recognizing small- bandwidth graphs in polynomial time, SIAM J. Alg. Discr. Meth. 1 (1980), 363-369.
5. P. A. MacMahon, Combinatory Analysis, Cambridge University Press, 1915 (reprinted by Chelsea, 1960).
6. F. R. Gantmacher, Applications of the theory of matrices, Interscience Publishers, New York, 1959.
7. A. Meir and J. W. Moon, On the altitude of nodes in random trees, Can. J. Math. 30 (1978), 997-1015.
8. V. E. Stepanov, The distribution of the number of vertices in the strata of a random tree, *Teor. Verojatnost. i Primenen.* 14 (1969), 64-77 (see also Theor. Probability Appl. 14 (1969), 65-78.
9. J. W. Moon, Counting labelled trees, Canadian Mathematical Monographs No. 1, Canadian Mathematical Congress, 1970.
10. A. Renyi and G. Szekeres, On the height of trees, J. Austral. Math. Soc. 7 (1967), 497-507.
11. P. Flajolet and A. Odlyzko, The average height of binary trees and other simple trees, J. Comp. Syst. Sci. 25 (1982), 171-213.

12. J. Turner, Probabilistic analysis of bandwidth minimization algorithms, Proc. 15th Annual ACM Symposium on Theory of Computation (1983), 467-476.

13. V. Chvatal, A remark on a problem of Harary, Czechoslovak Math. J. 20 (1970), 95.

2-SUPER-UNIVERSAL GRAPHS

János Pach
Mathematical Institute of the
Hungarian Academy of Sciences
Budapest, H-1053, Hungary

László Surányi
Mathematical Institute of the
Hungarian Academy of Sciences
Budapest, H-1053, Hungary

ABSTRACT

A graph G is called k-super-universal if for every k element subset $A \subset V(G)$ and for every $B \subseteq A$ there exists a vertex (not in A) which is connected to all points in B but to none in $A \setminus B$. In general, the definition of k-super-universal given here is simply the definition of (k+1)-superuniversal given in [1, 2]. In [1, 2] the term used for these graphs was "superuniversal", not "super-universal". The following conjecture of P. Erdös is proved: If n is large enough, then every 2-super-universal graph on n vertices has at least 3n-9 edges. A characterization of the extremal graphs is also given.

1. Introduction

Let G be a simple finite graph whose vertex set and edge set are V(G) and E(G), respectively. Following S.H. Hechler [2], we call G <u>k-super-universal</u> if for every k element subset $A \subseteq V(G)$ and every $B \subseteq A$ there exists a vertex $x \in V(G) \setminus A$ which is joined to all points in B but to none in $A \setminus B$. Let f(n,k) denote the minimum number of edges of a k-super-universal graph on n vertices. For $k \geq 3$, $n \geq n_o(k)$, Erdös-Hechler-Kainen [1] and Pach [3] established the following bounds:

$$(1) \qquad 2^{k-5} n \log n < f(n,k) < k2^k n \log n \, .$$

In the case $k = 2$, it was shown in [3] that $f(n,2) = 3n-O(1)$, and several classes of 2-super-universal graphs with n vertices and $3n-9$ edges were constructed. Erdös conjectured that the following theorem is true.

<u>1.1. Theorem</u>. *If n is sufficiently large, then any 2-super-universal graph on n vertices has at least $3n-9$ edges, and this bound cannot be improved.*

The aim of the present note is to prove this conjecture and to characterize the extremal configurations (see Lemma 2.5).

2. <u>Preliminaries</u>

Next we summarize some earlier results which will be used in the rest of the paper.

<u>2.1. Proposition</u> ([5], [3]). Every vertex of a 2-super-universal graph is of degree at least 4.

<u>2.2. Proposition</u> ([4]). Let $\varepsilon > 0$, and let G be a graph of diameter 2 on n vertices whose maximal degree is at most $(1-\varepsilon)n$ and whose minimal degree is at least $d \geq 3$. Then for any sufficiently large n,

$$|E(G)| \geq \left\lceil \frac{d+2}{2}(n-3) \right\rceil$$

holds, with equality only if there are three vertices $x_1, x_2, x_3 \in V(G)$ such that all other vertices of G are joined to exactly two x_i's and all (but at most one) are of degree d.

2.3. Definition. A graph G with n vertices is said to be of type (i) if there exist three vertices x_1, x_2, $x_3 \in V(G)$ and a subset $V' \subseteq V(G)$, $|V'| < n/\log n$, such that every element of $V(G) \setminus V' \setminus \{x_1,x_2,x_3\}$ is joined to at least two x_i's and has degree at least 4.

If, in addition, this is valid with $V' = \phi$, and every element of $V(G) \setminus \{x_1,x_2,x_3\}$ is joined to exactly two x_i's and has degree exactly 4, then G is said to be of type (i*).

2.4. Proposition ([4]). Let $\varepsilon > 0$, and let G be a graph of diameter 2 with n vertices whose minimal degree is at least 4 and whose maximal degree is at most $(1-\varepsilon)n$. If n is sufficiently large, then either

$$|E(G)| > 3n + \frac{\varepsilon n}{2} - \frac{n}{\log n} ,$$

or G is of type (i).

Using the fact that every 2-super-universal graph is of diameter 2, 2.1-4 immediately yield.

2.5. Lemma. Let $\varepsilon > 0$, and let G be a 2-super-universal graph on n vertices whose maximal degree does not exceed $(1-\varepsilon)n$. If $|E(G)| \leq 3n$ and n is large enough, then

(a) $|E(G)| \geq 3n-9$ with equality only if G is of type (i*);

(b) G is of type (i).

Hence our theorem is true for graphs without points of 'large' degree.

3. Proof of Theorem 1.1

Throughout this section let $\varepsilon = 1/43$ be fixed, and let G denote a 2-super-universal graph on n vertices with a fixed vertex x of degree

$$(2) \qquad d(x) \geq (1-\varepsilon)n = \frac{42}{43} n ,$$

and suppose

$$(3) \qquad |E(G)| < 3n-3.$$

In what follows, we will try to describe the structure of G.

For any $y \in V(G)$, let

$$S(y) := \{v \in V(G) | yv \in E(G)\},$$
$$M(y) := V(G) \setminus S(y) \setminus \{y\} .$$

Thus we have $|S(y)| = d(y)$ and $|M(y)| = n-d(y) - 1$. Further, if $A \subseteq V(G)$, then let

$$d_A(y) := |\{v \in A | yv \in E(G)\}| .$$

3.1. Lemma. Let G, and $x \in V(G)$ be as above. Then there are two vertices $z_0, z_1 \in M(x)$ such that every other point of $M(x)$ is connected to both of them.

Proof. Put $S := S(x)$, $M := M(x)$, and $M := |M| \leq \frac{n}{43} - 1$. By the definition of the 2-super-universality and by 2.1, we know that $d(y) \geq 4$ and $d_M(y) \geq 1$ for any $y \in S$. Let

$$S_o := \{y \in S | d(y) > 4\} \cup \{y \in S | d_M(y) > 1\}.$$

Obviously,

$$(4)\quad \begin{aligned} 2|E(G)| &= d(x) + \sum_{y\in S} d(y) + \sum_{y\in S} d_M(y) + \sum_{z\in M} d_M(z) \\ &= 6d(x) + \sum_{y\in S_o} (d(y)-4) + \sum_{y\in S_o} (d_M(y)-1) + \sum_{z\in M} d_M(z). \end{aligned}$$

Thus, using (3), we get

$$(5)\quad \begin{aligned} |S_o| &\le \sum_{y\in S_o} (d(y)-4) + \sum_{y\in S_o} (d_M(y) - 1) \\ &< 6(n-d(x)-1) - \sum_{z\in M} d_M(z) \le 6m. \end{aligned}$$

Now let

$$S_1 := \{y \in S \setminus S_o | yv \in E(G) \ \text{ for some } \ v \in S_o\},$$

$$S_2 := \{y \in S \setminus S_o \setminus S_1 | yv \in E(G) \ \text{ for some } \ v \in S_1\}.$$

Since every point of S_1 has exactly 2 neighbors in S, at least one of them lying in S_o, we have $|S_2| \le |S_1|$. On the other hand,

$$\sum_{y\in S_o} (d(y)-4) + \sum_{y\in S_o} (d_M(y)-1)$$

$$\ge |S_o| + \sum_{y\in S_o} (d_{S_1}(y)-3) \ge |S_1|-2|S_o|,$$

and taking (5) into account, we obtain $|S_1| < 6m + 2|S_o| < 18m$. Thus, by (2), $|M \cup S_o \cup S_1 \cup S_2| < 43m < n - 1$, which means that there exists at least one vertex $u_o \in S \setminus S_o \setminus S_1 \setminus S_2$.

Clearly, u_o has exactly 2 neighbors in S (say, u_1 and u_2) and one neighbor in M (say, z_o). By the 2-super-universality of G, we get that u_o and z_o have a common neighbor (say, u_1), and there is a point $z_1 \in M$ which is not connected to z_o. The definition of u_o ensures that $u_1, u_2 \in S \setminus S_o \setminus S_1$, hence u_1 cannot be joined to any point of M different from z_o. In particular, $u_1 z_1 \notin E(G)$, so the only common neighbor of u_o and z_1 is necessarily u_2. Further, let u_3 denote the (only) point of G connected to both u_2 and z_1. Obviously, $u_3 \in S \setminus S_o$ and $u_3 \neq u_o, u_1$.

Now let z be any point in $M \setminus \{z_o, z_1\}$. We know, by the 2-super-universality of G, that u_o and z must have a common neighbor, and this cannot be x, u_1 or u_2. Hence, z is connected to z_o. Similarly, applying the condition to the pair (u_2, z), we obtain that z is connected to z_1, too. □

As a matter of fact, we have proved a somewhat stronger assertion about the structure of G.

<u>3.2. Lemma</u>. Let G, $x \in V(G)$, $M = M(x)$, $m = |M|$, $S = S(x)$, S_o, S_1 be as above. Then there exists a natural number t, $1 \leq t \leq 3$, such that M and S can be partitioned into ($t + 1$ and $2t + 1$, resp.) pairwise disjoint subsets

$$M = \{z_o^1, z_1^1\} \cup \{z_o^2, z_1^2\} \cup \cdots \cup \{z_o^t, z_1^t\} \cup M' ,$$

$$S = (S_o^1 \cup S_1^1) \cup (S_o^2 \cup S_1^2) \cup \cdots \cup (S_o^t \cup S_1^t) \cup S', \; (S' = S_o \cup S_1)$$

having the following properties:

(i) $S_j^i = S(z_j^i) \cap (S \setminus S_o \setminus S_1)$, $(1 \leq i \leq t,\ 0 \leq j \leq 1)$,

(ii) $S' = |S_o \cup S_1| < 24m$,

(iii) $z_o^i z_1^i \notin E(G)$, but both z_o^i and z_1^i are connected to all other elements of M $(1 \le i \le t)$,

(iv) every vertex $y \in S_j^i$ has degree 4: it is connected to x, z_j^i and to one-one vertex in both $S(z_j^i)$ and $S(z_{1-j}^i)$, $(1 \le i \le t,\ 0 \le j \le 1)$.

Observe that t cannot be larger than 3, for otherwise $2|E(G)| \ge 6(n-m-1) + 2t(m-2) \ge 6n - 6$ would hold, contradicting (3).

Put $z_o = z_o^1$, $z_1 = z_1^1$ and $M_o := M \setminus \{z_o, z_1\}$. By 3.2 (iii) and (4), we clearly have that

$$2|E(G)| = 6(n-m-1) + \sum_{y \in S} (d(y)-4) + \sum_{y \in S} (d_M(y)-1) + \sum_{z \in M_o} d_{M_o}(z) + 4(m-2). \tag{6}$$

Now let $N \subseteq M_o$ and $T \subseteq S$ be <u>maximal</u> subsets satisfying the following conditions:

(a) there are no edges running between $M_o \setminus N$ and T,

(b) $\sum_{y \in T} (d(y)-4) + \sum_{y \in T} (d_M(y)-1) + \sum_{z \in N} d_{M_o}(z) \ge 2|N|$.

(Note that $N = T = \phi$ satisfies both (a) and (b).)

In what follows, we are going to prove the following chain.

<u>3.3. Claim.</u> $N = M_o$.

From here, using (6) and property (b), we obtain $2|E(G)| \ge 6(n-m-1) + 6(m-2) = 6n - 18$, and Theorem 1.1 follows.

The proof of 3.3 is divided into several steps.

Let $N^* := M_o \setminus N$, $T^* := S \setminus T$.

3.4. Proposition.

(i) Every point of T^* has at least one neighbor in N^*.

(ii) There are no edges running between T^* and $\{z_o, z_1\}$.

(iii) There are no edges running between T^* and N.

(iv) Every point of N^* has exactly one neighbor in M_o.

(v) $T \supseteq S_o$.

Proof. Let $y \in T^*$. If $S(y) \cap N^* = \phi$, then $N' := N$ and $T' := T \cup \{y\}$ satisfy the conditions (a) and (b), contradicting the maximal choice of N and T. This proves (i).

To show (ii), suppose first that some $y \in T^*$ is joined to both z_o and z_1. Since, by the 2-super-universality of G, y must have a neighbor in S, we obtain

$$d(y)-4) + (d_M(y)-1) \geq 2|S(y) \cap N^*| + 1.$$

In this case, set $N' := N \cup (S(y) \cap N^*)$, $T' := T \cup \{y\}$ to get a contradiction.

Otherwise, for $j = 0, 1$, let

$$N_j^* := \{z \in N^* | z \text{ and } z_j \text{ have a common neighbor in } S\},$$

$$T_j^* := \{y \in T^* | y \text{ is a common neighbor of } z_j \text{ and some } z \in N^*\}.$$

We clearly have $T_o^* \cap T_1^* = \phi$. Put $N' := N \cup N_o^* \cup N_1^*$ and $T' := T \cup T_o^* \cup T_1^*$. Then (a) remains valid, and so does (b), because

$$\sum_{y \in T' \setminus T} (d_M(y)-1) + \sum_{z \in N' \setminus N} d_{M_o}(z)$$
$$\geq (|N_o^*|+|N_1^*|) + (|N_1^* \setminus N_o^* \setminus| + |N_o^* \setminus N_1^*|) = 2|N' \setminus N|.$$

This is a contradiction unless $T_o^* = T_1^* = \phi$, i.e. (ii) is true.

Next we prove (iv). Let $z \in N^*$. In view of (ii), any common neighbor of z and z_j must be in M_o, whence $d_{M_o}(z) \geq 1$. If $d_{M_o}(z) \geq 2$, then setting $N' := N \cup \{z\}$, $T' := T$, we get the desired contradiction.

To prove (iii), we have to note only that if $S(y) \cap N \neq \phi$ for some $y \in T^*$, then by (iv)

$$(d_M(y)-1) + \sum_{z \in S(y) \cap N^*} d_{M_o}(z) \geq 2|S(y) \cap N^*|,$$

is valid, and putting $N' := N \cup (S(y) \cap N^*)$, $T' := T \cup \{y\}$ we get a contradiction.

The proof of (v) is again indirect. Suppose first that $d(y') \geq 5$ for some $y' \in T^*$. Then, as above, the substitution $N' := N \cup (S(y') \cap N^*)$, $T' := T \cup \{y'\}$ easily leads to a contradiction. Assume next that y' is some vertex in T^* with $d_M(y') \geq 2$, i.e., with $d_M(y') = 2$. Then y' has one neighbor (say, y'') in S and two neighbors (say, z' and z'') in N^*. By the 2-super-universality of G, we have that $z'z'' \in E(G)$.

If $N^* = \{z',z''\}$, then

$$\Delta := \sum_{y \in S} (d(y)-4) + \sum_{y \in S} (d_M(y)-1) + \sum_{z \in M_o} d_{M_o}(z)$$

$$\geq 2|N| + (d_M(y')-1) + d_{M_o}(z') + d_{M_o}(z'') = 2|M_o| - 1.$$

Observe that in this case $\Delta = 2|E(G)| - 6|S| - 4|M_o|$ is an even number; thus we have $\Delta \geq 2|M_o|$, showing that $N' := M_o$, $T' := S$ fulfills the requirements of (a) and (b), a contradiction.

Hence, we may assume that there exists a vertex $z''' \in N^* \setminus \{z',z''\}$. Let u' (u'') be a point of G connected to z' (z'') but not to z'' (z', resp.). By (iv) we have that

$u', u'' \in T^*$, and at least one of them (say, u') is not connected to y', because $d(y') = 4$. Let $v' \in T^*$ be a common neighbor of u' and z'. We can suppose that $v'z'' \notin E(G)$, for otherwise $N' := N \cup \{z',z''\}$, $T'' := T \cup \{y',v'\}$ would contradict the maximum property of N and T. If v'' denote a common neighbor of u' and z'''. If v'' were in N^*, then $v''z' \in E(G)$ would follow, contradicting (iv). Hence, $v'' \in T^*$. Further, let v''' denote a common neighbor of v'' and z''. It is easy to see now that $v''' \in T^* \setminus \{u', v'\}$ and v''' is also the only common neighbour of v'' and z'''. But in this case both $z''z'$, $z''z''' \in E(G)$, which again contradicts (iv). This completes the proof of (v). □

<u>3.5. Proposition</u>. Let G, $x \in V(G)$, $S = T \cup T^*$, $M = \{z_o, z_1\} \cup N \cup N^*$ be defined as above, and let

$$R := \{v \in T | S(v) \supseteq N \text{ and } S(v) \cap T^* \neq \phi\}.$$

If $N^* \neq \phi$, then the following four statements hold:

(i) $|N^*| = 2$, and the 2 elements $z', z'' \in N^*$ are adjacent;

(ii) $|T^*| \geq 4$;

(iii) $S(y) \cap R \neq \phi$ for every $y \in T^*$;

(iv) $d(v) \geq 5$ for every $v \in R$.

<u>Proof</u>. By the 2-super-universality of G, every point of N^* has at least one neighbor in S (which is necessarily in T^*, as there are no edges between N^* and T). Hence, $N^* \neq \phi$ implies $T^* \neq \phi$.

Let $y \in T^*$. By 3.4 (i) and (v), we know that y has exactly one neighbor in N^* (say, z') and 2 neighbors (say, y' and y'') in S.

If $N^* = \{z'\}$, then by 3.4 (iv) we obtain

$$\Delta := \sum_{y \in S} (d(y)-4) + \sum_{y \in S} (d_M(y)-1) + \sum_{z \in M_o} d_{M_o}(z)$$

$$\geq \Big(\sum_{y \in T} (d(y)-4) + \sum_{y \in T} (d_M(y)-1) + \sum_{z \in N} d_{M_o}(z)\Big) + d_{M_o}(z')$$

$$\geq 2|N| + 1 = 2|M_o| - 1.$$

As Δ is even, $\Delta \geq 2|M_o|$ also follows. Thus, setting $N' := S$, $T' := M_o$, we get a contradiction with the maximality of N and T. So we may assume $N^* \setminus \{z'\} \neq \phi$.

Next we show that any element $z'' \in N^* \setminus \{z'\}$ is joined to z'. In view of 3.4 (iv), this implies (i).

Assume, without loss of generality, that $y'(\in T^*)$ is a common neighbor of y and z'. By 3.4 (v) we have that $y'z'' \notin E(G)$. If $z'z'' \notin E(G)$, then y'' must be the only common neighbor of y and z''. Thus y'' is also a member of T^*, and cannot be connected to any members of $M \setminus \{z''\}$. By superuniversality, there must be a member of G connected to both y'' and z''. By 3.4 (v) this vertex, which we call y''' must belong to $T^* \setminus (y,y')$. It is also easily seen that by 3.4 and superuniversality, neither T nor N, nor, therefore, $M_o \setminus \{z',z''\}$ can be empty. Now let z'' be any member of $M_o\{z',z''\}$. It is now easily seen that the only vertex of G which can be connected to both z''' and y is z'', and, similarly, the only vertex which can be connected to both z''' and y'' is z''. Hence we have shown that every member of $M_o \setminus \{z',z''\}$ is connected to both z' and z''. According to 3.4 (iv), this yields that $M_o \setminus \{z',z''\}$ contains only a single vertex z'''. Now, every point of G is connected either to x or to z''', contradicting the 2-super-universality of G. Hence $z'z'' \in E(G)$, which proves (i).

Let t be a common neighbor of x and z'', and let t' be a common neighbor of t and z''. To show (ii), observe only that y, y', t, t' are 4 distinct elements of T^*.

We are going to prove now that $y'' \in R$. Let z be an arbitrary element of N. It is enough to show that $y''z \in E(G)$. By 3.4 (iii) and (iv), z cannot be connected to either z' or y'. Hence, y'' is the only common neighbor of z and y, which establishes (iii).

Note that every vertex $v \in R$ is either connected to at least one of z_o and z_1, or it has at least 2 neighbors in T^* (one in $T^* \cap S(z')$ and one in $T^* \cap S(z''))$. Hence, if we prove $|N| \geq 2$, then (iv) follows.

We obviously have that $N \neq \phi$, otherwise every point of G would be connected either to x or to z', contradicting the fact that G is 2-super-universal. Suppose that $|N| = 1$, say $N = \{w\}$. Let $y_i \in S_o \subseteq T$ denote a common neighbor of w and z_i $(i = 0, 1)$.

If $y_o = y_1$, or $S_o \neq \{y_o, y_1\}$, or $d(y_i) \geq 5$ for some i, then

$$\Delta := \sum_{y \in S} (d(y)-4) + \sum_{y \in S} (d_M(y)-1) + \sum_{z \in M_o} d_{M_o}(z)$$

$$\geq \sum_{y \in S_o} (d(y)-4) + \sum_{y \in S_o} (d_M(y)-1) + d_{M_o}(z') + d_{M_o}(z'')$$

$$\geq 5 = 2|M_o| - 1$$

and, since Δ is even, putting $N' := M_o$ and $T' := S$ we get a contradiction.

So we may assume that none of the above 3 possibilities holds. From here, using that $|T^*| \geq 4$ and each vertex of T^* has at least one neighbor in R (cf. (ii), (iii)), it follows that there exists a vertex $u \in R \setminus \{y_o, y_1\} \setminus d(y_o) \setminus d(y_1)$. Then u is adjacent to exactly 4 points of G: to x, to w and to 2 elements of T^*. By 3.4 (iii), this yields that u and w have no common neighbors. This contradiction proves that $|N| \geq 2$, as desired. □

Now we are in a position to complete the proof of 3.3. Assume again that $N^* \neq \phi$. Then

$$(7) \qquad \sum_{v \in R} (d(v)-4) + \sum_{v \in R} (d_{M_o}(v)-1) \geq 2|N|.$$

To see this, suppose first that there is only one vertex v in R. Applying 3.5 (iii), we get that v is connected to all points of T^*. Thus, by 3.5 (ii), we have

$$d(v) \geq |N| + |T^*| + 1 \geq |N| + 5, \; d_{M_o}(v) \geq |N|,$$

which proves (7). On the other hand, by 3.5 (iv) we obtain

$$(d(v)-4) + (d_{M_o}(v)-1) \geq 1 + (|N|-1) = |N|$$

for any $v \in R$, so (7) holds in case $|R| \geq 2$, too.

Let w be an arbitrary element of N, and let y_i denote a common neighbor of w and z_i $(i = 0, 1)$. Let $Y = \{y_o, y_1\}$. Note that y_o and y_1 are not necessarily distinct and that Y must be a subset of the union of T and N. No matter how Y is distributed amount these two sets, it is clear that

$$\sum_{y \in Y \cap R} d_{\{z_o, z_1\}}(y) + \sum_{y \in Y \cap (T \setminus R)} (d_M(y)-1) + \sum_{z \in Y \cap M_o} d_M(z) \geq 2 .$$

Comparing this with (7), we obtain

$$\Delta := \sum_{y\in S} (d(y)-4) + \sum_{y\in S} (d_M(y)-1) + \sum_{z\in M_o} d_{M_o}(z)$$

$$\geq \sum_{y\in S} (d(y) - 1) + \sum_{y\in S} (d_M(y)-1) + \sum_{y\in Y\cap R} d_{\{z_o,z_1\}}(y)$$

$$+ \sum_{z\in Y\cap(T\setminus R)} (d_M(y)-1) + \sum_{z\in Y\cap M_o} d_N(z) + 2$$

$$\geq 2|N| + 2 + 2 \geq 2|M_o|.$$

contradicting the maximality of M and T. This completes the proof of 3.3 and, therefore, the theorem. □

3.6. Remark. We have some constructions with $S_o \neq \phi$, showing that S' in Lemma 3.2 cannot be omitted. However, our proof gives that $|S'| = |S_o \cup S_1|$ is bounded by an absolute constant.

References

1. P. Erdös, S. H. Hechler and P. Kainen, On finite superuniversal graphs, Discrete Math. 24 (1978), 235-249.
2. S. H. Hechler, Large superuniversal metric spaces, Israel J. Math 14 (1973), 115-148.
3. J. Pach, On super-universal graphs, Studia Sci. Math. Hung. 12 (1977), 19-27.
4. J. Pach and L. Surányi, On graphs of diameter 2, Ars Combinatoria 11 (1981), 61-78.
5. G. B. Purdy, Planarity of two-point universal graphs, Coll. Math. Soc. J. Bolyai 10, Infinite and finite sets, Keszthely, Hungary (1975), VOl. III., 1149-1157.

AN EFFICIENT ALGORITHM FOR EMBEDDING GRAPHS IN THE PROJECTIVE PLANE

Branislava Peruničić
University of Sarajevo
and IRIS-ENERGOINVEST

Zoran Durić
IRCA-ENERGOINVEST

ABSTRACT

The embeddings of the nonplanar graphs in the projective plane have at least one essential cycle. If an essential cycle is contracted to a vertex the considered graph becomes planar. Moreover, if it is replaced by another cycle and suitably connected to the remaining graph, a planar graph, termed equivalent planar graph (EPG) is produced. The embedding in the projective plane can be very simply made from the equivalent planar graph. The presented algorithm has three main steps. First, a finite set of the essential cycle candidates is obtained as the result of a trial to embed the graph in the plane. Next, each of the candidates is tested by an attempt to produce an EPG. Finaly, the embedding in the projective plane is obtained from the EPG. The input of the algorithm is the adjacency list and it has three outputs: a planar embedding a projective plane embedding, and an information that there are no embedding. The duration of the algorithm is proportional to maximal degree and the square of the number of vertices.

1. Introduction

The problem of embedding graphs on various surfaces has been treated in numerous papers. The majority of them were, however, devoted to embedding in the plane, the likely reason being practical value of such algorithms for the realization of printed circuits. The problem of embedding in other orientable and nonorientable surfaces, which at the same time may determine the genus of the graph, seems at present to be of theoretical interest only.

The first algorithm for embedding toroidal graphs in the torus |1| was published 1976, It describes the embedding of an arbitrary toroidal graph, and is based on an algorithm for planar embedding |2|, which is an extended version of |3|, In 1978 another algorithm |4| for embedding cubic toroidal graphs in the torus was published. The same author developed that algorithm in 1980 |7|, and presented later a generalized approach to determine the genus of an arbitrary graph |5|.

This paper presents an efficient algorithm for embedding suitable nonplanar graphs in the projective plane or in the Möbius band. The input of the algorithm is the adjacency matrix of the graph and there are three outputs: an embedding in the plane if the graph is planar; an embedding in the projective plane if it exists, and, finally, information that the graph cannot be embedded in projective plane, if that is the case.

2. The representation of the embedding

The embeding $g:G\to M$ of the graph G in the projective plane is a homeomorphism of G into M. The components of M-g(G) are the faces of embedding, However, in this paper, by the face sometimes will be understood the cycle that borders the face. An embedding in the projective plane may be presented by a map in the plane, as in Figure 1. All the faces of this representation but the outer one are equal to faces of a projective plane embedding. The outer face is bordered by a twice repeated cycle, denoted by c. This cycle is the essential cycle. If two images of the essential cycle are made different, by introducing different indices to equal nodes, a planar graph is produced, named the equivalent planar graph and denoted as EPG.

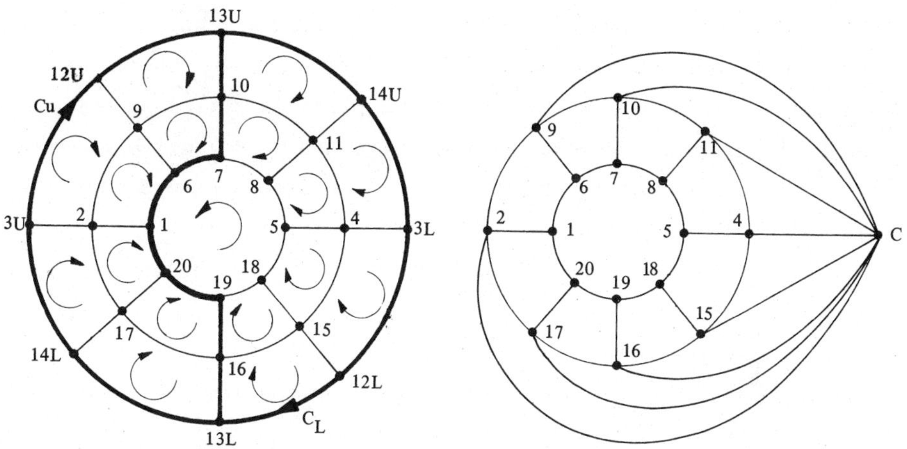

Figure 1. Figure 2.

The algorithm described in this paper has as a goal to produce an EPG, as this obviously solves the problem of embedding. Shortly, the algorithm starts with the search of a finite set of cycles which should contain at least one essential cycle if there is a projective plane embedding; then all the candidates are examined by trying to produce en EPG.

3. The essential cycles on the projective plane

In Figure 1. all the inner faces are oriented clockwise. In that case two faces with a common arc which belongs to the essential cycle c have the same direction along this arc whereas two faces that have a common arc which does not belong to c have the opposite direction along this arc. The other essential cycles can be found if two representations of a vertex are connected by a path that has no other common vetrices with essential cycle c. An example of such a cycle is 13-10-7-6-1-20-19-16-13 denoted by a thick line on Figure 1.

If the projective plane is cut along a drawing of an essential cycle, a sphere with a hole bordered by a twice repeated essential cycle is obtained. All the cycles that are not essential ones enclose some open discs. On such a disc a planar subgraph of the graph is embedded. If the projective plane is cut along the representation of such a cycle, an open disc and a projective plane with a hole-equivalent to Mobius band-are obtained. Each essential cycle of the projective plane has another important property: if it is contracted to a node, say C, the resulting graph, denoted by G(C), is planar (see Figure 2).

It is worthwhile to note that nonplanar graphs embeddable on the Mobius band have essential cycles with the same property. Their drawings are closed curves going around the Mobius band, A planar drawing of the Mobius band with a map is shown in Figure 3. If the faces bordering the cycle c are oriented in the same direction as c, all the other cycles may be oriented in the opposite direction on their common arcs. The cutting of the Mobius band along the drawing of c produces a doubly twisted band, having two distinct sides and two separate edges. One of these edges is equal to the Mobius boundary, whereas the other edge is bordered by the doubled essential cycle c.

If the band is cut, straightened, and joined again, a cylinder is obtained. The plane representation of such a cylinder is an annulus (Figure 4.). Note that the aperture of the annulus is the Mobius band produced by the Mobius boundary, whereas its outer edge is equal to the twice repeated cycle c. This drawing can not be distinguished from the projective plane map in Figure 1. That makes the embedding of the graphs on the projective plane and the Mobius band possible with a single algorithm.

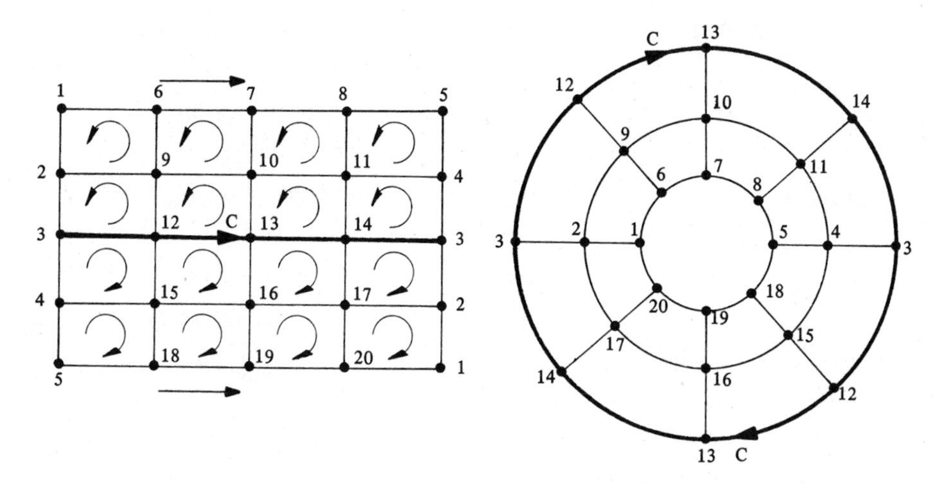

Figure 3.

Figure 4.

4. The search for a set of essential-cycle candidates

The algorithm begins by a trial to embed the considered graph in the plane by the algorithm described in |2| and briefly repeated here. First, a planar three-connected subgraph G_o is embedded. Next, the paths that connect already embedded vertices are searched for. If such a path has a unique embedding (a forced-choice path), it is embedded, and a new subgraph G_1 is obtained. The embedding of the multiple-choice paths is postponed, and nonplanar paths prove the nonplanarity. If only multiple-choice paths remain, the graph is two-connected. Then, one of such paths is embedded in one of the possible faces, and the procedure continues. The blocks are discovered and embedded in a similar manner.

The paths are obtained by a depth-first search that begins in an already embedded vertex P. When another embedded vertex Q is found, the possibility of the embedding of path PQ is tested.

This procedure stops when a nonplanar path between two nodes P and Q is detected in a subgraph G_k. A set of essentiel-cycle candidates may be found from the faces of G_k.

Let the set of cycles defining the faces of G_k be {F}. This set may be partitioned into the set of cycles that are essential in the projective plane, denoted by {EF}, and those that are not essential {NF}. Define by {F(P)} the elements of {F} that have arcs incident to P. The corresponding faces encompass P in the plane. Besides, the union of faces of {F(P)} has as a border a cycle b(P). If {F(P)}$\subset${NF}, then embeddings of the union of {F(P)} in the projective plane also encompasses the embedding of P. Namely, the embedding of this union is a closed surface S(P), with an embedding of b(P) as a border, and P is an inner point. Any path that connects P with a vertex Q not in S(P) must have a common vertex with b(P). So, no nonplanar paths can be found starting from P in the subgraph G_k. Any such path will meet first a vertex at the border b(P), will be recognized as planar, and then embedded in one of faces of {F(P)}. (See for ilustration Figure 5)

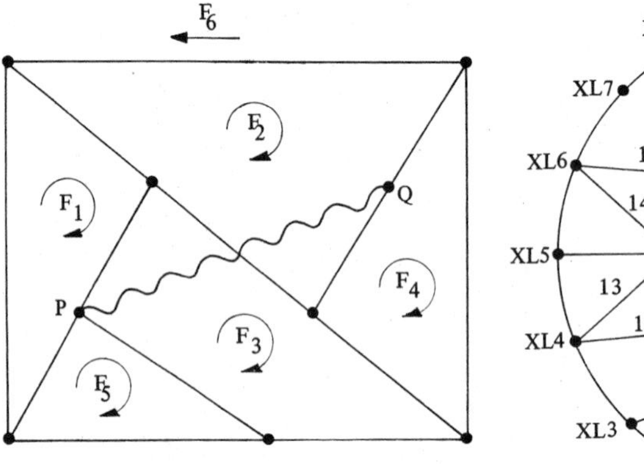

{F(P)}={F_1,F_3,F_5};{F(Q)}={F_2,F_4}

Figure 5.

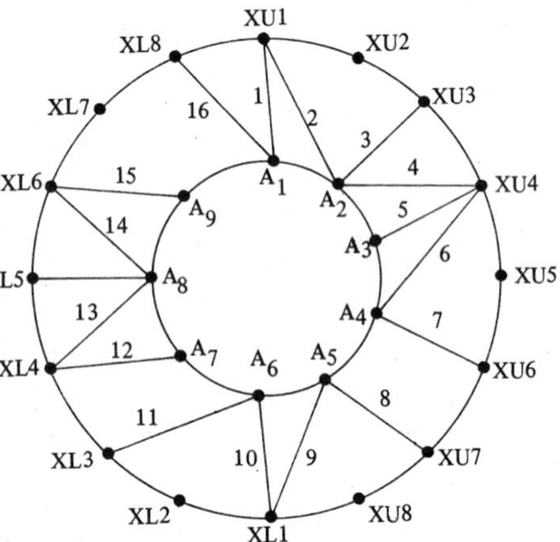

Figure 6.

The same reasoning applies for {F(Q)}. If {F(Q)}$\subset${NE}, any path starting from P will first meet a vertex at b(Q), so Q can not be reached by a nonplanar path.

Therefore, when a nonplanar path from P to Q is found, both {F(P)} and {F(Q)} must contain at least one essential cycle.

To make the examination of the set of candidates easier, the arcs that connect vertices in the same faces of {F(P)} are searched for, embedded, and then {F(P)} is consequently modified.

In the second stage of the algorithm, each of candidates is tested by trying to produce an embedding in the projective plane, via generating an equivalent planar graph.

5. <u>The equivalent planar graph generation</u>

The equivalent planar graph is obtained by replacing an essential cycle and all the arcs incident to it by another set of arcs. For each node X_i of the essential cycle two new nodes XL_i and XU_i are introduced. All the arcs X_i-X_j save one, say X_i-X_m, are replaced by two new arcs XL_i-XL_j and XU_i-XU_j. The arc X_1-X_m is replaced by XL_1-XU_m and XL_m-XU_1. The obtained cycle will be named the <u>doubled contour</u> and denoted by cUcL. The letters L and U stand for lower and upper. The arcs incident to the essential cycle, say A_j-X_i, are replaced by A_j-XL_i or A_j-XU_i. If a correct choice is made, an EPG is produced, and the problem of embedding is solved.

Each of the essential-cycle candidates is examined by a trial to produce an EPG, First, the candidate cycle $c \subset$ {F(P)} is contracted to a node C, and the contracted graph G(C) is tested for planarity. If a planar map is produced, the node C together with all the arcs stemming from it is deleted. So,the set of faces encopmassing C in the embedding of G(C), {F(C)}, disappear, and a new face, defined by b(C), and denoted FC, is created. If that face is taken as the outer one, it encloses the faces that appear in the projective plane map. To find other faces, b(C) is encircled by the doubled contour. The arcs A_j-XL_i and A_j-XU_i connect the vertices A_i from b(C) to vertices XL_i and XU_i of the doubled contour. The crucial part of the EPG generation is the making of the right choice between A_j-XL_i and A_j-XU_i. The procedure of the making of the choice is called <u>indexing</u> and is described below.

6. The Indexing

First, the vertices of b(C) that are connected via arcs with the doubled contour are denoted by A_1, A_2,...A_n, going along the b(C) in an arbitrary direction. They have a cyclic order; ..A_1, A_2,....A_n, A_1,.... Next, the essential cycle vertices are denoted by X_i, and their cyclic order is defined going along the essential cycle in an arbitrary direction. Finaly, the cyclic order of the doubled contour is established as... XU_1, XU_2,....XU_m,XL_1, XL_2,... XL_m,XU_1,...

The rules that govern a correct indexing can be envisaged from the Figure 6. On that picture the arcs A_j-XU_i and A_j-XL_i are embedded in an annulus between the b(C) and the cUcL. Since there are no crossing of the arcs, a cyclic order of the arcs can be found if the annulus is toured starting from an arbitrary arc and going along the annulus in the direction of b(C). It is obvious that vertices A_j and XL_i and XU_i that are on the ends of ordered arcs obey their respective cyclic order. More precisely, after A_i only the A_i or A_{i+1} can pursue. After XU_i may follow XU_i or some XU_{i+k} (k positive).

The ordering of the arcs is conveniently visualized in Figure 7. There, a block of the adjacency matrix is shown for a possible orientations of cUcL. The arcs are depicted by encireled X´s. The cyclic order of the arcs is represented bu the starcaise-like trajectory connecting the x. Taking an arc as a starting one, we construct the trajectory at Figure 7 according to the following rule:

Take as a starting arc the first arc in the first row. Go from the starting arc rightward and collect the arcs until the last one in this row is found. Go one step down the collumn of the last arc. Then go again rightward and collect again the arcs. If the right end of the row is reached, ond there are more arcs in the row, make a jump to the left end of the same row, and continue to collect the arcs until the last one is found. Go one step down the collumn of the last arc and continue to collect the arcs as described. When in the last row, replace the step down with a jump to the upper end of the column.

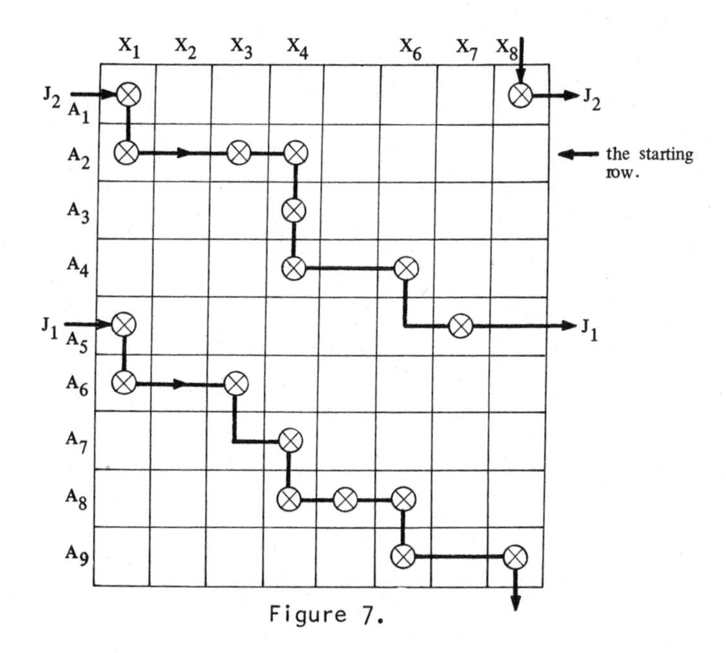

Figure 7.

Row jumps reflect passage from cU to cL and vice-versa, and may happen in two rows only. This two rows will be named the split rows. The split rows in Figure 7 are A_1 and A_5. The collumn jump reflects the cyclic order of A_{i-s}. All the arcs should be collected, and the trajectory closed.

This rule will produce a closed trajectory with two row jumps if the starting row is not a split row. If trajectory can not be closed, the described procedure has to be applied in the second and possibly in the third row. Namely, as there are only two split rows, it is enough to try three rows as starting ones.

If a closed trajectory is not obtained, the orientation of the essential cycle has to be reversed, and the procedure should be repeated.

If no trajectory is produced, the candidate cycle is not

an essential cycle in the projective plane. (It may however be essential on some other surface!) The next candidate cycle, if any, has to be tested. However, if a trajectory is produced, the arcs between two row jumps are indexed by U end L.

If b(C) is not a simple cycle, but has repeated vertices, there are blocks embedded in face FC. That makes the ordering of A_{i-s} an ambiguous one. Namely, a cut vertex A_k will appear if the graph G(C) has a two-connected component tied between the vertices A_k and C (See for ilustration Figure 8a). Such a component can be embedded in two ways, producing two different ordering of A_i-s between two appearances of A_k in cycle b(C) (Figure 8 b and 8c). To avoid this obstacle, the block attached to A_k is removed, and a set of dummy arcs is introduced. A dummy arc A_k-X_j is created, if a vertex in the deleted block is connected to X_j. This can be performed on the adjacency matrix by collecting all the arcs incident to the block in the row of the cut vertex A_k. If, for example the vertices 1,2,4 and 5 are connected to the vertices X_1, X_2, X_5 and X_6 of the essencial cycle, the dummy arc A_5-X_1, A_5-X_3, A_5-X_5 are introduced as represented at Figure 8d.

When indexing is finished, the letters L and U from the dummy arcs are transfered to real ones, and the other arcs of the block are returned to FC. The indexing will automatically determine the correct embedding of the block, if it exists.

After the indexing is finished, the embedding is completed, and the letters U and L are removed The faces of the obtained map, save the outer one, are the faces of the projective plane map. The algorithm may be resumed by four steps:

1. Start embedding the graph in the plane by expanding a three-connected subgraph by planar paths. If the embedding may be finished, the graph is planar, Else, found the set of candidates for essential cycles and go to step 2.

2. Contract the candidate cycles, and try to produce a planar map

of the resulting graph G(C). If such a graph is produced, go to step 3. Else,try next candidate. If no candidates are left, the graph cannot be embedded in projective plane.

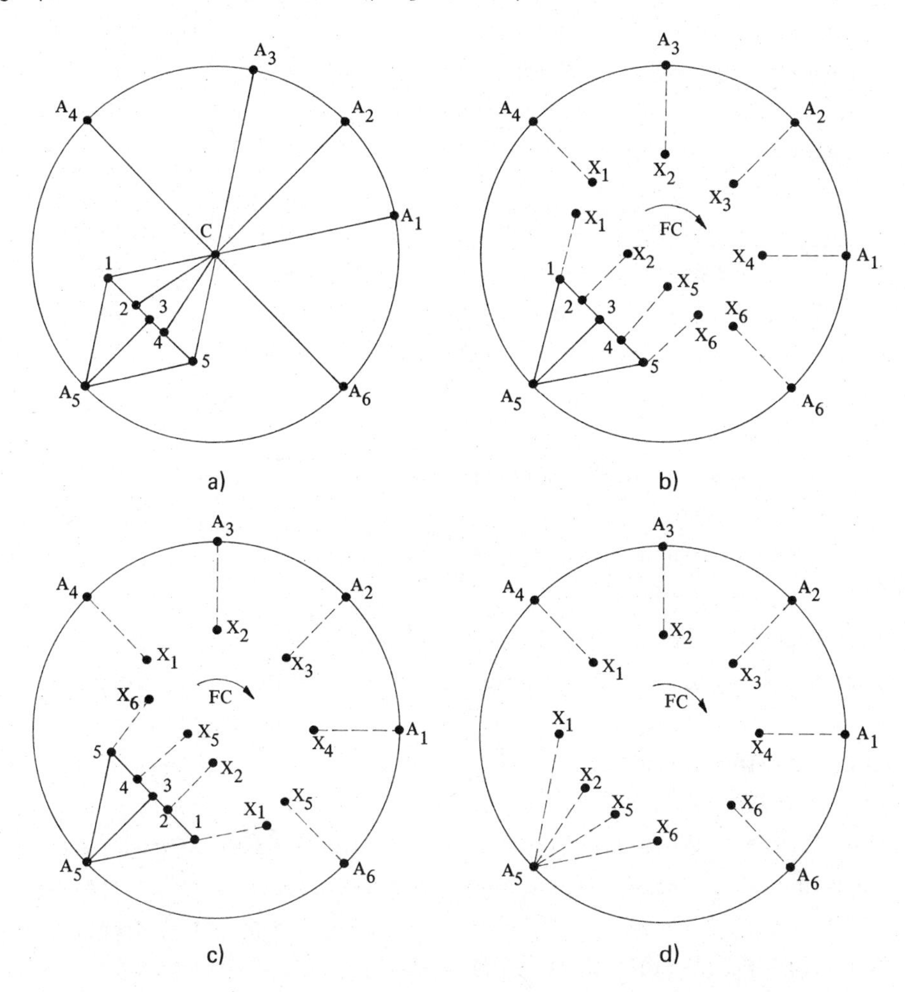

Figure 8.

3. Find the face FC. If necessery remove the blocks and introduce dummy arcs, Establish the cyclic order of A_{j-s} and X_{i-s}, of the adjacency matrix. Try the indexing procedure. If this is succesfully completed go to step 4. Else return to 2.

4. Delete the dummy arcs. Return the removed blocks and arcs incident to the blocks with indices transfered from the corresponding dummy arcs. Try to produce the planar map of the so-obtained EPG with the doubled contour as outer face. If this is possible, remove the indices. All the faces of the resulting graph, with exeption of outer one, are equal to the faces of the projective plane map. Else return to step 2.

7. The duration of the Algorithm

The duration of the algorithm deponds basically on the algorithm applied for planar embedding. In step 1, an algorithm which establishes nonplanarity by finding a nonplanar path in a three-connected planar subgraph as in |2| has to be used. Such an algorithm runs in $O(N^2)$, where N is the number of the vertices in the graph. However, in other steps any algorithm for planar embedding is suitable. For example, well known algorithm described in |6|, which runs in $O(N)$ may be used.

Let D be the maximal degree of the vertices in the considered graph, NX the number of the vertices of the doubled contour, and NA the number of the vertices of the face FC. Obviously, maximal value of (NX+NA) is equal to N. Execution of the first step takes $O(N^2)$. The number of candidates for the essential cycles is at most (D-1). The second step is run at most (D-1) times, and it needs $(D-1)O(N)$ for execution. In the step 3, the longest procedure is indexing. As it needs exploring of all the elements of a (NAxNX) matrix, and is repeated six times in a trial, the third step has a duration of $6(D-1)O(NA \times NX)$. As the maximal value of (NXxNA) is $N^2/2$ the worst time of step 3 is $6(D-1)O(N^2)$. Finally, in the step 4, the planarity algorithm is repeated at most (D-1) time. So its duration is $(D-1)O(N)$. The total duration is:

$$O(N^2) + (D-1)O(N) + 6(D-1)O(N^2) + (D-1)O(N)$$

The third step dominates the execution of the algorithm, which is proportional to the maximal degree of the graph, and the square of the number the vertices.

8. Concluding remarks

The possible extentions of the ideas presented in this paper may be looked for in the following directions. The technique for the search for essential cycles by a trial of planar embedding may be used for a search of essential cycles in other surfaces. The indexing may be regarded as a check of planarity in a cylindrical face. This technique may be adapted for connecting pieces of graphs in surgical approach for genus testing. The authors are presently trying to decrease further the complexity of the presented algorithm, and are looking for an algorithm which will find an embedding of a suitable nonplanar graph in projective plane, torus or Klein bottle in a unique procedure.

REFERENCES

|1| B. Kljuić, Z. Salčić, *An Algorithm for Testing Genus one, and Drawing of Toroidal Graphs,* IEEE International conference on Circuits and Systems. Munchen, 1976.

|2| B. Kljuić, Z. Salčić, *An effective Algorithm for Drawing of planar networks.* The European Conference on Circuit Theory and Dessign, London 1974.

|3| G. Demoucron, Y. Malgrange and R. Pertuiset, *Graphes planaires treconnaissence et construction de reperesentation planaires topologiques.* Rev. Francaise informat. Rech. operationelle, 8(1964) 33-47.

|4| I. S. Filloti, *An Efficient Algorithm for Determining wheather a Cubic Graph is toroidal.* Tenth Annual ACM Symposium on the Theory of Computing pp 133-142. 1978.

|5| I. S. Filotti and G. L. Miller, *On determining the Genus of a Graph in* $O(v^{O(g)})$ *steps.* ACM 1979, pp 27-37.

|6| Hopcroft and R. Tarjan, *Efficient planarity testing*. J. Assoc. Comp. Mach. 21(1974) 549-568.

|7| I. S. Filotti, *An Algorithm for imbedding Cubic Graphs in the Torus*. Journal of Computer and Sciences 20, 2550276 (1980).

TOWARD A MEASURE OF VULNERABILITY

I. THE EDGE-CONNECTIVITY VECTOR

Raymond E. Pippert
Indiana University-Purdue University
at Fort Wayne

Marc J. Lipman
Indiana University-Purdue University
at Fort Wayne

ABSTRACT:

In the process of determining a parameter which appropriately measures the vulnerability of a graph (as a communication network, for example), both relative parameters, which incorporate the size of the graph, and absolute parameters are considered. Arising naturally from these considerations is the parameter λ_i, which denotes the minimum number of edges which must be removed so as to separate at least i vertices of a graph from the rest of its vertices. In a given graph, the values of λ_i and λ_j $(i \neq j)$ may not be independent, in which case the values yield information about the structure of the graph.

I. Introduction.

In the use of graphs to model networks, especially in the case of communication networks, there is considerable interest in measuring the ease or difficulty of disrupting the network by the removal of some set of vertices or edges from the graph. A more or less generic term for such measurements is vulnerability.

A natural starting point in the search for an appropriate parameter is with connectivity or edge-connectivity. These are quickly found to be inadequate, since they give no indication of the relative sizes of the components obtained when the graph is disconnected. These measures are extended in a straightforward way to obtain both connectivity and edge-connectivity vectors. The connectivity vector proves to be of less interest than the edge-connectivity vector, and its discussion is left to the concluding section. In our work, G or $G(p,q)$ will denote a graph with p vertices and q edges. To simplify notation, we shall let $h = \lfloor p/2 \rfloor$ throughout. We are now ready for the following:

Definition: The edge-connectivity vector $\overline{\lambda}(G)$ of a graph G is $\overline{\lambda}(G) = (\lambda_1, \lambda_2, \lambda_3, \ldots, \lambda_h)$, where λ_i denotes the minimum number of edges G which must be removed in order to separate a set of at least i but not more than h vertices of G from the remaining vertices of G.

Some observations are in order. We need not consider values of λ_i for $i > h$, since separating $(p - i)$ vertices of a graph from the remaining i vertices is equivalent to separating i vertices from the remaining $(p - i)$ vertices. We have chosen not to have λ_i represent the number of edges which must be deleted to separate a set of exactly i vertices from the rest, but rather the number of edges which must be deleted to separate a set of at least i but not more than h vertices from the rest. This yields $\lambda_i \le \lambda_{i+1}$ $(1 \le i \le h - 1)$. We refer to a vector having this property as an increasing vector. As we shall see from the theorems, not every increasing vector is the edge-connectivity vector of a graph.

From the definition, it follows that λ_1 is simply the edge-connectivity λ, and hence does not exceed the minimum degree δ. Furthermore, successive entries of $\overline{\lambda}$ can differ by no more than the maximum degree Δ.

II. Examples

All entries of $\overline{\lambda}$ may be equal. If $m \geq n + 1$ in the graph G of Figure 1, then $\lambda_i(G) = n$ for $1 \leq i \leq h$.

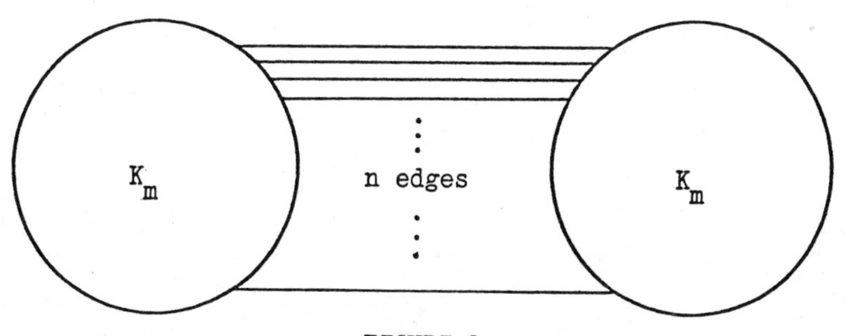

FIGURE 1

On the other hand, $\overline{\lambda}$ may have no two entries equal. The three graphs of Figure 2 have, respectively, $\overline{\lambda}(K_{1,n}) = (1,2,3, \ldots, h)$, $\overline{\lambda}(H) = (2,3,4, \ldots, h + 1)$, and $\overline{\lambda}(W_{n+1}) = (3,4,5, \ldots, h + 2)$.

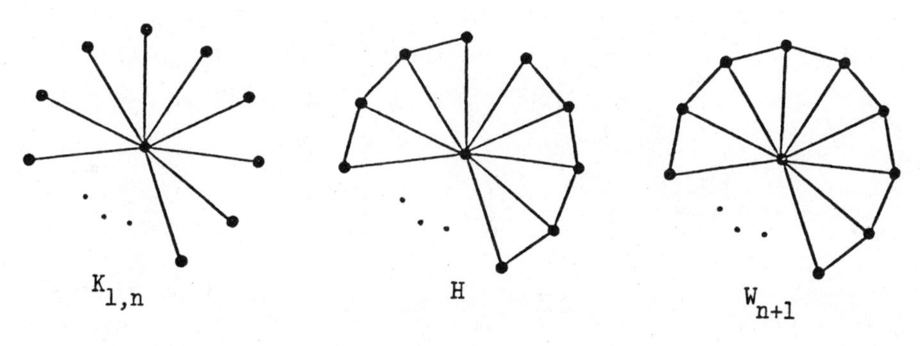

FIGURE 2

That no example of a graph with $\overline{\lambda} = (4,5,6 \ldots, h + 3)$ has been provided is not an oversight; in fact, there is no graph having $\overline{\lambda} = (4,5,6, *, *, \ldots, *)$. This leads us to consider the types of restrictions which must be satisfied by an increasing vector if it is to be the edge-connectivity vector of some graph.

III. Restrictions on edge-connectivity vectors.

If the first entry of an edge-connectivity vector is relatively large (at least four), and the second is larger, then succeeding entries may be restricted.

Theorem 1. Let $G(p,q)$ be a graph with edge-connectivity vector $\overline{\lambda} = (m,n,*,*, \ldots, *)$, where $m \geq 4$, $m + 1 \leq n \leq 2m - 3$, and $p \geq 2m$. Then $\overline{\lambda} = (m,\underbrace{n,n, \ldots, n}_{(m-1) \text{ terms}}, *,*, \ldots, *)$,

i.e. $\lambda_i = n$ for $2 \leq i \leq m$.

Proof: Since one vertex of G can be separated from the rest by the removal of m edges, but more edges must be removed to separate two or more vertices, we see that $m = \lambda = \delta$ (the minimum degree of G). Since $\lambda_2 = n$, there exists a set S containing k vertices of G, $2 \leq k \leq h$, which can be separated from the rest of the vertices of G by the removal of n edges. Because $\delta = m$, the total degree of S (sum of the degrees of the vertices of S) must be at least km. Now the n edges from S to the other vertices of G contribute n to the total degree of S, while the maximum possible contribution from edges in the subgraph induced by S is $k(k-1)$, so we have

$$k(k - 1) + n \geq km$$

or

$$k^2 - (m + 1)k + n \geq 0.$$

Since the graph of $f(k) = k^2 - (m + 1)k + n$ is a parabola, we can determine those values of k which satisfy the inequality simply by evaluating $f(k)$ at a few points.

Recalling that $k \geq 2$, we check

$$f(2) = 4 - 2m - 2 + n \leq 2 - 2m + (2m - 3) < 0 \quad [\text{since } n > m > 3].$$
$$f(m - 1) = (m - 1)^2 - (m^2 - 1) + n \leq -2m + 2 + (2m - 3) < 0$$
$$f(m) = m^2 - (m + 1)m + n = -m + n \geq -m + (m + 1) > 0.$$

Consequently, since k is an integer, we must have $k \geq m$, which indicates that at least m vertices of G can be separated from the rest of G by the removal of n edges of G,
i.e. $\overline{\lambda}(G) = m,\underbrace{n,n, \ldots, n,}_{(m-1)\text{ terms}} *,*, \ldots, *)$. □

The restriction $n \leq 2m - 3$ in Theorem 1 may appear to be rather arbitrary. Investigation of cases with relatively small values of m indicates that the restriction is appropriate. Clearly, some other inequalities may be derived for larger values of n, however they are of less interest than that obtained in the theorem.

The second type of restriction on an edge-connectivity vector involves the location of the first increase in the entries of the vector.

<u>Theorem 2</u>. Let G be a graph with
$m = \lambda_1 = \lambda_2 = \ldots = \lambda_k < \lambda_{k+1}$, where $k \geq 2$. Then $m \leq k$, i.e.
$\overline{\lambda} = (\underbrace{m,m, \ldots, m}_{\text{at least } m \text{ terms}} *, *, .., *)$, with equality only if $m = \delta$

<u>Proof</u>: There is a set S consisting of k vertices of G which can be separated from the rest of the vertices of G by the removal of m edges. Since $m \leq \delta$, the total degree of S is at least km. Consequently,

$$k(k - 1) + m \geq km$$
$$k^2 - k + m - km \geq 0$$
$$(k - m)(k - 1) \geq 0.$$

Since $k \geq 2$, we have $k \geq m$. If $m < \delta$, then the first inequality above is strict, so we must have $k > m$. []

IV. Conclusion.

Because of the restrictions imposed by the theorems, we see that there are whole classes of increasing vectors which cannot be the edge-connectivity vectors of any graphs. Examining especially the first increase in the entries of a graph's edge-connectivity vector, we obtain some information about the structure of the graph. Further analysis of the edge-connectivity vector may yield additional information.

The situation changes significantly when connectivity replaces edge-connectivity. We observe that if κ_n denotes the minimum number of vertices which must be removed from a graph G in order to separate a set of at least n vertices from the rest of G, we must decide how to categorize those vertices which have been removed. We may include them among the vertices which have been "separated," we may consider them part of "the rest of G," or we may exclude them from both categories. In the first case, $\kappa_n \leq n$ since the removal of n vertices clearly "separates" at least that many from the rest. In the latter two cases, given an arbitrary increasing vector $(v_1, v_2, \ldots, v_k)$, it is easy to construct a graph whose connectivity vector $\overline{\kappa} = (\kappa_1, \kappa_2, \kappa_3, \ldots, \kappa_h)$ is $(v_1 . v_2, \ldots, v_k, *, *, \ldots *)$, as shown in Figure 3.

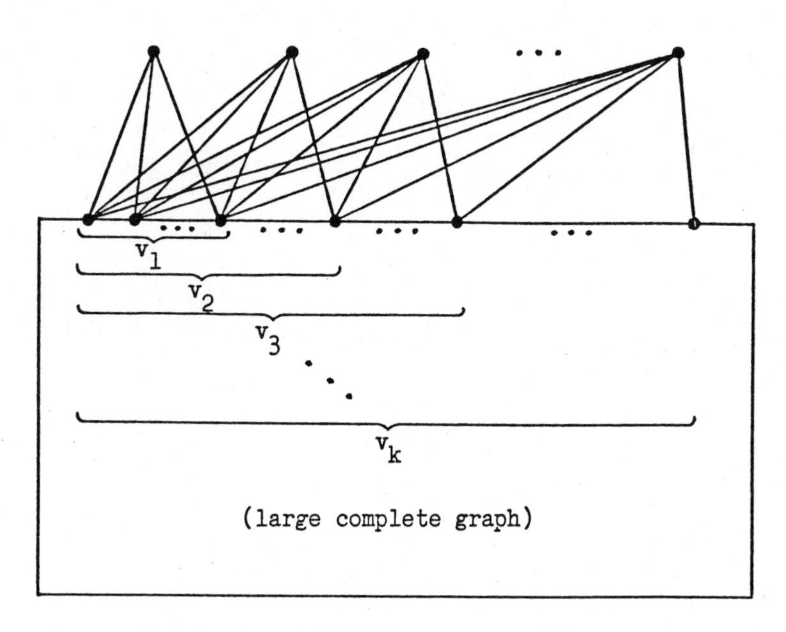

FIGURE 3

In any case, it is clear that $\overline{\kappa}$ does not contain the kind of information about the graph that is available in $\overline{\lambda}$.

SOME RESULTS ON AUTOMORPHISMS OF ORDERED RELATIONAL SYSTEMS AND THE THEORY OF SCALE TYPE IN MEASUREMENT

Fred S. Roberts
Department of Mathematics
Rutgers University
New Brunswick, New Jersey 08903

Zangwill Rosenbaum
Department of Mathematics
Rutgers University
New Brunswick, New Jersey 08903

ABSTRACT

This paper presents some theorems on the classification of scales of measurement by degree of homogeneity and degree of uniqueness, as begun by Narens and by Luce and Narens. The results generalize an earlier result of the first author's and relate to results on automorphisms of valued digraphs obtained by Roberts and Rosenbaum.

1. Introduction

The theory of measurement is concerned with putting measurement on a firm mathematical foundation. In the "representational" approach to measurement described in the books by Krantz, et al. [1971], Pfanzagl [1968], and Roberts [1979], measurement is looked at as the study of order-preserving mappings between relational systems. Such mappings define the scales of measurement, and a major emphasis is put on defining the scale type by considering transformations which map one scale into another. In the modern point of view as expressed by Narens [1981 a, b], Luce and Narens [1983, 1984 a, b], and Alper [1984], such transformations are looked at as automorphisms of a relational system. The scale type is then defined by classifying the group of automorphisms according to two parameters called degree of homogeneity and degree of uniqueness. In this paper, we study these parameters, and show how, under certain assumptions, the problem of computing the degree of homogeneity and the degree of uniqueness of an arbitrary relational system can be reduced to computing these parameters for a simple relational system, a strict weak order (a linear order with ties). The results generalize earlier results obtained by the first author for arbitrary relational systems and by Roberts and Rosenbaum [1984] for automorphisms of valued digraphs.

2. Definitions

For the purpose of this paper, a relational system is a (p+1)-tuple $\mathcal{B} = (B, R_1, R_2, \ldots, R_p)$, where B is a set and R_i is an m_i-ary relation on B. An automorphism of $\mathcal{B}$ is a function $\phi: B \to B$, one-to-one and onto, such that for all i and all $a_1, a_2, \ldots, a_{m_i}$ in B,

$$R_i(a_1, a_2, \ldots, a_{m_i}) \longleftrightarrow R_i(\phi(a_1), \phi(a_2), \ldots, \phi(a_{m_i})). \quad (1)$$

Following the papers by Narens and by Luce and Narens, we shall study just relational systems $\mathcal{B}$ where $R_1 = \succ$ is a strict weak order on B. (For definitions of undefined terms, see Roberts [1979] or Roberts [1984].) Such a relational system will be called ordered. If $(B, \succ)$ is a strict weak order, $a \sim b$ will mean

$\sim(a \succ b)$ and $\sim(b \succ a)$. Then $(B, \sim)$ is an equivalence relation. B^* will denote the collection of equivalence classes.

If $\mathcal{B}$ is an ordered relational system, Roberts and Rosenbaum [1984] call an (ordinary) automorphism a <u>loose automorphism</u>, and call an automorphism ϕ <u>tight</u> if in addition, $\phi(x) \sim x$, for all x in B.

Whereas Narens and also Luce and Narens only define one notion of homogeneity and one of uniqueness, Roberts and Rosenbaum define several. We shall use two of each here. Suppose $\mathcal{B}$ is an ordered relational system, and $M \leq |B^*|$. We say that $\mathcal{B}$ is <u>M-loose homogeneous</u> if whenever

$$x_1 \succ x_2 \succ \cdots \succ x_M \quad \text{and} \quad y_1 \succ y_2 \succ \cdots \succ y_M \tag{2}$$

for $x_1, x_2, \ldots, x_M, y_1, y_2, \ldots y_M$ in B, there is a loose automorphism ϕ of $\mathcal{B}$ such that $\phi(x_i) = y_i$, $i = 1, 2, \ldots, M$. We say that $\mathcal{B}$ is <u>M-tight homogeneous</u> if (2) and $x_i \sim y_i$, $i = 1, 2, \ldots, M$, imply there is a tight automorphism ϕ of $\mathcal{B}$ such that $\phi(x_i) = y_i$, $i = 1, 2, \ldots, M$. Corresponding to each type of homogeneity, there is a parameter h called the <u>degree of homogeneity</u>. Thus, $h_L(\mathcal{B})$ is defined to be the largest M such that $\mathcal{B}$ is M-loose homogeneous, and $h_T(\mathcal{B})$ is defined similarly using M-tight homogeneous. We take $h_L(\mathcal{B})$ or $h_T(\mathcal{B})$ to be 0 if there is no M and ∞ if there are arbitrarily large M.

We say that $\mathcal{B}$ is <u>N-loose unique</u> if whenever $x_1, x_2, \ldots, x_N$ are distinct elements of B and ϕ and ϕ' are two loose automorphisms of $\mathcal{B}$ such that $\phi(x_i) = \phi'(x_i)$, $i = 1, 2, \ldots, N$, then $\phi(x) = \phi'(x)$ for all x in B. <u>N-tight uniqueness</u> is defined similarly using tight automorphisms. Corresponding to each type of uniqueness, there is a parameter u called the <u>degree of uniqueness</u>. Thus, $u_L(\mathcal{B})$ is defined to be the smallest N such that $\mathcal{B}$ is N-loose unique, and $u_T(\mathcal{B})$ is defined similarly using N-tight unique. We take $u_L(\mathcal{B})$ or $u_T(\mathcal{B})$ to be ∞ if there is no such N. Note that $u_L(\mathcal{B})$ or $u_T(\mathcal{B})$ is 0 if the only loose or tight automorphism is the identity.

Narens [1981 a, b] and Luce and Narens [1983, 1984 a, b] begin the investigation of the possible <u>scale types</u> (h_L, u_L), and Roberts and Rosenbaum [1984] introduce the <u>scale types</u> (h_T, u_T). If $\mathcal{B}$ is an ordered relational system with $(B, \succ)$ isomorphic onto $(Re, >)$, there is evidence that the pair (h_L, u_L) cannot take on many values. For instance, Narens [1981a, b] and Alper [1984] show that in this case, if (h_L, u_L) is (M, M) or $(M, M+1)$, $M \geq 1$, then (h_L, u_L) is $(1, 1)$, $(2, 2)$ or $(1, 2)$. The results in Section 3 suggest that many scale types will be (∞, ∞) in this situation. That begins to explain why so few scale types (h_L, u_L) seem to occur. Indeed, for $0 < M, N < \infty$, no (M, N) scale types are known to occur except $(1, 1)$, $(2, 2)$ and $(1, 2)$. The results in Section 4 address the scale types (h_T, u_T).

3. <u>The Loose Case</u>

<u>Theorem 1.</u> Suppose $\mathcal{B} = (B, \succ, R_2, R_3, \ldots, R_p)$ is an ordered relational system and $(B, \succ)$ is a strict simple (linear, total) order. Suppose R_i is an m_i-ary relation and $m_i \leq M$, all i, and suppose $\mathcal{B}$ is M-loose homogeneous. Then the (loose) automorphisms of $\mathcal{B}$ are exactly the (loose) automorphisms of $(B, \succ)$.

<u>Proof.</u> Clearly every (loose) automorphism of $\mathcal{B}$ is a (loose) automorphism of $(B, \succ)$. Conversely, suppose ϕ is a loose automorphism of $(B, \succ)$. Then ϕ is one-to-one and onto. It suffices to show (1) for $i = 2, 3, \ldots, p$. Since $\mathcal{B}$ is M-loose homogeneous, M is at most $|B^*|$. Thus, since $m_i \leq M$, we can find a set X containing $\{a_1, a_2, \ldots, a_{m_i}\}$ and consisting of M distinct elements. (Two a_i can be the same but each a_i is listed at most once in X.) Since ϕ is one-to-one, $\phi(X)$ must consist of M distinct elements. Since $(B, \succ)$ is strict simple, no two elements in X are equivalent under $\sim$, and no two elements of $\phi(X)$ are equivalent under $\sim$. Also, $a_i \succ a_j$ holds if and only if $\phi(a_i) \succ \phi(a_j)$. Thus, by M-loose homogeneity of $\mathcal{B}$, there is

a loose automorphism ψ of $\mathcal{B}'$ such that $\psi(x) = \phi(x)$ for all x in X, i.e., such that $\psi(a_i) = \phi(a_i)$, $i = 1, 2, \ldots, m_i$. It follows that

$$(a_1, a_2, \ldots, a_{m_i}) \in R_i \longleftrightarrow (\psi(a_1), \psi(a_2), \ldots, \psi(a_{m_i})) \in R_i$$

$$\longleftrightarrow (\phi(a_1), \phi(a_2), \ldots, \phi(a_{m_i})) \in R_i.$$

Q.E.D.

Corollary 1. Under the hypotheses of Theorem 1, $h_L(\mathcal{B}') = h_L(B, \succ)$ and $u_L(\mathcal{B}') = u_L(B, \succ)$.

Corollary 2. Suppose $\mathcal{B}' = (B, \succ, R_2, R_3, \ldots, R_p)$ is an ordered relational system with $(B, \succ)$ isomorphic onto $(Re, >)$. Suppose R_i is an m_i-ary relation and $m_i \leq M$, all i, and suppose $\mathcal{B}'$ is M-loose homogeneous. Then $h_L(\mathcal{B}') = u_L(\mathcal{B}') = \infty$.

Corollary 2 was originally obtained by the first author in August 1982, and was first published (with acknowledgements) in Luce and Narens [1984b]. It begins to explain why scale types (M, N) with M, N < ∞ might not be easy to find.

We now discuss how to eliminate the hypothesis that $(B, \succ)$ is strict simple. Suppose $\mathcal{B}'$ is an ordered relational system and $M \leq |B|$. We say that $\mathcal{B}'$ is M-loose perfect homogeneous if whenever $x_1, x_2, \ldots, x_M$ are distinct, $y_1, y_2, \ldots, y_M$ are distinct, and for all i, j,

$$x_i \succ x_j \longleftrightarrow y_i \succ y_j, \tag{3}$$

then there is a loose automorphism ϕ of $\mathcal{B}'$ such that $\phi(x_i) = y_i$, $i = 1, 2, \ldots, M$. Note that for $M \leq |B^*|$, M-loose perfect homogeneity implies M-loose homogeneity, but not conversely, since we could have $x_i \sim x_j$, some $i \neq j$.

Theorem 2. Suppose $\mathcal{B}$ = (B, $\succ$, R_2, R_3, ..., R_p) is an ordered relational system, R_i is an m_i-ary relation, $m_i \leq M$, all i, and $\mathcal{B}$ is M-loose perfect homogeneous. Then the loose automorphisms of $\mathcal{B}$ are exactly the loose automorphisms of (B, $\succ$).

Proof. The proof is exactly the same as that of Theorem 1, except that it is not necessary to show that no two elements of X are equivalent under $\sim$ and no two elements of $\phi(X)$ are equivalent under $\sim$. Q.E.D.

4. The Tight Case

Results analogous to Theorem 1 and its corollaries are trivial in the case of tight homogeneity and tight automorphisms. Specifically, suppose $\mathcal{B}$ = (B, $\succ$, R_2, R_3, ..., R_p) is an ordered relational system and (B, $\succ$) is a strict simple (linear, total) order. Suppose R_i is an m_i-ary relation and $m_i \leq M$, all i, and suppose $\mathcal{B}$ is M-tight homogeneous. Then the tight automorphisms of $\mathcal{B}$ are exactly the tight automorphisms of (B, $\succ$). This is because in both cases, the only tight automorphism is the identity. (The one difference from the loose situation is that here, $u_T(\mathcal{B}) = u_T(B, \succ) = 0$.)

The following theorems show how to eliminate the hypothesis that (B, $\succ$) is strict simple. We say that two elements a and b of B are perfect substitutes, and write aEb, if for all i and for all j, $0 \leq j \leq m_i - 1$,

$$R_i(x_1, x_2, \ldots, x_j, a, x_{j+2}, \ldots, x_{m_i}) \Longleftrightarrow R_i(x_1, x_2, \ldots, x_j, b, x_{j+2}, \ldots, x_{m_i}).$$

Theorem 3. Suppose $\mathcal{B}$ = (B, $\succ$, R_2, R_3, ..., R_p) is an ordered relational system and $a \sim b$ implies aEb. Then the tight automorphisms of $\mathcal{B}$ are exactly the tight automorphisms of (B, $\succ$).

Proof. Clearly every tight automorphism of $\mathcal{B}'$ is a tight automorphism of $(B, \succ)$. Conversely, suppose ϕ is a tight automorphism of $(B, \succ)$. Then ϕ is one-to-one and onto and $\phi(x) \sim x$, all x. We show (1) for $i = 2, 3, \ldots, p$. But this follows trivially since $\phi(a_i) \sim a_i$, so $\phi(a_i)Ea_i$, so (1) follows.

Q.E.D.

Corollary. Under the hypotheses of Theorem 3, $h_T(\mathcal{B}') = h_T(B, \succ)$ and $u_T(\mathcal{B}') = u_T(B, \succ)$.

Theorem 3 does not have an analogue for the loose case, since $\phi(a_i) \sim a_i$ does not necessarily hold. However, if B^* is finite, every loose automorphism is tight, and the result holds for loose automorphisms as well as tight ones.

If $M \leqslant |B|$ let us say that the ordered relational system $\mathcal{B}'$ is M-tight perfect homogeneous if whenever $x_1, x_2, \ldots, x_M$ are distinct, $y_1, y_2, \ldots, y_M$ are distinct, and $x_1 \sim y_1, x_2 \sim y_2, \ldots, x_M \sim y_M$, there is a tight automorphism ϕ of $\mathcal{B}'$ so that $\phi(x_i) = y_i$, $i = 1, 2, \ldots, M$. Note that for $M \leqslant |B^*|$, M-tight perfect homogeneity implies M-tight homogeneity, but not conversely, since we could have $x_i \sim x_j$, some $i \neq j$. Note also that if B^* is finite, M-tight perfect homogeneity implies M-loose perfect homogeneity. To see this, note that $x_i \sim y_i$, $i = 1, 2, \ldots, M$, implies (3), and note that when B^* is finite, every loose automorphism is tight.

Theorem 4. Suppose $\mathcal{B}' = (B, \succ, R_2, R_3, \ldots, R_p)$ is an ordered relational system, R_i is an m_i-ary relation, $m_i \leqslant M$, all i, and $\mathcal{B}'$ is M-tight perfect homogeneous. Then the tight automorphisms of $\mathcal{B}'$ are exactly the tight automorphisms of $(B, \succ)$.

Proof. As in the proof of Theorem 3, it suffices to assume that ϕ is a tight automorphism of $(B, \succ)$ and to show that ϕ is a tight automorphism of $\mathcal{B}'$. Clearly ϕ is one-to-one and onto and $\phi(x) \sim x$. Thus, it remains to demonstrate (1) for $i = 2, 3, \ldots, p$.

We shall use the fact that M-tight perfect homogeneity implies k-tight perfect homogeneity for $k \leq M$, which is simple to verify.

Suppose $a_1, a_2, \ldots, a_{m_i}$ are elements of B. These are not necessarily distinct. But we can pick k distinct elements $a_{i_1}, a_{i_2}, \ldots, a_{i_k}$ from these so that every other a_r is equal to one of these a_{i_j}. Since ϕ is one-to-one, $\phi(a_{i_1})$, $\phi(a_{i_2}), \ldots, \phi(a_{i_k})$ are also distinct. Moreover, by tightness of ϕ, $a_{i_j} \sim \phi(a_{i_j})$, $j = 1, 2, \ldots, k$. Since $k \leq m_i \leq M$, M-tight perfect homogeneity implies that there is a tight automorphism ψ of $\mathcal{B}$ so that $\psi(a_{i_j}) = \phi(a_{i_j})$, $j = 1, 2, \ldots, k$. But then $\psi(a_i) = \phi(a_i)$, all i. Hence,

$$(a_1, a_2, \ldots, a_{m_i}) \in R_i \longleftrightarrow (\psi(a_1), \psi(a_2), \ldots, \psi(a_{m_i})) \in R_i$$

$$\longleftrightarrow (\phi(a_1), \phi(a_2), \ldots, \phi(a_{m_i})) \in R_i.$$

Q.E.D.

It should be remarked that Theorem 3 follows from Theorem 4, since the hypothesis $a \sim b \longrightarrow aEb$ implies the hypothesis M-tight perfect homogeneous for any $M \leq |B|$. To show this, suppose $M \leq |B|$, $x_1, x_2, \ldots, x_M$ are distinct, $y_1, y_2, \ldots, y_M$ are distinct, and $x_i \sim y_i$, $i = 1, 2, \ldots, M$. Let $\phi(x_i) = y_i$, all i. Then ϕ can be extended to a one-to-one, onto function from B into B which simply permutes elements in each equivalence class of B^*. It follows that $\phi(x) \sim x$, all x. The claim is that ϕ is a tight automorphism of $\mathcal{B}$. This is because $a_i \sim \phi(a_i)$, all i, and so $a_i E \phi(a_i)$, all i, and therefore (1) follows. The proof of Theorem 3 is not yet complete if by chance some $m_i > |B|$. But, as is easy to check, the proof of Theorem 4 really only needs the hypothesis k-tight perfect homogeneity for $k \leq |B|$ and $k \leq \max\{m_i\}$, and this hypothesis has been verified.

Corollary. Under the hypotheses of Theorem 4, $h_T(\mathcal{B}) = h_T(B, \succ)$ and $u_T(\mathcal{B}) = u_T(B, \succ)$.

This corollary is closely related to a result of Roberts and Rosenbaum [1984]. We say an ordered relational system $\mathcal{B} = (B, \succ, R)$ is a <u>valued digraph</u> if R is binary. Roberts and Rosenbaum (Proposition 5) show that if B^* is finite and $\mathcal{B}$ is a valued digraph which is 2-tight homogeneous, then $h_T(\mathcal{B}) = |B^*| = h_T(B, \succ)$. The above corollary gives this same conclusion under a stronger hypothesis, since 2-tight perfect homogeneity implies 2-tight homogeneity. Theorem 4 also gives a stronger conclusion under the stronger hypothesis, as 2-tight homogeneity does not imply that the tight automorphisms of $(B, \succ, R)$ are exactly the same as those of $(B, \succ)$. To give an example, consider $B = \{a, b, c, d\}$, $\succ = \{(d, a), (d, b), (d, c)\}$, and $R = \{(a, b), (b, c), (c, a)\}$. (In the language of Roberts and Rosenbaum [1984], $(B, \succ, R)$ consists of a 3-cycle together with an isolated vertex; the isolated vertex has one value and the other three vertices share a common lesser value.) Then $(B, \succ, R)$ is 2-tight homogeneous, but $\phi(a) = c$, $\phi(c) = a$, $\phi(b) = b$, $\phi(d) = d$ defines a tight automorphism of $(B, \succ)$ which is not a tight automorphism of $(B, \succ, R)$.

We close this section by noting that Theorem 3 also has an implication for valued digraphs. In a valued digraph, aEb means that $a \sim b$ (a and b have the same value) and a and b have the same in-neighbors and out-neighbors in the digraph (B, R). (Note that if (x, x) is a loop of (B, R), it is both an in- and an out-neighbor of x.) Thus, we have the following result: Suppose in a valued digraph $\mathcal{B}$, whenever a and b have the same value, they have the same in-neighbors and out-neighbors in the digraph (B, R). Then the tight automorphisms of $\mathcal{B}$ are exactly the tight automorphisms of $(B, \succ)$, i.e., the one-to-one, onto functions on B which map each point into another one of the same value.

Acknowledgements

This research was supported by NSF Grant number IST-83-01496 to Rutgers University.

REFERENCES

Alper, T.M., "A Note on Real Measurement Structures of Scale Type (m, m+1)," J. Math. Psychol., 1984, to appear.

Krantz, D.H., Luce, R.D., Suppes, P., and Tversky, A., Foundations of Measurement, Vol. I, Academic Press, New York, 1971.

Luce, R.D., and Narens, L., "Symmetry, Scale Types, and Generalizations of Classical Physical Measurement," J. Math. Psychol. 27 (1983), 44-85.

Luce, R.D., and Narens, L., "Classification of Real Measurement Representations by Scale Type," Measurement, 1984, to appear. (a)

Luce, R.D., and Narens, L., "Classification of Concatenation Measurement Structures According to Scale Type," J. Math. Psychol., 1984, to appear. (b)

Narens, L., "A General Theory of Ratio Scalability with Remarks about the Measurement-theoretic Concept of Meaningfulness," Theory and Decision, 13 (1981), 1-70. (a)

Narens, L., "On the Scales of Measurement," J. Math. Psychol., 24 (1981), 249-275. (b)

Pfanzagl, J., Theory of Measurement, Wiley, New York, 1968.

Roberts, F.S., Measurement Theory, with Applications to Decisionmaking, Utility, and the Social Sciences, Addison-Wesley, Reading, MA, 1979.

Roberts, F.S., Applied Combinatorics, Prentice-Hall, Englewood Cliffs, N.J., 1984.

Roberts, F.S., and Rosenbaum, Z., "Tight and Loose Value Automorphisms," submitted to F. Harary (ed.), Proc. First Hoboken Symposium on Graph Theory, Annals of Discrete Math., 1984.

COUNTING STRONGLY CONNECTED FINITE AUTOMATA

Robert W. Robinson*
University of Georgia

ABSTRACT

Recurrence formulas are derived for the numbers C(p) and S(p) of initially connected and strongly connected labeled finite automata with p states and an arbitrary input alphabet. For this purpose it was found necessary to determine the Möbius function of the lattice of quasi-orders of a set, ordered by reverse inclusion, with respect to the minimum element. Detailed asymptotic conjectures are included, but little progress has been made toward proving them.

*Research supported by the N.S.F. under grant MCS-8302282.

1. Introduction.

By a *finite automaton* we mean a triple $M = (Q,\Sigma,f)$ where Q is a finite set of *states*, Σ is a finite *input alphabet*, and $f: Q \times \Sigma \to Q$ is a *transition function*. One can think of M as a digraph with vertex set Q, multiple loops and arcs allowed, which is out-regular of degree $d = |\Sigma|$. The arcs are to be labeled with members of Σ so as to correspond to the transition function; that is, M has an arc from q to q' labeled with σ if and only if $f(q,\sigma) = q'$. Then for each finite string $\omega \in \Sigma^*$ and each state $q \in Q$, the transition function defines a unique directed path in M starting at q and labeled with the symbols of ω in sequence. The endpoint of this path is the *ω-successor* of q. We say that q' is *reachable* from q in M if q' is the *ω-successor* of q for some $\omega \in \Sigma^*$.

Sometimes one state is specified to be the initial state, or *start state*, of a finite automaton. If q is the start state for M, then M is *initially connected* if every state is reachable from q. If every state is reachable from every other state then M is said to be *strongly connected*, regardless of whether a start state has been specified. Clearly M is strongly connected if and only if it is strongly connected in the usual sense as a digraph. Following digraph terminology, M is said to be *weakly connected* if the underlying graph (ignoring directions on the arcs) is connected. Maximal weakly connected subgraphs of M are called *weak components* of M. However maximal strongly connected subgraphs of M are called *blocks*, in contradistinction to the usual terminology of graph theory.

As defined so far, there is no provision for output from our finite automata. There are two standard models for specifying output using some finite output alphabet Γ with $g = |\Gamma|$. For a Moore machine (state-assigned), output is given by a map $h: Q \to \Gamma$. There are g^p such maps possible, where $p = |Q|$. For a Mealy machine (transition - assigned), output is given by a map $h: Q \times \Sigma \to \Gamma$. For each of our automata, there are g^{pd} such maps. In the present paper interest is focused on connectivity rather than output, so finite automata will be enumerated without provision for output.

In counting labeled finite automata with p states we are simply counting those with the fixed state set {1,2,...,p}. In counting unlabeled finite automata we are counting isomorphism classes in which the input alphabet is held fixed but the states can be mapped in any way that preserves the transition function (and the start state if one is specified).

In this paper, initially connected and strongly connected labeled finite automata are counted. If a start state is specified, this allows the unlabeled automata to be counted by dividing by (p-1)! A method for counting strongly connected finite automata was published by Radke [7], but it is computationally unwieldy. Curiously, the much easier problem of counting initially connected finite automata was posed quite explicitly by Harary [3], but apparently has not been discussed in the literature. Unlabeled finite automata have been counted by Harrison [5], and by Harary and Palmer [4], allowing permutations of the input and output alphabets as well as permutations of the state set. Narushima [6] has solved the very difficult problem of counting reduced finite automata. Presumably Narushima's methods could be combined with those of the present paper to count reduced initially connected finite automata and reduced strongly connected finite automata, but this possibility has not yet been explored.

2. Initially connected finite automata.

The number of finite automata which are complete and deterministic with p states, d input symbols, and no provision for output is p^{pd}. Let C(p) denote the number of these which are initially connected from a specified start state. Without loss of generality we take {1,...,p} as the state set, and state 1 as the start state. Of course the number C(p) depends on d as well as p. However the point of view taken is that d is arbitrary but fixed, while p is allowed to vary. Thus it is convenient to suppress d in the notation.

A recurrence for C(p) is easily derived by considering the number k of states reachable from 1 in an arbitrary finite automaton. Since 1 is reachable from itself, $k \geq 1$ and the other $k - 1$ reachable states can be chosen from the other $p - 1$ states in $\binom{p-1}{k-1}$ different ways. The structure of the automaton on the reachable states is that of an arbitrary initially connected finite automaton, which can be chosen in C(k) ways. Transitions from the $p - k$ unreachable states can be chosen in $p^{(p-k)d}$ ways. Moreover these three sets of choices are independent of each other, which leads to the identity

$$\sum_{k=1}^{p} \binom{p-1}{k-1} C(k)p^{(p-k)d} = p^{pd}. \tag{2.1}$$

Solving for C(p) gives the recurrence

$$C(p) = p^{pd} - \sum_{k=1}^{p=1} \binom{p-1}{k-1} C(k)p^{(p-k)d} \tag{2.2}$$

for $p \geq 1$. This recurrence determines the numbers C(p) for all $p \geq 1$, and can be used to calculate these numbers efficiently.

One might hope to find a better recurrence in which all terms are positive. Such a recurrence can be found for the numbers of initially connected finite automata if an additional parameter is introduced, such as the number of states at the maximum distance from the start state. Unfortunately the additional parameter makes the resulting recurrences less efficient computationally without making asymptotic analysis more tractable, so the details are omitted.

An explicit expression for C(p) is

$$C(p) = \sum_{r=1}^{p} (-1)^{r-1} \sum_{\substack{k_i \geq 1 \\ \Sigma_1^r k_i = p}} \frac{k_1(p-1)!}{\Pi_1^r(k_i!)} \prod_{j=1}^{r} (\Sigma_1^j k_i)^{dk_j}. \tag{2.3}$$

This can be proved by verifying that the right side of (2.3) satisfies the identity (2.1). One can separate the factor p^{dk_r}, in which k_r plays the role of p - k in (2.1). Alternatively, (2.3) can be deduced directly from facts concerning the Möbius function on the lattice of quasi-orders on the state set which are proved in the next section. In any case, this explicit expression seems to be suitable neither for computing exact numbers efficiently nor for asymptotic calculations.

The number of unlabeled initially connected finite automata is just $C(p)/(p-1)!$ Here the input alphabet is held fixed, but the allocation of the labels 2,3, ..., p to states other than the designated start state is immaterial. Each unlabeled initially connected finite automaton corresponds to exactly $(p-1)!$ labeled versions because the unlabeled automaton has no nontrivial automorphisms. This is because each state q is the ω-successor of the initial state 1 for some sequence ω of input symbols. In any automorphism the initial state 1, the individual input symbols, and the successor relation are all left fixed, so each state q is mapped to itself by any automorphism.

Table 1 shows the first few values of $C(p)/(p-1)!$ for d = 2 and d = 3. Values for d = 1 are not shown because in that case $C(p) = p!$. The latter can be seen directly because the states of distance 1,2,3,..., p-1 respectively from the start state can be any permutation of the set {2,...,p}, of which there are $(p-1)!$ This determines the unique transition away from each state except the one farthest from the start state, which can go to any of the p states.

An additional divisibility condition satisfied by C(p) is that $C(p)/p^d(p-1)!$ is integral. Note that the set of states having maximum distance from the start state in not empty. If all transitions are specified except for those leaving one such state, say q, then the automaton can be completed in p^d different ways. This is because each of the d transitions from q can be sent to any of the p states, since none of these transitions is needed to ensure initial connectivity.

	C(p)/(p-1)!	p
	12	2
	216	3
	5248	4
	1 60675	5
	59 31540	6
	2561 82290	7
	1 26654 45248	8
d = 2	70 50680 85303	9
	4363 12502 29700	10
	2 97058 13455 16818	11
	220 64283 93429 06336	12
	17753 18168 75445 16980	13
	15 38156 94793 65241 72656	14
	1427 67837 72754 41137 83650	15
	56	2
	7965	3
	21 28064	4
	9149 29500	5
	57 66892 14816	6
d = 3	50075 01723 37212	7
	572 87912 63921 78688	8
	8 35007 87475 93938 78655	9
	15104 92370 20431 47773 45000	10
	332 04702 73536 65897 07397 63334	11

Table 1

Numbers of unlabeled initially connected finite automata

3. <u>The lattice of quasi-orders.</u>

Let M be a finite automaton with state set Q. The reachability relation of M, written $\rho(M)$, is the set of ordered pairs (i,j) of states such that j is reachable from i in M. For any M the relation $\rho(M)$ is reflexive and transitive, that is, it is a quasi-order. Clearly M is strongly connected if and only if $\rho(M) = Q \times Q$. In this section the lattice L(Q) of quasi-orders over Q is studied. The results will be applied in the next section to count strongly connected finite automata.

The order relation in L(Q) is reverse inclusion; i.e., $\alpha \leq \beta$ in L(Q) if and only if $\beta \supseteq \alpha$ as sets of ordered pairs. The universal relation $Q \times Q$ is denoted 0 since it is the minimum element of L(Q). In L(Q) $\alpha \cap \beta$ gives the join of α and β, while $(\alpha \cup \beta)^*$ gives the meet of α and β, where * denotes transitive closure. Our main interest will be in the value of the Möbius function of α over 0 in L(Q), which we denote simply by $\mu(\alpha)$. First we give a characterization of certain complements in L(Q). For i, j ε Q we use $i|j$ to indicate that i and j are incomparable in some $\alpha \in L(Q)$. Also, the sublattice between 0 and α is written L_α.

> Lemma. Suppose $i|j$ in α, $\alpha' = (\alpha \cup \{(i,j),(j,i)\})^*$, and β is a complement of α' in L_α. Then β consists of two incomparable blocks, one containing i and the other containing j. In particular, i and j are not weakly connected in β.

Proof. We will show that any state q is equivalent in β to i or else to j. Since $\beta \cap \alpha' = \alpha$ we have $i|j$ in β, so the two blocks containing i and j will be incomparable.

Since $(\beta \cup \alpha')^* = 0$, we have $(\beta \cup \{(i,j),(j,i)\})^* = 0$, so that $\beta \cup \{(i,j),(j,i)\}$ must contain a directed q - i path. This may contain (i,j) or (j,i) as an edge, but the first occurrence of i or j on this path marks off a q - i path or a q - j path contained in β. Thus, $(q,i) \in \beta$ or $(q,j) \in \beta$. Similarly, $(i,q) \in \beta$ or $(j,q) \in \beta$. Now β can't contain both (i,q) and (q,j) nor both (j,q) and (q,i), since $i|j$ in β. So either β contains both (i,q) and (q,i), or else both (j,q) and (q,j), from which q is equivalent to i or j in β.

A weak component γ of a quasi-order α is <u>total</u> if every two states of γ are comparable in γ. A total weak component is simply a linearly ordered chain of blocks. We can now present the main result of this section.

> <u>Theorem</u> If $\alpha \in L(Q)$ has a non-total weak component then $\mu(\alpha) = 0$. Otherwise $\mu(\alpha) = (n-1)!(-1)^{k-n}$ where n is the number of weak components of α and k is the number of blocks of α.

<u>Proof</u>. We induct on $|Q|$. It is trivial to check the cases $|Q| = 1$ and $|Q| = 2$.

<u>Case 1</u>. α has a weak component which is not total. In this case we will show that L_α is not complemented, so $\mu(\alpha) = 0$; see [1, Cor. to Thm 3]. Let i,j be incomparable in α but weakly connected in α. Then $\alpha' = (\alpha \cup \{(i,j),(j,i)\})^*$ has no complement in L_α. For by the Lemma, in any complement β the states i and j would not be weakly connected. On the other hand $\beta \supseteq \alpha$, so i and j would have to be weakly connected in β as well as in α.

<u>Case 2</u>. α is total. In this case $n = 1$, and we may assume that $k \geq 2$. Let $B_1, \ldots, B_k$ be the blocks, as ordered by α. If $\beta < \alpha$ then β is also total, and each block of β consists of a segment $B_i \ldots B_{i+r}$ of blocks of α. Such a β is uniquely specified by some choice of $v \geq 1$ of the $k - 1$ consecutive pairs B_i, B_{i+1} to join together in blocks of β. There are $\binom{k-1}{v}$ such choices, and each gives a total quasi-order with exactly $k - v$ blocks. By our induction hypothesis, then,

$$\sum_{\beta<\alpha} \mu(\beta) = \sum_{v=1}^{k-1} \binom{k-1}{v}(-1)^{k-v-1}$$

$$= (1-1)^{k-1} - (-1)^{k-1} = (-1)^k.$$

Thus

$$\mu(\alpha) = -\sum_{\beta<\alpha} \mu(\beta) = (-1)^{k-1},$$

as required when n = 1.

Case 3. $n \geq 2$ and each weak component of α is total. Let i and j lie in different weak components of α, so that $i|j$. Let $\alpha' = (\alpha \cup \{(i,j),(j,i)\})^*$, and let β be a complement of α' in L_α. By the Lemma, β must consist of two incomparable blocks B_1, B_2, say with $i \in B_1$ and $j \in B_2$. Conversely, it is clear that any such β with $\beta \supseteq \alpha$ is a complement of α' in L_α. To satisfy these conditions, B_1 must consist of a union of weak components of α including the weak component containing i but not that containing j. Then B_2 is the union of the remaining weak components of α.

It is also clear that no two complements of α' in L_α are comparable. This allows us to simplify the standard expression for $\mu(\alpha)$ in terms of complements [1, Thm. 3] to the form

$$\mu(\alpha) = \sum_{\beta} \mu(\beta)\mu(\beta,\alpha) \tag{3.1}$$

where β ranges in the sum over all complements of α'. Since β consists of two blocks B_1, B_2 which are incomparable, $\mu(\beta) = 1$. If α_t denotes the restriction of α to B_t for t = 1, 2 then the segment of L_α between β and α is isomorphic to the direct product $L_{\alpha_1} \times L_{\alpha_2}$. Thus $\mu(\beta,\alpha) = \mu(\alpha_1)\,\mu(\alpha_2)$ by the product law for Möbius functions; see [8, Prop. 5 of §3].

To compute the right side of (3.1), suppose that a complement β of α' is to have a block B_1 containing i which is the union of r weak components. There are $\binom{n-2}{r-1}$ choices for forming such a complement β for α'. If B_1 contains m blocks of α then B_2 contains the remaining k - m blocks of α. By induction,

$$\mu(\alpha_1) = (r-1)!(-1)^{m-r}$$

and

$$\mu(\alpha_2) = (n-r-1)!(-1)^{(k-m)} - (n-r).$$

Thus

$$\mu(\beta,\alpha) = (r-1)!(n-r-1)!(-1)^{k-n} .$$

Summing over r gives

$$\mu(\alpha) = \sum_{r=1}^{n-1} \binom{n-2}{r-1} (r-1)!(n-r-1)!(-1)^{k-n}$$

$$= \sum_{r=1}^{n-1} (n-2)!(-1)^{k-n}$$

$$= (n-1)!(-1)^{k-n} ,$$

completing the proof of Case 3 and the Theorem.

It should be noted that segments of L(Q) which lie above partial orderings of Q have been studied elsehwhere; see [2].

4. Strongly connected finite automata.

For $p \geq 1$ let L_p denote $L(\{1,2,\ldots,p\})$. For $\alpha \in L_p$ let $H(\alpha)$ be the number of finite automata M with $\rho(M) = \alpha$, and let $E(\alpha)$ be the number with $\rho(M) \subseteq \alpha$. Then

$$E(\alpha) = \sum_{\alpha \leq \beta} H(\beta)$$

for every $\alpha \in L_p$, so by Möbius inversion we have

$$H(0) = \sum_{\alpha \in L_p} \mu(\alpha) E(\alpha). \tag{4.1}$$

The number S(p) of strongly connected finite automata with state set $\{1,2,\ldots,p\}$ is precisely H(0), so (4.1) gives an expression for S(p). This will be solved to give a recurrence relation for S(p) using the values of the Möbius function which were determined in the previous section.

As an auxiliary we will need the numbers A(p) defined by

$$A(p) = \sum_{\text{total } \alpha \varepsilon L_p} \mu(\alpha)E(\alpha).$$

To derive a recurrence satisfied by A(p), for each total $\alpha \varepsilon L(p)$ let $m(\alpha)$ be the number of states in the minimum block of α. If $m(\alpha) = p$ then $\alpha = 0$ so $\mu(\alpha)E(\alpha) = p^{pd}$. If $m(\alpha) = i$ for $i < p$, then α is obtained from some total $\beta \varepsilon L_{p-i}$ by adding a new minimum block of i states, for which there are $\binom{p}{i}$ choices. By the Theorem of Section 3, $\mu(\alpha) = -\mu(\beta)$. All possible transitions from states in the minimum block are consistent with α, so $E(\alpha) = E(\beta)p^{id}$. Summing the possibilities, we have

$$\sum_{\substack{\text{total } \alpha \varepsilon L_p \\ m(\alpha) = i}} \mu(\alpha)E(\alpha) = -\sum_{\text{total } \beta \varepsilon L_{p-i}} \binom{p}{i} \mu(\beta)E(\beta)p^{id} ,$$

$$= -\binom{p}{i} A(p-i)p^{id} .$$

Summing over i then gives the recurrence

$$A(p) = p^{pd} - \sum_{i=1}^{p-1} \binom{p}{i} A(p-i)p^{id} \tag{4.2}$$

for all $p \geq 1$.

By the Theorem of Section 3, if $\mu(\alpha) \neq 0$ then α is a state-disjoint union of total quasi-orders, say $\beta_1, \ldots, \beta_n$. The Theorem then gives

$$\mu(\alpha) = (n-1)! \prod_{i=1}^{n} \mu(\beta_i).$$

Also it is immediate that

$$E(\alpha) = \prod_{i=1}^{n} E(\beta_i),$$

so we have

$$\mu(\alpha)E(\alpha) = (n-1)! \prod_{i=1}^{n} \mu(\beta_i)E(\beta_i).$$

Let A(x) be the exponential generating function

$$A(x) = \sum_{p=1}^{\infty} A(p)x^p/p!.$$

Then $A(x)^n$ is the exponential generating function for products of the form $\prod_{i=1}^{n} \mu(\beta_i)E(\beta_i)$ for ordered sequences $\beta_1, \ldots, \beta_n$ of total quasi-orders. The state sets of $\beta_1, \ldots, \beta_n$ are disjoint, so the product $\prod_{i=1}^{n} \mu(\beta_i)E(\beta_i)$ is counted n! times in $A(x)^n$. Dividing by n and summing gives an exponential generating function for the right side of (4.1). Thus if we let

$$S(x) = \sum_{p=1}^{\infty} S(p)x^p/p!$$

we have

$$S(x) = \sum_{n=1}^{\infty} \frac{1}{n} A(x)^n = -\ln(1-A(x)).$$

Differentiating, we find

$$S'(x) = A'(x) + S'(x)A(x).$$

Equating coefficients of $x^{p-1}/(p-1)!$ then gives the recurrence

$$S(p) = A(p) + \sum_{j=1}^{p-1} \binom{p-1}{j} S(p-j)A(j) \qquad (4.3)$$

for $p \geq 1$.

The first few values of $S(p)/(p-1)!$, computed using (4.2) and (4.3), are given in Table 2 for d = 2 and d = 3. Values for d = 1 are not given because in this case A(1) = 1, A(p) = 0 for $p \geq 2$,

	S(p)/(p-1)!	p
	9	2
	148	3
	3493	4
	1 06431	5
	39 50832	6
	1723 25014	7
d = 2	86170 33285	8
	48 52670 03023	9
	3036 36917 15629	10
	2 08869 80406 37242	11
	156 61253 92154 05732	12
	12709 74531 99471 41220	13
	11 09746 20939 04795 79732	14
	49	2
	6877	3
	18 54545	4
	8074 78656	5
d = 3	51 47982 04147	6
	45118 23237 94896	7
	519 96186 47032 59753	8
	7 62210 14796 13304 21167	9
	13849 45048 77450 01470 47194	10

Table 2

Numbers of unlabeled strongly connected finite automata with a start state specified and $S(p) = (p-1)!$ for $p \geq 1$. The latter can be seen at once from the fact that for $d = 1$ any strongly connected finite automaton is a single directed cycle. That $S(p)/(p-1)!$ is an integer for any d follows just as for $C(p)/(p-1)!$ in Section 2. In fact $S(p)/(p-1)!$ is the number of unlabeled strongly connected finite automata with p states and a specified start state. It is not hard to count these without specifying a start state, using Burnside's Lemma, but that is outside the scope of the present paper.

5. Conjectured asymptotic behavior.

When $d = 1$, $C(p) = p!$ and $S(p) = (p-1)!$ so Stirling's formula gives the asymptotic behavior. When $d \geq 2$, let c be the unique positive solution to $c = 1 - e^{-dc}$, and let γ and K be defined by

$$\gamma = \frac{c(1-c)^{\frac{1-c}{c}}}{c^{d-1}} ,$$

$$K = \frac{1}{c}\sqrt{dc-(d-1)} .$$

Then it is conjectured that $C(p) \sim K\gamma^p p^{pd}$ and $A(p) \sim S(p) \sim cC(p)$.

These conjectures result from assuming a general form of $a \cdot p^b \cdot \delta^p \cdot p^{pd}$, from which the constants a, b, and δ can be determined in each case. The conjectured behavior agrees with the exact values which have been computed. To date the only progress on the conjectures is a proof that $\log(C(p)/p^{pd}) \sim p\log\gamma$ (joint work with L. B. Richmond). For $d = 2$ approximate values of the constants are $c = 0.796812$, $\gamma = 0.835906$, and $K = 0.966941$. For $d = 3$ the values are $c = 0.940480$, $\gamma = 0.945700$, and $K = 0.963692$.

There is a striking contrast between the asymptotic effects of weak connectivity versus initial or strong connectivity. As indicated by Harrison [5, Thm. 7.3], for fixed $d \geq 2$ and large p, almost all finite automata are weakly connected. However even the weak form of the asymptotic conjecture proved jointly with L. B.

Richmond shows that the proportion of finite automata which are initially (or strongly) connected tends to 0 geometrically as $p \to \infty$.

REFERENCES

1. H.H. Crapo, The Möbius function of a lattice, J. Combin. Theory 1 (1966), 126-131.

2. P.H. Edelman and P. Klingsberg, The subposet lattice and the order polynomial, European J. Combin. 3 (1982) 341-346.

3. F. Harary, Unsolved problems in the enumeration of graphs, Publ. Math. Inst. Hungar. Acad. Sci. 5 (1960), 63-95.

4. F. Harary and E.M. Palmer, Enumeration of finite automata, Inform. and Control 10 (1967), 499-508.

5. M.A. Harrison, A census of finite automata, Canad. J. Math. 17 (1965), 100-113.

6. H. Narushima, Principles of Inclusion-Exclusion on Semilattices and its Applications. Doctoral Thesis, Waseda Univ., Tokyo, 1977.

7. C.E. Radke, Enumeration of strongly connected sequential machines, Inform. and Control 8 (1965), 377-389.

8. G.-C. Rota, On the foundations of combinatorial theory I. Theory of Möbius functions, Z. Wahrscheinlichkeitstheorie 2 (1964), 340-368.

SOME EXTENSIONS OF ORE'S THEOREM

Edward Schmeichel

San Jose State University

David Hayes

San Jose State University

Our starting point is the following generalization of a well-known theorem of Ore [4] (for a well-written proof, see [2, p. 282]):

Theorem 1. *Let G be a 2-connected graph on p vertices. Suppose that for $k \geq 0$, we have $\deg v + \deg w \geq p - k$ for every nonadjacent pair of vertices v,w. Then G contains a cycle of length at least $p - k$, and this lower bound on the cycle length is best possible.*

One can see that the bound on the cycle length is best possible by considering $K_{\frac{p-k}{2}, \frac{p+k}{2}}$.

For $k \leq 4$, the conclusion of the above theorem can be substantially strengthened if we include one additional hypothesis with the degree sum condition.

Let c(G) denote the number of components of a graph G.

Definition. A graph G is said to have a _component problem_ if there exists a set $X \subseteq V(G)$ such that $c(G - X) > |X|$.

It is well known that a graph with a component problem cannot be hamiltonian. If we add the assumption that G does not have a component problem to the hypotheses of Theorem 1, we will be able to conclude that G is almost always hamiltonian when k is sufficiently small. The precise statement is the following theorem of Jung [1]:

Theorem 2. _Let G be a 2-connected graph on $p \geq 11$ vertices. Suppose that $\deg v + \deg w \geq p - 4$ for every nonadjacent pair of vertices v,w, and that G does not have a component problem. Then G is hamiltonian._

Jung also pointed out that there exists an infinite family of graphs which are nonhamiltonian, have no component problem, and satisfy $\deg v + \deg w \geq p - 5$ for every nonadjacent pair v,w. One such family (due to Karel Zikan) is exhibited in Figure 1.

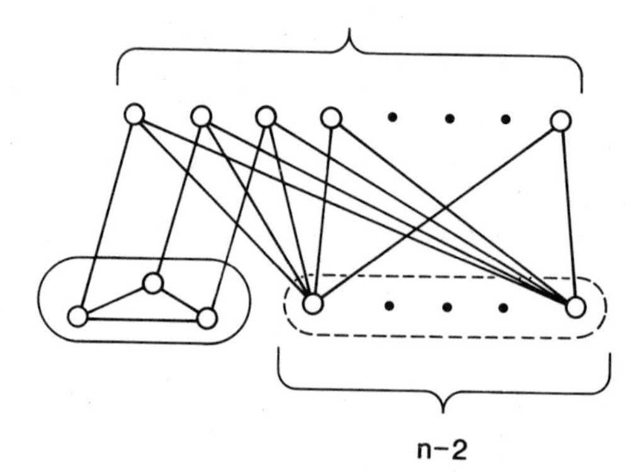

Fig. 1

The broken circle indicates that an edge between any pair of these n - 2 vertices is optional. The symbol = indicates the join of the two circled subgraphs.

There are some important questions not answered by Theorem 2. In particular, what are the exceptional graphs on $p \leq 10$ vertices. The search for these graphs is made somewhat difficult by the fact that the degree sum condition $\deg v + \deg w \geq p - 4$ is not very helpful when p is small (e.g., *every* cubic graph on 10 vertices satisfies it). The authors did perform the search, however, and obtained the following strengthening of Theorem 2.

Theorem 3. Let G *be a* 2-*connected graph on* p *vertices. Suppose that* $\deg v + \deg w \geq p - 4$ *for every non-adjacent pair* v,w, *and that* G *does not have a component problem. Then* G *is hamiltonian unless* G *is one of the graphs in Figure* 2.

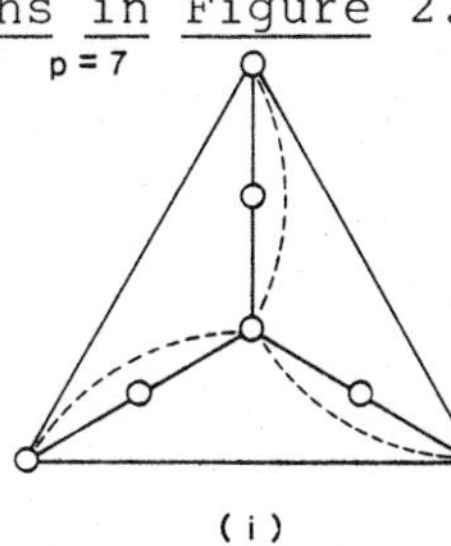

(i)

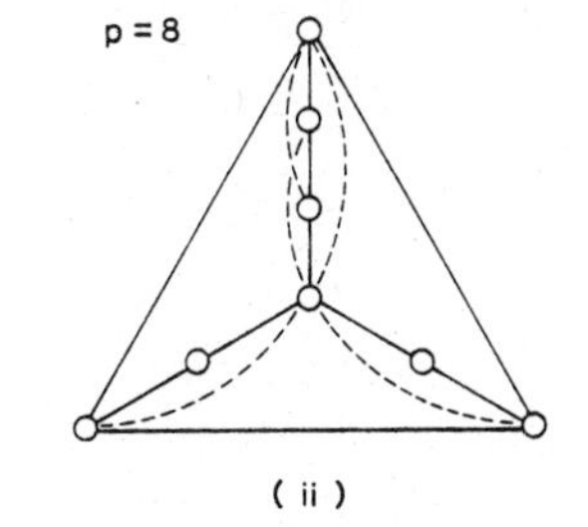

(ii)

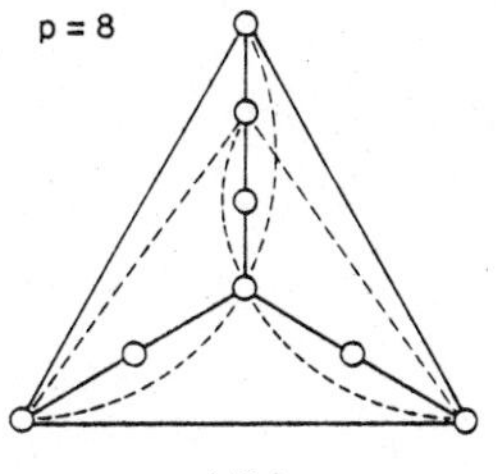

(iii)

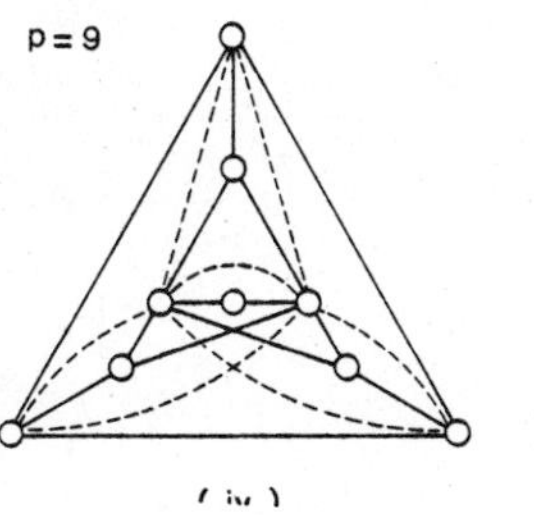

(iv)

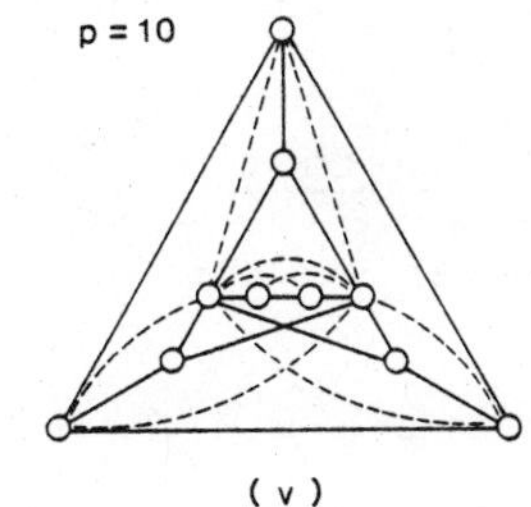

(v)

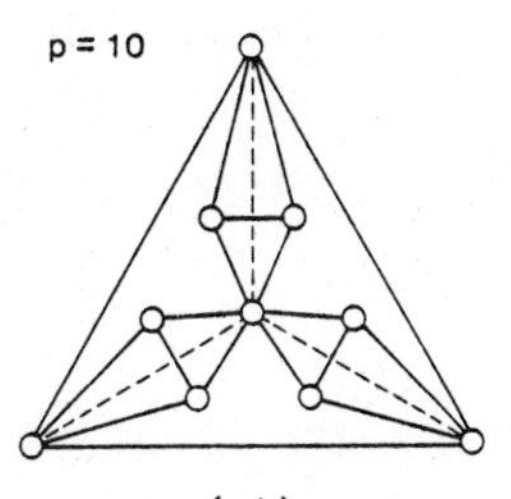

(vi)

Fig. 2

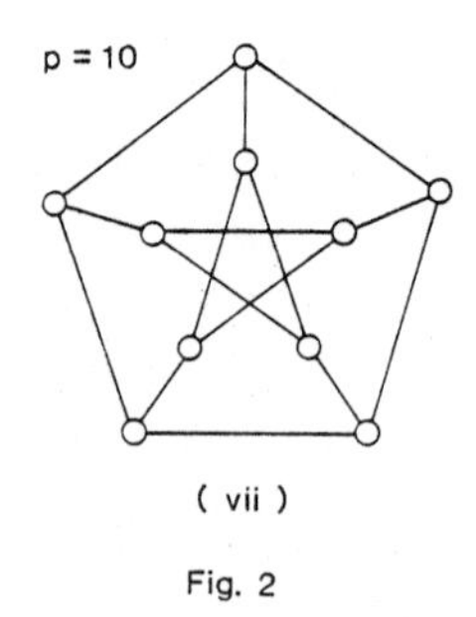

(vii)

Fig. 2

(The regular lines represent mandatory edges, while the broken lines represent optional edges. Any subset of the optional edges may be added without making the resulting graph hamiltonian.)

Observe that each of the graphs in (i) contains a nonadjacent pair with degree sum $p - 3$, while the remaining graphs in Figure 2 all contain a nonadjacent pair with degree sum $p - 4$. This immediately yields

Corollary 1. *Let G be a 2-connected graph on p vertices. If* $\deg v + \deg w \geq p - 2$ *for every nonadjacent pair v,w, then G is hamiltonian if and only if G does not have a component problem.*

We now consider the question of precisely which graphs satisfying the hypotheses of Corollary 2 have a component problem. The following lemma will prove helpful.

Lemma 1. *Let G be a 2-connected graph on p vertices, with* $\deg v + \deg w \geq p - k$ $(k \geq 1)$ *for every nonadjacent pair v,w. Suppose also that* $c(G - X) > |X|$ *for some* $X \subseteq V(G)$. *Then we must have* $\frac{p-k}{2} \leq |X| \leq \frac{p-1}{2}$, *if* $p \geq 5k$.

Proof. The upper bound for $|X|$ is trivial. For the lower bound, let $x = |X|$, and suppose $G - X$ has n components, where $n > x \geq 2$. Let H_i be the vertex set of the ith component, and suppose $|H_1| \leq |H_2| \leq \cdots \leq |H_n|$. Choose any $v_i \in H_i$, for $i = 1,2$. Since $(v_1, v_2) \notin G$, since v_i could be adjacent only to the vertices in X and the other vertices in H_i, and since

$|H_1|, |H_2| \leq \frac{1}{n} \sum_{j=1}^{n} |H_j|$, we have

$$p - k \leq d(v_1) + d(v_2) \leq \sum_{i=1}^{2} (x + |H_i| - 1) \leq 2(x+\frac{p - x}{n}-1)$$

$$\leq 2(x + \frac{p - x}{x + 1} - 1), \text{ or } f(x) = 2x^2 - (p - k + 2)x + (p + k - 2) \geq 0.$$

Letting $0 < r_1 < r_2$ be the (distinct, real) roots of $f(x)$, we see that $f(x) \geq 0$ precisely if $x \leq r_1$ or $x \geq r_2$. But if $k \geq 1$ and $p \geq 5k$, then $r_1 < 2$, and so $|X| = x \leq r_1$ is impossible, since G is 2-connected. Hence we must have $x \geq r_2 > \frac{p - k - 1}{2}$ (if $p \geq 5k$), or $|X| \geq \frac{p - k}{2}$. ∎

It is now a simple matter to derive the following extension of Ore's Theorem.

Theorem 4. *Let G be a 2-connected graph on p vertices, and suppose* $\deg v + \deg w \geq p - 1$, *for every nonadjacent pair of vertices* v,w. *Then G is hamiltonian unless G is one of the graphs in Figure* 3.

Proof. By Corollary 1 and Lemma 1, G will be hamiltonian unless there is a set X of $\frac{p - 1}{2}$ vertices whose removal leaves at least $\frac{p - 1}{2}$ components. Hence these

p odd, ≥ 5

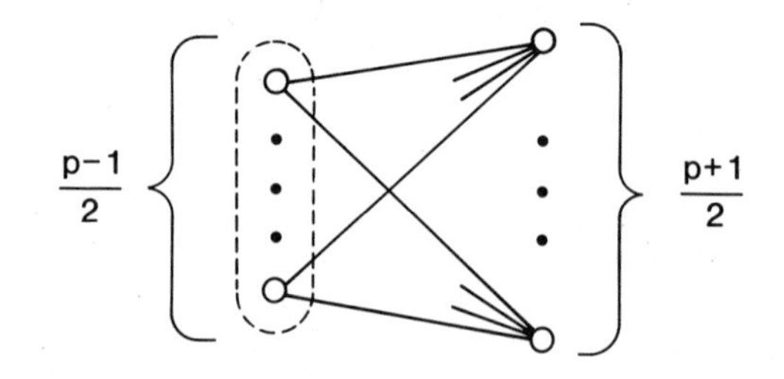

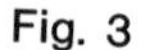
Fig. 3

components are independent single vertices, each of which is adjacent to every vertex in X (if the degree sum condition is to hold). Thus G is one of the graphs in Figure 3, since edges between the vertices of X are optional. ∎

It is a simple matter to recognize whether a graph is one of the exceptional graphs in Theorem 4. In fact, G is one of these if and only if p is odd and the $\frac{p+1}{2}$ vertices of smallest degree in G are independent and each of degree $\frac{p-1}{2}$. Thus Theorem 4 provides an effective extension of Ore's Theorem.

Corollary 2. (Nash-Williams [3]) *An* n-*regular graph* G *on* 2n + 1 *vertices is hamiltonian*.

Proof. It is easy to verify that G satisfies the hypotheses of Theorem 4. Since none of the exceptional graphs in Theorem 4 are regular, we have that G is hamiltonian. ∎

It is interesting to note that we can give a simple proof of Theorem 4 without using Corollary 1 or Lemma 1 by appealing to Theorem 1. If G satisfies the conditions of Theorem 4 but is not hamiltonian, then by Theorem 1 G contains a (p - 1) -cycle C. Let x be the vertex not on C.

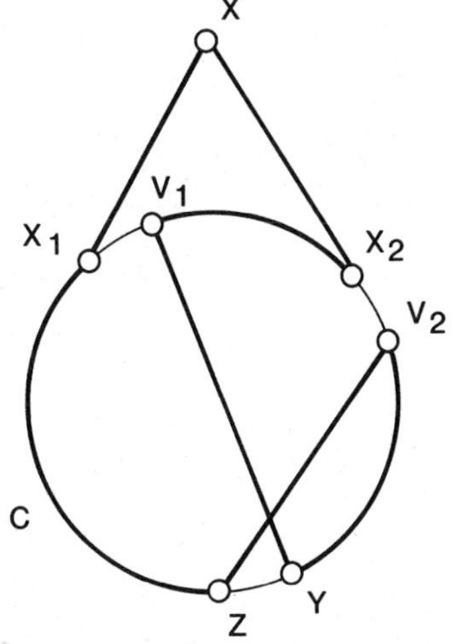

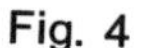
Fig. 4

Clearly deg x $\leq \frac{p-1}{2}$ (or x would be adjacent to two consecutive vertices of C, and G would be hamiltonian). If $2 \leq \deg x < \frac{p-1}{2}$, let x_1, x_2 be any two vertices in N(x) (the set of neighbors of x). Let v_i and x_i be consecutive vertices on C, for i = 1,2, as shown in Figure 4. Since $\deg x + \deg v_i \geq p - 1$, we have $\deg x_i \geq \frac{p-1}{2}$, for i = 1,2. But then G contains edges (v_1,y), (v_2,z) as shown, and hence a hamiltonian cycle, a contradiction. Hence $\deg x = \frac{p-1}{2}$.

But then x is adjacent to alternate vertices on C. If $v \in V(C) - N(x)$, then $(v,w) \notin G$ for every $w \in V(C) - N(x)$, or G would be hamiltonian. On the other hand, since $(x,v) \notin G$, we must have $\deg v \geq \frac{p-1}{2}$, and so v must be adjacent to every vertex in N(x). As we argued previously, G is one of the exceptional graphs in Figure 3. This completes the alternate proof of Theorem 4.

Using Corollary 1 and Lemma 1, we can obtain (much as we did Theorem 4) the following result; the proof is omitted.

Theorem 5. *Let G be a 2-connected graph on p vertices. If $\deg v + \deg w \geq p - 2$ for every nonadjacent pair of vertices v,w, then G is hamiltonian unless G is one of the graphs in Figure 5.*

(In (i), there is one optional edge incident at v. In (iii), there are optional edges incident at both v_1 and v_2, and these edges are not necessarily independent.)

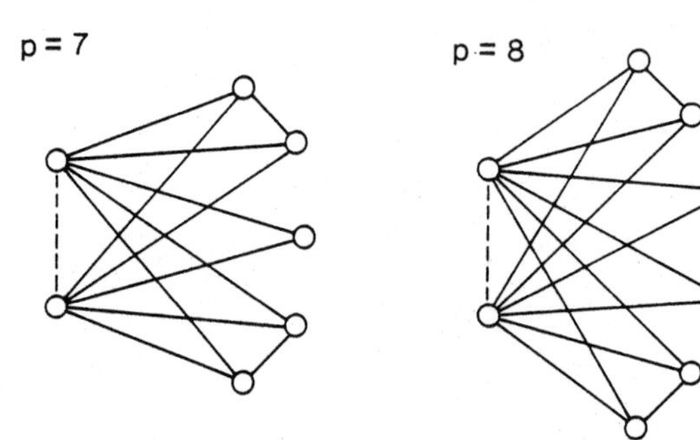

Fig. 5

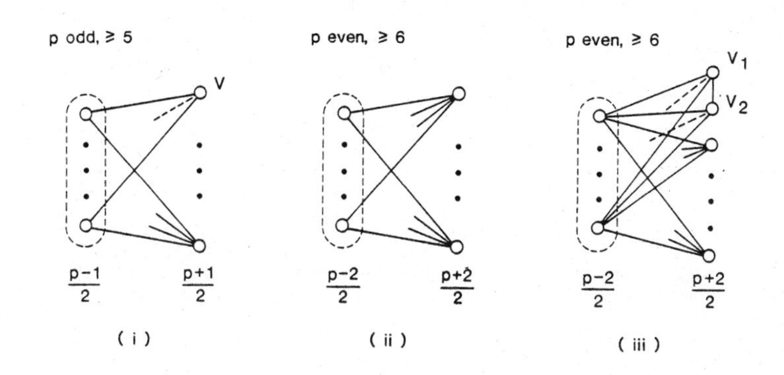

Fig. 5

Using Theorem 3 and Lemma 1, one could prove the analouge of Theorem 4 and 5 when $\deg v + \deg w \geq p - k$, for k equals 3 or 4.

REFERENCES

1. H.A. Jung. On maximal circuits in finite graphs, Annals of Discrete Math. 3(1978), 129-144.

2. L. Lovasz. Combinatorial Problems and Exercises, Elsevier North-Holland, New York, 1979.

3. C.St.J.A. Nash-Williams. Valency sequences which force a graph to have hamiltonian circuits, Report of the Univ. of Waterloo, 1972.

4. O. Ore. Note on hamiltonian circuits, Amer. Math. Monthly 67(1960), 55.

PACKING A TREE OF ORDER p WITH A (p,p) GRAPH

Dedicated to Ronald M. Foster

Seymour Schuster
Carleton College

ABSTRACT

A packing of graph G_1 with graph G_2 is an isomorphic embedding of G_1 in the complement of G_2. The problem of packing a pair of trees of the same order and that of packing a tree with a graph of the same size have recently been solved: the former by S. M. Hedetniemi, S. T. Hedetniemi and P. J. Slater and the latter by P. J. Slater, S. K. Teo and H. P. Yap. The present paper completely determines the conditions under which there exists a packing of a tree of order p with a (p,p) graph.

1. Introduction. The notion of packing a pair of graphs seems to have first appeared in 1977 and 1978 in several disguises (see [1], [2], [5], and [11]), all equivalent to the following:

Definition. Let G_1 and G_2 be two graphs of the same order p. A packing of G_1 with G_2 is an isomorphic embedding of G_1 in $\overline{G_2}$, the complement of G_2.

The terminology has its origin in the fact that there exists an edge-disjoint isomorphic embedding of the two graphs in the complete graph K_p. When a packing exists, we often say, "G_1 packs with G_2."

Problems of packing graphs have received a modest amount of attention, as can be seen in [3], [4], [6], [8], [12], and [13] as well as in the aforementioned references. Apart from concern and curiosity about the structure of abstract graphs, there has been additional motivation to study packing because of its application to computational complexity (see [1]) and to Ramsey Theory (see [9] and [10]).

Recently, particular attention has been given to packing trees with other graphs. H. Joseph Straight [14] was apparently the first to observe that every non-trivial tree other than a star packs with itself, a fact that is an immediate corollary of the Principal Theorem of D. Burns and S. Schuster in [3]. S. M. Hedetniemi, S. T. Hedetniemi and P. J. Slater, in [8], went beyond packing a tree with itself by proving the following:

Theorem 1. There is a packing of any two trees of the same order $p \geq 4$, provided neither tree is a star.

This result was extended significantly by P. J. Slater, S. K. Keo, and H. P. Yap in [13], where they proved:

Theorem 2. Let T be a tree of order $p \geq 5$ and let G be a graph of order p and size $p - 1$. If neither T nor G is a star, then there is a packing of T with G.

The purpose of the present paper is to continue the study of packing trees by determining the conditions under which a tree T of order p packs with a graph G of order p and size p.

2. Preliminaries. We begin by establishing an important convention that shall be maintained throughout the paper: T will denote a tree of order p and the accompanying G will denote a

(p,p) graph. Although most of the notation and terminology used will be standard, we selectively review some of it that may not be well known to all readers.

The maximum degree occurring among the vertices of G is denoted by $\Delta(G)$ and the minimum by $\delta(G)$. The <u>neighborhood</u> of vertex v, written $N(v)$, is the set of vertices adjacent to v. The star, the path, and the cycle (all of order n) have the standard notations $K_{1,n-1}$, P_n, and C_n, respectively. Apart from these, we shall require notation for other frequently encountered graphs. Suppose H and K are graphs having no vertices in common. Then $H \cup K$, the <u>union</u> of H and K, is the graph whose vertex-set is $V(H) \cup V(K)$ and whose edge-set is $E(H) \cup E(K)$. The union of m copies of H will be denoted by mH. The graph obtained from G by deleting the vertex v is denoted by $G-v$; if vertices $v_1, v_2, \ldots, v_i$ are deleted, the resulting graph will be written $G - \{v_1, v_2, \ldots, v_i\}$. If an edge e is inserted in G by joining two of its vertices, the resulting graph is $G + e$; of course, $G + e$ is well-defined only if G is so symmetric that joining any pairs of vertices results in the same graph. $G - e$ is defined analogously. Further, as shown in Figure 1, $P_2 \cdot K_{1,p-2}$ will denote the tree of order p obtained by inserting a new vertex in an edge of the star $K_{1,p-2}$; similarly, $P_3 \cdot K_{1,p-3}$ will denote the tree of order p obtained by inserting two vertices in a single edge of $K_{1,p-3}$; and $2P_2 \cdot K_{1,p-3}$ will denote the tree obtained from $K_{1,p-3}$ by inserting a vertex in each of two different edges of $K_{1,p-3}$. (The notation $P_2 \cdot K_{1,p-2}$ stems from the idea that we are identifying an endvertex (darkened in Figure 1) of P_2 with one of the vertices of $K_{1,p-2}$. The situation is analogous in the other cases.)

Finally, we review some simple facts about the structure of (p,p) graphs, for the proof of our packing theorem rests on an

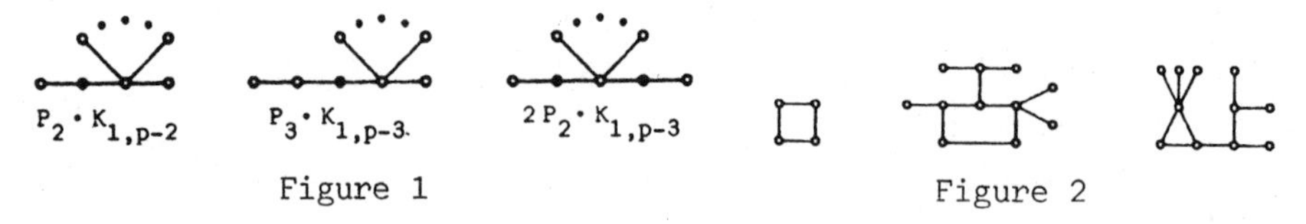

Figure 1

Figure 2

exhaustive consideration of the possible structures of a (p,p) graph.

A measure of the number of cycles in a graph H is its <u>cyclomatic number</u> $\gamma(H)$, which is defined precisely as the minimum number of edges that must be deleted from H in order to reduce it to a forest. It is easy to see that if H is an (m,n) graph having k components, then

$$\gamma(H) = n - m + k.$$

We shall be interested only in the cyclomatic number of a component of a graph, in which case $k = 1$.

If the cyclomatic number of some component is 1, then the component in <u>unicyclic</u>, possessing the same number of vertices as edges; it is either a cycle or a cycle with one or more trees attached to vertices of the cycle. (See Figure 2.) Thus, if a (p,p) graph has no components that are trees, then it must be the union of unicyclic components. On the other hand, if the (p,p) graph contains a tree then one of its cyclic components must have cyclomatic number greater than one. Figure 3 exhibits four different types of (p,p) graphs, each with one component having cyclomatic number 2, and showing how such components can vary in structure.

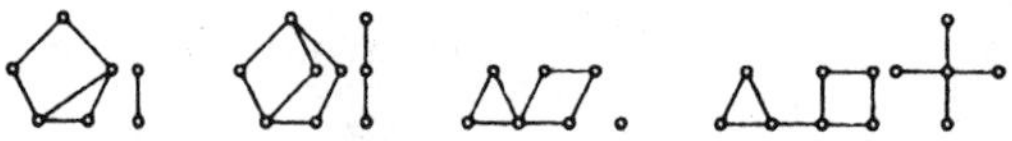

Figure 3

If a cycle has one or more trees attached at a single vertex, then we shall say that <u>the cycle has one appendage</u>; if the cycle has trees attached at two of its vertices, then we shall say that <u>the cycle has two appendages</u>.

3. The Packing Theorems. We begin by disposing of two simple cases in which one of the two graphs to be packed possesses a vertex adjacent to all other vertices.

Lemma 1. (i) If $\Delta(T) = p - 1$, then T packs with G if and only if G has an isolated vertex.

(ii) If $\Delta(G) = p - 1$, then no tree packs with G.

Though these two statements and their proofs are obvious or nearly so, we have elevated them to lemma status because they enable us to dispense with further consideration of the packing problem if $T = K_{1,p-1}$ or if $G = K_{1,p-1} + e$.

Lemma 2 (Induction Anchor). If T is of order $p = 5$ or 6, then there is a packing of T with G except in the following cases:

1. T is a star and G has no isolated vertex;
2. T is any tree and $G = K_{1,p-1} + e$;
3. $T = P_2 \cdot K_{1,p-2}$ and G is a union of cycles;
4. $T = P_3 \cdot K_{1,3}$ and $G = 2K_3$;
5. $T = P_5 = 2P_2 \cdot K_{1,2}$ and $G = K_4 - e \cup K_1$;
6. $T = P_6$ and $G = K_4 \cup 2K_1$;
7. $T = 2P_2 \cdot K_{1,3}$ and $G = K_4 \cup 2K_1$.

The proof of this theorem requires exhibiting a packing of over 100 pairs of graphs, which clearly cannot be done within these pages. It is actually an easy--but somewhat tedious--task that can be carried out systematically by utilizing the graph diagrams in the appendix of [7].

We shall complete our study of packing T with G through a sequence of additional lemmas that exhaust all the possible structures for G. Lemma 5 is the first of two major results, but its proof must await the elimination of the special cases dealt with in the following two results.

Lemma 3. Let T be a tree of order $p > 6$ different from a star and let G be the graph shown in Figure 4(a). Then T packs with G.

Proof. Since the ramsey number $r(3,3) = 6$, T has a set of three independent vertices. By examining the trees of order 7 in [7], it is easy to see that such a set can always (for all $p > 6$) be chosen to include the neighborhood of an endvertex. Let us denote such an independent set by $\{v_1, v_2, v_3\}$ so that $v_1 = N(w)$, where w is an endvertex of T. Then an isomorphic embedding $\varphi : T \to \overline{G}$ is exhibited in Figure 4(b), where the unlabelled vertices may be assigned arbitrarily as the images of $V(T) - \{v_1, v_2, v_3, w\}$.

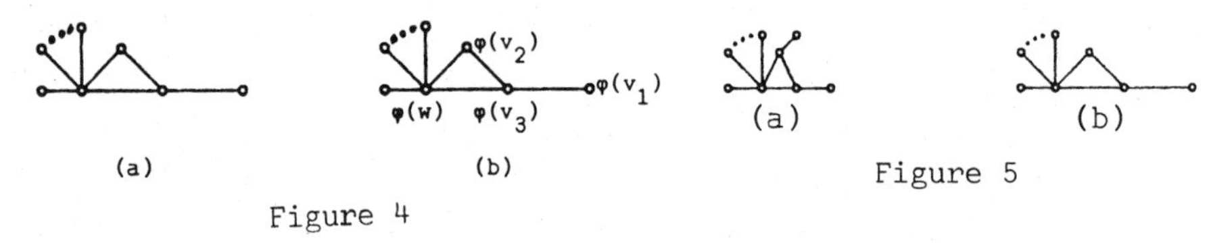

Figure 5

Figure 4

The third of the forbidden cases of Lemma 2 arises if $T' = P_2 \cdot K_{1,p-2}$ while G' is a union of cycles for $p = 5$ and 6. The possibilities for T and G in this case are shown in Figures 6 and 7. It is somewhat time-consuming but easy to show that each such T can be packed with each respective G.

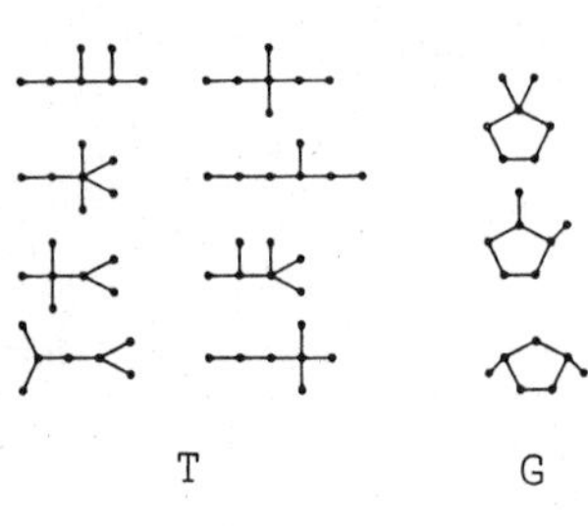

Figure 6

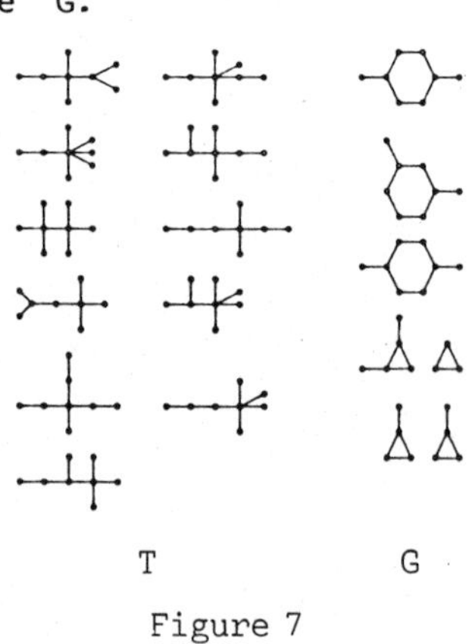

Figure 7

In the event that $T' = P_3 \cdot K_{1,3}$ and $G' = 2K_3$, the possibilities for T and G are shown in Figure 8 and, again, it is easy to exhibit a packing of every T with every G.

If $T' = P_5$, and $G' = K_4 - e \cup K$, the possibilities for T and G are shown in Figure 9. As in the previous cases, each T packs with each G.

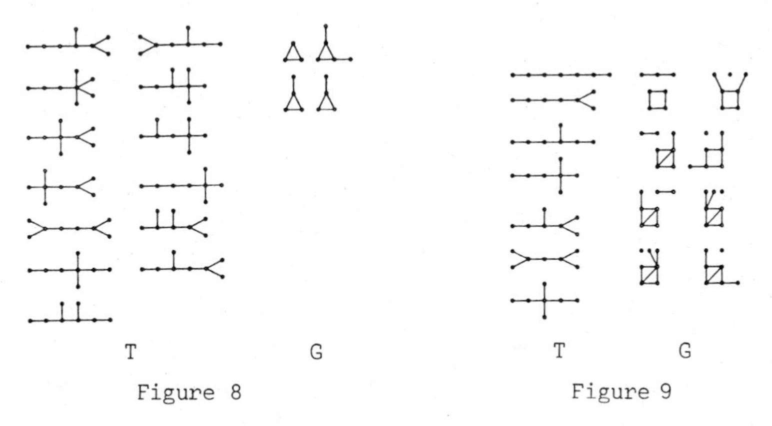

Figure 8

Figure 9

If $T' = P_6$ and $G' = K_4 \cup 2K_1$, then the T and G from which they might arise are shown in Figure 10. Once again, it is easy to verify that a packing of T with G exists in all these cases.

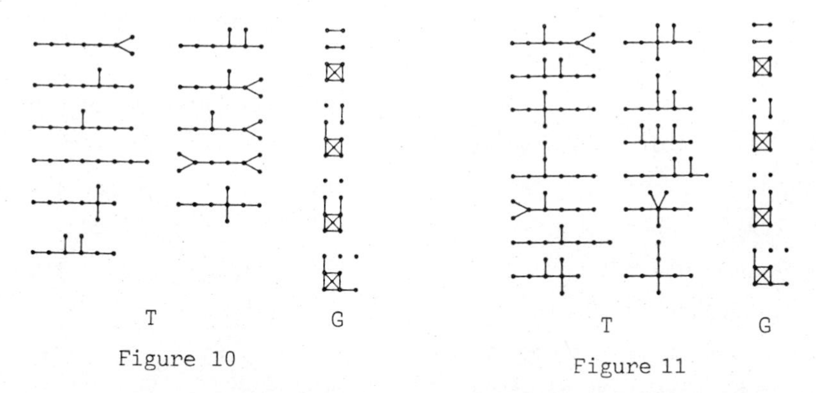

Figure 10

Figure 11

Finally, the case in which $T' = 2P_2 \cdot K_{1,3}$ and $G' = K_4 \cup 2K_1$ requires that we consider the pairs T and G of Figure 11.

As in all the previous cases, every T packs with every G.

Now, the induction follows trivially. By the induction hypothesis, we have an isomorphism $\varphi : T' \longrightarrow \overline{G}'$, which we extend by defining $\varphi(v_1) = x_1$ and $\varphi(v_2) = x_2$. This extension is a packing of T with G.

In consequence of Lemma 5 and the earlier discussion in §2 on the nature of (p,p) graphs, it is seen that we have left to consider only the following possibilities for G:

(A) G is a union of cycles;

(B) G is a union of cyclic components with only one appendage;

(C) G contains, as its acyclic components, only isolated vertices (while not satisfying the endvertex hypothesis of Lemma 5).

(D) G has exactly one non-trivial tree, a star $K_{1,n}$ $(n \geq 1)$, among its components while the cyclic components have no appendages.

The forbidden case $T = P_2 \cdot K_{1,p-2}$ with G a union of cycles was noted in Lemma 2. Our next result shows that in all other cases, where $\Delta(G) \neq p - 1$, this tree does pack with G.

Lemma 4. If $T = P_2 \cdot K_{1,p-2}$, G not a union of cycles, and $\Delta(G) \neq p - 1$, then T packs with G.

Proof. The presence of an endvertex or isolated vertex in G (since G contains an acyclic component) makes it is an easy matter to construct a packing of T with G in both these cases.

Lemma 5. Let $p \geq 5$, and $\Delta(T), \Delta(G) < p - 1$. If G contains two endvertices x_1, x_2 such that $d(x_1,x_2) \geq 3$, then there is a packing of T with G. (N.B. If endvertices x_1 and x_2 are in different components of G, then $d(x_1,x_2) = \infty$, so such a pair certainly satisfies the hypothesis $d(x_1,x_2) \geq 3$.)

Proof. The argument is by induction on p. Since the lemma

is true for $p = 5$ and 6, we assume it true for all p with $5 \leq p < n$, where $n \geq 7$. We then consider the possibility of packing any tree T (of order n) with an arbitrary (n,n) graph G where $\Delta(T)$, $\Delta(G) < n - 1$.

First, we note that T contains a pair of endvertices v_1 and v_2 such that $d(v_1,v_2) \geq 3$. Let $u_1 = N(v_1)$, $u_2 = N(v_2)$, $y_1 = N(x_1)$ and $y_2 = N(x_2)$. Now, let $T' = T - \{v_1,v_2\}$ and $G' = G - \{x_1,x_2\}$. In order to proceed with the induction argument we must consider those cases in which T' and G' do not satisfy the induction hypothesis, namely those cases in which T' and G' may be among the seven forbidden cases of Lemma 2.

First, we observe that since T itself is not a star, and since we may exclude $T = P_2 \cdot K_{1,p-2}$ from consideration by virtue of Lemma 4, there exist endvertices v_1, $v_2 \in V(T)$ such that $T' = T - \{v_1,v_2\}$ is not a star. Next we ask: What is the structure of G that leads to $\Delta(G') = n - 3$? Since $\Delta(G) < n - 1$, the only two possibilities for G are shown in Figure 5. In the first case (Fig. 5(a)), we may delete x_2' rather than x_2 so that $G' = G - \{x_1,x_2'\}$, in which case $\Delta(G') < n - 3$. The second case (Fig. 5(b)) is eliminated from consideration by virtue of Lemma 3.

The third of the forbidden cases of Lemma 2 arises if $T' = P_2 \cdot K_{1,p-2}$ while G' is a union of cycles for $p = 5$ and 6. The possibilities for T and G in this case are shown in Figures 6 and 7. It is somewhat time-consuming but easy to show that each such T can be packed with each respective G.

In the event that $T' = P_3 \cdot K_{1,3}$ and $G' = 2K_3$, the possibilities for T and G are shown in Figure 8 and, again, it is easy to exhibit a packing of every T with every G.

If $T' = P_5$, and $G' = K_4 - e \cup K$, the possibilities for T and G are shown in Figure 9. As in the previous cases, each T packs with each G.

If $T' = P_6$ and $G' = K_4 \cup 2K_1$, then the T and G from which they might arise are shown in Figure 10. Once again, it is easy to verify that a packing of T with G exists in all these cases.

Finally, the case in which $T' = 2P_2 \cdot K_{1,3}$ and $G' = K_4 \cup 2K_1$ requires that we consider the pairs T and G of Figure 11. As in all the previous cases, every T packs with every G.

Now, the induction follows trivially. By the induction hypothesis, we have an isomorphism $\varphi : T' \rightarrow \overline{G'}$, which we extend by defining $\varphi(v_1) = x_1$ and $\varphi(v_2) = x_2$. This extension is a packing of T with G.

In consequence of Lemma 5 and the earlier discussion in §2 on the nature of (p,p) graphs, it is seen that we have left to consider only the following possibilities for G:

(A) G is a union of cycles;

(B) G is a union of cyclic components with only one appendage;

(C) G contains, as its acyclic components, only isolated vertices (while not satisfying the endvertex hypothesis of Lemma 5).

(D) G has exactly one non-trivial tree, a star $K_{1,n}$ $(n \geq 1)$, among its components while the cyclic components have no appendages.

Our next result, the second of the major lemmas, eliminates from consideration all but a few types of cyclic components that may be present in G.

Lemma 6. Let G, of order $p > 6$, contain two adjacent vertices x_1, x_2 such that $\deg x_1 = \deg x_2 = 2$ and x_1, x_2 not on a 3-cycle. Then T packs with G with only the following exceptions: (1) T is a star and (2) $T = P_2 \cdot K_{1,p-2}$ and G is a union of cycles.

Proof. Let v_1, v_2 be endvertices of T such that $d(v_1, v_2) \geq 3$. Then $T' = T - \{v_1, v_2\}$ is a tree (possibly a star) and $G' =$

$G - \{x_1, x_2\}$ is a $(p-2, p-3)$ graph. If neither T' nor G' is a star, then Theorem 2 implies that T' packs with G'. The graph T' is a star only if $T = P_2 \cdot K_{1,p-2}$, but Lemma 4 tells us that this tree packs with every G other than a union of cycles.

The graph G' is a star only if G is a star with a 4-cycle attached at its center, as in Figure 12. If this is the case, let v_1, u_0, u_1 $V(T)$ such that v_1 is an endvertex, $u_1 = N(v_1)$ and u_0 arbitrary. Defining $\varphi(v_1)$, $\varphi(u_0)$, $\varphi(u_1)$ as in Figure 12 shows that T packs with G.

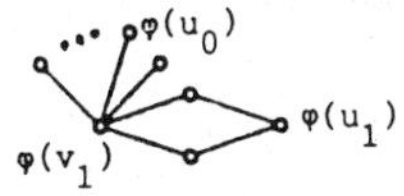

Figure 12

Figure 13

The remainder of the proof follows by the usual induction.

In order to complete consideration of (A), we need only consider the case in which $G = nK_3$, $n > 2$.

Lemma 7. Let $p = 3n \geq 6$ and $G = nK_3$. If T is any tree different from $K_{1,3n-1}$, $P_2 \cdot K_{1,3n-2}$ and $P_3 \cdot K_{1,3n-3}$, then there exists a packing of T with G.

Proof. First, we observe that this lemma is obviously true if T is a path, so we exclude this case from consideration. The remainder of the proof is by induction on n. The case of $n = 2$ was taken case of in Lemma 2. Assuming the result for $n = k$, we suppose that $G = (k + 1)K_3$ and that T is not a path. Noting that $p \geq 9$, we know that T possesses three independent vertices v_1, v_2 and v_3 with one, say v_1, adjacent to an endvertex. We now form $T' = T - \{v_1, v_2, v_3\}$ and $G' = G - \{x_1, x_2, x_3\}$, where x_1, x_2, x_3 are the vertices of a 3-cycle of G. Since T' is disconnected and smaller in size than G', it obviously packs with G'. We extend this packing by mapping v_i onto x_i, for $i = 1$, 2 and 3.

Now we turn to a consideration of (B), in which G is the union of cyclic components with only one appendage. Lemma 6 tells us that we need consider only the case in which the cycles are of order 3. Further, the appendage cannot contain two endvertices x_1, x_2 such that $d(x_1, x_2) \geq 3$.

Lemma 8. Let $\Delta(T) < p - 1$, $\Delta(G) < p - 1$, and $\Delta G = (T_1 \cdot K_3) \cup nK_3$ where $n \geq 0$ and the appendage T_1 is a tree. Then there exists a packing of T with G.

Proof. From the structure of G, we see that $p \geq 7$. Also, by virtue of Lemma 6, we know that G can only be among the structures suggested in Figure 13, where n has been chosen as 2. Observe that in each of these, T_1 contains a star. We proceed by induction on n.

Let $n = 0$. Since $p \geq 7$, T possesses three independent vertices v_1, v_2 and v_3, so that one, say v_1, is adjacent to an endvertex u. By mapping u to the center of the star in the appendage and judiciously mapping the v_i onto the vertices of the 3-cycle to which T_1 is appended, we can obtain a packing of T with G. If the center of the star is actually on this 3-cycle then $n > 0$, for otherwise $\Delta(G) = p - 1$; hence, the lemma is vacuously true in this case.

The remainder of the induction is precisely the argument given in the proof of Lemma 7.

We may now turn to a consideration of (C).

Lemma 9. If T is any tree and G has an isolated vertex, then T packs with G.

Proof. Let v_1 be an endvertex of T and $u_1 = N(v_1)$. Let x_1 be the isolated vertex of G and $x_2 \in V(G)$ such that $\deg x_2 > 2$. Then there is a packing φ of $T - \{v_1, u_1\}$ with $G - \{x_1, x_2\}$, by Theorem 2. We can extend φ by writing $x_1 = \varphi(u_1)$ and $x_2 = \varphi(v_1)$. This extension is a packing of T

with G.

Finally, we treat (D), the case in which G has exactly one non-trivial acyclic component, namely a star. In this case, G must also contain one component whose cyclomatic number is 2. As a consequence of Lemma 6, we realize that this component must be among the five exhibited in Figure 14. Also, from Lemma 6 and Lemma 5, we conclude that any other component of G must be a K_3.

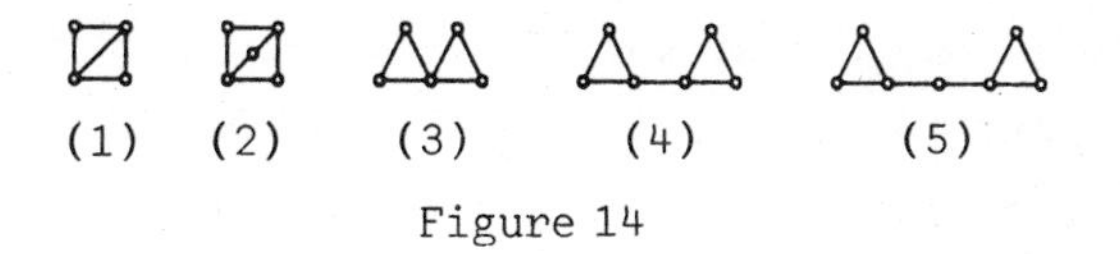

Figure 14

Lemma 10. Let T be any tree of order p different from a star and let G have $K_{1,m}$ $(m \geq 1)$ as its only acyclic component. Then T packs with G.

Proof. We dispense, first, with the case $m = 1$. Form $T' = T - v$ and $G' = G - x$, where v is an endvertex of T and x is a vertex of the $K_{1,1}$ component of G. By Lemma 9, we have a packing of T' with G', which we extend in the obvious way. Observe that such an extendable packing exists because the contruction of φ in Lemma 9 shows that we may choose the mapping φ in the present theorem so that $\varphi(N(v)) \neq N(x)$.

Next, we consider $m = 2$. Let v_1, v_2 be endvertices of T such that $d(v_1, v_2) \geq 3$, $u_1 = N(v_1)$, and $u_2 = N(v_2)$. Also, let the vertex set of the $K_{1,2}$ component of G be $\{x_1, x_2, x_3\}$ with $\deg x_2 = 2$. Again, by Lemma 9, there is a packing φ of $T' = T - \{v_1, v_2\}$ with $G' = G - \{x_1, x_2\}$. If $\varphi(u_1) = x_3$ then we extend φ by defining $\varphi(v_1) = x_1$, and $\varphi(v_2) = x_2$; otherwise, we write $\varphi(v_1) = x_2$ and $\varphi(v_2) = x_1$. Thus we have a packing of T with G.

Now we are ready to consider the general case $m \geq 3$. We observe that $p > 8$ if the cyclic component of G is one of (2) - (5) in Figure 14. Excluding $T = P_2 \cdot K_{1,p-2}$ from consideration, which may be done by virtue of Lemma 4, we have that every tree of order $p \geq 9$ possesses two endvertices v_1, v_2 such that $d(v_1, v_2) > 3$. Hence, their respective neighbors u_1 and u_2 are not adjacent. In the star component of G, we let x_0 be the center, x_1, x_2 two endvertices, and x be a vertex of degree 2 in G. We then form $T' = T - \{v_1, v_2, u_1, u_2\}$ and $G' = G - \{x, x_0, x_1, x_2\}$.

Surely, there exists a packing φ of T' with G', for G' has at least one isolated vertex. If, in this packing, both vertices adjacent to x are images of neighbors of u_1, we extend φ as follows: $\varphi(u_1) = x_1$, $\varphi(u_2) = x_0$, $\varphi(v_1) = x_2$, $\varphi(v_2) = x$. If both vertices adjacent to x are images of neighbors of u_2, we extend φ similarly. If one of the vertices adjacent to x is the image of a neighbor of u_1 and the other is the image of a neighbor of u_2, then we extend φ by writing: $\varphi(v_1) = x$, $\varphi(u_1) = x_0$, $\varphi(v_2) = x_1$, $\varphi(u_2) = x_2$. Thus, there exists a packing if the cyclic component of G is among the graphs (2) - (5) of Figure 14.

The foregoing argument also holds in the remaining case if T possesses two endvertices v_1, v_2 such that $d(v_1, v_2) > 3$. Such v_1, v_2 exist if $p > 8$; in fact, the only case in which there do not exist such v_1, v_2 is T : 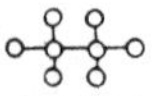 , which obviously packs with the union of $K_{1,3}$ and graph (1) of Figure 14.

Having examined all possible structures for the (p,p) graph G, we summarize our conclusion in the following theorem.

Theorem 3. Let T be a tree of order $p \geq 5$ and G a (p,p) graph. Then there exists a packing of T with G if and only if the graphs are not among the following cases:

1. T is a star and G has no isolated vertex;
2. T is any tree and $G = K_{1,p-1} + e$;
3. $T = P_2 \cdot K_{1,p-2}$ and G is a union of cycles;
4. $T = P_3 \cdot K_{1,3n}$ and $G = (n + 1)K_3$, $(p = 3(n + 1))$;

5. $T = P_5 = 2P_2 \cdot K_{1,2}$ and $G = K_4 - e \cup K_1$;
6. $T = P_6$ and $G = K_4 \cup 2K_1$;
7. $T = 2P_2 \cdot K_{1,3}$ and $G = K_4 \cup 2K_1$.

The author has recently learned that Theorem 3 was discovered independently by S. K. Teo and H. P. Yap.

Note on Dedicatee. R. M. Foster is a creative mathematician who, for over six decades, has been somewhat allergic to publishing. Although he has contributed to several branches of mathematics and enjoys a towering reputation in electrical network theory, he is too little known and recognized by graph theorists and combinatorists (even though some of the results that bear his name in network theory were proved with the aid of graph theory). H.S.M. Coxeter and R. Frucht have dedicated their recent volume to him, no doubt in appreciation of his unpublished--privately communicated--work on geometry, graphs, and groups. The present author wishes to join in honoring this combinatorist *extraordinaire*, Ronald M. Foster.

Bibliography

[1] B. Bollobas and S. E. Eldridge, Packing of graphs and applications to computational complexity, Proc. Fifth British Combinatorics Conference (Aberdeen, 1978), Utilitas Mathematica Publishing, Inc., Winnipeg.

[2] D. Burns and S. Schuster, Every (p,p-2) graph is contained in its complement, J. Graph Theory 1 (1977), 277-279.

[3] D. Burns and S. Schuster, Embedding (p,p-1) graphs in their complements, Israel J. of Math., 30 (1978), 313-320.

[4] P. A. Catlin, Embedding subgraphs under extremal degree conditions, Proc. 8th S.E. Conf. on Combinatorics, Graph Theory, and Computing, (1977), 139-145.

[5] P. A. Catlin, Subgraphs of graph, I, Discrete Math., 10, (1974), 225-233.

[6] R. J. Faudree, C. C. Rousseau, R. H. Schelp and S. Schuster, Embedding graphs in their complements, Czechoslovak Math J., 31 (106), (1981), 53-62.

[7] F. Harary, Graph Theory, Addison Wesley, Reading, Massachusetts, 1969.

[8] S. M. Hedetniemi, S. T. Hedetniemi and P. J. Slater, A note on packing two trees in K_n, Ars Combinatoria, 11, (1981), 149-153.

[9] A. D. Polimeni, H. J. Straight and J. Yellen, Some arrowing results for trees versus complete graphs, J. of Graph Theory, 5, (1981), 363-369.

[10] A. D. Polimeni, H. J. Straight and J. Yellen, Arrowing results for trees versus complete graphs, Theory of Applications of Graphs, ed. G. Chartrand, et al., John Wiley & Sons, Inc., 1981, 493-506.

[11] N. Sauer and J. Spencer, Edge-disjoint placement of graphs, J. of Combinatorial Theory, Ser. B., 25 (1978), 295-302.

[12] S. Schuster, Fixed-point-free embeddings of graphs in their complements, International Journal of Mathematics and Mathematical Sciences, 1, No. 3 (1978), 335-338.

[13] P. J. Slater, S. K. Teo, and H. P. Yap, Packing a tree with a graph of the same size, J. Graph Theory, accepted for publication.

[14] H. J. Straight, private communication, September 1976.

HOW MANY RINDS CAN A FINITE SEQUENCE OF PAIRS HAVE?

Allen J. Schwenk
U. S. Naval Academy and
Office of Naval Research

ABSTRACT

A rind of the finite sequence $a_1, a_2, \ldots, a_n$ is a new sequence formed by peeling one element at a time from the original sequence. If the the a_i's are distinct, there are 2^{n-1} different rinds that can appear, but if repetitions occur among the a_i's the number of rinds is reduced. In particular, we consider sequences of length $n = 2m$ comprised of m pairs. We attempt to determine the starting arrangement that permits the maximum number of rinds. A combination of analysis and computer search identifies the optimal arrangement through m = 14 pairs and suggests a conjecture for all m.

1. Introduction.

In [2] Fritz Göbel introduced the concept of peeling a finite sequence $S = a_1, a_2, \ldots, a_n$ to produce a new sequence $b_1, b_2, \ldots, b_n$ called a rind. The first element of the rind, b_1, is either a_1 or a_n which we imagine being peeled from one end of the sequence. Then b_2 is selected as an element stripped from either the left or right end of the remaining list of n-1 elements. After n-1 elements have been peeled the original sequence is reduced to a single element which must be chosen as b_n. Notice that every rind is a rearrangement of S, but not all rearrangements are legitimate rinds. For example, b_1 is never a_2 when $n \geq 3$. Evidently, if S contains n distinct elements, 2^{n-1} different rinds can be formed, but if S has some repeated elements the number of rinds is reduced. Let $\rho(S)$ denote the number of different rinds that can be peeled from S. We have just observed that $\rho(S) \leq 2^{n-1}$ with equality if and only if S is comprised of n distinct values.

Göbel [2] raised the following difficult question: Define $R(S) = \max \{\rho(T)$: T is any rearrangement of $S\}$. For a given sequence S, what is the value of R(S) and which arrangements T attain this maximum? Clearly the answer depends on the multiplicities of the entries in S. In particular, when all are distinct we have $R(S) = \rho(T) = 2^{n-1}$ for every T. At the other extreme, if S consists of n identical entries, all rearrangements are identical and $R(S) = \rho(S) = 1$. If S has b occurrences of 1 and $c = n-b$ occurrences of 2, there are $\binom{b+c}{b}$ ways to rearrange S. But any of these sequences can be peeled from the "grouped" sequence $T = 1,1,\ldots,1,2,2,\ldots,2$ by the simple expedient of peeling from the left whenever we need a 1 and from the right whenever we need a 2. Thus $R(S) = \rho(T) = \binom{b+c}{b}$. Moreover, only T and its reversal T^{-1} can achieve this maximum. Any other sequence U either contains a string of 1's surrounded by 2's (in which case T can never be peeled from U) or a string of 2's surrounded by 1's (in which case T^{-1} can never occur). Thus, the maximum number of rinds for a sequence having only two distinct entries occurs if and only if the like entries are grouped together to form either T or T^{-1}.

Based on this evidence, Gobel conjectured that for any sequence S with m distinct values occurring among its n entries, at least one of the m! possible grouped sequences will be included among those sequences T that attain $\rho(T) = R(S)$, the maximum number of rinds. Now for S having b 1's, c 2's, and $d = n - b - c$ 3's, we shall find the number of rinds for the grouped sequence $T = 1,1,\ldots,1,2,2,\ldots,2,3,3,\ldots,3$. Let us call an entry a_i in T <u>buried</u> if it equals both its neighbors $a_{i-1} = a_i = a_{i+1}$, and otherwise let us call a_i <u>exposed</u>, that is if i=1 or n or $a_i \neq a_{i-1}$ or $a_i \neq a_{i+1}$. Evidently, if b, c, and d are at least 2, then T has 6 exposed entries.

Lemma 1. For any sequence S, every possible rind of S can be obtained by terminating with an exposed entry of S.

Proof. If the last entry of the rind U is $b_n = j$ which came from a buried j in S, then there were two exposed j's that were peeled earlier. But at the time the latter of these was peeled, the remaining subsequence of S contained only j's. We might as well presume that these j's are all peeled from the (originally) buried side to terminate with the exposed j.

This lemma enables us to count the number of rinds of T by counting rinds ending with each possible exposed entry, a_i. Since in the group sequence each entry to the left of a_i is strictly less than (and hence distinguishable from) each entry to the right of a_i , the number of rinds is precisely $\binom{n-1}{i-1}$, the number of ways to arrange i - 1 "left" choices among n - 1 selections of left or right. Unfortunately, those rinds that have all occurrences of a given value j appearing at the end of the rind have been counted twice - once for each exposed j. We correct by subtracting the number of such rinds to get

$$\rho(T) = \binom{n-1}{0}+\binom{n-1}{b-1}+\binom{n-1}{b}+\binom{n-1}{d}+\binom{n-1}{d-1}+\binom{n-1}{0}-\binom{c+d}{0}-\binom{b+d}{b}-\binom{b+c}{0}. \quad (1)$$

This simplifies to give

$$\rho(T) = \binom{n}{b} + \binom{n}{d} - \binom{b+d}{b} . \tag{2}$$

Even though this formula was derived by assuming b, c, and d were at least 2, a careful reexamination shows the same formula holds when one or more of the variables equal 1. This formula for rinds of a grouped sequence T gives three possible values as we try all 3! orders for the three groups. We find the largest of these values occurs when the smallest group is placed in the center, that is, $c = \min\{b,c,d\}$.

Now Göbel conjectured that (2) gives the maximum value for all sequences with three values. However, Guy [3] reported that Peck [4] and Gibson and Slater [1] independently found
$\rho(1221333) = 49 > 46 = \rho(1122333)$.
Incidentally, the present author found this same example but did not report it. Thus, the grouped sequence theorem for two values does not hold for three values, and we might well expect it to fail for all larger numbers of values. Moreover, determining general formulas for R(S) is probably much harder than originally thought since we have no obvious candidate for the optimal starting arrangement T.

2. Sequences of pairs.

Since the general problem appears to be too ambitious, we shall limit our goal to finding R(S) for S = 1122 ... mm, a sequence of m pairs. Our first theorem generalizes the counting formula (2) to a grouped sequence T = 11...1 22...2 33...3...mm ...m with b_j occurrences of j. Let $S_j = \sum_{i=1}^{j} b_i$ with $S_o = 0$ and $S_m = n$.

Theorem 2. $\rho(T) = \rho(11...122...2...mm...m)$

$$= \sum_{i=o}^{m} \binom{n}{s_i} - \binom{n-b_i}{s_{i-1}} .$$

Proof. Just as in the previous section, we count all rinds ending with an exposed entry and subtract the number of rinds in which all entries with common value j occur at the end. Upon simplification we obtain the desired formula. Reflection on what happens when various $b_j = 1$ leads to the same formula in all cases.

Corollary 3. $\rho(1122...mm) = 2^{2m-1} - 2^{2m-3}$.

Proof. Substituting $b_j = 2$ for all j into Theorem 2 gives

$$\rho(1122...mm) = \sum_{i=0}^{m} \binom{2m}{2i} - \binom{2m-2}{2i-2} = 2^{2m-1} - 2^{2m-3} . \qquad (3)$$

Although we can combine these two terms if we wish, we prefer to leave them as given because the first term happens to be the number of rinds of a sequence of 2m distinct values. Thus, the second term tells us that changing to m grouped pairs has cost us 2^{2m-3}, exactly one quarter of the original rinds. We shall call the difference $2^{n-1} - \rho(S) = d(S)$ the <u>deficit</u> of S. Obviously finding a rearrangement T of S to maximize $\rho(T)$ is equivalent to finding T to minimize d(T). To gain familiarity, we derive two other counting formulas.

<u>Theorem 4</u>. For $m \geq j$

$$\rho(m\ldots321123\ldots j) = 2^{m+j-1} - \sum_{i=0}^{j-1} \binom{m+j}{i} .$$

<u>Proof</u>. We prove this by induction on the length m+j. The formula for m=1, j=1 is trivial. To prove the inductive step for length m+j, assume the formula has already been verified for all shorter sequences. There are two cases. If j=m, every rind begins with $b_1 = m$ and then follows with a rind of m...321123...m-1. By induction, this gives

$$\rho(m\ldots321123\ldots m) = \rho(m\ldots321123..m-1) \tag{4}$$

$$= 2^{2m-2} - \sum_{i=0}^{m-2} \binom{2m-1}{i} = \binom{2m-1}{m-1} .$$

But the claimed formula reduces to

$$2^{2m-1} - \sum_{i=0}^{m-1} \binom{2m}{i} = \tfrac{1}{2}\binom{2m}{m} = \binom{2m-1}{m-1} \quad . \tag{5}$$

Thus the case j=m has been verified.

For j<m, each rind begins either with an m or a j, so

$$\rho(m\ldots321123\ldots j) = \rho(m-1\ldots321123\ldots j)+\rho(m\ldots321123\ldots j-1) \tag{6}$$

$$= 2^{m+j-2} - \sum_{i=0}^{j-1} \binom{m+j-1}{i} + 2^{m+j-2} - \sum_{i=0}^{j-2} \binom{m+j-1}{i}$$

$$= 2^{m+j-1} - \sum_{i=0}^{j-1} \binom{m+j}{i} \quad .$$

Incidentally, the completely nested sequence in (4) having $\binom{2m-1}{m-1}$ rinds is almost certainly the starting arrangement of m pairs yielding the fewest rinds, but we have not proved this conjecture.

<u>Theorem</u> <u>5</u>. For $m \geq 3$ and $m \geq j \geq 1$,

$$\rho(123\ldots m12\ldots j) = 3 \cdot 2^{m+j-3} \quad . \tag{7}$$

For m=2, $\rho(121) = 2$ and $\rho(1212) = 4$.

<u>Proof</u>. The case m=2 is irregular and is found by exhaustive listing of all rinds. For $m \geqslant 3$ we use induction on j. For j=1 we find each rind begins either with the left 1 (in 2^{m-1} ways)

or with the right 1 (in 2^{m-1} ways). However, rinds beginning 11 have been counted twice so we subtract 2^{m-2} to get $3 \cdot 2^{m-2}$. For $j>1$ there are $3 \cdot 2^{m+j-4}$ rinds starting with 1 and an equal number starting with j. Adding these totals completes the proof.

Observe that for $j=m \geq 3$ the formula in Theorem 5 matches the grouped formula of Corollary 3. However, we shall soon see that for $m \geq 4$ neither of these sequences has the maximum possible number of rinds. In an attempt to avoid becoming too tedious, we shall omit the proof of our final formula of this sort.

Corollary 6. For $m \geq 3$ and $m \geq j$, $\rho(12\ldots m\ 12\ldots m\ 12\ldots j) =$

$$9 \cdot 2^{2m+j-5}. \tag{8}$$

In order to refine further our ability to count rinds, we shall determine how often each particular rind can be produced when peeling S. The analysis that follows is only valid when S has no element occurring more than twice. To illustrate terminology, consider the starting sequence S = 21132443. We shall call the 1's and 2's nested in S because they occur in the relative order 2112. The order 1221 would also be called nested. The 1's and 3's are disjoint because their order is 1133. The 2's and 3's occur in relative order 2323 and are called linked. Every collection of four entries consisting of two pairs is either nested, disjoint, or linked. An ambiguous element of a rind is one that might have been peeled from either end of S. For example, in the rind U peeled from S above and beginning 3442..., clearly the 3 and both 4's had to be peeled from the right of S. But the 2 could have been peeled from either end. It is "potentially" ambiguous. Potentially, because if it is followed by a 1 or 3 the ambiguity is removed. Only if it is followed by the other 2 is the result irrevocably ambiguous. No matter how we complete the rind 34422 ... it is impossible to decide which 2 came from the left and which came from the right. Any sequence of consecutive elements such as 22 whose peeling history can not be determined is called an _ambiguous block_.

What form can ambiguous blocks take? First, consider non terminal blocks of U. The simplest and shortest ambiguous blocks are a double entry jj. To obtain a longer ambiguous block, the entries nested immediately inside the j's in S must also match, i.e., S = ...jk...kj..., for otherwise the ambiguity of the j's would be removed when the next entry is peeled. This nested pattern in S allows jkjk to be a single ambiguous block of U. On the other hand, if U contains jjkk these are viewed as two consecutive ambiguous blocks since peeling the second j made it certain that both j's are ambiguous and then peeling both k's created a second ambiguous block. If s has the pattern ...jkl...lkj... we can form the ambiguous block jkljkl in U.

But note that jkjlkl is also ambiguous. At first it may appear that the structure of long ambiguous blocks will grow hopelessly complicated, but in fact we find that each block of length 2i was produced by making i left hand choices and i right hand choices. Moreover, if at any earlier stage the number of left choices equals the number of right choices, we would close one ambiguous block at that point, and possibly begin another. Thus, it happens that the number of ambiguous blocks of length 2i that can be formed from i nested pairs of S is just the Catalan number $\frac{i}{i+1} \binom{2i}{i}$.

Now what about a terminal block in U? Such a block is ambiguous if and only if the substring in S it comes from is equal to its own reversal. That is, the substring has the form 12...j...21 in S with the central j either a double or single entry. Thus, only a terminal ambiguous block may have odd length. Our observation about the structure of ambiguous blocks are summarized in the following theorem:

Theorem 7. For any sequence S having at most two occurrences of each entry, each possible rind has a collection of $k \geq 0$ disjoint ambiguous blocks. Consequently, that particular rind can be obtained in precisely 2^k different ways.

Let $N_k(S) = N_k$ stand for the number of rinds of S having k specified ambiguous blocks (and possibly others).

Corollary 8. $$\rho(S) = \sum_{k=0}^{m} N_k(S)(-1)^k \qquad (9)$$

and $$d(S) = \sum_{k=1}^{m} N_k(S)(-1)^{k-1} \qquad (10)$$

Proof. Each particular rind U with j ambiguous blocks will be counted $\binom{j}{k} 2^{j-k}$ times in $N_k(S)$ because there are $\binom{j}{k}$ way to select k specified blocks and then 2^{j-k} ways to peel the other unidentified ambiguous blocks. Thus U is counted

$$\sum \binom{j}{k} 2^{j-k}(-1)^k = (2-1)^j = 1$$ time in (9). Since

$N_0 = 2^{n-1}$ and $d(S)=2^{n-1} - \rho(S)$ we also have (10).

It appears that this formula ought to allow the direct evaluation of $\rho(S)$ for any starting sequence S. However, the possibilities of long ambiguous blocks and many ambiguous blocks in the same rind seriously complicate the evaluation process. But if the ambiguous blocks in a given rind contain 2j or 2j+1 entries, these must include j pairs which were mutually nested j levels deep in the original sequence. Thus, if S has no nesting, no rind can have more than one single short ambiguous block. If S has no nesting three levels deep, then each rind has at most two short ambiguous blocks or one block of length four. Thus, the corollary is most effective when S has only limited nesting.

For each pair of like elements in S, we define C(jj) as the "cost" of the ambiguous block jj among rinds of S. This is precisely the contribution of the ambiguous block jj to N_1 in (10). Assuming b entries precede the first j in S, c entries lie between the j's, and $d = n - 2 - b - c$ entries follow the second j, the number of ways to produce the block jj is

$$C(jj) = \begin{cases} \binom{b+d}{b} 2^{c-1} & \text{for } c \geq 1 \\ \binom{b+d}{b} & \text{for } c = 0 \, . \end{cases} \qquad (11)$$

In addition, when c=1 the possible terminal ambiguous block jxj contributes $C(jxj) = \binom{b+d}{b}$ to the count N_1. If S has no nesting these are all the terms in (10) and in this case

$$d(S) = \sum_{j=1}^{m} C(jj) + C(jxj) \qquad (12)$$

where the latter term is nonzero only when c=1.

To illustrate the use of (12), let $S_k = 1,2,\ldots,k$ denote the first k integers in sequence. Let $k+S_k = k+1, k+2, \ldots, 2k$. Consider the sequence $S = S_k, S_k, k + S_k, k + S_k$ of length n=4k having m=2k pairs. Since S has no nesting, (12) yields

$$d(S_k, S_k, k+S_k, k+S_k) = 2^{k-1} \sum_{j=1}^{k} \binom{3k-1}{j-1} . \tag{13}$$

For all $k \geq 3$ this deficit is smaller than the deficit for the grouped sequence $d(1,1,2,2,\ldots,2k,2k) = \sum_{j=1}^{2k} \binom{4k-2}{2j-2} = 2^{4k-3}$. (14)

Thus, the grouped sequence will never give the most rinds when $k \geq 3$, as already observed by Peck [4]. Peck conjectured that this sequence S would give the most rinds, but Gibson and Slater [1] found that for $k \geq 4$ altering S by interchanging a_{k-1} with a_k and a_{3k+1} with a_{3k+2} yields more rinds. They cautiously avoided conjecturing that their sequence would be the ultimate. Using our cost function approach to devise a computer aided exhaustive search, we have found the optimal arrangements for all $m \leq 14$ pairs. The results are shown in Table 1 along with their deficits. The actual number of rinds is not shown, but is simply 2^{2m-1} minus the deficit. We observe that the grouped sequences are best only for $m \leq 3$ and at m=3 three other sequences do just as well. Four pairs is the last case found with more than one pattern. For $5 \leq m \leq 7$, there is no apparent pattern to the optimal sequences but for $8 \leq m \leq 14$ the optimal sequences all have the same pattern. To describe it, let T(k) = k, k-1,... 4,2,3,1,1,2,3,4,...,k . That is, $T(k) = S_k^{-1}, S_k$ with

m PAIRS	BEST SEQUENCE	DEFICIT
1	11	1
2	1122	2
3	112233	8
	211233	
	212313	
	123123	
4	21124334	18
	21142334	
5	1231235445	55
6	231123564456=T(3),3+T(3)	138
7	41231234675567	470
8	T(4),4+T(4)	1 164
9	T(4),4+T(5)	4 055
10	T(5),5+T(5)	10 140
11	T(5),5+T(6)	35 609
12	T(6),6+T(6)	89 782
13	T(6),6+T(7)	316 513
14	T(7),7+T(7)	803 040

TABLE 1. All sequences of $m \leq 14$ pairs having the maximum number of rinds.

the 3,2 pair in positions k-2 and k-1 reversed to form 2,3. For $8 \leq m \leq 14$, let $k = [m/2]$. Then the optimal pattern is $T(k), k+T(m-k)$. This also happens to be the pattern for m=6, but at m=7 the listed irregular pattern has deficit 470 to edge out $T(3), 3+T(4)$ which has a deficit of 471. It seems reasonable to conjecture that:

Conjecture: The sequence $T(k), k+T(m-k)$ with $k = [m/2]$ has the most rinds of any sequence of m pairs for all $m \geq 8$.

For $m \leq 5$ we counted the number of rinds of every possible sequence of m pairs, but for $m \geq 6$ a more sophisticated approach was required. We shall illustrate the technique for m=6. At this size we can carry out the analysis by hand without any computer search. As m increased to 14 the computer dependence becomes increasingly heavy. Beyond m=14 the amount of run time grows excessive, and round off error for the large integers involved is a problem.

To find the optimal sequences for a given value of m, say m=6, we start by constructing a 2mx2m "cost matrix" as shown in Figure 1. The i j entry of this matrix is the cost, or contribution to the deficit, that results from placing a pair of identical entries in positions i and j while keeping all other entries distinct. In other words, it counts the number of rinds in which this pair comprise an ambiguous block. The diagonal

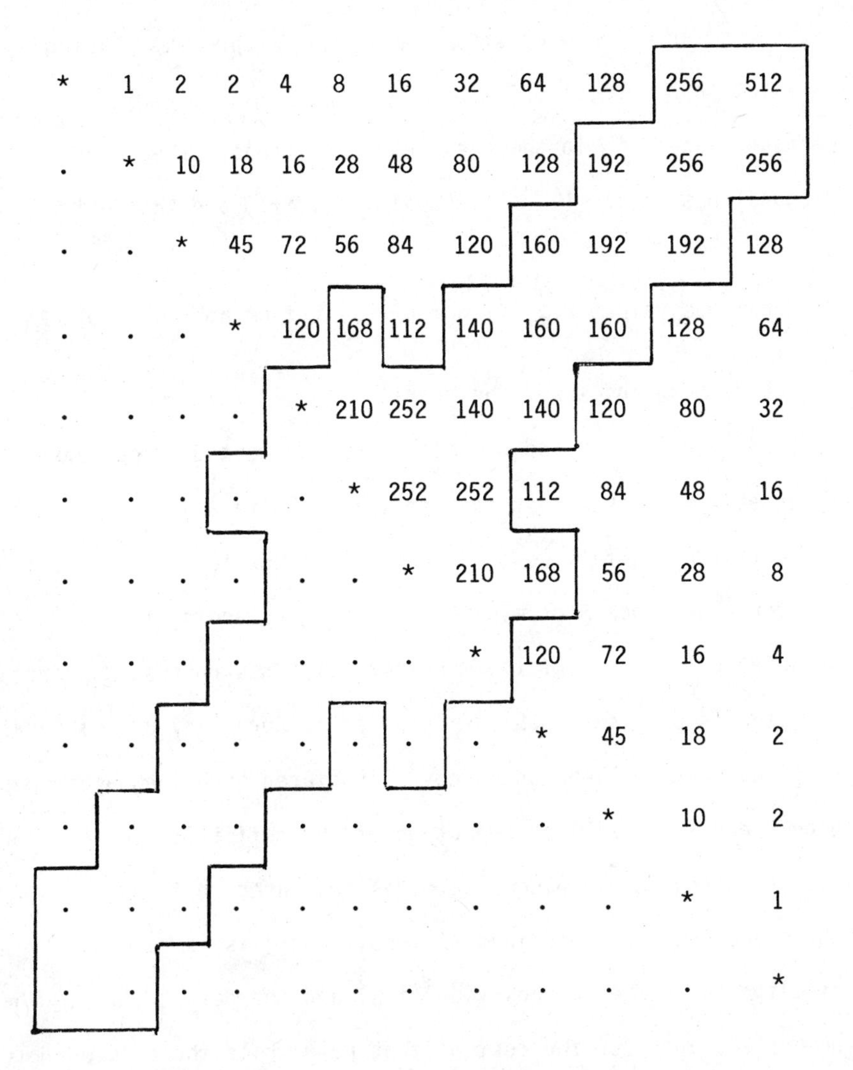

Figure 1. The cost matrix for m=6 pairs and its region of large entries.

positions cannot be used since these would represent placing a pair into a single slot.

We have placed *'s in these excluded positions. The super diagonal has values $\binom{2m-2}{i-1}$ in position i,i+1, and the next diagonal has values $2\binom{2m-3}{i-1}$ which result from ambiguous blocks of the form xx and xyx. The entries for $j-i \geq 2$ are $\binom{2m+i-j-1}{i-1} 2^{j-i-1}$. The entries below the main diagonal are shown as • since the cost in position i,j with i<j already accounts for the full cost of that pair. Now to find the deficit for a sequence S of m pairs, we start by adding the costs associated with each of the m pairs. If S has no nested pairs, this sum is precisely d(S) as detailed in Corollary 8. If S has nesting, there will be added cost associated with long ambiguous blocks and there will be reduction of the deficit to correct for over counting double blocks, etc. as indicated in formula (10). The placement of a pair in positions i and j is represented by circling the number at position (i,j) and the dot at (j,i). The number accounts for the cost of that pair while the circled dot in row j reminds us that position j in the sequence has been filled. Placing all m pairs inserts 2m circles in a pattern symmetric under transposition having one circle in every row and column.

The group sequence 112233445566 gives a deficit of 1+45+210+210+45+1 = 512. Since there is no nesting, there are no correction terms. This sequence does not give the most rinds because 123123456456 gives a deficit of 2+16+56+56+16+2 = 148. Is it possible to find a sequence with an even smaller deficit? A little inspection of the matrix, trying to avoid the large entries, suggests 231123564456 whose costs sum to 4+28+45+45+28+4 = 154, but now nesting allows a deduction of 2(1+7) for double blocks while no long blocks have been added to yield the deficit 138.

To conclude that this is indeed the minimum deficit, we must show that no sequence can have a smaller deficit. First, we must not place any pair to use an entry $\geq$ 138 for the placement of each successive pair can never decrease the deficit already incurred by the earlier pairs. Thus we forbid the entries inside the irregular region shown in Figure 1. Now each pair produces either two circles above the region or two below. If four pairs were used above the region, we would have to circle one number from each of the first four rows. In particular, in row 4 we suffer a cost of either 120 or 112 for our first pair. The second pair, using a number from row 3, already gives a deficit exceeding 138 even when correction terms are computed. Thus, four pairs must not appear above the region. Since the same argument applies below the region, we conclude that there must be three upper pairs and three lower. The only two ways to nest an upper pair with a lower pair involve (1,10) nesting (6,9) or (8,9). Both choices give an excessive deficit (even after the correction term), so no upper pair may nest with a lower pair. Consequently, the deficit for the three uppers adds to that for the three lowers to give the total deficit. If we hope to undercut 138, at least one of these two deficits must be less than 69. This excludes further entries $\geq$ 69 and so we are forced to select three values from the first three rows. In particular, in row 3 we must choose either 45 or 56. This leaves very few possibilities, the least being 45+16+2-1-7=69 as already found. Thus, sequence 231123564456 is the unique sequence with minimum deficit 138. We have just verified its inclusion in Table 1.

This instance for m=6 illustrates the method used to obtain the values through m=14. The computer was used in an interactive mode to exclude large entries one by one. In some cases dozens of trial choices were extended to four or five pairs until it was found that all possible choices would exceed the deficit already computed for a good candidate sequence.

3. Other ways to count rinds.

While the cost/deficit approach is quite efficient for restricting the placement of individual pairs, we also found a recursive approach to determine $\rho(S)$ for an arbitrary sequence $S = a_1, a_2, \ldots, a_n$ with no restrictions on the frequencies of individual values in S. To describe this approach, we let $S[i,j]$ denote the segment of S of length $j+1-i$ starting with a_i and ending with a_j. Thus, $S = S[1,n]$ and $a_i = S[i,i]$. Call S reversible if $S = S^{-1} = a_n, \ldots, a_2, a_1$. The number of rinds of S can be determined from one of three equations relating it to the number of rinds of segments of S.

Lemma 9. If $a_1 \neq a_n$, then $\rho(S) = \rho(S[2,n]) + \rho(S[1,n-1])$. If $S = S^{-1}$ is reversible, then $\rho(S) = \rho(S[2,n]) = \rho(S[1,n-1])$. Otherwise let $k < n/2$ be the maximum value with $S[1,k]^{-1} = S[n+1-k,n]$. Then $\rho(S) =$

$$\rho(S[2,n]) + \rho(S[1,n-1]) - \sum_{j=1}^{k} \frac{1}{j}\binom{2j-2}{j-1} \rho(S[j+1,n-j]). \quad (15)$$

Proof. For $a_1 \neq a_n$, there are $\rho(S[2,n])$ rinds starting with a_1 and $\rho(S[1,n-1])$ different rinds starting with a_n. For $S=S^{-1}$, every rind of $S[2,n]$ is also a rind of $S[1,n-1]$ obtained by always making the opposite left/right choice as made in $S[2,n]$. Thus $S[2,n] = S[1,n-1]$, and every rind of S consists of one of these rinds of length $n-1$ preceded by a_1. Finally, if k is the length of the matching outer segments, then $\rho(S[2,n])$ counts rinds starting with a_1 and $\rho(S[1,n-1])$ counts those starting with a_n. We must reduce this sum by the number of rinds that begin with an ambiguous block. Such a block must have even length 2j with $1 \leq j \leq k$, consist of j pairs, and have no initial segment of even length also perfectly paired. The number of ways to peel an initial ambiguous block of length 2j is precisely the Catalan number $\frac{1}{j}\binom{2j-2}{j-1}$. The number of ways to peel the rest of S is just $\rho(S[j+1,n-j])$. The product of these two expressions counts the number of rinds of S having an initial ambiguous block of length 2j.

The lemma can be used recursively to find $\rho(S)$. Typically, $\rho(S)$ equals the sum of the numbers of rinds for two sequences of length n-1, but occasionally it is just one sequence, and sometimes we have the k extra terms in (15). The lemma can be reapplied to each subsequence to produce more terms of shorter length. Proceeding blindly in this fashion until only singletons remain computes $\rho(S)$ in about 2^n steps. But as elements are stripped from S the segment S[i,j] may be produced $\binom{n+i-j-1}{i-1}$ times. Thus, the blind recursive approach recomputes the same $\rho(S[i,j])$ over and over again. Instead we only need to compute each of the $\binom{n}{2}$ possible $\rho(S[i,j])$'s once. These segments may be arranged in a pattern resembling Pascal's triangle. Starting from bottom level, it is easy to apply Lemma 9 once for each position in the triangle. The final entry found is $\rho(S)$. This process is illustrated in Figure 2 for the sequence S = 12345321. Note particularly the use of equation (15) for the center column. Thus, $\rho(S) = 50+50-1\cdot 20-1\cdot 6-2\cdot 2 = 70$.

Further inspection of Table 1 reveals that each optimal sequence for $m \geq 5$ can be partitioned into two segments, one comprised of $k = \lfloor m/2 \rfloor$ pairs and the other with m-k pairs. For such partitioned sequences, it is possible to develop another formula for the number of different rinds.

Theorem 10. If T = S[1,t] and U = S[t+1,n] partitions S into two segments having no element in common, then

$$\rho(S) = \sum_{j=0}^{t-1} \binom{n-t-1+j}{j} \rho(S[j+1,t]) + \sum_{j=0}^{n-t-1} \binom{t-1+j}{j} \rho(S[t+1,n-j]) \qquad (16)$$

or equivalently, the deficit of S is

$$d(S) = \sum_{j=0}^{t-1} \binom{n-t-1+j}{j} d(S[j+1,t]) + \sum_{j=0}^{n-t-1} \binom{t-1+j}{j} d(S[t+1,n-j]). \qquad (17)$$

Proof. Since T and U have no common elements, no ambiguous block can appear until one of T or U is exhausted. The jth term of the first sum in (16) counts rinds with a_{t+1} appearing before a_t and in position n-t+j. The jth term of the second sum

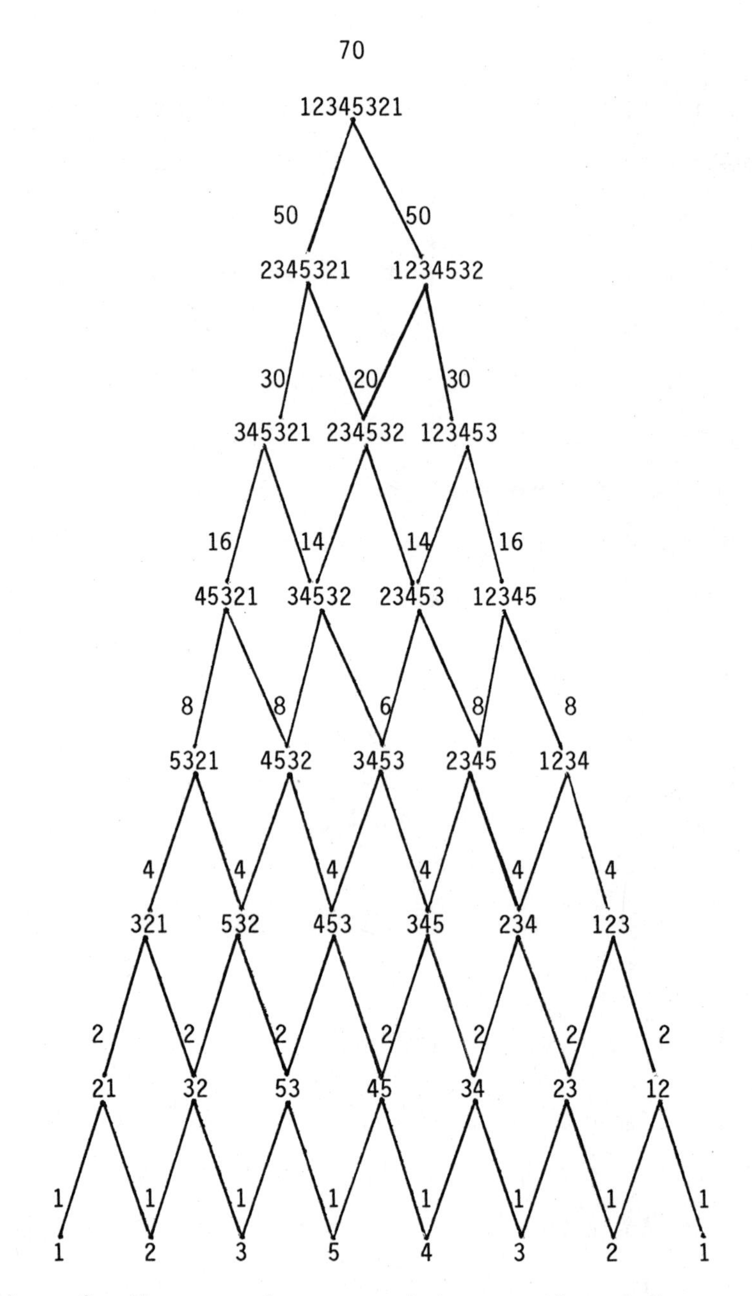

Figure 2. The recursive approach to computing rinds.

counts rinds with a_t appearing before a_{t+1} and in position t+j. Every rind is counted once by this process, verifying (16). To obtain the deficit formula, first observe that if S has n distinct entries (16) becomes

$$2^{n-1} = \sum_{j=0}^{t-1} \binom{n-t-1-j}{j} 2^{t-1-j} + \sum_{j=0}^{n-t-1} \binom{t-1+j}{j} 2^{n-t-1-j} \tag{18}$$

Now subtracting (16) from (18) gives the deficit formula (17).

4. Related problems.

Gibson and Slater [1] have suggested a nice generalization of the rind process. For an arbitrary labeled connected graph, a rind of G is a sequence of vertices obtained by removing vertices in succession so that the remaining subgraph continues to be connected. The rinds of sequences we have been studying can be viewed as rinds of a path in which each vertex label appears twice. Considering the difficulty of computing the number of rinds of a sequence, we can expect evaluating the number of rinds of an arbitrary graph to be very difficult.

We close with an easy exercise that has a pretty answer.

<u>Problem</u>. $S = a_1\ a_2,\ldots,a_n$ consists of the first n positive integers in some scrambled order. The peeling process can be used to produce a rind, S_1. Now repeel S_1 to form a rind S_2, and so on. How many times must a particular sequence S be

peeled in order to obtain the natural order 1,2,...,n? What is the maximum number of peels that might be needed for any S?

Solution. Call a_j a local maximum if $a_{j-1} < a_j > a_{j+1}$. Of course a_1 is a local maximum if $a_1 > a_2$ and similarly for a_n. Let m(S) be the number of local maxima in S. We shall find that any rind S_1 of S has at least ½m(S) local maxima. To see this, suppose that a_j is the first local maximum peeled from S. If a_j is not a local maximum in S_1 then it must be followed by elements from the opposite end. At the latest we will have a local maximum in S_1 when the second local maximum of S is peeled. In this way every two local maxima of S yield at least one in S_1. Finally, if m(S) is odd the last maximum to be peeled will automatically be a local maximum in S_1.

Furthermore, $\lceil \frac{1}{2}m(S) \rceil$ can be attained. Let $b_1, b_2, \ldots b_m$ be the positions of the local maxima in S, and let $c_1, c_2, \ldots, c_{m-1}$ be the intervening local minima. First peel $S[1,b_1] \cup S[b_m,n]$ into increasing order. Then peel $S[b_1+1,c_1] \cup S[c_{m-1},b_m-1]$ into decreasing order, and so on, alternating increasing and decreasing segments. Thus, S must be peeled $\lceil \log_2 m(S) \rceil$ times to reduce it to one maximum, and then one more time to move that maximum to the last position. Obviously, this means the maximum over all sequences is

$1 + \lceil \log_2 \lceil n/2 \rceil \rceil$ which happens to equal $\lceil \log_2 n \rceil$.

REFERENCES

1. P. M. Gibson and P. J. Slater, Rinds of a graph, preprint.

2. F. Göbel, How many different rinds can you peel from a sequence? Amer. Math. Monthly 89 (1982), 113-114.

3. R. K. Guy, Unsolved problems, Amer. Math. Monthly 90 (1983), 686-687.

4. G. W. Peck, How many different rinds can you peel from a sequence?, preprint.

ITERATIVE ALGORITHMS FOR CALCULATING NETWORK RELIABILITY

D. R. Shier
Clemson University

ABSTRACT

This paper focuses on determining the reliability of a network in which edges fail with known probabilities. In particular, we wish to calculate the probability that an operative path exists between two specified nodes of the network. An underlying algebraic structure is identified and exploited to produce a class of iterative algorithms for exact and approximate calculation of network reliability.

1. Introduction.

The reliability of a complex communication or distribution system depends critically on the structure and reliability of its individual components. A number of deterministic and probabilistic reliability measures have been studied when the underlying system can be modeled as a network. Deterministic measures [4, 5, 20] typically focus on the topology of the network and emphasize the worst-case performance of the system, while probabilistic measures [2, 6, 14] often incorporate nonuniform failure rates for the various components and emphasize the average-case performance of the system. A specific instance of the latter type of measure is the s-t reliability: namely, the probability that two distinguished nodes can communicate via a path of operative edges in the network.

In what follows we suppose $G=(N,E)$ is a directed graph with node set N and edge set E. Moreover, there is associated with every $e \in E$ a reliability p_e, indicating the probability that edge e is operative. It is well known [9, 11] that the s-t reliability of such a (directed) network can be easily expressed in terms of its *states* $\delta = (\delta_1, \ldots, \delta_m)$, where $m=|E|$. Here δ is simply the characteristic vector for a set of (functioning) edges in the network: $\delta_k=1$ if edge k is operative and $\delta_k=0$ otherwise. If the edges are assumed to fail independently, then the s-t reliability of G is given by

$$\sum_{\delta \in \Omega} I(\delta) \prod_{k=1}^{m} p_k^{\delta_k}(1-p_k)^{1-\delta_k},$$

where Ω is the set of states and $I(\delta)$ is an indicator variable for the existence of a path from s to t when the system is in state δ. Because there are 2^m such states, calculation of network reliability by this approach is clearly infeasible even for rather small networks. As a result, a number of alternative approaches have been studied which require enumerating simple s-t paths [8, 16], simple s-t cutsets [6, 13] or "factoring" the network into two smaller networks by conditioning on the state of some selected edge [12, 15].

Despite the various approaches and proposed algorithms for calculating s-t network reliability, all known general procedures exhibit exponential behavior in the worst case. That is, an unacceptably large amount of time (exponential in the problem size) will be required to compute reliability for certain types of networks. In fact, it has been recently demonstrated [1, 19] that the s-t reliability problem (as well as many other such network reliability problems) is NP-hard.

2. An Algebraic Structure.

In order to gain additional insight, we shall study an algebraic structure underlying the s-t reliability problem. Associate with each edge e_k of the given network a *variable* x_k; two edges that always fail together are assigned the same variable. For example, in Figure 1 the pair of edges (2, 3) and (3, 2) both fail simultaneously, so these edges are labeled with the same variable x_3. All other edges fail independently of one another and independently of the edges labeled x_3.

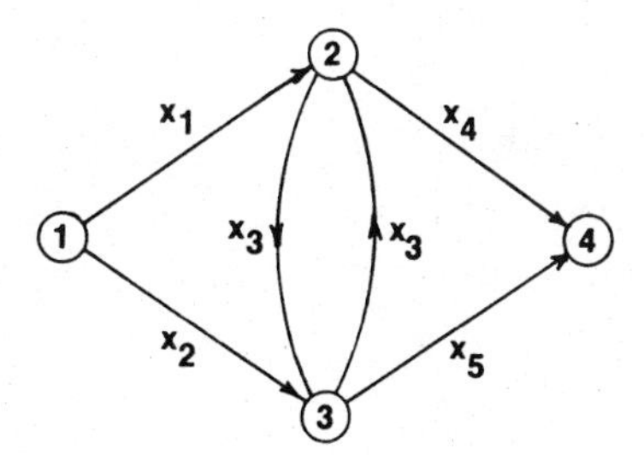

Figure 1.

Suppose that the edges of G have been labeled using $x_1, \ldots, x_r$. Then we would like to obtain the *reliability polynomial* $F(x_1,\ldots,x_r)$ associated with the pair (s,t). Namely, $F(x_1,\ldots,x_r)$ is a polynomial in $x_1,\ldots,x_r$ such that when the numerical values $p_1,\ldots,p_r$ are substituted for the corresponding variables $x_1,\ldots,x_r$ then the resulting value is the probability that an operative path exists from node s to node t. It will be shown how this polynomial can be concisely expressed in terms of two operations $\oplus$, $\otimes$ defined on polynomials.

The operation $\otimes$ when applied to single terms $x_1^{a_1} x_2^{a_2} \ldots x_r^{a_r}$ and $x_1^{b_1} x_2^{b_2} \ldots x_r^{b_r}$ yields $x_1^{c_1} x_2^{c_2} \ldots x_r^{c_r}$ where $a_i, b_i, c_i \varepsilon \{0,1\}$ and $c_i = \max\{a_i, b_i\}$. This operation is extended to arbitrary polynomials by distributivity. For example

$$x_1x_3x_7x_8 \otimes x_2x_3x_8 = x_1x_2x_3x_7x_8,$$
$$(x_1x_3 + x_7x_8) \otimes x_2x_3x_8 = x_1x_2x_3x_8 + x_2x_3x_7x_8.$$

The operation $\oplus$ is defined on polynomials $f(x_1,\ldots,x_r)$ and $g(x_1,\ldots,x_r)$ using

$$f(x_1,\ldots,x_r) \oplus g(x_1,\ldots,x_r) = f(x_1,\ldots,x_r) + g(x_1,\ldots,x_r) - f(x_1,\ldots,x_r) \otimes g(x_1,\ldots,x_r).$$

For example, $x_1x_2x_3 \oplus x_2x_4 = x_1x_2x_3 + x_2x_4 - x_1x_2x_3x_4$. Operations related to $\oplus$ and $\otimes$ have been previously stated in a rudimentary form, without further development, by Mine [9] and by Kim et al. [7]. It turns out that the operations $\oplus$ and $\otimes$ defined above enjoy a number of useful properties:

$$x \oplus y = y \oplus x \qquad x \otimes y = y \otimes x$$
$$x \oplus (y \oplus z) = (x \oplus y) \oplus z \qquad x \otimes (y \otimes z) = (x \otimes y) \otimes z$$
$$x \oplus 0 = x \qquad x \otimes 1 = x$$
$$x \otimes (y \oplus z) = (x \otimes y) \oplus (x \otimes z) \qquad x \oplus (y \otimes z) = (x \oplus y) \otimes (x \oplus z)$$
$$x \oplus (x \otimes y) = x \qquad x \otimes (x \oplus y) = x$$
$$x \oplus x = x \qquad x \otimes x = x$$

Moreover, we can now express the reliability polynomial $F(x_1,\ldots,x_r)$, relative to nodes s and t, as a certain sum with respect to operation $\oplus$: namely,

$$F(x_1,\ldots,x_r) = \oplus \Sigma \{v(P) : P \varepsilon \mathcal{P}_{st}\} \tag{1}$$

where $\mathcal{P}_{st}$ is the set of simple paths from node s to node t in G and v(P) denotes the *value* of path P, defined as the product (with respect to $\otimes$) of the edge variables along the path:

$$v(P) = \otimes \Pi \{x_k : e_k \varepsilon P\}.$$

As an illustration, consider the network in Figure 1 with s=1 and t=4. Since there are four simple paths extending from s to t, equation (1) becomes

$$F(x_1,\ldots,x_5) = x_1x_4 \oplus x_2x_5 \oplus x_2x_3x_4 \oplus x_1x_3x_5.$$

This relation just expresses (in an inclusion/exclusion sense) the s-t reliability as the probability that at least one of the simple s-t paths is operative. By expanding the above expression using the definitions of $\oplus$ and $\otimes$ and then substituting the numerical values p_i for the corresponding variables x_i, one obtains the s-t reliability of the network.

3. Iterative Procedures.

Equation (1) bears a striking resemblance to an algebraic formulation [3, 10] of the classical shortest path problem, in which a path of minimum length from node s to node t is sought. Specifically, if the x_i denote real numbers and $\otimes$ is ordinary addition, then the value v(P) of path P is simply its (additive) length. As a result, if $\oplus$ is the minimum operation then equation (1) represents the minimum path length from node s to node t, assuming there are no (directed) circuits of negative length in the network. Indeed, there are other network optimization problems which can be formulated in a similar algebraic way, by judiciously defining two binary operations that satisfy certain properties and then evaluating the expression in (1). The problems of determining maximum capacity paths, most reliable paths, minimax control paths, and k^{th} shortest paths can all be formulated in this manner by appropriate choices for $\oplus$ and $\otimes$ [3, 10, 17].

In this more general algebraic setting, a number of techniques have been developed to calculate the quantity in (1). Such techniques amount to different methods for solving a certain system of "linear" equations--equations which are now linear in $\oplus$ and $\otimes$. This system is given by $z = zA \oplus e_s$, where A is the (weighted) adjacency matrix for the network and e_s is the sth "unit" row vector. More concretely, for the example given in Figure 1 with s=1, the appropriate equations to be solved for z are the following:

$$\begin{aligned} z_1 &= 1 \\ z_2 &= (x_1 \otimes z_1) \oplus (x_3 \otimes z_3) \\ z_3 &= (x_2 \otimes z_1) \oplus (x_3 \otimes z_2) \\ z_4 &= (x_4 \otimes z_2) \oplus (x_5 \otimes z_3) \end{aligned} \qquad (2)$$

The first observation to be made is that by solving the above system of equations, one obtains more than just the s-t reliability for a fixed s and t. Namely, the solution component z_j gives the s-j reliability, and so we obtain here simultaneously the 1-j reliability for *all* $j \varepsilon N$. Moreover, unlike other techniques for calculating s-t reliability, it is not necessary to determine explicitly the set of simple s-t paths in the network; by solving the above linear system, such paths will be *automatically* accounted for during the solution process.

Given the above form of the system (2), a fairly natural way of obtaining a solution is by means of an iterative scheme. Specifically, we can apply a generalized Jacobi method [3, 18] to (2), whereby the current estimate $z^{(i)}$ at iteration i is substituted into the right-hand side of (2) in order to obtain the new estimate $z^{(i+1)}$ at iteration i+1. As shown in [3, 18], this generalized Jacobi method will converge in at most $|N|-1$ iterations, at which point component z_j of z represents the reliability polynomial from node s=1 to node j.

Table 1 details the progress of the iterative scheme applied to the example of Figure 1 when s=1; the initial estimate $z^{(0)}=(1,0,0,0)$ has been used to start the Jacobi scheme. For example, the estimate for z_2 at iteration 3 is obtained using the estimates for z_1 and z_3 at iteration 2:

$$\begin{aligned} z_2 &= (x_1 \otimes 1) \oplus (x_3 \otimes [x_2 \oplus x_1x_3]) \\ &= x_1 \oplus x_2x_3 \oplus x_1x_3 \\ &= [x_1 \oplus x_1x_3] \oplus x_2x_3 \\ &= x_1 \oplus x_2x_3. \end{aligned}$$

Notice that the properties given earlier for $\oplus$ and $\otimes$ are useful here in simplifying the estimate for z_2. Convergence is achieved after three iterations, and so the correct reliability polynomial (see Table 1) from node 1 to node 4 is

$$\begin{aligned} z_4 &= x_1x_4 \oplus x_2x_5 \oplus x_2x_3x_4 \oplus x_1x_3x_5 \\ &= x_1x_4 + x_2x_5 + x_2x_3x_4 + x_1x_3x_5 - x_1x_2x_4x_5 - x_1x_2x_3x_4 \\ &\quad - x_2x_3x_4x_5 - x_1x_3x_4x_5 - x_1x_2x_3x_5 + 2x_1x_2x_3x_4x_5. \end{aligned}$$

In a similar manner z_2 and z_3 yield the reliability polynomials to nodes 2 and 3 respectively.

	ITERATION			
	0	1	2	3=4
z_1	1	1	1	1
z_2	0	x_1	$x_1 \oplus x_2x_3$	$x_1 \oplus x_2x_3$
z_3	0	x_2	$x_2 \oplus x_1x_3$	$x_2 \oplus x_1x_3$
z_4	0	0	$x_1x_4 \oplus x_2x_5$	$x_1x_4 \oplus x_2x_5 \oplus x_2x_3x_4 \oplus x_1x_3x_5$

Table 1. Successive iterations of the Jacobi method

4. Approximations.

Application of this generalized Jacobi method to any network yields a sequence of estimates $z^{(i)}$ converging to the exact network reliability vector z. Notice that the estimate at iteration i includes all terms involving paths of length $\leq i$, so convergence is assured in at most $|N|-1$ iterations (no simple path in G has length exceeding $|N|-1$). It also follows that better and better approximations to the true reliability are generated as the algorithm progresses. This suggests that fairly good approximations to network reliability can be obtained by terminating the iterative process after relatively few iterations.

To illustrate the utility of such approximations, the Jacobi scheme was applied to the undirected network in Figure 2, with s=1. (Each undirected edge represents two oppositely directed edges that fail together.) For ease of comparison, the sequence of approximating polynomials from node s=1 to node t=3 has been evaluated using the same reliability $p_e=p$ for each edge (Table 2). These approximations $z^{(1)},\ldots,z^{(4)}$ have been plotted, as a function of the common edge reliability p, in Figure 3. Notice that these approximations form a nondecreasing sequence of functions over [0,1] that converges to the exact reliability polynomial $F(p)=z^{(4)}$. The approximation $z^{(3)}$ is, however, virtually indistinguishable in Figure 3 from this exact reliability polynomial. Indeed even the approximation $z^{(2)}$ provides a fairly close approximation over the entire interval [0,1] to the exact network reliability.

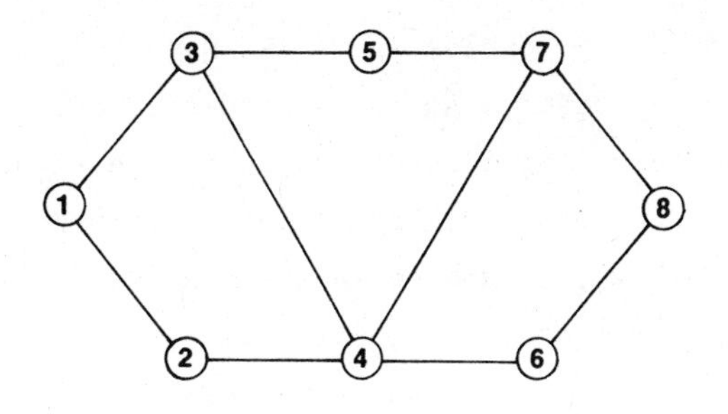

Figure 2.

$$z^{(0)} = 0$$
$$z^{(1)} = p$$
$$z^{(2)} = p + p^3 - p^4$$
$$z^{(3)} = p + p^3 - p^4 + p^5 - 2p^6 + p^7$$
$$z^{(4)} = p + p^3 - p^4 + p^5 - 2p^6 + 2p^7 - 3p^8 + 3p^9 - p^{10}$$

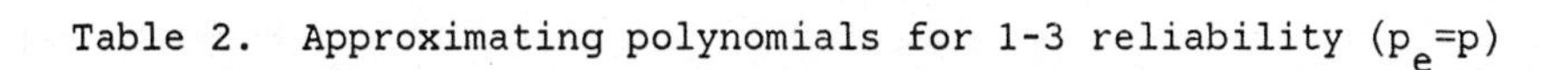

Table 2. Approximating polynomials for 1-3 reliability (p_e=p)

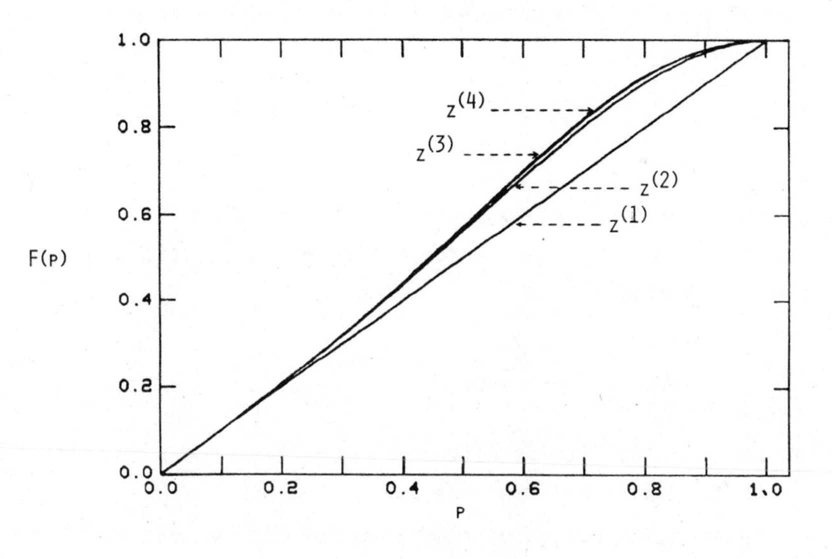

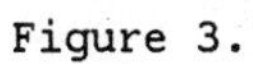

Figure 3.

Other, more complex, networks have been analyzed using this iterative approach. Again, a nondecreasing sequence of approximating polynomials is generated by the algorithm. Each approximation provides a "conservative" estimate for the true network reliability, in the sense that the exact network reliability is guaranteed to be at least as large as the value specified by the approximation. In these larger examples the computation time to determine the exact reliability polynomial grows quite rapidly; however, reasonably good approximations can be found in only a small fraction of this time.

While the previous discussion has concentrated on one particular iterative scheme (the Jacobi method), there are several other well-known iterative procedures that can be adapted to solve the linear system $z = zA \oplus e_s$. Further research is proceeding to develop such schemes and associated data structures for efficient implementation. Finally, it should be emphasized that all such iterative schemes provide an unexpected bonus: the s-t reliability is available from a given source node s to *every* other node t in the network.

Acknowledgement

This research was supported in part under NSF Grant ISP-8011451 (EPSCOR) and Air Force Office of Scientific Research Grant AFOSR-84-0154.

REFERENCES

1. M. Ball, Complexity of network reliability computations, *Networks*, 10 (1980), 153-165.
2. J. Buzacott, A recursive algorithm for finding reliability measures related to the connection of nodes in a graph, *Networks*, 10 (1980), 311-327.

3. B. Carré, An algebra for network routing problems, *J. Inst. Math. Appl.*, 7 (1971), 273-294.
4. H. Frank and I. Frisch, Analysis and design of survivable networks, *IEEE Trans. Commun. Technol.*, COM-18 (1970), 501-519.
5. H. Frank and I. Frisch, *Communication, Transmission, and Transportation*, Addison-Wesley, Reading, Mass., 1971.
6. E. Hänsler, G. McAuliffe and R. Wilkov, Exact calculation of computer network reliability, *Networks*, 4 (1974), 95-112.
7. Y. Kim, K. Case and P. Ghare, A method for computing complex system reliability, *IEEE Trans. Rel.*, R-21 (1972), 215-219.
8. M. Locks, Recursive disjoint products: A review of three algorithms, *IEEE Trans. Rel.*, R-31 (1982), 33-35.
9. H. Mine, Reliability of physical system, *IRE Trans. Circuit Theory*, CT-6 (1959), 138-151.
10. E. Minieka and D. Shier, A note on an algebra for the k best routes in a network, *J. Inst. Math. Appl.*, 11 (1973), 145-149.
11. P. Mirchandani, Shortest distance and reliability of probabilistic networks, *Comput. & Ops. Res.*, 3 (1976), 347-355.
12. F. Moskowitz, The analysis of redundancy networks, *AIEE Trans. Commun. Electron.*, 39 (1958), 627-632.
13. J. Provan and M. Ball, *Computing network reliability in time polynomial in the number of cuts*, MS/S 82-023, Management Science and Statistics, University of Maryland, College Park, Md., 1982.
14. A. Satyanarayana, A unified formula for analysis of some network reliability problems, *IEEE Trans. Rel.*, R-31 (1982), 23-32.
15. A. Satyanarayana and M. Chang, Network reliability and the factoring theorem, *Networks*, 13 (1983), 107-120.
16. A. Satyanarayana and A. Prabhakar, A new topological formula and rapid algorithm for reliability analysis of complex networks, *IEEE Trans. Rel.*, R-27 (1978), 82-100.

17. D. Shier, A decomposition algorithm for optimality problems in tree-structured networks, *Discrete Math.*, 6 (1973), 175-189.
18. D. Shier, Iterative methods for determining the k shortest paths in a network, *Networks*, 6 (1976), 205-229.
19. L. Valiant, The complexity of enumeration and reliability problems, *SIAM J. Comput.*, 8 (1979), 410-421.
20. R. Wilkov, Analysis and design of reliable computer networks, *IEEE Trans. Commun.*, COM-20 (1972), 660-678.

SHORTEST PATH ALGORITHMS

Robert E. Tarjan
AT&T Bell Laboratories
Murray Hill, New Jersey 07974

ABSTRACT

This paper surveys recent work on Dijkstra's algorithm and related methods of computing shortest paths. Included is a discussion of recent work by M. L. Fredman and the author on a new data structure called Fibonacci heaps and its use in shortest path calculations.

1. Dijkstra's Algorithm and its Variants

Let $G = (V,E)$ be a directed graph with vertex set V and edge set E. We shall denote the number of vertices of G by n and the number of edges by m. For purposes of stating time bounds simply, we assume that $n \geq 2$ and $m \geq n$. Let each edge (v,w) of G have a non-negative *length* $a(v,w)$. The *length* of a path in G is the sum of the lengths of its edges. A *shortest path* from a source vertex s to a target vertex t is a path from s to t of minimum possible length. The *distance* from s to t is the length of such a shortest path.

In this paper we shall discuss the *single-pair shortest path problem*: given a source s and target t, find the distance from s to t. The classical algorithm, for this problem, due to Dijkstra [4], is a form of *shortest-first search*. The algorithm maintains a tentative distance $u(v)$ for each vertex v, which is an upper bound on the length of a shortest path from s to v. Initially $u(s) = 0$ and $u(v) = \infty$ for

$v \neq s$. Every vertex is in one of three stages: <u>unlabeled</u>, <u>labeled</u>, or <u>scanned</u>. Initially s is labeled and all other vertices are unlabeled. The algorithm consists of repeating the following step until t becomes scanned:

<u>Scanning Step</u>. Select a labeled vertex v with $u(v)$ minimum. Convert v to the scanned state. For every edge (v,w) such that $u(v) + a(v,w) < u(w)$, replace $u(w)$ by $u(v) + a(v,w)$, and if $u(w)$ was previously infinite, convert w to the labeled state.

The non-negativity of the edge lengths implies that when a vertex v becomes scanned, $u(v)$ is the length of a shortest path from s to v. By changing the terminating condition of the algorithm, we can solve the <u>single-source shortest path problem</u>, which is that of computing distances from a given source s to all other vertices. We merely continue scanning until there are no labeled vertices (any remaining unlabeled vertices are not reachable from s). We can augment the algorithm to compute the actual shortest paths as well as their lengths by maintaining for each vertex v a <u>predecessor</u> $p(v)$, which precedes v on a path from s to v of length $u(v)$. Further discussion of this and other shortest path algorithms can be found in [14].

Dijkstra's algorithm, even when it is used to solve the single-pair problem, scans many, if not most of the vertices, since the order of scanning is unaffected by the location of the target vertex. If we can easily compute an estimate of the distance to the target, we can bias the algorithm so that it preferentially scans vertices likely to be close to the shortest path. Let $b(v)$ be an estimate of the distance from vertex v to the target vertex t. We modify the scanning step so that it selects a labeled vertex v with $u(v) + b(v)$ minimum. This method, called <u>heuristic search</u>, was studied by Hart, Nilsson, and Raphael [6].

For the algorithm to be correct, the estimate function b must satisfy the following consistency condition: $b(v) \le a(v,w) + b(w)$ for every edge (v,w).

Given the consistency condition, we can drop the requirement that the length function a be non-negative. Heuristic search is equivalent to running Dijkstra's algorithm using transformed edge lengths $a'(v,w) = a(v,w) + b(w) - b(v)$ for every edge (v,w). The consistency condition is equivalent to $a'(v,w) \ge 0$ for every edge (v,w), hence the correctness of heuristic search is equivalent to the correctness of Dijkstra's algorithm. The consistency condition implies that $b(v)$ is a lower bound on the distance from v to t.

The performance of heuristic search depends on the accuracy of the estimating function b. The more accurate b is, the fewer vertices will be scanned. A particularly promising class of graphs for heuristic search are the Euclidean graphs. A Euclidean graph consists of a set of vertices in k-dimensional space and a subset of the possible edges, with the length of an edge equal to the Euclidean distance between its end vertices. As an estimate $b(v)$, we use the Euclidean distance from v to t. This example is mentioned by Hart, Nilsson, and Rafael; Sedgewick and Vitter [13] have done some average-case analyses of the behavior of heuristic search on Euclidean graphs.

Instead of solving a shortest path problem by searching forward from s, we can alternatively search backward from t. Or, we can search forward from s and backward from t concurrently. This method is called bidirectional search [11,12]. The formulation of the correct termination condition is tricky, especially when a distance estimate is incorporated to produce bidirectional heuristic search [3]. Lawler, Luby, and Parker have done a careful study of heuristic and bidirectional search [10].

2. Implementation and Analysis of Dijkstra's Algorithm

Let us call the set of labeled vertices in Dijkstra's algorithm the *frontier set*. The hardest part of implementing the algorithm is in maintaining the frontier set. We must be able to perform the following operations in this set.

make: Initialize the frontier set to be empty.

insert (v,x): Insert a new vertex v into the frontier set with $u(v) = x$.

delete min: Delete from the frontier set a vertex v with $u(v)$ minimum.

decrease (v,x): For vertex v in the frontier set with $u(v) > x$, redefine $u(v)$ to be x.

Dijkstra's algorithm requires the performance of one *make* operation, at most $n - 1$ *delete min* operations, at most n *insert* operations, and at most m *decrease* operations. Aside from operations on the frontier set, the algorithm takes $O(m)$ time. Thus the algorithm is dominated by the time spent on these operations. A data structure supporting *make*, *insert*, and *delete min* is called a *priority queue* or a *heap* [14]. The *decrease* operation, though non-standard, is crucial in implementing Dijkstra's algorithm.

Classical implementations of heaps include the implicit heaps of Williams [16], the leftist heaps of Crane [2,9], and the binomial heaps of Vuillemin [1,15]. All these data structures require $O(1)$ time for a *make* operation and $O(\log n)$ time for each of the other heap operations, resulting in an $O(m \log n)$ running time for Dijkstra's algorithm. Dijkstra's original implementation, which stores the label function u in an array and uses no special data structure to store the frontier set, runs in $O(n^2)$ time and has the virtue of being extremely simple. Nevertheless, on sufficiently sparse graphs the heap-based implementations are faster, both in theory and in practice.

E. L. Johnson [8] and D. B. Johnson [7] adapted the implicit heaps of Williams to the shortest path problem. With a proper choice of parameters, their data structure reduces the time for an insert or decrease operation to $O(\log_{(m/n+1)} n)$ while increasing the time of a delete min operation to $O(\frac{m}{n} \log_{(m/n+1)} n)$. This results in a running time of $O(m \log_{(m/n+1)} n)$ for Dijkstra's Algorithm, matching Dijkstra's $O(n^2)$ bound on very densegraphs $(m = \Omega(n^2))$ and matching the $O(m \log n)$ bound for classical heap-based methods on sparse graphs. Details of this algorithm can be found in [14].

Recently Fredman and Tarjan [5] proposed a new implementation of heaps called Fibonacci heaps that further reduce the running time of Dijkstra's algorithm. Fibonacci heaps have a time bound of $O(\log n)$ for delete min and $O(1)$ for all the other heap operations. These time bounds are amortized; that is, the running time of an arbitrary sequence of heap operations is bounded by the sum of their amortized time bounds, though individual operations in the sequence may exceed their bounds. With Fibonacci heaps, Dijkstra's algorithm runs in $O(n \log n + m)$ time.

The $O(n \log n + m)$ time bound for Dijkstra's algorithm is best possible assuming only binary comparisons of path lengths are used, since in the worst case every edge must be examined and Dijkstra's algorithm can be used to sort $n - 1$ numbers. However, this does not preclude the existence of a faster algorithm for the shortest path problem; it only means that such an algorithm cannot be obtained by divising an even more clever implementation of Dijkstra's algorithm.

Fibonacci heaps can also be used to reduce the time necessary to solve other network optimization problems, most notably the minimum spanning tree problem [14]. Fredman and Tarjan obtained a time bound for solving this problem of $O(m\beta(m,n))$, improved from $O(m \log \log_{(m/n+1)} n)$, where $\beta(m,n) = \min\{i \mid \log^{(i)} n \leq m/n\}$. Gabow and Galil (private communication) have further reduced the time to $O(m \log \beta(m,n))$. Whether there is a linear-time algorithm for the minimum spanning tree problem remains unknown.

REFERENCES

1. M. R. Brown, "Implementation and analysis of binomial queue algorithms", SIAM J. Comput. 7 (1978), 298,319.
2. C. A. Crane, "Linear lists and priority queues as balanced binary trees," Techn. Rep. STAN-CS-72-259 Computer Science Dept., Stanford Univ., Stanford, CA, (1972).
3. D. de Champeaus, "Bidirectional search again," J. Assoc. Comput. Mach. 30 (1983), 22-32.
4. E. W. Dijkstra, "A note on two problems in connexion with graphs," Numer. Math.. 1 (1959), 269-271.
5. M. L. Fredman and R. E. Tarjan, "Fibonacci heaps and their uses in improved network optimization algorithms," Proc. 25th Annual IEEE Symp. on Found. of Comp. Sci. (1984), to appear.
6. P. E. Hart, N. J. Nilsson, and B. Rafael, "A formal basis for the heuristic determination of minimum cost paths," IEEE Trans. Sys. Sci. and Cyb. SSC-4 (1968), 100-107.
7. D. B. Johnson, "Efficient algorithms for shortest paths in sparse networks, J. Assoc. Comput. Mach. 24 (1977), 1-13.
8. E. L. Johnson, "On shortest paths and sorting," Proc. ACM Annual Conf., Boston, AM (1972), 510-517.
9. D. E. Knuth, The Art of Computer Programming, Vol. 3: Sorting and Searching, Addison-Wesley, Reading, MA, (1973).
10. E. L. Lawler, M. G. Luby, and B. Parker, "Finding shortest paths in very large networks," Proc. WG '83 International Workshop on Graph-theoretic Concepts in Computer Science, Osnabruck, West Germany, (1983), 184-199.

11. T.A.J. Nicholson, "Finding the shortest route between two points in a network," Computer J. 9 (1966), 275-280.

12. I. Pohl, "Bi-directional search," Machine Intelligence 6, B. Meltzer and D. Michie, eds., Edinburgh Univ. Press, Edinburgh, (1971), 124-140.

13. R. Sedgewick and J. S. Vitter, "Shortest paths in Euclidean graphs," submitted for publication.

14. R. E. Tarjan, Data Structures and Network Algorithms, CBMS 44, Society for Industrial and Applied Mathematics, Philadelphia, PA, (1983).

15. J. Vuillemin, "A data structure for manipulating priority queues," Comm. ACM 21 (1978), 309-314.

16. J.W.J. Williams, "Algorithm 232: Heapsort," Comm. ACM 7 (1964), 347-348.

THE BINDING NUMBER OF LEXICOGRAPHIC PRODUCTS OF GRAPHS

Jianfang Wang
Institute of Applied Mathematics,
Academia Sinica, Beijing, China

Songlin Tian
Institute of Applied Mathematics,
Academia Sinica, Beijing, China

Jiuqiang Liu
Shandong College of Oceanography,
Qiangdao, China

ABSTRACT

In this paper, the binding numbers of lexicographic products of graphs are obtained.

1. Introduction

In [5] Woodall introduced the concept of the binding number of a graph and presented some relations between the binding number and other invariants of a graph. In [2] Kane, Mohanty and Males determined the binding numbers of some products of graphs. In this paper, we obtain some formulas for the binding numbers of lexicographic products of graphs.

1. Terminology and Preliminaries

We follow the terminology and notation of [1], unless otherwise specified.

Definition 1 For graphs G and H, the lexicographic product of G and H, denoted by G(H), is defined by

$$V(G(H)) = V(G) \times V(H),$$
$$E(G(H)) = \{(u_i,v_j)(u_k,v_\ell) \mid u_iu_k \in E(G) \text{ or } u_i = u_k,\ v_jv_\ell \in E(G)\}.$$

Definition 2. For a graph G and $x \in X \subseteq V(G)$, let

$$\Gamma_G(x) = \{v \mid xv \in E(G)\}, \quad \Gamma_G(X) = \bigcup_{x \in X} \Gamma_G(x),$$

$$F(G) = \{X \mid \phi \neq X \subseteq V(G) \text{ and } \Gamma_G(X) \neq V(G)\}.$$

If there is no confusion, we denote these sets by $\Gamma(x)$, $\Gamma(X)$ and F, respectively.

Definition 3 The binding number bind(G) of a graph G is defined by $\text{bind}(G) = \min_{X \in F} \frac{|\Gamma(X)|}{|X|}$. The set X is called a realizing set of G if $x \in F$ and $\frac{|\Gamma(X)|}{|X|} = \text{bind}(G)$. Further, X is called a maximum realizing set of G if

$$|X| = \max\{|Y| \mid Y \text{ is a realizing set of } G\}.$$

Let $G = G_1(G_2)$ where $V(G_1) = \{v_1, v_2, \cdots, v_m\}$. For every $X \in F$, let $V_i = \{v_i\} \times V(G_2)$, $X_i = X \cap V_i$, $A = \{i | X_i = \phi, \ 1 \le i \le m\}$, $B = \{i | X_i \ne \phi, \ 1 \le i \le m\}$, $\tilde{B} = \{i | i \in B, X_i \ne V_i\}$ and

$$S = \{X | X \in F, \ \text{if} \ i \in B \ \text{and} \ v_j \in \Gamma_{G_1}(v_i) \ \text{then} \ j \in A\}.$$

If G_2 is a bipartite graph with maximal independent sets C and D, $|C| \ge |D|$, let $C_i = \{v_i\} \times C$, $D_i = \{v_i\} \times D$ and

$$T = \{X | X \in F, \ \text{if} \ i \in B \ \text{then} \ X_i = C_i\}.$$

Proposition 1 ([3, Lemma 3.1])

If G is a graph with minimum degree ρ, $X \subseteq V(G)$ and $\Gamma(X) \ne V(G)$, then $|X| \le |V(G)| - \rho$.

Proposition 2 (Woodall (5), Corollary 7.1)

If G is a graph on n vertices with minimum degree ρ, then $\text{bind}(G) \le \frac{n-1}{n-\rho}$.

Proposition 3

If $G = G_1(G_2)$ and X is a realizing set of G, then $X \in S$.

Proof:

Suppose $i \in B$, $j \in B$ and $v_j \in \Gamma_{G_1}(V_i)$. Then $V_i \subseteq \Gamma(X)$. Set $Y = (X - X_i) \cup V_i$. Then $\Gamma(Y) = \Gamma(X)$ and $|Y| = |X| + |V_i - X_i| > |X|$. Therefore $\frac{|\Gamma(Y)|}{|Y|} < \frac{|\Gamma(X)|}{|X|}$, which is a contradiction. □

Proposition 4

If $G = G_1(K_{a,b})$, $a \geq b$, and X is a realizing set of G, then $X \in T$.

The proof of Proposition 4 is straightforward and is consequently omitted.

3. Main Results

Theorem 1. Let G be a graph with $|V(G)| = m \geq 2$, $1 \leq \rho(G) \leq 2$, and $\text{bind}(G) \geq 1$. Further, let L_n be a path on $n \geq 2$ vertices. If X_o is a maximum realizing set of G, then

$$\text{bind}(L_n(G)) = \begin{cases} \dfrac{mn-1}{mn-(m+1)}, & \text{if } \rho(G) = 1 \\[2ex] \dfrac{2m-1}{m-2}, & \text{if } \rho(G) = 2, \text{bind}(G) > 1, \\ & n = 2 \\[2ex] \min\{\dfrac{|X_o|+m}{|X_o|}, \dfrac{2m-1}{m-2}\}, & \text{if } \rho(G) = 2, \text{bind}(G) = 1, \\ & n = 2 \\[2ex] \min\{\dfrac{n-1}{n-2}, \dfrac{mn-1}{mn-(m+2)}\}, & \text{if } \rho(G) = 2, \text{bind}(G) > 1, \\ & n \geq 3 \\[2ex] \min\{\dfrac{mn-1}{mn-(m+2)}, \dfrac{|X_o| + (n-1)m}{|X_o|+(n-2)m}\}, & \text{if } \rho(G) = 2, \\ & \text{bind}(G) = 1, \\ & n \geq 3 \end{cases}$$

Proof:

Let $L_n = v_1v_2 \cdots v_n$ and let $X \in S$. We consider two cases.

Case 1 $\rho(G) = 1$.

If $\tilde{B} = \phi$, then, for any $j \in B$, $X_j = V_j$. Therefore, there exists an integer $i \in A$ such that $V_i \subseteq \Gamma(X)$. So $|\Gamma(X) \geq |X| + |V_i| = |X| + m$.

Thus assume $B \neq \phi$, where, say, $j \in B$. Since $X \in S$, without loss of generality, we assume that $j - 1 \in A$. Then $V_{j-1} \subseteq \Gamma(X)$. Since $\text{bind}(G) \geq 1$, this implies that $|\Gamma(X) \cap V_i| \geq |X_i|$, if $i \in B$. Thus

$$|\Gamma(X)| \geq |X| + |V_{j-1}| = |X| + m.$$

By Proposition 1,

$$\frac{|\Gamma(X)|}{|X|} \geq \frac{|X|+m}{|X|} \geq \frac{mn-1}{mn-(m+1)} .$$

Propositions 2 and 3 imply that $\text{bind}(L_n(G)) = \dfrac{mn-1}{mn-(m+1)}$.

Case 2 $\rho(G) = 2$.

We distinguish three subcases.

Subcase (i) $\tilde{B} = \phi$.

In this subcase, for any $j \in B$, $|X_j| = |V_j|$. Hence $n \geq 3$ and

$$\frac{|\Gamma(X)|}{|X|} \geq \frac{(|B|+1)m}{|B|m} \geq \frac{n-1}{n-2} .$$

Subcase (ii) $\phi \neq \tilde{B} \subseteq \{1,n\}$.

Suppose $j \in B - \{1,n\}$. Since $X \in S$, then $j - 1 \in A$, $j + 1 \in A$ and $V_{j-1} \cup V_{j+1} \subseteq \Gamma(X)$. Since $\text{bind}(G) \geq 1$, for any $i \in B$, $|\Gamma(X) \cap V_i| \geq |X_i|$. Therefore

$$|\Gamma(X)| \geq |X| + 2m > |X| + m + 1.$$

By Proposition 1,

$$\frac{|\Gamma(X)|}{|X|} \geq \frac{|X| + m + 1}{|X|} \geq \frac{mn - 1}{mn-(m+2)} .$$

Subcase (iii) $\phi \neq \tilde{B} \subseteq \{1,n\}$.

(a). $\phi \neq B \neq \{1,n\}$.

Suppose $B = \{1\}$. If $|\Gamma(X_1) \cap V_1| \geq |X_1| + 1$, since bind(G) ≥ 1, then $|\Gamma(X)| \geq |X| + m + 1$ and so $\frac{|\Gamma(X)|}{|X|} \geq \frac{mn-1}{mn-(m+2)}$.

If $|\Gamma(X_1 \cap V_1| = |X_1|$, then bind(G) = 1 and X_1 is a realizing set of G.

Therefore

$$\frac{|\Gamma(X)|}{|X|} = \begin{cases} \dfrac{|X_1| + |\Gamma(X - X_1)| + m}{|X_1| + |X - X_1|}, & \text{if } n \geq 3, \ 3 \in A \\ \dfrac{|X_1| + |\Gamma(X-X_1)|}{|X_1| + |X - X_1|}, & \text{if } n \geq 3, \ 3 \in B \\ \dfrac{|\Gamma(X_1)|}{|X_1|}, & \text{if } n = 2 \end{cases}$$

$$\geq \begin{cases} \dfrac{|X_1|+|X-X_1|+2m}{|X_1|+|X-X_1|}, & \text{if } n \geq 3, \ 3 \in A \\ \dfrac{|X_1|+|X-X_1|+m}{|X_1|+|X-X_1|}, & \text{if } n \geq 3, \ 3 \in B \\ \dfrac{|X_1|+m}{|X_1|}, & \text{if } n = 2 \end{cases}$$

$$\geq \begin{cases} \dfrac{|X_1|+(n-3)m+2m}{|X_1|+(n-3)m}, & \text{if } n \geq 3, \quad 3 \in A \\[2ex] \dfrac{|X_1|+(n-2)m+m}{|X_1|+(n-2)m}, & \text{if } n \geq 3, \quad 3 \in B \\[2ex] \dfrac{|X_1|+m}{|X_1|}, & \text{if } n = 2 \end{cases}$$

$$\geq \quad (|X_o|+(n-1)m)/(|X_o|+(n-2)m).$$

(b). $\tilde{B} = \{1,n\}$,

If there exists an integer $i \in \{1,n\}$ such that $|\Gamma(X_i) \cap V_i| \geq |X_i| + 1$, then, similar to (a), we have

$$\frac{|\Gamma(X)|}{|X|} \geq \frac{mn-1}{mn-(m+2)} .$$

Otherwise,

$$\frac{|\Gamma(X)|}{|X|} = \begin{cases} \dfrac{|X_1|+|X_n|+(n-2)m}{|X_1|+|X_n|}, & \text{if } n = 3,4 \\[3ex] \dfrac{|X_1| + |X_n|+2m+|\Gamma(X-X_1-X_n) \cap (\bigcup_{i=3}^{n-2} V_i)|}{|X_1| + |X_n| + |X - (X_1 X_n)|}, & \text{if } n \geq 5 \end{cases}$$

Similar to (a),

$$\frac{|\Gamma(X)|}{|X|} \geq \begin{cases} \dfrac{2|X_o|+m}{2|X_o|}, & \text{if } n = 3 \\[2ex] \dfrac{2|X_o|+m(n-2)}{2|X_o|+(n-4)m}, & \text{if } n \geq 4. \end{cases}$$

Thus, by the definition for Proposition 3, Theorem 1 holds.

□

We now consider $C_n(G)$ for a cycle C_n of length $n(\geq 3)$.

Theorem 2. Let G be a graph, $|G| = m \geq 2$, $1 \leq \rho(G) \leq 2$, $\text{bind}(G) \geq 1$ and X_o be a maximum realizing set of G. Then

$$\text{bind}(C_n(G)) = \begin{cases} \dfrac{mn-1}{mn-(2m+1)}, & \text{if } \rho(G) = 1 \\[2ex] \dfrac{3m-1}{m-2}, & \text{if } \rho(G) = 2, \ \text{bind}(G) > 1, \ n = 3 \\[2ex] \min\{\dfrac{|X_o|+2m}{|X_o|}, \dfrac{3m-1}{m-2}\}, & \text{if } \rho(G) = 2, \\ & \quad \text{bind}(G) = 1, \ n = 3 \\[1ex] \min\{\dfrac{n-1}{n-3}, \dfrac{mn-1}{mn-(2m+2)}\}, & \text{if } \rho(G)=2, \\ & \quad \text{bind}(G) > 1, \ n \geq 4 \\[1ex] \min\{\dfrac{(n-1)m+|X_o|}{(n-3)m+|X_o|}, \dfrac{mn-1}{mn-(2m+2)}\}, & \text{if } \rho(G) = 2, \\ & \quad \text{bind}(G) = 1, \ n \geq 4. \end{cases}$$

The proof of Theorem 2 is similar to that of Theorem 1 and hence omitted.

Theorem 3. If $m, n \geq 2$ then

$$\mathrm{bind}(L_n(L_m)) = \begin{cases} \dfrac{mn-1}{mn-1(m+1)}, & \text{if } m \text{ is even} \\[2ex] \min\{\dfrac{3m-1}{m+1}, \dfrac{mn-1}{mn-(m+1)}, \dfrac{m(n-1)+\frac{m-1}{2}}{m(n-2)+\frac{m+1}{2}}\}, & \text{if } m \text{ is odd, } n \neq 3,5 \\[2ex] \min\{\dfrac{5m-1}{4m-1}, \dfrac{7m-3}{3m+3}, \dfrac{4m+\frac{m-1}{2}}{3m+\frac{m+1}{2}}\}, & \text{if } m \text{ is odd } n = 5 \\[2ex] \min\{\dfrac{2m-1}{m+1}, \dfrac{5m-1}{3m+1}, \dfrac{3m-1}{2m-1}\}, & \text{if } m \text{ is odd } n = 3 \end{cases}$$

Proof: Let $X \in S$, $B_1 = \{i | i \in \tilde{B} \text{ and } |\Gamma(X_i) \cap V_i| < |X_i|\}$ and $B_2 = B - \tilde{B}$. It is clear that, if $i \in B_1$, then m is odd and $|X_i| = \frac{m+1}{2}$ and $|\Gamma(X_i) \cap V_i| = \frac{m-1}{2}$.

If $B_1 = \phi$, then $|\Gamma(X)| \geq |X| + m$. By Proposition 1,

$$\frac{|\Gamma(X)|}{|X|} \geq \frac{|X|+m}{|X|} \geq \frac{mn-1}{mn-(m+1)} .$$

If $B_1 \neq \phi$, we distinguish two cases.

Case 1. $B_2 \neq \phi$.

If $B - (B_1 \cup B_2) \neq \phi$ or $|B - B_2| = |B_1| \geq 2$, then

$$|\Gamma(X)| \geq |B_2|m + \sum_i |B - B_2||X_i| - |B_1| + |B - B_2|m$$

$$\geq |X| + |B-B_2|m - |B_1| \geq |X| + m.$$

By Proposition 1,

$$\frac{|\Gamma(X)|}{|X|} \geq \frac{|X|+m}{|X|} \geq \frac{mn-1}{mn-(m+1)}$$

If $B = B_1 \cup B_2$ and $|B_1| = 1$, then

$$\frac{|\Gamma(X)|}{|X|} \geq \frac{|(B_2|+1+\frac{m-1}{2}}{|B_2|m+\frac{m+1}{2}} \geq \frac{(n-1)m+\frac{m-1}{2}}{(n-2)m+\frac{m+1}{2}}$$

Case 2. $B_2 = \phi$.

In this case,

$$|\Gamma(X)| \geq \begin{cases} (|B|-1)m+|X| - |B_1|, & \text{if } n \text{ is odd, } B = \{1,3,\cdots,n\} \\ |B|m+|X| - |B_1|, & \text{otherwise.} \end{cases}$$

Therefore $(|B|-1)m + |X| - |B_1| \geq |X| + m$. If $n \geq 7$, $B = \{1,3, \cdots , n\}$ or if $n = 5$, $B = \{1,3,5\} \neq B_1$.

Now $|B|m + |X| - |B_1| \geq |X| + m$, if $|B| \neq |B_1|$ or $|B_1| \geq 2$.

If $n = 5$, then $B = \{1,3,5\} = B_1$ and $\frac{|\Gamma(X)|}{|X|} = \frac{7m-3}{3m+3}$.

If $n = 3$, then $B = \{1,3\} = B_1$ and $\frac{|\Gamma(X)|}{|X|} = \frac{2m-1}{m+1}$.

If $n = 3$, $B = \{1,3\}$, $\phi \neq B_1 \neq B$, say $B_1 = \{1\}$, then

$|\Gamma(X_3) \cap V_3| \geq |X_3| + 1$, and so $|\Gamma(X)| \geq |X| + m$ and

$$\frac{|\Gamma(X)|}{|X|} \geq \frac{|X|+m}{|X|} \geq \frac{3m-1}{2m-1}.$$

If $|B| = |B_1| = 1$, then $\frac{|\Gamma(X)|}{|X|} \geq \frac{3m-1}{m+1}$.

Thus, by Proposition 3 and the definition of binding number, the theorem holds. □

We now consider the lexicographic product of a cycle with a path.

Theorem 4. If $m \geq 2$, $n \geq 3$, then

$$\text{bind}(C_n(L_m)) = \begin{cases} \frac{mn-1}{mn-(2m+1)}, & \text{if } m \text{ is even} \\ \min\{\frac{5m-1}{m+1}, \frac{mn-1}{mn-(2m+1)}, \frac{m(n-1)+\frac{m-1}{2}}{m(n-3)+\frac{m+2}{2}}\}, & \text{if } m \text{ is odd } n \neq 4,6. \\ \min\{\frac{3m-1}{m+1}, \frac{4m-1}{2m-1}, \frac{7m-1}{3m+1}\}, & \text{if } m \text{ is odd } n = 4. \\ \min\{\frac{3m-1}{m+1}, \frac{6m-1}{4m-1}, \frac{11m-1}{7m+1}\}, & \text{if } m \text{ is odd } n = 6. \end{cases}$$

The proof of Theorem 4 is similar to that of Theorem 3 and hence omitted.

Theorem 5. If $n \geq 2$, $a \geq b \geq 1$, then

$$\text{bind}(L_n(K_{a,b})) = \begin{cases} \frac{(n-1)(a+b)+b}{(n-2)(a+b)+a}, & \text{if } n \text{ is even} \\ \min\{\frac{(n-1)a+2nb}{(n+1)a}, \frac{(n-1)(a+b)+b}{(n-2)(a+b)+a}\}, & \text{if } n \text{ is odd.} \end{cases}$$

Proof: Let $L_n = v_1v_2 \cdots v_n$, $X \in S \cap T$, $\ell = |B|$ and $h = |B - B|$. By the definition of T, if $i \in B$, then $|X_i| = |C_i| = a$. By definition of S, if $i \in B$, then $i - 1$, if $i+1 \in A$ (if $i = 1$, then $i+1 \in A$, if $i = n$, then $i-1 \in A$). Therefore,

$$\frac{|\Gamma(X)|}{|X|} \geq \begin{cases} \frac{(h+1)(a+b)}{h(a+b)} \geq \frac{n-1}{n-2}, & \text{if } \ell = 0 \\ \frac{(n-1)a+2nb}{(n+1)a}, & \text{if } h = 0, \ n \text{ is odd } B = \{1,3,\cdots,n\}. \\ \frac{\ell(a+b)+\ell b}{\ell a} = \frac{a+2b}{a}, & \text{if } h = 0, \\ & n \text{ is even or } n \text{ is odd } B \neq \{1,3,\cdots,n\}. \\ \frac{\ell(a+b)+h(a+b)+\ell b}{h(a+b)+\ell a}, & \text{if } \ell \geq 1, \ h \geq 1, \ 2\ell + h \leq n \end{cases}$$

We can prove that

$$\frac{\ell(a+b)+h(a+b)+\ell b}{h(a+b)+\ell a} \geq \frac{(n-1)(a+b)+b}{(n-2)(a+b)+a}, \quad \text{if } \ell \geq 1, h \geq 1, 2\ell + h \leq n.$$

Hence, by Propositions 3 and 4 and the definition of binding number, Theorem 5 follows.

Theorem 6. If $n \geq 3$ and $a \geq b \geq 1$, then

$$\text{bind}(C_n(K_{a,b})) = \begin{cases} \frac{(n-1)(a+b)+b}{(n-3)(a+b)+a}, & \text{if } n \text{ is even} \\ \min\{\frac{(n-1)(a+b)+b}{(n-3)(a+b)+a}, \frac{(n-1)b+(n+1)(a+b)}{(n-1)a}\}, & \\ & \text{if } n \text{ is odd.} \end{cases}$$

The proof of Theorem 6 is similar to that of Theorem 5.

Theorem 7. Let G be a graph with $|V(G)| = n \geq 2$, $1 \leq \rho(G) \leq 2$ and $\text{bind}(G) \geq 1$. If X_o is a maximum realizing set of G and $a \geq b \geq 1$, then

$$\text{bind}(K_{a,b}(G)) = \begin{cases} \min\{\frac{a+b-1}{a-1}, \frac{n(a+b)-1}{n(a+b)-(nb+\rho(G))}\} & \text{if } \text{bind}(G) > 1 \\ & \text{or } \text{bind}(G) = 1 \text{ and } \rho(G) = 1. \\ \min\{\frac{n(a+b)-1}{n(a+b)-(bn+2)}, \frac{|X_o|+bn+(a-1)n}{|X_o|+(a-1)n}\}, & \\ & \text{if } \text{bind}(G) = 1,\ \rho(G) = 2. \end{cases}$$

Proof: Let $X \in S$. We consider two cases.

Case 1: $B = \phi$.

In this case

$$\frac{|\Gamma(X)|}{|X|} \geq \min_{\substack{1 \leq \ell \leq a-1 \\ 1 \leq t \leq b-1}} \{\frac{\ell n+bn}{\ell n}, \frac{tn+an}{tn}\} = \frac{a+b-1}{a-1}.$$

Case 2: $|B| = \ell \geq 1$.

(i). There exists an integer $j \in \tilde{B}$ such that $|\Gamma(X_j) \cap V_j| \geq |X_j| + 1$.

Since $\rho(K_{a,b}(G)) = bn + \rho(G)$, $\text{bind}(G) \geq 1$. Therefore, $|\Gamma(X)| \geq |X| + bn + 1 \geq |X| + \rho(K_{a,b}(G)) - 1$. By Proposition 1,

$$\frac{|\Gamma(X)|}{|X|} \geq \frac{|X|+\rho(K_{a,b}(G))-1}{|X|} \geq \frac{n(a+b)-1}{n(a+b)-(nb+\rho(G))}.$$

(ii). $|\Gamma(X_j) \cap V_j| = |X_j|$, if $j \in \tilde{B}$.

In this subcase, $\text{bind}(G) = 1$. Thus

$$\frac{|\Gamma(X)|}{|X|} \le \min_{\substack{1 \le K_1+\ell \le a \\ 1 \le K_2+\ell \le b}} \left\{ \frac{\sum_{j=1}^{\ell} |X_j|+bn+K_1 n}{\sum_{j=1}^{\ell} |X_j|+K_1 n}, \frac{\sum_{j=1}^{\ell} |X_j|+an+K_2 n}{\sum_{j=1}^{\ell} |X_j|+K_2 n} \right\}$$

$$= \frac{|X_o|+bn+(a-1)n}{|X_o|+(a-1)n}.$$

Hence, by Proposition 3, Theorem 7 follows. □

Corollary 1. If $m \ge 3$, n is even and $a \ge b \ge 1$, then

(i)

$$\text{bind}(K_{a,b}(C_m)) = \begin{cases} \min\{\frac{a+b-1}{a-1}, \frac{m(a+b)-1}{m(a+b)-(bm+2)}\}, & \text{if } m \text{ is odd} \\ \min\{\frac{2a+2b-1}{2a-1}, \frac{m(a+b)-1}{m(a+b)-(bm+2)}\}, & \text{if } m \text{ is even.} \end{cases}$$

(ii)

$$\text{bind}(K_{a,b}(L_n)) = \frac{n(a+b)-1}{n(a+b)-(bn+a)}.$$

We conclude the paper by presenting a formula for the binding number of the lexicographic product of this complete bipartite graphs.

Theorem 8. Let $G = K_{a,b}(K_{c,d})$, $a \ge b \ge 1$, and $c \ge d \ge 1$. Then

$$\text{bind}(G) = \begin{cases} \min\{\frac{a+b-1}{a-1}, \frac{bc+bd+ad}{ac}\}, & \text{if } c \ge \frac{a+b}{a} d \\ \min\{\frac{a+b-1}{a-1}, \frac{(a+b)(c+d)-C}{a(c+d)-d}\}, & \text{if } c < \frac{a+b}{a} d. \end{cases}$$

Proof: Suppose Y_a, Y_b are two maximal independent sets of G, and $|Y_a| = a$ and $|Y_b| = b$. Set $W = \{X | X \in F, X \subseteq Y_a \times V(K_{c,d})\}$. If X_o is a realizing set of G, then $X_o \in W$.

Let $X \in S \cap T \cap W$. We consider two cases.

Case 1. $\tilde{B} = \phi$.

$$\text{Then} \quad \frac{|\Gamma(X)|}{|X|} \geq \frac{|B|(c+d)+b(c+d)}{|B|(c+d)} \geq \frac{a+b-1}{a-1} \quad .$$

Case 2. $\tilde{B} \neq \phi$.

Let $|\tilde{B}| = l$ and $|B-\tilde{B}| = k$. Then

$$\frac{|\Gamma(X)|}{|X|} \geq \min_{\substack{l\geq k+\ell\leq a \\ \ell\geq 1 \\ k\geq o}} \left\{\frac{\ell d+k(c+d)+b(c+d)}{\ell c+k(c+d)}\right\}$$

$$\geq \begin{cases} \dfrac{bc+bd+ad}{ac}, & \text{if } c \geq \dfrac{a+b}{a} d \\[2ex] \dfrac{(a+b)(c+d)-c}{a(c+d)-d}, & \text{if } c < \dfrac{a+b}{a} d. \end{cases}$$

Hence, by Proposition 3 and 4, the theorem follows. □

REFERENCE

[1] F. Harary, Graph Theory. Addison-Wesley, Reading, MA (1969).

[2] V. G. Kane, S. P. Mohanty and R. S. Males, Product graphs and binding number. Ars Combinatoria 11(1981) 201-224.

[3] S. Tian, J. Wang and J. Liu, The binding number of the tension product of some graphs (to appear).

[4] J. Wang, S. Tian and J. Liu, The binding number of product graphs (to appear).

[5] D. R. Woodall, The binding number of a graph and its Anderson number. _J. Combinatorial Theory Ser. B._ 15(1973) 225-255.

SPANNING TREES IN PROGRAM FLOWGRAPHS

R.W. Whitty
South Bank Polytechnic
Borough Road, London SE1 0AA, England

ABSTRACT

Sequential programs are modelled as directed graphs. The problem is discussed of representing programs by 2 directed trees which describe respectively the True and False outcomes of the program predicates. It is shown to be a special case of Edmonds' disjoint branching theorem and we give a new proof and algorithm for this special case.

1) INTRODUCTION

We tackle the following problem: when can the control flow of a sequential program be represented by 2 directed trees or branchings so that all True outcomes of program predicates are contained in one tree and all False outcomes are contained in the other. This problem has applications in adapting sequential programs to run concurrently on 2 processors and, in view of the well-known Level Algorithm of Hu [6], to multiprocessor-scheduling the tasks in algorithm flowcharts. Of course, such applications depend very much on the data-flow properties of the programs or algorithms but we shall concentrate on the purely structural aspect of the problem. We shall deal with a rather abstract graph theoretic model of program control flow, the *CGK-graph*, first developed in [1]. For details of the role of this model in mathematical structured programming and control flow analysis see [1,4,5,10].

We now give some necessary definitions and notation:

Let G be a directed graph. We allow G to have multiple arcs and self-loops. Arcs from node x to node y are denoted by the pair <x,y> and x is called the *tail* and y the *head* of these arcs. The number of arcs of which x is the head is the *indegree* of x, denoted id(x); the number of arcs of which x is the tail is the *outdegree* of x, denoted od(x). G\x denotes the set of all nodes of G except x. For a subgraph G' of G, those nodes $x \in G$ with id(x) smaller in G' than in G are called *entry nodes* of G'; arcs in G\G' with head x are called *entry arcs* of G'. Similarly, nodes $x \in G$ with od(x) smaller in G' than in G are *exit nodes* of G' and arcs in G\G' with tail x are *exit arcs* of G'.

For any node x in G, a _spanning tree to x_ in G is a spanning tree of G in which all nodes have a path to x. More generally, for a set X of n nodes of G, a _branching to X_ in G is a spanning collection of n trees in G such that exactly one tree is directed to each node of X.

2) A MODEL OF SEQUENTIAL PROGRAMS

The CGK-graph is a formalisation and abstraction of the normal program or algorithm flowchart. It contains only the most essential information about the flow of control, namely the relationship between the program predicates.

2.1 Definition.

A CGK-graph is a triple F=(G,a,z) consisting of a directed graph G, the _graph_ of F, together with 2 distinguished nodes of G. a and z, satisfying:

1) The nodes of G\z (the _predicate nodes_) have outdegree 2; od(z) = 0.
2) The distinguished node a (the _start node_) has a path to all other nodes of G; all nodes of G have a path to the distinguished node z (the _stop node_).

As will be illustrated, analysis of CGK-graphs F=(G,a,z) can often be reduced to a study of the minimal _subflowgraphs_ of F, i.e. subgraphs of G which are themselves the graphs of CGK-graphs and which are minimal with respect to this property. Such subgraphs must necessarily have a single exit node in G.

2.2 Definition.

A CGK-graph F=(G,a,z) is called a CGK-irreducible if G has no proper subgraph which is the graph of a CGK-graph.

Figure 1 shows the graphs of all CGK-irreducibles with up to 3 predicate nodes. By symmetry, there is exactly 1 triple (G,a,z) corresponding to each of the graphs with up to 2 predicate nodes. The graphs with 3 predicate nodes collectively produce 6 non-isomorphic triples, so there are 10 CGK-irreducibles with 3 predicate nodes or less.

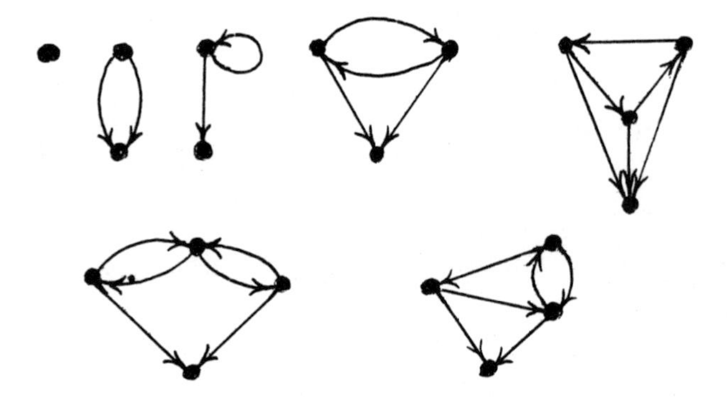

Figure 1

A second characterisation of the CGK-irreducibles, which will be used in the next section, can be made in terms of an explicit construction of all the graphs of CGK-irreducibles of a certain order.

2.3 Definition.

Let F=(G.a,z) be a CGK-irreducible. Define the _generation operation_ as follows:

let <x,y> be any arc in G and let v be any node in G\y. Then a new graph G' is _generated_ from G by adding a new node u in <x,y> and a new arc <u,v> (see Figure 2).

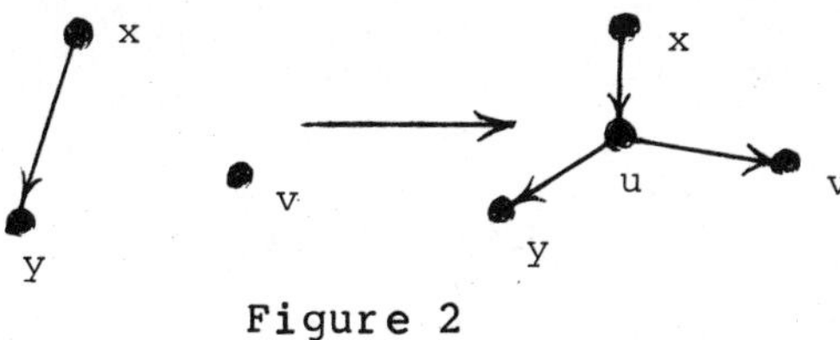

Figure 2

The following important result is proved in [10]:

2.4 Theorem.

Let F=(G,a,z) be a CGK-graph with n predicate nodes, $n \geqslant 2$. Then F is a CGK-irreducible if and only if G is generated by the graph of some CGK-irreducible with n-1 predicate nodes.

3) TREE FACTORISATIONS

We can now give a formal statement of the problem proposed in Section 1:

3.1 Definition.

Let F=(G,a,z) be a CGK-graph and let T_1, T_2 be spanning trees of G directed to z. We say $T=\{T_1,T_2\}$ is a _tree factorisation_ of F if every arc of G lies in one and only one of T_1 and T_2.

As we shall see, finding tree factorisations in CGK-graphs reduces essentially to finding tree factorisations in CGK-irreducibles.

3.2 Proposition.
If $F=(G,a,z)$ is a CGK-irreducible with at least 2 predicate nodes then there exists a tree factorisation of F.

Proof
We argue by induction on n, the number of predicate nodes of G. For $n = 2$ the result is true from Figure 1. So suppose that the result holds for some N, $N \geqslant 2$ and let $n = N+1$. Now by Theorem 2.4, G is generated by the graph G' of some CGK-irreducible F' with N predicate nodes by adding $\langle u,v \rangle$ in arc $\langle x,y \rangle$ as in Definition 2.3. By inductive hypothesis, F' has a tree factorisation $T'=\{T_1',T_2'\}$. Without loss of generality, assume $\langle x,y \rangle \in T_1'$. Now G differs from G' only in as much as a new node u has been inserted in arc $\langle x,y \rangle$ so that $\langle x,u \rangle$ and $\langle u,y \rangle$ replace $\langle x,y \rangle$ as arcs in G and a new arc $\langle u,v \rangle$ is in G. thus we may define

$T_1 = T_1'$ with $\langle x,y \rangle$ deleted and u and $\langle x,u \rangle$, $\langle u,y \rangle$ added

T_2 consists of all other arcs of G.

It is clear that $\{T_1,T_2\}$ is the required tree factorisation.

It is easy to see how the proof of Proposition 3.2 implies a polynomial-time algorithm for constructing a tree factorisation in any CGK-irreducible.

3.3 Theorem.

Let F=(G,a,z) be any CGK-graph. Then the following are equivalent:

1) There exists a tree factorisation of F.
2) If S is any proper subset of the set of predicate nodes of G and $z \in S$ then S has at least 2 entry arcs in G.
3) Every predicate node in G has at least 2 arc-disjoint paths to z.
4) Every cycle of G has at least 2 exit nodes.

Proof

$1 \Rightarrow 2$ is clear, as noted by Edmonds [3] for a much more general theorem on edge-disjoint branchings.
$2 \Rightarrow 3$ is a reformulation by Lovasz [7] of Edmonds' theorem.
$3 \Rightarrow 4$ is clear.
$4 \Rightarrow 1$: we shall prove this by induction on n, the number of predicate noes of G, using Proposition 3.2. This will illustrate how properties and algorithms for CGK-irreducibles can be extended to apply to arbitrary CGK-graphs. For n=1, the result follows from Figure 1. So suppose $n \geq 2$ and the result holds for $N < n$. If F is a CGK-irreducible then F has a tree factorisation by Proposition 3.2. So suppose G has a proper subgraph G' which is the

graph of a CGk-irreducible F'. By Definition 2.2, G' has at most one exit node in G and, by Proposition 3.2, F' has a tree factorisation $T'=\{T_1',T_2'\}$. Now derive the graph G^* of a new CGK-graph F^* by contracting G' to a single node p in G, where entry arcs and exit arcs of G' are replaced by arcs with head and tail p, respectively. By inductive hypothesis, F^* has a tree factorisation $T^*=\{T_1^*,T_2^*\}$. Define T_1 to consist of all arcs of $G' \in T_1'$ together with the following:

For each arc $\langle x,y\rangle \in G^* \cap T_1^*$ either:

1) $x,y \neq p$ in which case $\langle x,y\rangle \in G$, so add $\langle x,y'\rangle$ to T_1.
2) $x = p$ in which case there is some node $y' \in G'$ such that $\langle x,y'\rangle \in G$, so add $\langle x,y'\rangle$ to T_1.
3) $y = p$ in which case there is some node $x' \in G'$ such that $\langle x',y\rangle \in G$, so add $\langle x',y\rangle$ to T_1

Define T_2 to consist of all other arcs of G. T_1 and T_2 are clearly disjoint spanning subgraphs and by Definition 2.1, applied to F' and F^*, and the assumption that every cycle in G has at least 2 exit nodes, T_1 and T_2 are trees directed to z in G. So $\{T_1,T_2\}$ is a tree factorisation as required.

We conclude with some remarks concerning Proposition 3.2 and Theorem 3.3:

a) Edmonds' theorem on disjoint branchings was given in [2] as an example of the intersection of 2 matroids. Min-max and weighted versions were also

given. The directional dual of this theorem may be stated as follows:

let G be a directed graph and let R_i, $i \in I$, be a collection of proper subsets of the nodes of G. There exist mutually edge-disjoint branchings B_i directed to R_i in G, $i \in I$, if and only if, for every proper subset U of the nodes of G

entry arcs of U $\geqslant$ # sets R_i contained in U.

The equivalence of 1 and 2 in Theorem 3.3 is now just the special case when I={1,2} and $R_1=R_2=z$, the stop node. In [8], Schrijver has given a result of which Edmonds' theorem is itself a corollary.

b) Edmonds [3] gave an algorithmic proof of his theorem but without considering its time complexity. In [9], Tarjan derived a new algorithm from this proof which was polynomial-time; in fact, for the special case of Theorem 3.3 it runs in O(nlogn) time. The construction given in the proot of Proposition 3.2 can be extended to a polynomial-time algorithm for Theorem 3.3 (a description of the necessary decomposition algorithms is given in [4]) which is is less efficient than Tarjan's but is very simple conceptually.

c) In programming terms, the equivalence of 1 and 4 in Theorem 3.3 means that we can represent sequential programs as proposed in Section 1 exactly when a program contains no WHILE loops or their equivalent.

d) In [10] it is proved that all predicate nodes in a CGK-irreducible have 2 node-disjoint paths to the stop node. Thus Proposition 3.2 follows trivially from Theorem 3.3(3) and it would seem that a much stronger result would hold in this important case. In fact we prove in [11] that, in any CGK-irreducible, we can find a tree factorisation $\{T_1, T_2\}$ such that, for any 2 predicate nodes x and y, there are node disjoint paths from x to z and y to z lying in T_1 and T_2 , or in T_2 and T_1 , respectively. The proof is to long to include here.

4) REFERENCES

1) D.F. Cowell, D.F. Gillies and A.A. Kaposi, Synthesis and Structural Analysis of Abstract Programs. Computer Journal 23 (1980), pp 243-247.

2) J. Edmonds, Submodular Functions, matroids and Certain Polyhedra. in Combinatorial Structures and Their Applications, R. Guy et al (eds.), Gordon and Breach, New York 1969, pp 69-87.

3) J. Edmonds, Edge-Disjoint Branchings. in Combinatorial Algorithms, R. Rustin (ed.), Algorithmic Press, New York, 1972, pp 91-96.

4) N.E. Fenton, R.W. Whitty and A.A. Kaposi, A Generalised Mathematical Theory of Structured Programming. Theor. Comput. Sci. (to appear Vol. 37, March 1985)

5) N.E. Fenton, The Structural Complexity of Flowgraphs. in Proc. Fifth Int. Conf. on the Theory and Applications of Graphs, 1984.

6) T.C. Hu, Parallel Sequencing and Assembly Line Problems. Oper. Res. 9 (1961), pp 841-848.

7) L. Lovasz, Connectivity in Digraphs. J. Combin. Theory (B) (1973) 15. pp 174-177.

8) A. Schrijver, Min-max Relations for Directed Graphs. in Bonn Workshop on Combinatorial Optimisation, A. Bachem et al (eds.) Annals of Discrete Maths 16, North Holland, 1982.

9) R.E. Tarjan, A Good Algorithm for Edge-Disjoint Branchings. Inform. Processing. Lett. 3 (1974), pp 51-53.

10) R.W. Whitty, On an Important Class of Program Flowgraphs. Submitted to SIAM J. Computing, 1984.

11) R.W. Whitty, PhD Thesis, South Bank Polytechnic, London, 1984.

ANALYSIS SITUS[†]

Robin J. Wilson
The Open University, Milton Keynes, England,
and The Colorado College, Colorado Springs, CO 80903.

ABSTRACT

In this paper we trace the development of 'analysis situs' from its origin in 1679 until the middle of the 19th century, and discuss some of the interpretations that were placed on this phrase during this period.

1. Leibniz and 'Analysis Situs'

On 8 September 1679 the philosopher and mathematician Gottfried Wilhelm Leibniz wrote a letter [1] to Christiaan Huygens in the following terms:

> I am not content with algebra, in that it yields neither the shortest proofs nor the most beautiful constructions of geometry. Consequently, in view of this I consider that we need yet another kind of analysis, geometric or linear, which deals directly with position, as algebra deals with magnitude...

Leibniz introduced the phrase *analysis situs*, meaning the analysis of position or situation, to denote this area of study. Although it has usually been used to refer to problems of a topological nature, this phrase has been interpreted in several different ways, and it

---

† A talk based on this material was presented at the special Conference Dinner.

now seems certain that topology was not what Leibniz had in mind. In this paper we consider the diverse interpretations put on the phrase 'analysis situs' by mathematicians in the 18th and 19th centuries, and attempt to explain what Leibniz intended when he introduced this phrase.

2. Euler and the Königsberg Bridges

On 26 August 1735 Leonhard Euler presented a paper [2] to the Academy of Sciences in St. Petersburg (now Leningrad) on 'the solution of a problem relating to the geometry of position'. This problem, which he believed to be widely known, was the famous *Königsberg bridges problem* which asks whether it is possible to find a route which crosses each of the seven bridges of Königsberg once and once only (see Figure 1). More generally, given any division of the river into branches and any arrangement of bridges, can one find a general method for determining whether such a route is possible?

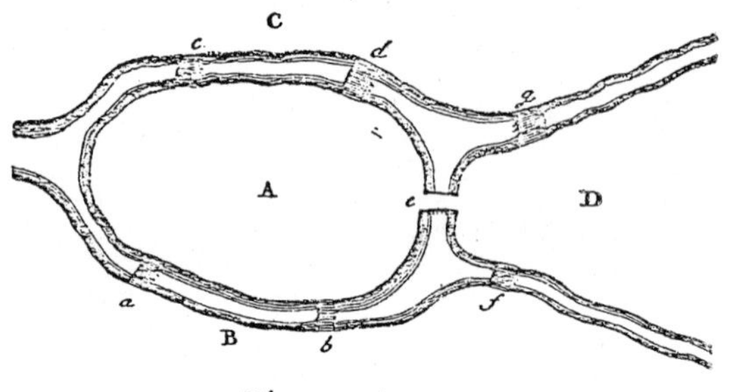

Figure 1

Euler began his paper by referring to Leibniz as follows:

> In addition to that branch of geometry which is concerned with magnitudes and which has always received the greatest attention, there is another branch, previously almost unknown, which Leibniz first mentioned, calling it the *geometry of position* [geometria situs]. This branch is concerned only with the determination of position and its properties; it does not involve measurements, or calculations made with them...

Euler went on to remark that it was not yet clear exactly which problems are relevant to the geometry of position, or what methods should be used in solving them, but he certainly regarded the Königsberg bridges as such a problem, especially since its solution 'involved only position, and no calculation was of any use'.

In order to solve the Königsberg bridges problem, Euler attempted to find a sequence of eight letters (each A, B, C or D) representing the areas of the city that one would visit while crossing the seven bridges in turn. However, since area A has five bridges emanating from it, and the other areas each have three bridges, simple reasoning shows that any such sequence of letters must involve three As, two Bs, two Cs and two Ds—a total of nine letters altogether. This shows that the Königsberg bridges problem has no solution. It is worth noting that Euler never used a graph such as that in Figure 2 to settle the Königsberg bridges problem. As far as is known, the first appearance of such a graph was in W. Rouse Ball's *Mathematical Recreations and Problems* [3] in 1892, more than 150 years after Euler's original paper.

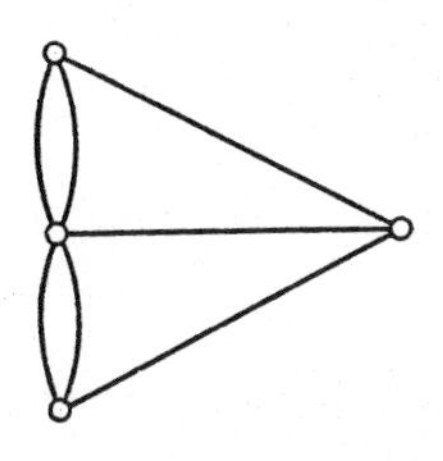

Figure 2

The rest of Euler's paper is devoted to the description of a method for determining whether such a route is possible for a general arrangement of rivers and bridges.

3. Early References to Euler's Paper

Euler's paper on the Königsberg bridges first appeared in the *Commentarii Academiae Scientiarum Imperialis Petropolitanae* [2]. Although dated 1736, it was not actually published until 1741 and was later republished in the new edition of the *Commentarii* which appeared in 1752. A commentary on his paper appeared in the *Nova Acta Eruditorum*, published in Leipzig in 1751. After describing the 'quite singular problem', it continues:

> Since this problem seems to belong to Geometry, and yet is so formulated that it does not require the determination of quantities, nor does it allow a solution by means of a calculation of quantities, it is referred to by this learned scholar as the geometry of position. In this work he unravels the method whereby the problem can thus be solved, but it is impossible to explain it in a few words.

Euler's paper was also mentioned in Diderot's *Encyclopédie* [4] of 1765. In an article on 'Situation', Jean D'Alembert described the analysis of situation in terms of geometrical problems which 'appear to have more solutions than they ought to have under the conditions of the problem'. He concluded by referring to Euler's paper, remarking that:

> One sees nothing in this memoir which relates to the analysis of situation that we have mentioned. It concerns only the finding of a route...

After this, there seem to be no references to Euler's solution of the Königsberg bridges problem until 1804 when it was mentioned by the Swiss mathematician Simon-Antoine-Jean Lhuilier in a footnote in his book *Élémens Raisonnés d'Algèbre* [5].

Finally, in 1810, Louis Poinsot referred to Euler's solution in a paper entitled *Sur les polygones et les polyèdres* [6]. In this memoir he introduced the four non-convex 'star polyhedra' (two of which had already been known to Johannes Kepler) and then discussed a variety of geometrical problems involving the tracing of diagrams. In addition, he included an extended footnote on the geometry of position, describing it as 'still less advanced than the theory of numbers, and we scarcely know, up to now, either the principles of the subject or the analysis'. After referring to Leibniz, he described Euler's Königsberg bridges paper and two papers by Euler and Vandermonde on the knight's-tour problem as 'nearly everything which has been done on the geometry of position'. It is to the knight's-tour problem that we now turn our attention.

4. The Knight's-Tour Problem

The knight's-tour problem—that of finding a sequence of knight's moves on a chessboard so that each square is visited exactly once—is another geometrical problem which came to be regarded as pertaining to the area of 'analysis situs'. Chessboard problems of this type have a long history (see [3]), and solutions are known to exist as far back as the fourteenth century. But the first systematic approaches to the solution of such problems did not appear, as far as is known, until the eighteenth century, when Euler [7] and Vandermonde [8] discussed the knight's-tour problem in great detail.

Euler had been interested in chess problems for some time, and had corresponded with Christian Goldbach and Louis Bertrand on the subject. His paper, written for the Academy of the Royal Society of Berlin in 1759, discussed the knight's-tour problem for various sizes of chessboard. At no point did Euler ascribe the problem to the geometry of position, but he did mention the non-geometrical form of reasoning involved.

Vandermonde's paper, on the other hand, explicitly refers to the geometry of position in his title *Remarques sur les problèmes de situation*. Written for the Académie Royale des Sciences in Paris, it is concerned with the development of an unambiguous mathematical notation for describing such objects as the interlacing of the threads in a knot or a braid. In order to illustrate his ideas, Vandermonde considered the knight's-tour problem and described a general method for solving such problems. He referred to Euler's paper on the subject, but remarked that whereas 'that great geometer presupposes that one has a chessboard to hand, I have reduced the problem to simple arithmetic'.

The knight's-tour problem continued to attract attention, and two later works on the subject—de Laisement's 1782 *Essai sur les Problèmes de Situation* and de Lavernède's *Problèmes de Situation* (1839)

— explicitly described such problems as belonging to the study of position.

5. Euler's Polyhedron Formula

In November 1750, Leonhard Euler wrote a letter to Christian Goldbach describing some 'general results in solid geometry'. Of these results, the most important was an equation

$$H + S = A + 2,$$

relating the number of faces (H), solid angles (S) and edges (A) of a solid enclosed by plane faces. This result is now known as *Euler's polyhedron formula*.

In 1752, Euler wrote two papers about his polyhedron formula. In the first [9] he verified the result for several families of solids but admitted that he was unable to prove it in general. In the second [10] he gave a non-metrical proof which involves slicing tetrahedral pieces from the polyhedron in such a way that $H + S - A$ is unchanged. Unfortunately, it is not clear that this procedure can always be carried out, and Euler's proof must be regarded as incomplete.

The first correct proof of the polyhedron formula was a metrical proof given by Adrien Marie Legendre in his celebrated textbook *Éléments de Géométrie* [11], published in 1794. His ideas were somewhat similar to those of René Descartes who in 1620 had obtained an expression for the sum of the angles of a polyhedron. It is now realized that Euler's formula can easily be deduced from this expression, but such a deduction was not made by Descartes himself.

Significant progress was made in the years 1811-13 with the appearance of papers by S.-A.-J. Lhuilier [12], [13] and Augustin-Louis Cauchy [14]. Both authors used Euler's formula to prove that there are only five regular solids, but Cauchy showed how the formula applies equally well to what are now called planar graphs. In addition, Cauchy gave the first non-metrical proof of the polyhedron formula, using a process of triangulation.

Lhuilier's papers are of great importance since they introduced several new ideas. In particular, he discussed the duality of particular polyhedra (such as the dodecahedron and the icosahedron), and he investigated some exceptions to Euler's formula, as when (for example) the faces are not simply-connected or the polyhedron has an interior cavity. But the most important exception discussed by Lhuilier is when the polyhedron has a number of holes through it—that is, it is embedded on a closed surface of positive genus. Lhuilier showed that, for surfaces of genus g (that is, homeomorphic to a doughnut with g holes), Euler's formula becomes

$$(\text{vertices}) - (\text{edges}) + (\text{faces}) = 2 - 2g.$$

This formula was the starting point of a major investigation by Johann Benedict Listing in 1861-62. Already well known for his book *Vorstudien zur Topologie* [15] in which he discussed a number of topics relating to the geometry of position, he now produced a pioneering paper *Der Census räumlicher Complexe* [16] on the topological properties of 'complexes'. This work influenced several other mathematicians and led eventually to the fundamental work of Henri Poincaré [17] on the foundations of the subject then universally called analysis situs but now known as algebraic topology. It is interesting to note that although the word 'topology' had been coined by Listing as far back as 1836, the name 'algebraic topology' did not replace the phrase 'analysis situs' until the 1920s.

6. Gauss and the Geometry of Position

Gauss' influence on the study of 'geometria situs', although minor, is also of interest. He first became aware of the geometry of position in 1794 at the age of 17. Five years later, in his doctoral dissertation on the fundamental theory of algebra, he remarked in a footnote:

> Moreover it follows from this reasoning which pertains to the geometry of position, whose principles are no less useful than those of the geometry of magnitudes, that...

Gauss did not state explicitly which principles he had in mind, but they seem to have involved the ordering of points on a curve.

Gauss also mentioned the geometry of position in his correspondence with other mathematicians, and Sartorius von Waltershausen, in his celebrated biography of Gauss [18], stated that Gauss

> ... placed extraordinary hope in the development of *geometria situs*, a field still completely unexplored...

However, by 1833 Gauss had become rather dismissive about the subject. In [19] he wrote:

> Of the geometry of position, which Leibniz initiated and in which only a few geometers (Euler, Vandermonde) have expressed a feeble interest, we know, after 150 years, scarcely more than nothing.

It was left to Gauss' student, Johann Benedict Listing, to develop the geometry of position. His contributions in this area were outlined in Section 5 of this paper.

7. Other Interpretations of 'Geometria Situs'

We have already seen how a number of different areas came to be described as 'geometria situs' or 'the geometry of position'. Another type of geometry which attracted the same description was *projective geometry* whose foundations were laid during the early nineteenth century.

Among the seminal works which led to the development of projective geometry was Lazare Carnot's *Géométrie de Position* of 1803 [20] in which Carnot helped to lay the foundations of modern synthetic geometry. This book referred to Leibniz' conception of 'analysis of situation, an idea which has not been followed although it merits the attention of savants'. Carnot also quoted from D'Alembert's article in Diderot's *Encyclopédie*, remarking that his aims differed from those of D'Alembert, but were analogous. In particular, he noted that

> Leibniz wished to have the diversity of position of parts of figures enter into the conditions of a geometric problem. Now this diversity is often expressed by changes of

> signs, and it is precisely the theory of these changes which is the essential object of the studies I have in view, and which I have named the *Geometry of Position*...

The mathematician most influential in the development of projective geometry was Jean-Victor Poncelet. After his pioneering book *Traité des Propriétés Projectives des Figures* [21], the parallel areas of synthetic and projective geometry advanced very rapidly. In 1827 August Ferdinand Möbius introduced homogeneous coordinates in his book *Der Barycentrische Calcul*, later described by Listing [16] as being related to the geometry of position. In 1846-47 Arthur Cayley wrote two papers (in French) on projective geometry [22], also referring to it as the geometry of position. Synthetic geometry was developed by a number of mathematicians, including Möbius, Plücker and Steiner, and in 1847 appeared Christian von Staudt's celebrated book *Geometrie der Lage* [23] (Lage = position). As the subject developed, more textbooks appeared with similar titles, including Karl Theodor Reye's *Geometrie der Lage* (1868). Even as late as 1891 there appeared a textbook on projective geometry by R. H. Graham, with the title *Geometry of Position*.

8. Leibniz and Vector Analysis

In his *Vorstudien zur Topologie*, already mentioned in Section 5, Listing cited a number of works as relating to the geometry of position. In addition to the papers of Euler and Vandermonde on the knight's-tour problem, he named Carnot's *Géométrie de Position*, Möbius' barycentric calculus and Grassmann's *Ausdehnungslehre*.

Hermann Grassmann's celebrated *Ausdehnungslehre* (the Theory of Extension) [24] appeared in 1844 and contained an extended treatment of the geometry of n dimensions, where $n > 3$. More importantly it introduced the basic ideas of vector analysis. At a meeting in the following year, Grassmann asserted:

> We still possess some remains of a characteristic geometry discovered by Leibniz, in which the reciprocal situation of places is determined by simple symbols as well as by their relations, without taking into consideration the

> notions of the size of an angle or the length of a line segment. Because of this, it should be completely different from either our analytic geometry or our algebraic geometry. We should reconstitute and extend this geometry — a task which does not appear at all impossible...

In an important prize essay published in 1847 [25], Grassmann said that in developing Leibniz' ideas he had been led to vector analysis and not to topology and he strongly expressed the opinion that vectors were what Leibniz had originally intended in his 1679 letter:

> It is also through [vector analysis] that Leibniz' idea is realized so that, it seems to me, no new light can now be shone on what he originally meant by 'geometrical analysis'.

Support for this viewpoint was later presented by M. Dehn and P. Heegaard in their extended article Analysis Situs [26] in the *Encyklopädie der Mathematischen Wissenschaften*

> ... Leibniz is thinking much more of a geometrical algorithm which yields a genuine method of solution for a particular geometrical problem than the methods of the usual analytic geometry,

and in the same *Encyklopädie* Alfred Lotze [27] remarked that:

> Grassmann's 'Ausdehnungslehre' represents the first extensive implementation of 'geometric analysis' (already desired by Leibniz) and at the same time the first generalization of such to n-dimensional space which can also be adapted to non-Euclidean space.

Similar views were expressed by Henri Lebesgue in a letter [28] to J. Itard dated 14 February 1939. In consequence, we can now claim with some confidence that *vector analysis, and not topology, was what Leibniz had in mind when he coined the phrase 'Analysis Situs'.*

REFERENCES

In addition to the specific references listed below, material relating to all of the topics in this paper can be found in the following three books:

N. L. Biggs, E. K. Lloyd and R. J. Wilson, *Graph Theory 1736-1936*, Clarendon Press, Oxford, 1976.

J. C. Pont, *La Topologie Algébrique des Origines à Poincaré*, Bibl. de Philos. Contemp., Presses Universitaires de France, Paris, 1974.

P. J. Federico, *The Origins of Graph Theory* (unpublished).

1. G. W. Leibniz, *Mathematische Schriften* (1) Vol. 2, Berlin, 1850, pp. 18-19.
2. L. Euler, Solutio problematis ad geometriam situs pertinentis, *Commentarii Academiae Scientiarum Imperialis Petropolitanae*, 8 (1736), 128-140.
3. W. W. Rouse Ball, *Mathematical Recreations and Problems of Past and Present Times*, (later entitled *Mathematical Recreations and Essays*), MacMillan, London, 1892.
4. D. Diderot, *Encyclopédie ou Dictionnaire Raisonné des Sciences, des Arts et des Métiers*, Vol. 15, Paris, 1751-1772, p. 232.
5. S.-A.-J Lhuilier, *Éléments Raisonnés d'Algebre*, Geneva, 1804.
6. L. Poinsot, Sur les polygones et les polyèdres, *J. École Polytech.*, 4 (Cah. 10) (1810), 16-48.
7. L. Euler, Solution d'une question curieuse qui ne paroit soumise à aucune analyse, *Mem. Acad. Sci. (Berlin)*, 15 (1759), 310-337.
8. A.-T Vandermonde, Remarques sur les problèmes de situation, *Mem. Acad. Roy. Sci. (Paris)*, (1771), 566-574.
9. L. Euler, Elementa doctrinae solidorum, *Novi Comm. Acad. Sci. Imp. Petropol.*, 4 (1752-3), 109-140.
10. L. Euler, Demonstratio nonnullarum insignium proprietatum quibus solida hedris planis inclusa sunt praedita, *Novi Comm. Acad. Sci. Imp. Petropol.*, 4 (1752-3), 140-160.
11. A. M. Legendre, *Éléments de Géométrie*, Firmin Didot, Paris, 1794.
12. S. L'Huilier, Demonstration immediate d'un theorème fondamental d'Euler sur les polyhèdres, et exceptions dont ce theorème est susceptible, *Mem. Acad. Imp. Sci. St. Petersb.*, 4 (1811), 271-301.
13. S. Lhuilier, Mémoire sur la polyèdrometrie, *Ann. de Math.*, 3 (1812-3), 169-189.
14. A.-L. Cauchy, Recherches sur les polyèdres, *J. École Polytech*, 9 (Cah. 16) (1813), 68-86.
15. J. B. Listing, Vorstudien zur Topologie, *Göttinger Studien*, 1 (1847), 811-875.

16. J. B. Listing, Der Census räumlicher Complexe oder Verallgemeinerung des Euler'schen Satzes von den Polyëdern, *Abh. K. Wiss. Göttingen Math. Cl.*, 10 (1861-2), 97-182.

17. H. Poincaré, Analysis situs, *J. École Polytech.* (2d. series), (1895), 1-211.

18. S. von Waltershausen, *Gauss zum Gedächtnis*, Leipzig, Hirzel, 1856.

19. C. F. Gauss, *Werke*, Vol. 5, Göttingen, 1867, p. 605.

20. L. Carnot, *Géométrie de Position*, Paris, 1803.

21. J.-V. Poncelet, *Traité des Propriétés Projectives des Figures*, Paris, 1822.

22. A. Cayley, Sur quelques théorèmes de la géométrie de position, *Crelle*, 31 (1846), 213-227, and 34 (1847), 270-275.

23. K. G. C. von Staudt, *Geometrie der Lage*, Nuremberg, 1847.

24. H. Grassmann, *Die Lineale Ausdehnungslehre, ein neuer Zweig der Mathematik*, Leipzig, 1844.

25. H. Grassmann, *Geometrische Analyse geknüpft an die von Leibniz erfundene geometrische Charakteristik*, Weidmann'sche Buchhandlung, Leipzig, 1847.

26. M. Dehn and P. Heegaard, Analysis situs, *Encyklopädie der Mathematischen Wissenschaften III*, 1 (1907), 153-220.

27. A. Lotze, Die Grassmannsche Ausdehnungslehre, *Encyklopädie der Mathematischen Wissenschaften III*, 1 (1907), 1425-1550.

28. H. Lebesgue, Notices d'Histoire des Mathématiques, *L'Enseignement Mathématique*, Geneva, 1958, p. 111.

ON GRAPHS WHICH ARE METRIC SPACES OF NEGATIVE TYPE

Peter M. Winkler
Emory University

Abstract

The vertices of a connected graph G constitute a metric space, with the distance between two vertices defined to be the number of edges in a shortest path between them. A graph is of <u>negative type</u> if it can be embedded in Euclidean space in such a way that each distance in the graph is the <u>square</u> of the corresponding Euclidean distance.

We give a new characterization of the graphs of negative type, based on relationships between edges of a spanning tree; and a graph is exhibited which is <u>not</u> of negative type but whose distance matrix has only one positive eigenvalue, confirming a conjecture of R.L. Graham.

I. Introduction

Embeddability of metric spaces in Euclidean space was studied extensively in the 'twenties and 'thirties. Menger [9] reduced the problem (for separable metric spaces) to embeddability of finite subspaces, and then I.J. Schoenberg [11] obtained the following pretty characterization in the finite case:

Theorem: Let M be a metric space with points $v_0,\ldots,v_n$ and distances $d_{ij} = d(v_i,v_j)$. Then M is (isometrically) embeddable in the Euclidean space $\mathbb{R}^n$ if and only if the quadratic form

$$Q(x,x) = \sum_{i,j=1}^{n} (d_{0i}^2+d_{0j}^2-d_{ij}^2)x_i x_j$$

is positive semi-definite, i.e. is always ≥ 0. Moreover $Q(x,x)$ will be positive definite exactly when the embedded points are the vertices of an n-simplex, that is, when M cannot be embedded in $\mathbb{R}^{n-1}$.

A more symmetrical characterization, found in [5], is easily derived from the above theorem by setting $x_0 = -\Sigma x_i$: M is embeddable in R^n iff the (n+1)-variable form

$$P(x,x) = \sum_{i,j=0}^{n} d_{ij}^2 x_i x_j$$

is negative semi-definite on the hyperplane $x_0+ x_1 +\ldots+x_n = 0$.

A (finite, connected) graph G with vertices $v_0,\ldots,v_n$ becomes a metric space with the usual graph distance, d_{ij} = the number of edges in a shortest path from v_i to v_j. It is not hard to show that G is (isometrically) embeddable in $\mathbb{R}^n$ just when G is either complete or a path, but this turns out to be the wrong question. In the important class of graphs (see e.g. [3], [6], [7] or [8]) isometrically embeddable in hypercubes, and in hypercubes themselves, it is the squares of the Euclidean distances which become equal to the graph distances. We are thus led via the above characterization to the following definition (see e.g. [8]):

Def. A graph G is of negative type if for any real numbers $x_0, x_1, \ldots, x_n$ which sum to 0,

$$\sum_{i,j=0}^{n} d_{ij} x_i x_j \leq 0.$$

In that case we say also that the distance matrix $D = (d_{ij})$ is of negative type.

The graphs of negative type fall neatly into a hierarchy of graphs, which we call the "metric hierarchy", studied partly or wholly in [1], [2], [3], [6], [7], [8], and [10]. We have:

G is isometrically embeddable in a hypercube

==> G is isometrically embeddable in ℓ_1

==> G is "hypermetric"

==> G is of negative type

==> D has only one positive eigenvalue.

Definitions and proofs can be found in the references listed above, as can the following facts: all the classes are identical for bipartite graphs; and for general graphs all implications are strict except possibly the last.

In what follows we show (by example) that the last implication is also strict. We develop also a new characterization of the graphs of negative type, using the approach of "edge relationships" which was employed to good effect in [7], [8] and [10]. The new characterization is computationally useful and in fact led us to the example mentioned above.

II. Notation

In what follows G will always be a connected graph on vertices $v_0,\dots,v_n$ without loops or multiple edges.

An <u>orientation</u> G' of G is a directed graph on the same vertex set which has, for each edge $\{u,v\}$ of G, exactly one of the arcs (u,v) or (v,u). We define the <u>inner product</u> $\langle e,f\rangle$ of two arcs $e=(s,t)$ and $f=(u,v)$ of G' as follows:

$$\langle e,f\rangle = \tfrac{1}{2}(d(s,v)+d(r,u)-d(s,u)-d(t,v))$$

where d is the ordinary path-length distance in G. (Clearly, this definition can be extended to any pair of point-pairs in a metric space; much of what follows can be generalized accordingly.)

Notice that if G is in fact embedded in $\mathbb{R}^n$ with d equal to the square of Euclidean distance, then $\langle e,f\rangle$ corresponds to the Euclidean inner product of the arcs e and f regarded as vectors. In any graph G the inner product can take only the values -1, $-\tfrac{1}{2}$, 0, $+\tfrac{1}{2}$ or $+1$, and is easy to evaluate.

If $e=\langle s,t\rangle$ is a fixed arc of G', the vertices of G are naturally partitioned into sets Nst, sNt and stN where Nst is the set of vertices nearer to s than t, stN is the set of vertices nearer to t than s, and sNt is the set (possibly empty) of vertices equidistant from both. If the arc f is contained in one of these sets, then $\langle e,f\rangle=0$; if it goes from Nst to sNt or from sNt to stN, then $\langle e,f\rangle=\tfrac{1}{2}$; if it goes from Nst to stN, then $\langle e,f\rangle=1$. Arcs going in the reverse direction have the corresponding negative inner product. This is illustrated in Fig. 1 below, with each arc f labelled by $\langle e,f \rangle$:

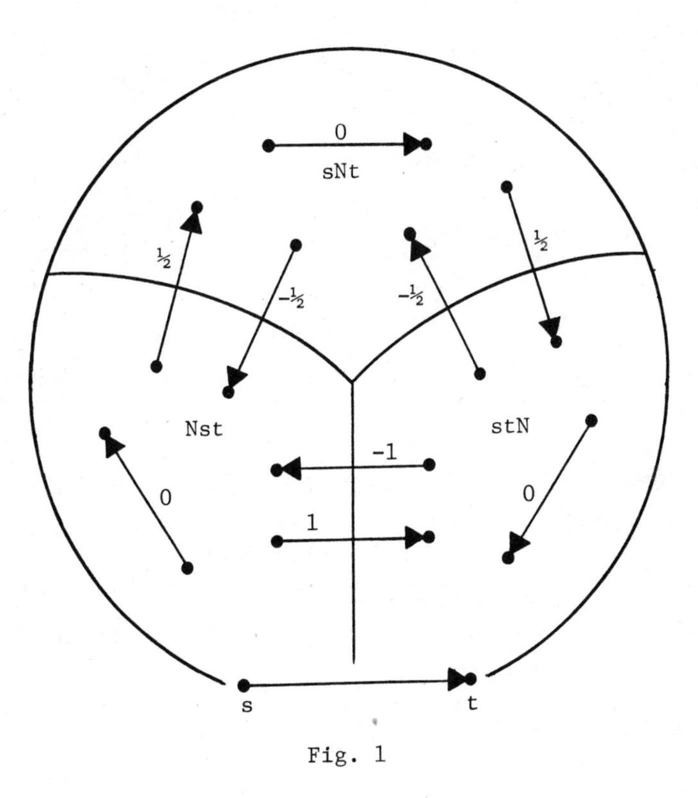

Fig. 1

III. Characterization

Theorem: Let G be a connected graph on vertices $v_0, v_1, \ldots, v_n$, and let G' be an arbitrary orientation of G with inner product $\langle \; , \; \rangle$ defined as above. Let T be an arbitrary spanning tree for G, with corresponding arcs $e_1, e_2, \ldots, e_n$ in G'. Then G is of negative type if and only if the quadratic form

$$R(y,y) = \sum_{i,j=1}^{n} \langle e_i, e_j \rangle y_i y_j$$

is positive semi-definite. Moreover, as in Schoenberg's theorem, G embeds as the vertices of an n-simplex just when $R(y,y)$ is positive definite.

In practice, to determine whether G is of negative type, we make a judicious choice of orientation and spanning tree, then compute the matrix $M = (m_{ij}) = (2\langle e_i, e_j \rangle)$ whose entries are integers between -2 and 2. It then remains to check that the upper left determinants of M are all non-negative.

Proof: The theorem will be proved if we can show that there is a 1-1 correspondence between unrestricted n-tuples $y = (y_1, \ldots, y_n)$ and $(n+1)$-tuples $x = (x_0, \ldots, x_n)$ on the hyperplane $x_0 + \ldots + x_n = 0$ such that origins correspond and $R(y,y) = -\frac{1}{2}P(x,x)$.

Given $y = (y_1, \ldots, y_n)$ and $0 \leq j \leq n$, let $A(j)$ be the set of indices i such that the arc e_i ends at v_j, and let $B(j)$ be the set of indices i such that e_i begins at v_j. Then, set

$$x_j = \sum_{i \in A(j)} y_i - \sum_{i \in B(j)} y_i.$$

This gives a linear transformation between two n-dimensional real vector spaces, and $x = (0, \ldots, 0)$ implies $y = (0, \ldots, 0)$ since the spanning tree T has no cycles. Hence the correspondence is 1-1 as claimed.

To check equality, regard $R(y,y)$ and $-\frac{1}{2}P(x,x)$ each as linear homogeneous polynomials in the $(n+1)^2$ variables d_{ij}, $0 \leq i, j \leq n$, with the x_i's and y_i's as constants related as above. Then it is easily seen that the coefficient of d_{ij} in each is

$$-\frac{1}{2}\left(\sum_{k \in A(i)} y_k - \sum_{k \in B(i)} y_k\right)\left(\sum_{k \in A(j)} y_k - \sum_{k \in B(j)} y_k\right)$$

and the proof is thus completed.

IV. Example

We define a sequence H_n of graphs, $n = 3,4,\ldots,$ as follows: H_n has vertices $v_0, v_1, \ldots, v_n$ and edges $\{v_0, v_1\}$, $\{v_0, v_i\}$, and v_1, v_1 with $i = 2,3,\ldots,n$. Thus H_n consists of $n-1$ triangles with a common base.

We now choose a spanning tree T_n for H_n and orient edges so that the arcs of T_n are $e_i = (v_0, v_i)$, $i = 1,\ldots,n$. The result for $n = 6$ is pictured in Fig. 2.

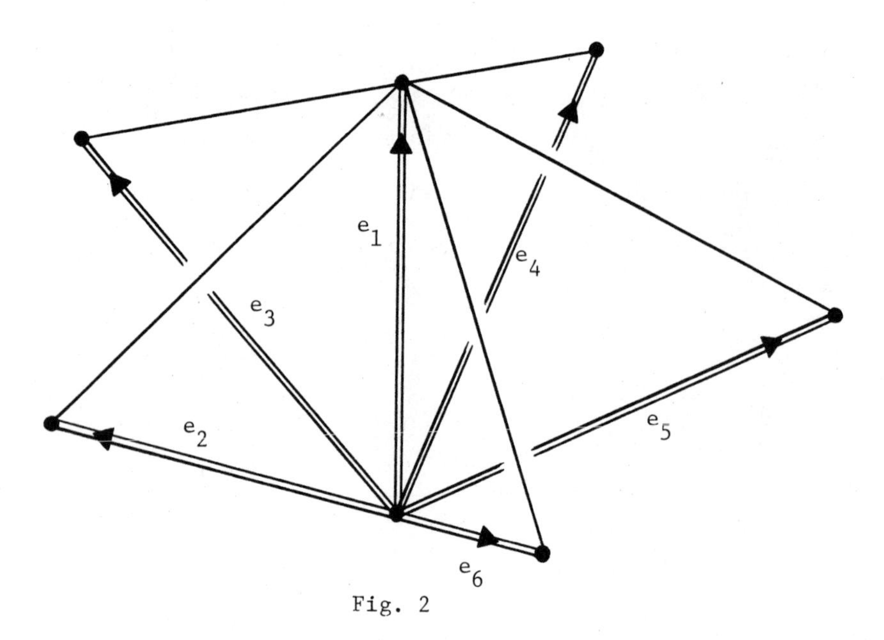

Fig. 2

It is now easily verified that for any n and any $k \leq n$, the upper left k-by-k submatrix of the matrix $M = (2\langle e_i, e_j\rangle)$ for H_n has the following form:

$$\underbrace{\left.\begin{matrix} 2 & 1 & 1 & 1 & \cdots & 1 \\ 1 & 2 & 0 & 0 & \cdots & 0 \\ 1 & 0 & 2 & 0 & \cdots & 0 \\ 1 & 0 & 0 & 2 & \cdots & 0 \\ \cdot & \cdot & \cdot & \cdot & & \cdot \\ \cdot & \cdot & \cdot & \cdot & & \cdot \\ \cdot & \cdot & \cdot & \cdot & & \cdot \\ 1 & 0 & 0 & 0 & \cdots & 2 \end{matrix}\right\}k}_{k}$$

This matrix has determinant $2^{k-1}(2-(n-1)/2) = 2^{k-2}(5-n)$. It follows that the matrix M is positive definite for $n < 5$, positive semi-definite but singular for $n = 5$, and neither for $n > 5$. It follows from the theorem, then, that H_n is of negative type for $n < 5$ and requires n dimensions for the embedding; H_5 is of negative type and has an appropriate embedding in R^4; and H_n is not of negative type for $n \geq 6$.

The distance matrix D of H_n looks like this:

$$\underbrace{\left.\begin{matrix} 0 & 1 & 1 & 1 & 1 & \cdots & 1 \\ 1 & 0 & 1 & 1 & 1 & \cdots & 1 \\ 1 & 1 & 0 & 2 & 2 & \cdots & 2 \\ 1 & 1 & 2 & 0 & 2 & \cdots & 2 \\ 1 & 1 & 2 & 2 & 0 & \cdots & 2 \\ \cdot & \cdot & \cdot & \cdot & \cdot & & \cdot \\ \cdot & \cdot & \cdot & \cdot & \cdot & & \cdot \\ \cdot & \cdot & \cdot & \cdot & \cdot & & \cdot \\ 1 & 1 & 2 & 2 & 2 & \cdots & 0 \end{matrix}\right\} n+1}_{n+1}$$

D has eigenvalues $2n-1$, -1, and -2 with multiplicities 1, 1, and $n-1$ respectively. Thus H_n for $n \geq 6$ has one positive eigenvalue but is not of negative type, as desired.

REFERENCES

[1] P. Assouad, Un espace hypermétrique non plongeable dans un espace L^1, C.R. Acad. Sci. Paris 285 (ser A) (1977), 361-363.

[2] P. Assouad and C. Delorme, Distances sur les graphes et plongements dans L^1, I, II (preprints).

[3] P. Assouad and M. Deza, Espaces métriques plongeables dans un hypercube, Annals of Disc. Math. 8 (1980), 197-210.

[4] D. Avis, Hypermetric spaces and the Hamming cone, Canad. J. Math. 33 (1981), 795-802.

[5] L.M. Blumenthal, Distance Geometry, Oxford University Press, London 1953.

[6] M. Deza (as M.E. Tylkin), On Hamming geometry of unitary cubes, Doklady Akad. Nauk S.S.R. 134 (1960), 1037-1040.

[7] D.Z. Djoković, Distance preserving subgraphs of hypercubes, J. Comb. Theory (B) 14 (1973), 263-267.

[8] R.L. Graham and P.M. Winkler, On isometric embeddings of graphs, submitted for publication.

[9] K. Menger, Die Metrik des Hilbertschen Raumes, Anzeiger de Akad. der Wissenschaften in Wien, Math. Nat. Kl., vol 65 (1928), 159-160.

[10] R.L. Roth and P.M. Winkler, Collapse of the metric hierarchy for bipartite graphs, submitted for publication.

[11] I.J. Schoenberg, Remarks to Maurice Frechet's article. . ., Annals of Math. 36 (1935), 724-732.